ETHICAL ISSUES IN HUMAN STEM CELL RESEARCH

VOLUME I

Report and
Recommendations
of the National
Bioethics Advisory
Commission

University Press of the Pacific
Honolulu, Hawaii

Ethical Issues in Human Stem Cell Research
(Three Volumes in One)

by
National Bioethics Advisory Commission

ISBN: 1-4102-1895-3

Copyright © 2005 by University Press of the Pacific

Reprinted from the 2000 edition

University Press of the Pacific
Honolulu, Hawaii
http://www.universitypressofthepacific.com

All rights reserved, including the right to reproduce
this book, or portions thereof, in any form.

Table of Contents

Letter of Transmittal to the President
National Bioethics Advisory Commission
National Bioethics Advisory Commission Staff and Consultants
Executive Summary i

Chapter 1: Introduction 1
Introduction 1
Human Stem Cells: An Overview 1
Ethical Issues 2
Framework for This Report 3
Definitions Used in This Report 4
Organization of This Report 5
Notes 6
References 6

Chapter 2: Human Stem Cell Research and the Potential for Clinical Application 7
Introduction 7
Stem Cell Types 8
Animal Models 14
Human Models 16
Growth and Derivation of Embryonic Stem Cells 19
Potential Medical Applications of Human Embryonic Stem Cell and Embryonic Germ Cell Research 20
Summary 23
Notes 24
References 24

Chapter 3: The Legal Framework for Federal Support of Research to Obtain and Use Human Stem Cells 29
Introduction 29
The Law Relating to Aborted Fetuses as Sources of Embryonic Germ Cells 29
The Law Relating to Embryos as Sources of Embryonic Stem Cells 33
The Law Relating to Deriving Stem Cells from Organisms Created Through Cloning 36
Summary 37
Notes 38
References 43

Chapter 4: Ethical Issues in Human Stem Cell Research 45
Ethical Issues Relating to the Sources of Human Embryonic Stem or Embryonic Germ Cells 45
The Arguments Relating to Federal Funding of Research Involving the Derivation and/or Use of Embryonic Stem and Embryonic Germ Cells 57
Ethical Issues in Adopting Federal Oversight and Review Policies for Embryonic Stem and Embryonic Germ Cell Research 61
Summary 61
Notes 62
References 62

Chapter 5: Conclusions and Recommendations 65
Introduction 65
Scientific and Medical Considerations 65
Ethical and Policy Considerations 66
Conclusions and Recommendations 67
Summary 81
Notes 81
References 81

Appendices
Appendix A: Acknowledgments 83
Appendix B: Glossary 85
Appendix C: Letters of Request and Response 87
Appendix D: The Food and Drug Administration's Statutory and Regulatory Authority to Regulate Human Stem Cells 93
Appendix E: Summary of Presentations on Religious Perspectives Relating to Research Involving Human Stem Cells, May 7, 1999 99
Appendix F: Points to Consider in Evaluating Basic Research Involving Human Embryonic Stem Cells and Embryonic Germ Cells 105
Appendix G: Public and Expert Testimony 109
Appendix H: Commissioned Papers 111

NATIONAL BIOETHICS ADVISORY COMMISSION

6100 Executive Blvd
Suite 5B01
Rockville, MD
20892-7508

Telephone
301-402-4242
Facsimile
301-480-6900
Website
www.bioethics.gov

Harold T. Shapiro, Ph.D.
Chair

Patricia Backlar

Arturo Brito, M.D.

Alexander M. Capron, LL.B.

Eric J. Cassell, M.D.

R. Alta Charo, J.D.

James F. Childress, Ph.D.

David R. Cox, M.D., Ph.D.

Rhetaugh G. Dumas, Ph.D., R.N.

Laurie M. Flynn

Carol W. Greider, Ph.D.

Steven H. Holtzman

Bette O. Kramer

Bernard Lo, M.D.

Lawrence H. Miike, M.D., J.D.

Thomas H. Murray, Ph.D.

Diane Scott-Jones, Ph.D.

Eric M. Meslin, Ph.D.
Executive Director

The President
The White House
Washington, DC 20500

September 7, 1999

Dear Mr. President:

    On November 14, 1998, you wrote to the National Bioethics Advisory Commission requesting that we "conduct a thorough review of the issues associated with...human stem cell research, balancing all medical and ethical issues." Your request came in response to reports of the successful isolation and culture of these specialized cells, which have simultaneously offered hope of new cures to debilitating and even fatal illness while renewing an important national debate about the ethics of research involving human embryos and cadaveric fetal material. After nine months of careful study, I am pleased to inform you that we have completed our deliberations and now provide you with the Commission's report, *Ethical Issues in Human Stem Cell Research*.

    The Commission considered a broad set of scientific, medical, and legal issues, but focused particular attention on the ethical questions relevant to federal sponsorship of research involving human embryonic stem (ES) cells and embryonic germ (EG) cells. In our deliberations, we benefited greatly from the testimony of scientists, religious scholars, ethicists, lawyers, and the public, which testimony fully reflected the wide diversity of moral perspectives on these issues that characterizes our nation. Although wide agreement exists that human embryos deserve respect as a form of human life, there is disagreement both on the form such respect should take and on the level of protection owed at different stages of embryonic development. Moreover, it was clear from the outset that no public policy or set of recommendations could fully bridge these disagreements and satisfy all the thoughtful moral perspectives that are held by members of the American public.

    The Commission proposes 13 recommendations in several areas. Perhaps the most important recommendations reflect the Commission's view that federal sponsorship of research that involves the derivation and use of human embryonic stem (ES) cells and human embryonic germ (EG) cells should be limited in two ways. First, such research should be limited to using only two of the current sources of such cells; namely, cadaveric fetal material and embryos remaining after infertility treatments. Second, that such sponsorship be contingent on an appropriate and open system of national oversight and review.

    Other recommendations address the requirements for consent from women or from couples who donate cadaveric fetal tissue or embryos remaining following infertility treatments; restrictions on the sale of fetal tissue or embryos and limits on the designation of those who may benefit from their use; the role of federal agencies in the review of research; and the encouragement of the private sector to comply with the same requirements recommended for federally funded researchers.

Taken together, we believe that these recommendations offer a policy framework that will provide the public with the assurance that important, potentially life-saving research can be conducted with federal sponsorship within a publicly accountable and rigorous system of oversight and review.

I would like to thank my fellow Commissioners, whose spirit of public service enabled them to work tirelessly to ensure that our report and its recommendations fully respected the moral worth of a wide variety of thoughtful viewpoints on the issues before us. We appreciate the opportunity to submit this report to you.

Sincerely,

Harold T. Shapiro
Chair

National Bioethics Advisory Commission

Harold T. Shapiro, Ph.D., Chair
President
Princeton University
Princeton, New Jersey

Patricia Backlar
Research Associate Professor of Bioethics
Department of Philosophy
Portland State University
Assistant Director
Center for Ethics in Health Care
Oregon Health Sciences University
Portland, Oregon

Arturo Brito, M.D.
Assistant Professor of Clinical Pediatrics
University of Miami School of Medicine
Miami, Florida

Alexander Morgan Capron, LL.B.
Henry W. Bruce Professor of Law
University Professor of Law and Medicine
Co-Director, Pacific Center for Health Policy and Ethics
University of Southern California
Los Angeles, California

Eric J. Cassell, M.D., M.A.C.P.
Clinical Professor of Public Health
Cornell University Medical College
New York, New York

R. Alta Charo, J.D.*
Professor of Law and Medical Ethics
Schools of Law and Medicine
The University of Wisconsin
Madison, Wisconsin

James F. Childress, Ph.D.
Kyle Professor of Religious Studies
Professor of Medical Education
Co-Director, Virginia Health Policy Center
Department of Religious Studies
The University of Virginia
Charlottesville, Virginia

David R. Cox, M.D., Ph.D.
Professor of Genetics and Pediatrics
Stanford University School of Medicine
Stanford, California

Rhetaugh Graves Dumas, Ph.D., R.N.
Vice Provost Emerita, Dean Emerita, and
  Lucille Cole Professor of Nursing
The University of Michigan
Ann Arbor, Michigan

Laurie M. Flynn
Executive Director
National Alliance for the Mentally Ill
Arlington, Virginia

Carol W. Greider, Ph.D.**
Professor of Molecular Biology and Genetics
Department of Molecular Biology and Genetics
The Johns Hopkins University School of Medicine
Baltimore, Maryland

Steven H. Holtzman
Chief Business Officer
Millennium Pharmaceuticals Inc.
Cambridge, Massachusetts

Bette O. Kramer
Founding President
Richmond Bioethics Consortium
Richmond, Virginia

Bernard Lo, M.D.
Director
Program in Medical Ethics
The University of California, San Francisco
San Francisco, California

Lawrence H. Miike, M.D., J.D.
Kaneohe, Hawaii

Thomas H. Murray, Ph.D.
President
The Hastings Center
Garrison, New York

Diane Scott-Jones, Ph.D.
Professor
Department of Psychology
Temple University
Philadelphia, Pennsylvania

*To avoid the appearance of a conflict of interest, Commissioner Charo recused herself from all Commission deliberations as of February 1, 1999. She neither dissents from nor endorses this report and its recommendations.

**To avoid the appearance of a conflict of interest, Commissioner Greider recused herself from Commission deliberations as of July 19, 1999.

The National Bioethics Advisory Commission (NBAC) was established by Executive Order 12975, signed by President Clinton on October 3, 1995. NBAC's functions are defined as follows:

a) NBAC shall provide advice and make recommendations to the National Science and Technology Council and to other appropriate government entities regarding the following matters:

1) the appropriateness of departmental, agency, or other governmental programs, policies, assignments, missions, guidelines, and regulations as they relate to bioethical issues arising from research on human biology and behavior; and

2) applications, including the clinical applications, of that research.

b) NBAC shall identify broad principles to govern the ethical conduct of research, citing specific projects only as illustrations for such principles.

c) NBAC shall not be responsible for the review and approval of specific projects.

d) In addition to responding to requests for advice and recommendations from the National Science and Technology Council, NBAC also may accept suggestions of issues for consideration from both the Congress and the public. NBAC also may identify other bioethical issues for the purpose of providing advice and recommendations, subject to the approval of the National Science and Technology Council.

National Bioethics Advisory Commission Staff and Consultants

Executive Director
Eric M. Meslin, Ph.D.

Research Staff

Kathi E. Hanna, M.S., Ph.D., *Research Director*
Emily C. Feinstein, *Research Analyst**
Melissa Goldstein, J.D., *Research Analyst**
E. Randolph Hull, Jr., *Research Analyst**
J. Kyle Kinner, J.D., M.P.A., *Presidential Management Intern*
Kerry Jo Lee, *Intern*
Debra McCurry, M.S., *Information Specialist*
Daniel J. Powell, *Intern*
Andrew Siegel, Ph.D., J.D., *Staff Philosopher***
Sean A. Simon, *Research Analyst**
Robert Tanner, J.D., *Research Analyst*

Consultants

Burness Communications, *Communications Consultant*
Sara Davidson, M.A., *Editor*
Elisa Eiseman, Ph.D., *Science Consultant*
Jeffrey P. Kahn, Ph.D., M.P.H., *Bioethics Consultant*
Tamara Lee, *Graphic Designer*
LeRoy Walters, Ph.D., *Bioethics Consultant*

Administrative Staff

Jody Crank, *Secretary to the Executive Director***
Evadne Hammett, *Administrative Officer*
Patricia Norris, *Public Affairs Officer*
Lisa Price, *Secretary***
Margaret C. Quinlan, *Office Manager*
Sherrie Senior, *Secretary*

*Until May 1999
**Until July 1999

ETHICAL ISSUES IN HUMAN STEM CELL RESEARCH

EXECUTIVE SUMMARY

Rockville, Maryland
September 1999

Executive Summary

Introduction

In November 1998, President Clinton charged the National Bioethics Advisory Commission with the task of conducting a thorough review of the issues associated with human stem cell research, balancing all ethical and medical considerations. The President's request was made in response to three separate reports that brought to the fore the exciting scientific and clinical prospects of stem cell research while also raising a series of ethical controversies regarding federal sponsorship of scientific inquiry in this area. Scientific reports of the successful isolation and culture of these specialized cells have offered hope of new cures for debilitating and even fatal illness and at the same time have renewed an important national debate about the ethics of research involving human embryos and cadaveric fetal material.

Scientific and Medical Considerations

The stem cell is a unique and essential cell type found in animals. Many kinds of stem cells are found in the body, with some more differentiated, or committed, to a particular function than others. In other words, when stem cells divide, some of the progeny mature into cells of a specific type (e.g., heart, muscle, blood, or brain cells), while others remain stem cells, ready to repair some of the everyday wear and tear undergone by our bodies. These stem cells are capable of continually reproducing themselves and serve to renew tissue throughout an individual's life. For example, they constantly regenerate the lining of the gut, revitalize skin, and produce a whole range of blood cells. Although the term *stem cell* commonly is used to refer to the cells within the adult organism that renew tissue (e.g., hematopoietic stem cells, a type of cell found in the blood), the most fundamental and extraordinary of the stem cells are found in the early stage embryo. These *embryonic stem (ES) cells*, unlike the more differentiated adult stem cells or other cell types, retain the special ability to develop into nearly any cell type. *Embryonic germ (EG) cells*, which originate from the primordial reproductive cells of the developing fetus, have properties similar to ES cells.

It is the potentially unique versatility of the ES and EG cells derived, respectively, from the early stage embryo and cadaveric fetal tissue that presents such unusual scientific and therapeutic promise. Indeed, scientists have long recognized the possibility of using such cells to generate more specialized cells or tissue, which could allow the generation of new cells to be used to treat injuries or diseases, such as Alzheimer's disease, Parkinson's disease, heart disease, and kidney failure. Likewise, scientists regard these cells as an important—perhaps essential—means for understanding the earliest stages of human development and as an important tool in the development of life-saving drugs and cell-replacement therapies to treat disorders caused by early cell death or impairment.

The techniques for deriving these cells have not been fully developed as standardized and readily available research tools, and the development of any therapeutic application remains some years away. Thus, ES and EG cells are still primarily a matter of intense research interest.

At this time, human stem cells can be derived from the following sources:

- human fetal tissue following elective abortion (EG cells),

Executive Summary

- human embryos that are created by *in vitro* fertilization (IVF) and that are no longer needed by couples being treated for infertility (ES cells),
- human embryos that are created by IVF with gametes donated for the sole purpose of providing research material (ES cells), and
- potentially, human (or hybrid) embryos generated asexually by somatic cell nuclear transfer or similar cloning techniques in which the nucleus of an adult human cell is introduced into an enucleated human or animal ovum (ES cells).

In addition, although much promising research currently is being conducted with stem cells obtained from adult organisms, studies in animals suggest that this approach will be scientifically and technically limited, and in some cases the anatomic source of the cells might preclude easy or safe access. However, because there are no legal restrictions or new ethical considerations regarding research on adult stem cells (other than the usual concerns about consent and risks), important research can and should go forward in this area. Moreover, because important biological differences exist between embryonic and adult stem cells, this source of stem cells should not be considered an alternative to ES and EG cell research.

Ethical and Policy Considerations

The scientific reports of the successful isolation and culture of ES and EG cells have renewed a longstanding controversy about the ethics of research involving human embryos and cadaveric fetal material. This controversy arises from sharply differing moral views regarding elective abortion or the use of embryos for research. Indeed, an earnest national and international debate continues over the ethical, legal, and medical issues that arise in this arena. This debate represents both a challenge and an opportunity: a challenge because it concerns important and morally contested questions regarding the beginning of life, and an opportunity because it provides another occasion for serious public discussion about important ethical issues. We are hopeful that this dialogue will foster public understanding about the relationships between the opportunities that biomedical science offers to improve human welfare and the limits set by important ethical obligations.

Although we believe most would agree that human embryos deserve respect as a form of human life, disagreements arise regarding both what form such respect should take and what level of protection is required at different stages of embryonic development. Therefore, embryo research that is not therapeutic to the embryo is bound to raise serious concerns and to heighten the tensions between two important ethical commitments: to cure disease and to protect human life. For those who believe that the embryo has the moral status of a person from the moment of conception, research (or any other activity) that would destroy the embryo is considered wrong and should not take place. For those who believe otherwise, arriving at an ethically acceptable policy in this arena involves a complex balancing of a number of important ethical concerns. Although many of the issues remain contested on moral grounds, they co-exist within a broad area of consensus upon which public policy can, at least in part, be constructed.

For most observers, the resolution of these ethical and scientific issues depends to some degree on the source of the stem cells. The use of cadaveric fetal tissue to derive EG cell lines—like other uses of tissues or organs from dead bodies—is generally the most accepted, provided that the research complies with the system of public safeguards and oversight already in place for such scientific inquiry. With respect to embryos and the ES cells from which they can be derived, some draw an ethical distinction between two types of embryos. One is referred to as the *research embryo*, an embryo created through IVF with gametes provided solely for research purposes. Many people, including the President, have expressed the view that the federal government should not fund research that involves creating such embryos. The second type of embryo is that which was created for infertility treatment, but is now intended to be discarded because it is unsuitable or no longer needed for such treatment. The use of these embryos raises fewer ethical questions because it does not alter their final disposition. Finally, the recent demonstration of cloning techniques (somatic cell nuclear transfer) in nonhuman animals suggests that transfer of a human somatic cell nucleus into an oocyte might create an embryo that could be used as a source of ES cells. The creation of a human organism using this technique raises questions similar to those raised by the creation of research embryos through IVF, and at this time federal funds may not be used for such research. In addition, if the enucleated oocyte that was to be combined with a human somatic cell nucleus came from an animal other than a human being, other issues would arise about the nature of the embryo produced. Thus, each source of material raises ethical questions as well as scientific, medical, and legal ones.

Conscientious individuals have come to different conclusions regarding both public policy and private actions in the area of stem cell research. Their differing perspectives by their very nature cannot easily be bridged by any single public policy. But the development of public policy in a morally contested area is not a novel challenge for a pluralistic democracy such as that which exists in the United States. We are profoundly aware of the diverse and strongly held views on the subject of this report and have wrestled with the implications of these different views at each of our meetings devoted to this topic. Our aim throughout these deliberations has been to formulate a set of recommendations that fully reflects widely shared views and that, in our view, would serve the best interests of society.

Most states place no legal restrictions on any of the means of creating ES and EG cells that are described in this report. In addition, current Food and Drug Administration regulations do not apply to this type of early stage research. Therefore, because the public controversy surrounding such activities in the United States has revolved around whether it is appropriate for the federal government to sponsor such research, this report focuses on the question of whether the scientific merit and the substantial clinical promise of this research justify federal support, and, if so, with what restrictions and safeguards.

Conclusions and Recommendations

This report presents the conclusions that the Commission has reached and the recommendations that the Commission has made in the following areas: the ethical acceptability of federal funding for research that either derives or uses ES or EG cells; the means of ensuring appropriate consent of women or couples who donate cadaveric fetal tissue or embryos remaining after infertility treatments; the need for restrictions on the sale of these materials and the designation of those who may benefit from their use; the need for ethical oversight and review of such research at the national and institutional level; and the appropriateness of voluntary compliance by the private sector with some of these recommendations.

The Ethical Acceptability of Federal Funding of ES and EG Cell Research by the Source of the Material

A principal ethical justification for public sponsorship of research with human ES or EG cells is that this research has the potential to produce health benefits for individuals who are suffering from serious and often fatal diseases. We recognize that it is possible that the various sources of human ES or EG cells eventually could be important to research and clinical application because of, for example, their differing proliferation potential, differing availability and accessibility, and differing ability to be manipulated, as well as possibly significant differences in their cell biology. **At this time, therefore, the Commission believes that federal funding for the use and derivation of ES and EG cells should be limited to two sources of such material: cadaveric fetal tissue and embryos remaining after infertility treatments.** Specific recommendations and their justifications are provided below.

Recommendation 1: EG Cells from Fetal Tissue
Research involving the derivation and use of human EG cells from cadaveric fetal tissue should continue to be eligible for federal funding. Relevant statutes and regulations should be amended to make clear that the ethical safeguards that exist for fetal tissue transplantation also apply to the derivation and use of human EG cells for research purposes.

Considerable agreement exists, both in the United States and throughout the world, that the use of fetal tissue in therapy for people with serious disorders, such

Executive Summary

as Parkinson's disease, is acceptable. Research that uses tissue from aborted fetuses is analogous to the use of fetal tissue in transplantation. The rationales for conducting EG research are equally strong, and the arguments against it are not persuasive. The removal of fetal germ cells does not occasion the destruction of a live fetus, nor is fetal tissue intentionally or purposefully created for human stem cell research. Although abortion itself doubtless will remain a contentious issue in our society, the procedures that have been developed to prevent fetal tissue donation for therapeutic transplantation from influencing the abortion decision offer a model for creating such separation in research to derive human EG cells. Because the existing statutes are written in terms of tissue transplantation, which is not a current feature of EG cell research, changes are needed to make it explicit that the relevant safeguards will apply to research to derive EG cells from aborted fetuses. At present, no legal prohibitions exist that would inhibit the use of such tissue for EG cell research.

Recommendation 2: ES Cells from Embryos Remaining After Infertility Treatments
Research involving the derivation and use of human ES cells from embryos remaining after infertility treatments should be eligible for federal funding. An exception should be made to the present statutory ban on federal funding of embryo research to permit federal agencies to fund research involving the derivation of human ES cells from this source under appropriate regulations that include public oversight and review. (See Recommendations 5 through 9.)

The current ban on embryo research is in the form of a rider to the appropriations bill for the Department of Health and Human Services (DHHS), of which the National Institutes of Health (NIH) is a part. The rider prohibits use of the appropriated funds to support any research "in which a human embryo [is] destroyed, discarded, or knowingly subjected to risk of injury greater than that allowed for research on fetuses *in utero*" (Pub. L. No. 105-78, 513(a)). The term "human embryo" in the statute is defined as "any organism...that is derived by fertilization, parthenogenesis, cloning, or any other means from one or more human gametes or human diploid cells." The ban is revisited each year when the language of the NIH appropriations bill is considered.

The ban, which concerns only federally sponsored research, reflects a moral point of view either that embryos deserve the full protection of society because of their moral status as persons or that there is sufficient public controversy to preclude the use of federal funds for this type of research. At the same time, however, some effects of the embryo research ban raise serious moral and public policy concerns for those who hold differing views regarding the ethics of embryo research. In our view, the ban conflicts with several of the ethical goals of medicine and related health disciplines, especially healing, prevention, and research. These goals are rightly characterized by the principles of beneficence and nonmaleficence, which jointly encourage pursuing social benefits and avoiding or ameliorating potential harm.

Although some may view the derivation and use of ES cells as ethically distinct activities, we do not believe that these differences are significant from the point of view of eligibility for federal funding. That is, we believe that it is ethically acceptable for the federal government to finance research that both derives cell lines from embryos remaining after infertility treatments and that uses those cell lines. Although one might argue that some important research could proceed in the absence of federal funding for research that derives stem cells from embryos remaining after infertility treatments (i.e., federally funded scientists merely using cells derived with private funds), we believe that it is important that federal funding be made available for protocols that also derive such cells. Relying on cell lines that might be derived exclusively by a subset of privately funded researchers who are interested in this area could severely limit scientific and clinical progress.

Trying to separate research in which human ES cells are used from the process of deriving those cells presents an ethical problem, because doing so diminishes the scientific value of the activities receiving federal support. This separation—under which neither biomedical researchers at NIH nor scientists at universities and other research institutions that rely on federal support could participate in some aspects of this research—rests on the mistaken notion that the two areas of research are so distinct that participating in one need not mean participating in the other. We believe that this is a misrepresentation of the new field of human stem cell research, and this misrepresentation could adversely affect scientific progress for several reasons.

First, researchers using human ES cell lines will derive substantial scientific benefits from a detailed understanding of the process of ES cell derivation, because the properties of ES cells and the methods for sustaining the cell lines may differ depending on the conditions and methods that were used to derive them. Thus, scientists who conduct basic research and are interested in fundamental cellular processes are likely to make elemental discoveries about the nature of ES cells as they derive them in the laboratory. Second, significant basic research needs to be conducted regarding the process of ES cell derivation before cell-based therapies can be realized, and this work must be pursued in a wide variety of settings, including those exclusively devoted to basic academic research. Third, ES cells are not indefinitely stable in culture. As these cells are grown, irreversible changes occur in their genetic makeup. Thus, especially in the first few years of human ES cell research, it is important to be able to repeatedly derive ES cells in order to ensure that the properties of the cells that are being studied have not changed.

Thus, anyone who believes that federal support of this important new field of research should maximize its scientific and clinical value within a system of appropriate ethical oversight should be dissatisfied with a position that allows federal agencies to fund research using human ES cells but not research through which the cells are derived from embryos. Instead, recognizing the close connection in practical and ethical terms between derivation and use of the cells, it would be preferable to enact provisions applicable to funding by all federal agencies, provisions that would carve out a narrow exception for funding of research to use or to derive human ES cells from embryos that are being discarded by infertility treatment programs.

Recommendation 3: ES Cells from Embryos Made Solely for Research Purposes Using IVF
Federal agencies should not fund research involving the derivation or use of human ES cells from embryos made solely for research purposes using IVF.

ES cells can be obtained from human research embryos created from donor gametes through IVF for the sole purpose of deriving such cells for research. The primary objection to creating embryos specifically for research is that there is a morally relevant difference between generating an embryo for the sole purpose of creating a child and producing an embryo with no such goal. Those who object to creating embryos for research often appeal to arguments about respecting human dignity by avoiding instrumental use of human embryos (i.e., using embryos merely as a means to some other goal does not treat them with appropriate respect or concern as a form of human life).

In 1994, the NIH Human Embryo Research Panel argued in support of federal funding of the creation of embryos for research purposes in exceptional cases, such as the need to create banks of cell lines with different genetic make-ups that encoded various transplantation antigens—the better to respond, for example, to the transplant needs of groups with different genetic profiles. This would require the recruitment of embryos from genetically diverse donors.

In determining how to deal with this issue, a number of points are worth considering. First, it is possible that the creation of research embryos will provide the only way in which to conduct certain kinds of research, such as research into the process of human fertilization. Second, as IVF techniques improve, it is possible that the supply of embryos for research from this source will dwindle. Nevertheless, we have concluded that, either from a scientific or a clinical perspective, there is no compelling reason at this time to provide federal funds for the creation of embryos for research. At the current time, cadaveric fetal tissue and embryos remaining after infertility treatment provide an adequate supply of research resources for federal research projects.

Recommendation 4: ES Cells from Embryos Made Using Somatic Cell Nuclear Transfer into Oocytes
Federal agencies should not fund research involving the derivation or use of human ES cells from embryos made using somatic cell nuclear transfer into oocytes.

Somatic cell nuclear transfer of the nucleus of an adult somatic cell into an enucleated human egg likely has the potential of creating a human embryo. To date, although little is known about these embryos as potential sources of human ES cells, there is significant reason to believe that their use may have therapeutic potential. For example, the potential use of matched tissue for autologous cell replacement therapy from ES cells may require the use of somatic cell nuclear transfer. The use of this technique to create an embryo arguably is different from all the other cases we considered—due to the asexual origin of the source of the ES cells—although oocyte donation is necessarily involved. The Commission concludes that, at this time, federal funding should not be provided to derive ES cells from this source. Nevertheless, scientific progress and the medical utility of this line of research should be monitored closely.

Requirements for the Donation of Cadaveric Fetal Tissue and Embryos for Research

Potential donors of embryos for ES cell research must be able to make voluntary and informed choices about whether and how to dispose of their embryos. Because of concerns about coercion and exploitation of potential donors, as well as societal controversy about the moral status of embryos, it is important, whenever possible, to separate donors' decisions to dispose of their embryos from their decisions to donate them for research. Potential donors should be asked to provide embryos for research only if they have decided to have those embryos discarded instead of donating them to another couple or storing them. If the decision to discard the embryos precedes the decision to donate them for research purposes, then the research determines only how their destruction occurs, not whether it occurs.

Recommendation 5: Requirements for Donation to Stem Cell Research of Embryos That Would Otherwise Be Discarded After Infertility Treatment

Prospective donors of embryos remaining after infertility treatments should receive timely, relevant, and appropriate information to make informed and voluntary choices regarding disposition of the embryos. Prior to considering the potential research use of the embryos, a prospective donor should have been presented with the option of storing the embryos, donating them to another woman, or discarding them. If a prospective donor chooses to discard embryos remaining after infertility treatment, the option of donating to research may then be presented. (At any point, the prospective donors' questions—including inquiries about possible research use of any embryos remaining after infertility treatment—should be answered truthfully, with all information that is relevant to the questions presented.)

During the presentation about potential research use of embryos that would otherwise be discarded, the person seeking the donation should

a) **disclose that the ES cell research is not intended to provide medical benefit to embryo donors,**

b) **make clear that consenting or refusing to donate embryos to research will not affect the quality of any future care provided to prospective donors,**

c) **describe the general area of the research to be carried out with the embryos and the specific research protocol, if known,**

d) **disclose the source of funding and expected commercial benefits of the research with the embryos, if known,**

e) **make clear that embryos used in research will not be transferred to any woman's uterus, and**

f) **make clear that the research will involve the destruction of the embryos.**

To assure that inappropriate incentives do not enter into a woman's decision to have an abortion, we recommend that directed donation of cadaveric fetal tissue for EG cell derivation be prohibited. Although the ethical considerations supporting a prohibition of the directed donation of human fetal tissue are less acute for EG cell research than for transplantation, certain concerns remain. Potential donors of cadaveric fetal tissue for EG cell derivation would not receive a direct therapeutic incentive to create or abort tissue for research purposes in the same way that such personal interest might arise in a transplant context. However, we agree that the prohibition remains a prudent and appropriate way of assuring that inappropriate incentives, regardless of how remote they may be, are not introduced into a woman's decision to have an abortion. Any suggestion of personal benefit to the donor or to an individual known to the donor would be untenable and possibly coercive.

Recommendation 6: No Promises to Embryo Donors That Stem Cells Will Be Provided to Particular Patient-Subjects
In federally funded research involving embryos remaining after infertility treatments, researchers may not promise donors that ES cells derived from their embryos will be used to treat patient-subjects specified by the donors.

Existing rules prohibit the practice of designated donation, the provision of monetary inducements to women undergoing abortion, and the purchase or sale of fetal tissue. We concur in these restrictions and in the earlier recommendation of the 1988 Human Fetal Tissue Transplantation Research Panel that the sale of fetal tissue for research purposes should not be permitted under any circumstances. The potential for coercive pressure is greatest when financial incentives are present, and the treatment of the developing human embryo or fetus as an entity deserving of respect may be greatly undermined by the introduction of any commercial motive into the donation or solicitation of fetal or embryonic tissue for research purposes.

Recommendation 7: Commerce in Embryos and Cadaveric Fetal Tissue
Embryos and cadaveric fetal tissue should not be bought or sold.

If and when sufficient scientific evidence and societal agreement exist that the creation of embryos specifically for research or therapeutic purposes is justified (specifically through somatic cell nuclear transfer), prohibitions on directed donation should be revisited. For obvious reasons, the use of somatic cell nuclear transfer to develop ES cells for autologous transplantation might require that the recipient be specified.

The Need for National Oversight and Review

The need for national as well as local oversight and review of human stem cell research is crucial. No such system currently exists in the United States. A national mechanism to review protocols for *deriving* human ES and EG cells and to monitor research using such cells would ensure strict adherence to guidelines and standards across the country. Thus, federal oversight can provide the public with the assurance that research involving stem cells is being undertaken appropriately. Given the ethical issues involved in human stem cell research—an area in which heightened sensitivity about the very research itself led the President to request that the Commission study the issue—the public and the Congress must be assured that oversight can be accomplished efficiently, constructively, and in a timely fashion, with sufficient attention to the relevant ethical considerations.

Recommendation 8: Creation and Duties of an Oversight and Review Panel
DHHS should establish a National Stem Cell Oversight and Review Panel to ensure that all federally funded research involving the derivation and/or use of human ES or EG cells is conducted in conformance with the ethical principles and recommendations contained in this report. The panel should have a broad, multidisciplinary membership, including members of the general public, and should

a) **review protocols for the derivation of ES and EG cells and approve those that meet the requirements described in this report,**

b) **certify ES and EG cells lines that result from approved protocols,**

c) **maintain a public registry of approved protocols and certified ES and EG cell lines,**

d) **establish a database—linked to the public registry—consisting of information submitted by federal research sponsors (and, on a voluntary basis, by private sponsors, whose proprietary information shall be appropriately protected) that includes all protocols that derive or use ES or EG cells (including any available data on research outcomes, including published papers),**

e) **use the database and other appropriate sources to track the history and ultimate use of certified cell lines as an aid to policy assessment and formulation,**

f) **establish requirements for and provide guidance to sponsoring agencies on the social and ethical issues that should be considered in the review of research protocols that derive or use ES or EG cells, and**

g) report at least annually to the DHHS Secretary with an assessment of the current state of the science for both the derivation and use of human ES and EG cells, a review of recent developments in the broad category of stem cell research, a summary of any emerging ethical or social concerns associated with this research, and an analysis of the adequacy and continued appropriateness of the recommendations contained in this report.

The Need for Local Review of Derivation Protocols

For more than two decades, prospective review by an Institutional Review Board (IRB) has been the principal method for assuring that federally sponsored research involving human subjects will be conducted in compliance with guidelines, policies, and regulations designed to protect human beings from harm. This system of local review has been subject to criticism, and, indeed, in previous analyses we have identified a number of concerns regarding this system. In the course of preparing this report, we considered a number of proposals that would allow for the local review of research protocols involving human stem cell research, bearing in mind that a decision by the Commission to recommend a role for IRBs might be incorrectly interpreted as endorsing the view that human ES or EG cells or human embryos are human subjects and therefore would be under the purview of the Common Rule.

We adopted the principle, reflected in these recommendations, that for research to derive human ES and EG cells, a system of national oversight and review supplemented by local review would be necessary to ensure that important research could proceed—but only under specific conditions. We recognized that for research proposals involving the derivation of human ES or EG cells, many of the ethical issues associated with these protocols could be considered at the local level, that is, at the institutions at which the research would be taking place. For protocols using but not deriving ES cells (i.e., generating the cells elsewhere), a separate set of ethical deliberations would have occurred. In general, the IRB is an appropriate body to review protocols that aim to derive ES or EG cells. Although few review bodies (including IRBs) have extensive experience in reviewing protocols of this kind, they remain the most visible and expert entities available. It is for this reason, for example, that we make a number of recommendations (8, 9, 10, 11, and 12) that discuss the importance of developing additional guidance for the review of such protocols.

For proposals involving the derivation of human ES or EG cells, particular sensitivities require attention through a national review process. This process should, however, begin at the local level, because institutions that intend to conduct research involving the derivation of human ES cells or EG cells should continue to take responsibility for assuring the ethical conduct of that research. More importantly, however, IRBs can play an important role, particularly by reviewing consent documents and by assuring that collaborative research undertaken by investigators at foreign institutions has satisfied any regulatory requirements for sharing research materials.

Recommendation 9: Institutional Review of Protocols to Derive Stem Cells

Protocols involving the *derivation* of human ES and EG cells should be reviewed and approved by an IRB or by another appropriately constituted and convened institutional review body prior to consideration by the National Stem Cell Oversight and Review Panel. (See Recommendation 8.) This review should ensure compliance with any requirements established by the panel, including confirming that individuals or organizations (in the United States or abroad) that supply embryos or cadaveric fetal tissue have obtained them in accordance with the requirements established by the panel.

Responsibilities of Federal Research Agencies

Federal research agencies have in place a comprehensive system for the submission, review, and approval of research proposals. This system includes the use of a peer review group—sometimes called a study section or initial review group—that is established to assess the scientific merit of the proposals. In addition, in some agencies, such as NIH, staff members review protocols prior to their transmittal to a national advisory council for final approval. These levels of review provide an opportunity to consider ethical issues that arise in the proposals.

When research proposals involve human subjects, federal agencies rely on local IRBs to review and approve the research in order to assure that it is ethically acceptable. (See Recommendation 9.) A grant application should not be funded until ethical issues that are associated with research involving human subjects have been resolved fully. Therefore, at every point in this continuum—from the first discussions that a prospective applicant may have with program staff within a particular institution to the final decision by the relevant national advisory council—ethical and scientific issues can be addressed by the sponsoring agency.

Recommendation 10: Sponsoring Agency Review of Research Use of Stem Cells

All federal agencies should ensure that their review processes for protocols using human ES or EG cells comply with any requirements established by the National Stem Cell Oversight and Review Panel (see Recommendation 8), paying particular attention to the adequacy of the justification for using such cell lines.

Research involving human ES and EG cells raises critical ethical issues, particularly when the proposals involve the derivation of ES cells from embryos remaining after infertility treatments. We recognize that these research proposals may not follow the paradigm usually associated with human subjects research. Nevertheless, research proposals being considered for funding by federal agencies must, in our view, meet the highest standards of scientific merit and ethical acceptability. To that end, the recommendations made in this report, including a proposed set of *Points to Consider in Evaluating Basic Research Involving Human ES Cells and EG Cells*, constitute a set of ethical and policy considerations that should be reflected in the respective policies of federal agencies conducting or sponsoring human ES or EG cell research.

Attention to Issues for the Private Sector

Although this report primarily addresses the ethical issues associated with the use of federal funds for research to derive and use ES and EG cells, we recognize that considerable work in both of these areas will be conducted under private sponsorship. Thus, our recommendations may have implications for those working in the private sector. First, for cell lines to be eligible for use in federally funded research, they must be certified by the National Stem Cell Oversight and Review Panel described in Recommendation 8. Therefore, if a private company aims to make its cell lines available to publicly funded researchers, it must submit its derivation protocol(s) to the same oversight and review process recommended for the public sector, i.e., local review (see Recommendation 9) and for certification that the cells have been derived from embryos remaining after infertility treatments or from cadaveric fetal tissue.

Second, we hope that nonproprietary aspects of protocols developed under private sponsorship will be made available in the public registry, as described in Recommendation 8. The greater the participation of the private sector in providing information on stem cell research, the more comprehensive the development of the science and related public policies in this area.

Third, and perhaps most relevant, in an ethically sensitive area of emerging biomedical research it is important that all members of the research community, whether in the public or private sectors, conduct the research in a manner that is open to appropriate public scrutiny. The last two decades have witnessed an unprecedented level of cooperation between the public and private sectors in biomedical research, which has resulted in the international leadership position of the United States in this arena. Public bodies and other authorities, such as the Recombinant DNA Advisory Committee, have played a crucial role in enabling important medical advances in fields such as gene therapy by providing oversight of both publicly and privately funded research efforts. We believe that voluntary participation by the private sector in the review and certification procedures of the proposed national panel, as well as in its deliberations, can contribute equally to the socially responsible development of ES and EG cell technologies and accelerate their translation into biomedically important therapies that will benefit patients.

Recommendation 11: Voluntary Actions by Private Sponsors of Research That Would Be Eligible for Federal Funding

For privately funded research projects that involve ES or EG cells that would be eligible for federal funding, private sponsors and researchers are

encouraged to adopt voluntarily the applicable recommendations of this report. This includes submitting protocols for the derivation of ES or EG cells to the National Stem Cell Oversight and Review Panel for review and cell line certification. (See Recommendations 8 and 9.)

In this report, we recommend that federally funded research to derive ES cells be limited to those efforts that use embryos remaining after infertility treatment. Some of the recommendations made in this context—such as the requirement for separating the decision by a woman to cease such treatment when embryos still remain and her decision to donate those embryos to research—simply do not apply to efforts to derive ES cells from embryos created (whether by IVF or somatic cell nuclear transfer) solely for research purposes, activities that might be pursued in the private sector. Nevertheless, other ethical standards and safeguards embodied in the recommendations, such as provisions to prevent the coercion of women and the commodification of human reproduction, remain vitally important, even when embryos are created solely for research purposes.

Recommendation 12: Voluntary Actions by Private Sponsors of Research That Would Not Be Eligible for Federal Funding

For privately funded research projects that involve deriving ES cells from embryos created solely for research purposes and that are therefore not eligible for federal funding (see Recommendations 3 and 4)

a) **professional societies and trade associations should develop and promulgate ethical safeguards and standards consistent with the principles underlying this report, and**
b) **private sponsors and researchers involved in such research should voluntarily comply with these safeguards and standards.**

Professional societies and trade associations dedicated to reproductive medicine and technology play a central role in establishing policy and standards for clinical care, research, and education. We believe that these organizations can and should play a salutary role in ensuring that all stem cell and embryo research conducted in the United States, including that which is privately funded, conforms to the ethical principles underlying this report. Many of these organizations already have developed policy statements, ethics guidelines, or other directives addressing issues in this report, and the Commission has benefited from a careful review of these materials. These organizations are encouraged to review their professional standards to ensure not only that they keep pace with the evolving science of human ES and EG cell research, but also that their members are knowledgeable about and in compliance with them. For those organizations that conduct research in this area but that lack statements or guidelines addressing the topics of this report, we recommend strongly that they develop such statements or guidelines. No single institution or organization, whether in the public or the private sector, can provide all the necessary protections and safeguards.

The Need for Ongoing Review and Assessment

No system of federal oversight and review of such a sensitive and important area of investigation should be established without simultaneously providing an evaluation of its effectiveness, value, and ongoing need. The pace of scientific development in human ES and EG cell research likely will increase. Although one cannot predict the direction of the science of human stem cell research, in order for the American public to realize the promise of this research and to be assured that it is being conducted responsibly, close attention to and monitoring of all the mechanisms established for oversight and review are required.

Recommendation 13: Sunset Provision for National Panel

The National Stem Cell Oversight and Review Panel described in Recommendation 8 should be chartered for a fixed period of time, not to exceed five years. Prior to the expiration of this period, DHHS should commission an independent evaluation of the panel's activities to determine whether it has adequately fulfilled its functions and whether it should be continued.

There are several reasons for allowing the national panel to function for a fixed period of time and for evaluating its activities before continuing. First, some of the hoped-for results will be available from research projects that are using the two sources we consider to be ethically acceptable for federal funding. Five years is a reasonable period of time to allow some of this information to amass, offering the panel, researchers, members of Congress, and the public sufficient time to determine whether any of the knowledge or potential health benefits are being realized. The growing body of information in the public registry and database described above (particularly if privately funded researchers and sponsors voluntarily participate) will aid these considerations.

Second, within this period the panel may be able to determine whether additional sources of ES cells are necessary in order for important research to continue. Two arguments are evident for supporting research using embryos created specifically for research purposes: one is the concern that not enough embryos remain for this purpose from infertility treatments, and the other is the recognition that some research requires embryos that are generated particularly for research and/or medical purposes. The panel should assess whether additional sources of ES cells that we have judged to be ineligible for federal funding at this time (i.e., embryos created solely for research purposes) are needed.

Third, an opportunity to assess the relationship between local review of protocols using human ES and EG cells and the panel's review of protocols for the derivation of ES cells will be offered. It will, of course, take time for this national oversight and review mechanism to develop experience with the processes of review, certification, and approval described in this report. Fourth, we hope that the panel will contribute to the national dialogue on the ethical issues regarding research involving human embryos. A recurring theme of our deliberations, and in the testimony we heard, was the importance of encouraging this ongoing national conversation.

The criteria for determining whether the panel has adequately fulfilled its functions should be set forth by an independent body established by DHHS. However, it would be reasonable to expect that the evaluation would rely generally on the seven functions described above in Recommendation 8 and that this evaluation would be conducted by a group with expertise in these areas. In addition, some of the following questions might be considered when conducting this evaluation: Is there reason to believe that the private sector is voluntarily submitting descriptions of protocols involving the derivation of human ES cells to the panel for review? Is the panel reviewing projects in a timely manner? Do researchers find that the review process is substantively helpful? Is the public being provided with the assurance that social and ethical issues are being considered?

Summary

Recent developments in human stem cell research have raised hopes that new therapies will become available that will serve to relieve human suffering. These developments also have served to remind society of the deep moral concerns that are related to research involving human embryos and cadaveric fetal tissue. Serious ethical discussion will (and should) continue on these issues. However, in light of public testimony, expert advice, and published writings, we have found substantial agreement among individuals with diverse perspectives that although the human embryo and fetus deserve respect as forms of human life, the scientific and clinical benefits of stem cell research should not be foregone. We were persuaded that carrying out human stem cell research under federal sponsorship is important, but only if it is conducted in an ethically responsible manner. And after extensive deliberation, the Commission believes that acceptable public policy can be forged, in part, on widely shared views. Through this report, we not only offer recommendations regarding federal funding and oversight of stem cell research, but also hope to further stimulate the important public debate about the profound ethical issues regarding this potentially beneficial research.

Chapter One

Introduction

Introduction

Late in 1998, three separate reports brought to the fore the debate over the scientific and clinical prospects as well as the ethical implications of research using human stem cells—those cells from which the different types of cells in a developing organism grow and that generate new cells throughout an organism's life (Van Blerkom 1994). The initial two reports were published by two independent teams of scientists that had accomplished the isolation and culture of human *embryonic stem cells* (hereafter referred to as *ES cells*) and *embryonic germ cells* (hereafter referred to as *EG cells*). The first report described the successful isolation of EG cells in the laboratory of John Gearhart and his colleagues at The Johns Hopkins University. This team derived stem cells from primordial gonadal tissue obtained from cadaveric fetal tissue (Shamblott et al. 1998). The second described the work of James Thomson and his coworkers at the University of Wisconsin, who derived ES cells from the blastocyst (~100 cells) of an early human embryo donated by a couple who had received infertility treatments (Thomson et al. 1998). Finally, an article in the November 12, 1998, edition of the *New York Times* described work funded by Advanced Cell Technology of Worcester, Massachusetts. Although this work has not yet been verified fully or published in a scientific journal, the company claims that its scientists have caused human somatic cells to revert to the primordial state by fusing them with cow eggs to create a hybrid embryo. From this hybrid embryo, a small clump of cells resembling human ES cells appears to have been isolated (Wade 1998).

Human Stem Cells: An Overview

Although many kinds of stem cells exist within the human body, scientists recognize a hierarchy of types. Some stem cells are more committed—or differentiated—than others. At the earliest stage of embryonic development, the cells of the blastomere are identical to each other and are relatively undifferentiated. Each one is individually capable of generating a whole organism, a quality referred to as *totipotency*. In the next stage, ES cells, although they no longer are capable of producing a complete organism, remain undifferentiated and retain the ability to develop into nearly any cell type found in the human body, representing a type of biological plasticity referred to as *pluripotency*. (The terms *totipotency* and *pluripotency* will be discussed again later in this chapter.) At this point, the ES cells branch out into many types; from each differentiated line, all the specialized cells (e.g., heart, muscle, nerve, skin, or blood) that constitute the tissues and organs of the body will develop (Weiss et al. 1996).

The potential versatility of ES and EG cells derived from the early stage embryo or from cadaveric fetal tissue offers unusual scientific and therapeutic promise. Because these cells have the ability to proliferate and renew themselves over the lifetime of the organism, scientists have long recognized the possibility of using such cells to generate a certain number of specialized cells or tissues, which could permit the generation of new cells or tissue as a treatment for injury or for damage done by diseases such as Alzheimer's disease, Parkinson's disease, heart disease, and kidney failure. Furthermore, scientists regard these cells as an important, perhaps essential, medium for understanding the details of human development and thus for developing life-saving drugs and other therapies. At the same time, the current source of these cells (the early stage embryo or cadaveric fetal tissue) makes them the subject of significant ethical considerations. Thus, the scientific reports of the successful isolation of these versatile cells simultaneously have raised the prospect of the development of new treatments and perhaps cures for debilitating and even fatal illnesses, while also renewing the debate regarding the ethics of research involving human embryos and cadaveric fetal material.

Ethical Issues

Within days of the publication of these reports and the *New York Times* article, President Clinton wrote to the National Bioethics Advisory Commission with two requests: that the Commission consider the implications of the purported cow-human fusion experiment and report back to him and that it "undertake a thorough review of the issues associated with human stem cell research balancing all ethical and medical considerations." On November 20, 1998, we responded to the President's first request by stating that "any attempt to create a child through the fusion of a human cell and a nonhuman egg would raise profound ethical concerns and should not be permitted." (See Appendix C, which includes these letters of request and response.) Our response was based upon the same principles we relied on when preparing our report to the President entitled *Cloning Human Beings* (1997). We noted, however, that insufficient scientific evidence is available at this time to determine whether the fusing of a human cell with the egg of a nonhuman animal would result in a human embryo. In addition, if the resulting hybrid embryo were to be used as a source of ES cells, it is not clear that those cells would be the same in all respects to those obtained from a nonhybrid human embryo.

The reports of the successful isolation and culture of ES and EG cells have added a new dimension to the ongoing controversy regarding the ethics of research involving human embryos and cadaveric fetal material. This controversy arises from sharply differing moral views regarding elective abortion or the use of embryos for research, and it has fueled the national and international debate over the ethical, legal, and medical issues that arise in this arena. This debate represents both a challenge and an opportunity: a challenge because it concerns important and morally contested questions regarding the beginning of life, and an opportunity because it provides another occasion for serious public discussion about important ethical issues. We are hopeful that this report will contribute to a dialogue that will foster increased public understanding of the ethical issues underlying research on ES and EG cells and an appreciation of the complexity of making responsible public policy in the face of moral disagreement and in light of a realistic appraisal of the scientific and clinical promise of that research.

We believe that most Americans agree that human embryos should be respected as a form of human life, but that disagreement exists both about the form that such respect should take and about what level of protection is owed at different stages of embryonic development. Therefore, embryo research, the purpose of which is not therapeutic to the embryo itself, is bound to raise serious concerns for some about how to resolve the tensions between the ethical imperative to cure diseases and the moral obligation to protect human life. For those who believe that the embryo has the moral status of a person from the moment of conception, research (or any other activity) that would destroy it is considered wrong and should not take place. For others, arriving at an ethically acceptable policy involves a complex balancing of a number of important ethical concerns. Although this is a controversial area, we should not lose sight of a broad area of consensus on which public policy could—in part—be constructed.

In order to respond effectively and responsibly to the President's request to consider issues related to human stem cell research and to "balance all medical and ethical considerations," we determined that it also is necessary to consider certain aspects of the broader issues regarding research using embryonic and/or fetal material. One reason for this approach is that the nature of some of the ethical issues involved depends on the source of the stem cells. For example, ES cells can be derived from early embryos that are destroyed in the process of ES cell derivation, an act that some people find ethically unacceptable. The use of cadaveric fetal tissue to derive EG cell lines is somewhat less controversial because the fetus is deceased prior to the initiation of the research and because a well-developed system of public oversight for such research is already in place. In addition, the recent demonstration of nuclear transfer techniques (somatic cell nuclear transfer [SCNT]) suggests that transfer of an adult nucleus into an oocyte might under certain conditions create an embryo. However, the use of this technique to combine an animal oocyte with a human diploid nucleus raises additional issues regarding both the nature of the embryo produced and the ethical issues involved. In addition, each source of material bears a unique set of scientific, ethical, and legal distinctions.

We believed that it was especially important to take a broad view of the status of the human embryo and of fetal tissue in relation to biomedical research, because it is likely that science will uncover additional characteristics of the early *ex utero* human embryo or fetal tissues that will raise additional important and unique therapeutic possibilities, separate from those that derive from ES or EG cells. If these developments occur, all of the same ethical considerations that pertain to embryo research and fetal tissue research in general would arise once again.[1] In fact, the 1994 National Institutes of Health Human Embryo Research Panel designated 13 areas in which embryo research could advance scientific knowledge or could lead to important clinical benefits. Among these areas is "the isolation of pluripotential embryonic stem cell lines for eventual differentiation and clinical use in transplantation and tissue repair."[2]

Recent scientific developments require the updating and review of the important work of U.S. bodies that have met previously to address the role of the ethical complexities of human embryo and fetal tissue research, particularly as they relate to the role of federally funded research. In addition, new policy statements from other countries (such as Canada and the United Kingdom) suggest well-thought-out novel approaches that must be considered carefully. In responding to the President's request, therefore, we elected to take a comprehensive approach that built on the work of these reflective efforts, both in this country and abroad.

In our 1997 report, *Cloning Human Beings*, we addressed a specific aspect of cloning, namely where genetic material would be transferred from the nucleus of a somatic cell of an existing human being to an enucleated human egg with the intention of creating a child. At the time that we were preparing this report, the issues surrounding embryo research were not revisited, although we began our discussions recognizing that any effort in humans to transfer a somatic cell nucleus into an enucleated egg likely involves the creation of an embryo, which has the potential to be transferred to a uterus and developed to term. We recognized that ethical concerns surrounding issues of embryo research recently had received extensive analysis and noted that under current law, the use of SCNT to create an embryo solely for research purposes is prohibited in any project involving federal funds. The President's request—together with new developments concerning human ES and EG cell research using embryos remaining after infertility treatments or fetal tissue following elective abortion—requires that we reconsider the appropriateness of using these sources of cells for research purposes.

In this respect it is important to note that research on human embryos, or the creation of human embryos for research purposes, is not only legal in the United States but proceeds without any public oversight as long as 1) federal funds are not involved, 2) Food and Drug Administration regulations do not apply, and 3) the laws of the state in which the research is to be conducted do not forbid such activity. Consequently, most of the public controversy surrounding such activities in the United States has focused on whether it is appropriate for the federal government to sponsor such research when it has significant scientific merit and substantive clinical promise. This question is also the focus of this report.

Framework for This Report

As noted above, President Clinton directed the Commission to conduct a thorough review of the issues associated with human stem cell research balancing all ethical and medical considerations. This approach—balancing or weighing difficult issues—often is used in public policy discussions and has much to recommend

Chapter 1: Introduction

it, particularly when such balancing involves a serious consideration of different moral points of view, the state of scientific and medical developments, and other factors. As discussed more fully in Chapter 4, some of the issues associated with research on human stem cells—the moral status of the human embryo, for example—are especially sensitive and do not lend themselves easily to balancing. We did not, for example, deem the views of those who consider the fetus to have the moral status of a human person from the moment of conception to be of less (or more) moral weight than the views of those who consider the fetus to lack this moral status. Similarly, we did not come to our conclusions simply by balancing potential medical benefits against the potential harms, because the possibility of social benefits, by itself, is not a sufficient reason for federal support of such controversial research, particularly given the interest in stem cell research in the private sector. Nor did we approach this issue based simply upon an interpretation of the existing legal environment. Instead, we combined, as thoughtfully as we could, a number of different perspectives on and approaches to this topic.

Through ongoing discussion and dialogue—informed by scientists, philosophers, legal and religious scholars, members of the public, and others—we developed our moral perspectives on the appropriateness of federal sponsorship of stem cell research involving the derivation and/or use of ES and EG cells, principally focusing on the ethical and scientific issues. We considered the sources of human EG and ES cells and the relevant moral differences that should be evaluated in determining the acceptability of federal funding for the derivation and/or use of cells from each of these sources. In this regard, we were assisted by a number of commissioned papers each of which addressed different aspects of the problem.[3] We also benefited from the input of a group of religious scholars from diverse faith traditions whose views within and across traditions reflected the diversity found within the public as a whole. We then considered some associated ethical issues including voluntary informed consent, the just distribution of potential benefits from stem cell research, and the commodification and sale of the body and its parts. Finally, we considered how and to what extent a mechanism of national oversight and review would provide the necessary assurance that research, conducted responsibly and with accountability, could go forward while protecting and honoring a number of deeply held values. These shared values include

- securing the safety and efficacy of clinical and/or scientific procedures, especially when fundamental ethical and social issues are involved,
- respecting human life at all stages of development, and
- ensuring the responsible pursuit of medical and scientific knowledge.

Although this report primarily addresses the ethical issues associated with the use of federal funds for research to derive and use ES and EG cells, we recognize that considerable work in both of these areas will be conducted under private sponsorship. Thus, our recommendations also may have implications for those working in the private sector.

Definitions Used in This Report

We recognize the need to define clearly the terms that are central to an understanding of this report. Because certain terms, such as *embryo* and *totipotent*, are not always used consistently, it is important to explain how the Commission uses this terminology.

It is most important that the reader understand how the term embryo is used. The Canadian Royal Commission on New Reproductive Technologies elucidated the confusion surrounding the term well in its 1993 report entitled *Proceed with Care: Final Report of the Royal Commission on New Reproductive Technologies*:

> ...In the language of biologists, before implantation the fertilized egg is termed a 'zygote' rather than an 'embryo.' The term 'embryo' refers to the developing entity after implantation in the uterus until about eight weeks after fertilisation. At the beginning of the ninth week after fertilisation, it is referred to as a 'fetus,' the term used until time of birth. The terms embryo donation, embryo transfer, and embryo research are therefore inaccurate, since these all occur with zygotes, not embryos. Nevertheless, because the terms are still commonly used in the public debate,

we continue to refer to embryo research, embryo donation, and embryo transfer (607).

For the sake of consistency and accuracy, when referring to the details of the developmental stages of an entity, we use the following terminology: 1) the developing organism is a *zygote* during the first week after fertilization, 2) the organism is an *embryo* during the second through eighth weeks of development, and 3) the organism is a *fetus* from the ninth week of development until the time of birth. However, in other contexts, we will continue to use the broad terms *embryo research, embryo donation,* and *embryo transfer* to refer to zygotes, because this is how the public commonly uses them.

Because there are several sources of human stem cells, we decided that each type of stem cell should be named in a way that clarifies its original source. Therefore, as discussed earlier, cells derived from the inner cell mass of a blastocyst—those cells within the conceptus that form the embryo proper—are called *ES cells*, and cells that are derived from primordial germ cells of embryos and fetuses are called *EG cells*. In addition, cells derived from teratocarcinomas—malignant embryonic tumors— are called *embryonal carcinoma cells*, and stem cells found in the adult organism are called *adult stem (AS) cells*.

Two other terms that require explanation—because the scientific community disagrees about their meaning— are *totipotent* and *pluripotent*. Some differentiate between the two terms by defining totipotency as the ability to develop into a complete organism and pluripotency as the ability to develop into all of the various cell types of an organism without the capability of developing into an entire organism. Others define a totipotent cell as any cell that has the potential to differentiate into all cells of a developing organism, but that does not necessarily have the ability to direct the complete development of an entire organism. These scientists would then define a pluripotent cell as any cell that has the ability to differentiate into multiple (more than two) cell types. Rather than engage in this debate, for the sake of clarity, we decided to avoid using this terminology in this report, unless it refers directly to specific work or to the statements of others in which these words were included. Instead, this report uses descriptions of the stage of development and the differentiation potential of cells to make clear to the reader which types of cells are being discussed.

Organization of This Report

This report comes at a time when the Commission has completed deliberations regarding the use of human biological materials in research (1999). In that report, we recognized that in research involving such materials as DNA, hair, and skin biopsies, a number of significant ethical issues must be addressed by Institutional Review Boards, researchers, and others; these include issues of privacy and confidentiality, potential discrimination, and stigmatization. As important as these issues are—and they must be handled satisfactorily in order for research to proceed with appropriate protections for human subjects—research on human stem cells, whether they are obtained from fetal tissue following elective abortions or from tissue obtained from embryos remaining after infertility treatments, requires additional and perhaps even deeper ethical reflection.

The Commission's primary goal for this report was the development of a set of recommendations that would provide guidance on the appropriateness of permitting the federal government to fund human ES and EG cell research and on what sorts of constraints, if any, should be placed on such support. This report first presents a summary of some of the key scientific issues involved in stem cell research (Chapter 2). To place our analysis in context and to understand the implications of any new recommendations regarding the oversight and regulation of research using fetal tissue and embryos, Chapter 3 describes the historical and current status of law and regulation governing the research use of these materials. Chapter 4 explores the various ethical issues surrounding the moral status of the embryo and cadaveric fetal tissue and ethical concerns governing the acceptable use of these materials in research. Finally, Chapter 5 offers our conclusions and recommendations regarding federal sponsorship of research and appropriate oversight activities in these ethically controversial areas.

Chapter 1: Introduction

Notes

1 For example, it has been generally recommended by most governmental and professional bodies that have previously examined this issue that research on the *ex utero* pre-implantation embryo should not be conducted beyond the 14th day following fertilization. At 14 days, the first stages of organized development begin, leading over the next few days to the first appearance of differentiated tissues of the body. The Commission concurs with this time limit on research involving the *ex utero* human embryo.

2 The 1994 National Institutes of Health Human Embryo Research Panel was asked to consider various areas of research involving the *ex utero* pre-implantation human embryo and to provide areas that 1) are acceptable for federal funding, 2) warrant additional review, and 3) are unacceptable for federal support. The panel did not consider research involving *in utero* human embryos, or fetuses, because guidelines for such research already exist in the form of regulations.

3 See Appendix H for a list of the papers that were prepared for the Commission. These papers are available in Volume II of this report.

References

Canadian Royal Commission on New Reproductive Technologies. 1993. *Proceed with Care: Final Report of the Royal Commission on New Reproductive Technologies*. 2 vols. Ottawa: Minister of Government Services.

National Bioethics Advisory Commission (NBAC). 1997. *Cloning Human Beings*. 2 vols. Rockville, MD: U.S. Government Printing Office.

———. 1999. *Research Involving Human Biological Materials: Ethical Issues and Policy Guidance*. 2 vols. Rockville, MD: U.S. Government Printing Office.

National Institutes of Health (NIH). Human Embryo Research Panel. 1994. *Report of the Human Embryo Research Panel*. 2 vols. Bethesda, MD: NIH.

Shamblott, M.J., J. Axelman, S. Wang, E.M. Bugg, J.W. Littlefield, P.J. Donovan, P.D. Blumenthal, G.R. Huggins, and J.D. Gearhart. 1998. "Derivation of Pluripotent Stem Cells from Cultured Human Primordial Germ Cells." *Proceedings of the National Academy of Sciences USA* 95:13726–13731.

Thomson, J.A., J. Itskovitz-Eldor, S.S. Shapiro, M.A. Waknitz, J.J. Swiergiel, V.S. Marshall, and J.M. Jones. 1998. "Embryonic Stem Cell Lines Derived from Human Blastocysts." *Science* 282:1145–1147.

Van Blerkom, J. 1994. "The History, Current Status, and Future Direction of Research Involving Human Embryos." In NIH Human Embryo Research Panel, *Report of the Human Embryo Research Panel, Vol. II: Papers Commissioned for the Human Embryo Research Panel*, 1–25. Bethesda, MD: NIH.

Wade, N. 1998. "Researchers Claim Embryonic Cell Mix of Human and Cow." *New York Times* 12 November, A-1.

Weiss, S., B.A. Reynolds, A.L. Vescovi, C. Morshead, C.G. Craig, and D. van der Kooy. 1996. "Is There a Neural Stem Cell in the Mammalian Forebrain?" *Trends in Neurosciences* 19(9):387–393.

Chapter Two

Human Stem Cell Research and the Potential for Clinical Application

Introduction

The stem cell is a unique and essential cell type found in animals. Many kinds of stem cells are found in the human body, with some more differentiated—or committed—to a particular function than others. In other words, when stem cells divide, some of the progeny mature into cells of a specific type (e.g., heart, muscle, blood, or brain cells), while others remain stem cells, ready to repair some of the everyday wear and tear undergone by our bodies. These stem cells are capable of continually reproducing themselves and serve to renew tissue throughout an individual's life. For example, they continually regenerate the lining of the gut, revitalize skin, and produce a whole range of blood cells. Although the term *stem cell* commonly is used to refer to the cells within the adult organism that renew tissue (e.g., hematopoietic stem cells, a type of cell found in the blood), the most fundamental and extraordinary of the stem cells are found in the early stage embryo (Van Blerkom 1994). These *embryonic stem (ES) cells*, unlike the more differentiated *adult stem (AS) cells* or other cell types, retain the special ability to develop into nearly any cell type. *Embryonic germ (EG) cells*, which originate from the primordial reproductive cells of the developing fetus, have properties similar to ES cells.

Because stem cells are able to proliferate and renew themselves over the lifetime of the organism—while at the same time retaining all of their multilineage potential—scientists have long recognized that such cells could be used to generate a large number of specialized cells or tissue through amplification, a possibility that could allow the generation of new cells that would treat injury or disease.[1] In fact, if it were possible to control the differentiation of human ES cells in culture, the resulting cells could be used to repair damage caused by such conditions as heart failure, diabetes, and certain neurodegenerative diseases.

In late 1998, three separate reports brought to the fore not only these scientific and clinical prospects but also the controversies inherent in human stem cell research. The first two reports, published by two independent teams of scientists supported by private funds from Geron Corporation, a biotechnology company located in Menlo Park, California, describe the first successful isolation and culture in the laboratory of human ES and EG cells. One team, led by John Gearhart of The Johns Hopkins University School of Medicine in Baltimore, Maryland, derived human EG cells from primordial gonadal tissue, which was obtained from fetal tissue following elective abortion (Shamblott et al. 1998). The second team, led by James Thomson of the University of Wisconsin, derived human ES cells from the blastocyst stage of early embryos donated by couples who had undergone infertility treatment (Thomson et al. 1998). The ES and EG cells derived by each of these means appear to be similar in structure, function, and potential, although additional research is needed in order to verify this claim (Varmus 1998). Finally, an article in the November 12, 1998, edition of the *New York Times* described work funded by Advanced Cell Technology of Worcester, Massachusetts. Although this work has not yet been verified fully or published in a scientific journal, the company claims that its scientists have caused human somatic cells to revert to the primordial state by fusing them with cow eggs to create a hybrid embryo. From this hybrid embryo, a small clump of cells resembling human ES cells appears to have been isolated (Wade 1998).

Chapter 2: Human Stem Cell Research and the Potential for Clinical Application

The methodologies used by these investigators for deriving human ES and EG cells are based on techniques that have been used in mice since the early 1980s and, more recently, from nonhuman primates and other animals. The isolation and culturing of these cells, however, for the first time open certain avenues of important research and future clinical possibilities. At the most basic level, the isolation of these cells allows scientists to focus on how human ES and EG cells differentiate into specific types of cells, with the goal of identifying the genetic and environmental signals that direct their specialization into specific cell types. Such studies using mouse stem cells are ongoing, but comparable studies with human cells will be required in order to determine whether the signals are the same. This research might, for example, lead to the discovery of new ways to treat a variety of conditions, including degenerative diseases, birth defects, and cancer and would build on investigations conducted over the last decade, in which laboratory animals have been used to determine whether ES cells can be used to re-establish tissue in an adult organism (Corn et al. 1991; Diukman and Golbus 1992; Hall and Watt 1989; Hollands 1991). Through processes scientists are only beginning to understand, these primitive stem cells can be stimulated to specialize so that they become precursors to different cell types, which then may be used to replace tissues such as muscle, skin, nerves, or liver. For example, in mid-1999, scientists used mouse ES cells to successfully generate glial (myelin-producing) cells that when transplanted into a rat model of human myelin disease were able to efficiently myelinate axons in the rat's brain and spinal cord (Brustle et al. 1999).

Stem Cell Types

Scientists often distinguish between different kinds of stem cells depending upon their origin and their potential to differentiate. Cells derived from malignant embryonic tumors, or teratocarcinomas, are called *embryonal carcinoma (EC) cells*; cells derived from the inner cell mass of a blastocyst-stage embryo are ES cells, and cells that are derived from precursors of germ cells from a fetus are EG cells. In addition, stem cells can be found in the adult organism, for example, in bone marrow; they may possibly also be found in skin and intestine. These AS cells serve to replenish tissues in which cells often have limited life spans, such as the skin, intestine, and blood. Although interesting new data suggest that stem cells found in the adult organism are not restricted to producing cells from the tissue in which they reside (Bjornson et al. 1999), it is unlikely that these cells are capable of differentiating into all cell types. In contrast, because human ES and EG cells are believed to be capable of differentiating into all cell types, they are likely to be of clinical use in treating a variety of diseases, especially those for which organ-specific stem cells are difficult to isolate and/or use.

EG Cells

Primordial germ cells are the embryonic precursors of the sperm and ova of the adult animal (Donovan 1998). The establishment of the germline in the embryo involves the separation of primordial germ cells from the somatic cells, the proliferation of primordial germ cells, the migration of these cells to the gonads, and finally their differentiation into gametes (Donovan 1994). Primordial germ cells are the only cells in the body that can give rise to successive generations, while the somatic cells that form the body of the animal lack this capability as soon as they start to differentiate (Matsui 1998).

In culture, primordial germ cells can give rise to EG cells that are capable of differentiating into cells of multiple lineages (Donovan 1998). (See Figure 2-1.) Primordial germ cells normally give rise to gametes, but sometimes if the developmental process goes awry, they become EC cells, the stem cells of benign teratomas and malignant teratocarcinomas, which are tumors containing derivatives of the three primary germ layers (Donovan 1998).

EG cells form embryoid bodies in culture, give rise to teratomas when introduced into histocompatible animals, and form germline chimeras when introduced into a host blastocyst (Donovan 1998). The derivation of EG cells directly from primordial germ cells provides a mechanism to study some aspects of primordial germ cell development, such as imprinting and differentiation (Donovan 1994). At the same time, it may be difficult to obtain an adequate supply of appropriate fetal tissue to

Figure 2-1. Isolation and Culture of Human ES Cells from Embryonic/Fetal Tissue

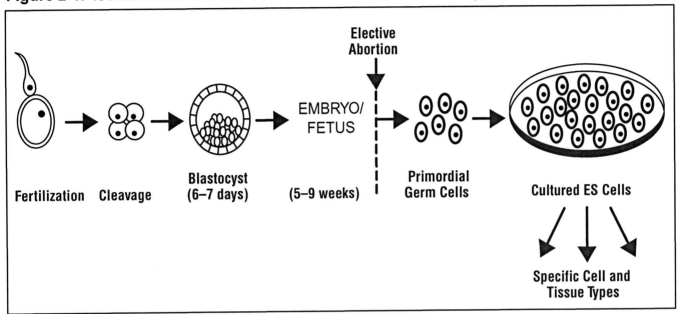

provide the relevant cell lines needed for both research and clinical uses.

ES Cells

In mammalian embryonic development, cell division gives rise to differentiated daughter cells that eventually comprise the mature animal. As cells become committed to a particular lineage or cell type, a progressive decrease in developmental potential presumably occurs. Early in embryonic development (until about 16 cells), each cell of the early cleavage-stage embryo has the developmental potential to contribute to any embryonic or extra embryonic cell type (Winkel and Pedersen 1988). However, by the blastocyst stage, the cells of the trophectoderm are irreversibly committed to forming the placenta and other trophectoderm lineages (Winkel and Pedersen 1988). By six to seven days postfertilization, the inner cell mass has divided to form two layers, which then give rise to the embryo proper and to extra embryonic tissues (Gardner 1982). (See Exhibit 2-A and Figure 2-2 for a description of early human embryonic development.)

Although the cells of the inner cell mass are precursors to all adult tissues, they can proliferate and replace themselves in the intact embryo only for a limited time before they become committed to specific lineages (Thomson and Marshall 1998). ES cells are derived from cells of the inner cell mass. Once they are placed in the appropriate culture conditions, these cells seem to be capable of extensive, undifferentiated proliferation *in vitro* and maintain the potential to contribute to all adult cell types (Evans and Kaufman 1981; Martin 1981). (See Figure 2-3.)

Even though these embryonic cells are stem cells, they differ substantially from the stem cells found within the fully developed, or adult, organism (see below). Most important, ES cells are highly proliferative, both in the embryo as well as in culture, while some stem cells of the adult can be nearly quiescent and may be more difficult to maintain and expand in culture (Van Blerkom 1994). Therefore, it appears that if stem cells were someday to be used for the treatment of disease, it might be advantageous to use ES cells to treat certain disorders.

Sources of Human ES Cells

We have distinguished between three sources of ES cells, which are derived from early embryos in culture: 1) embryos created by *in vitro* fertilization (IVF) for infertility treatments that were not implanted because they were no longer needed, 2) embryos created by IVF

Chapter 2: Human Stem Cell Research and the Potential for Clinical Application

> **Exhibit 2-A: Early Development of the Human Embryo**
>
> In humans, fertilization (the union of an oocyte [egg] and sperm) occurs in the fallopian tubes and results in the formation of the zygote. In the three to four days it takes for the zygote to travel down the fallopian tube to the uterus, several cell divisions (cleavages) occur.
>
> The first division occurs approximately 36 hours after fertilization, when the zygote begins to cleave into two cells called blastomeres. At about 60 hours following fertilization, the two blastomeres divide again to form four blastomeres. At three days postfertilization, the four blastomeres divide to form eight cells. Each blastomere becomes smaller with each subsequent division. In this early stage of development, all of the blastomeres are of equal size. These cells are unspecialized and have the capacity to differentiate into any of the cell types of the embryo as well as into the essential membranes and tissue that will support the development of the embryo. Therefore, one or more of the blastomeres can be removed without affecting the ability of the other blastomeres to develop into a fetus. In fact, if an embryo separates in half during this early stage of development, identical twins—two genetically identical individuals—will develop.
>
> When the cell division reaches approximately 16 cells, the zygote is called a morula. The morula leaves the fallopian tube and enters the uterine cavity three to four days following fertilization. After reaching the uterus, the developing zygote usually remains in the uterine cavity an additional four to five days before it implants in the endometrium (uterine wall), which means that implantation ordinarily occurs on the seventh or eighth day following fertilization.
>
> Cell division continues, creating a cavity known as a blastocele in the center of the morula. With the appearance of the cavity in the center, the entire structure is now called a blastocyst. This first specialization event occurs just before the zygote attaches to the uterus, approximately six to seven days after fertilization, when approximately 100 cells have developed. This specialization involves the formation of an outer layer of trophoblast cells, which will give rise to part of the placenta, surrounding a group of about 20 to 30 inner cells (the inner cell mass) that remain undifferentiated. At this stage, these cells no longer can give rise to all of the cells necessary to form an entire organism and therefore are incapable of developing into an entire human being. In general, as cells further differentiate, they lose the capacity to enter developmental pathways that were previously open to them.
>
> As the blastocyst attaches to the uterus, the outer layer of cells secretes an enzyme, which erodes the epithelial uterine lining and creates an implantation site for the blastocyst. Once implantation has taken place, the zygote becomes an embryo. The trophoblast and underlying cells proliferate rapidly to form the placenta and the various membranes that surround and nourish the developing embryonic cells.
>
> In the week following implantation, the inner cells of the blastocyst divide rapidly to form the embryonic disc, which will give rise to the three germ layers—the ectoderm, the mesoderm, and the endoderm. These three layers will eventually develop into the embryo. By 14 days, the embryonic disc is approximately 0.5 mm in diameter and consists of approximately 2,000 cells. It is at this time that the first stage of organized development, known as gastrulation, is initiated, leading over the next few days to the first appearance of differentiated tissues of the body, including primitive neural cells. Gastrulation is the process by which the bilaminar (two-layered) embryonic disc is converted into a trilaminar (three-layered) embryonic disc, and its onset at day 14 *in vivo* is marked by the appearance of the primitive streak, a region in which cells move from one layer to another in an organized way.
>
> During the third week, the embryo grows to 2.3 mm long, and the precursors of most of the major organ systems begin to form. At the beginning of the third month, the embryo becomes a fetus. During the third to ninth months, the organ systems and tissues of the fetus continue to develop, until birth.

expressly for research purposes, and 3) embryos resulting from somatic cell nuclear transfer (SCNT) or other cloning techniques. SCNT technology has, in fact, opened the door to a possible alternative approach to creating ES cells. (See Figure 2-4.) If the nucleus is removed from an immature egg (oocyte) and a mature diploid nucleus is inserted, the resulting cell will divide and develop with many characteristics of an embryo. In animal experiments in which a SCNT-derived embryo is transferred to a surrogate mother, a successful pregnancy may be established. (This was the technique used to generate the now-famous cloned sheep Dolly.) If, instead of being transferred to a surrogate, the SCNT-derived embryo is kept in culture, is allowed to divide, and is then dissociated, ES cells can

Figure 2-2. Stages of Development of the Human Embryo and Fetus

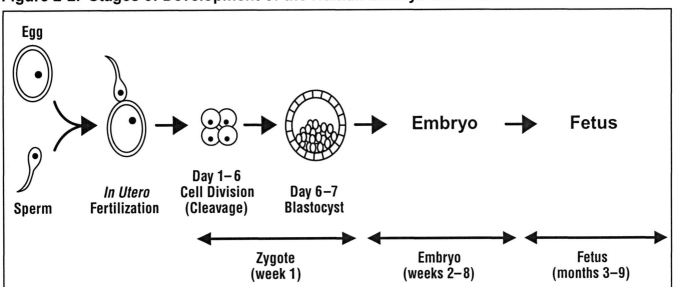

Figure 2-3. Isolation and Culture of Human ES Cells from Blastocysts

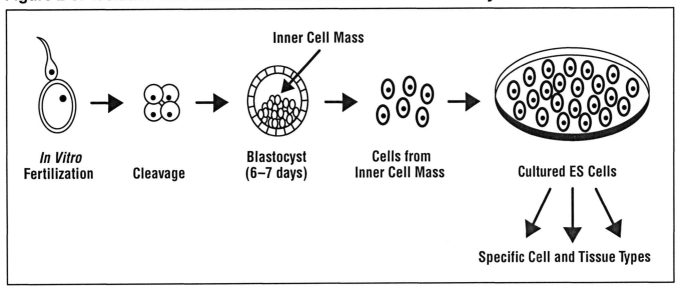

be derived. The potential advantage of using SCNT technology to create ES cells is that a somatic cell from an individual can be used to create ES cells that are completely compatible with that individual's tissue type. If cells or tissues are generated from these ES cells for transplant into a person, this tissue type compatibility may avoid many of the problems associated with tissue graft rejection that are currently encountered in the treatment of a variety of diseases.

The use of SCNT into an oocyte has been criticized as an asexual or "unnatural" way of creating a human embryo. However, it is important to distinguish the technique of SCNT from the type of cell that is created; in other words, SCNT techniques also might be used with recipient cells other than oocytes. For example, ES cells with matched tissue types for transplant might be generated by SCNT into an enucleated ES cell.[2] This possibility has not yet been explored, but it may be less morally

Figure 2-4. Isolation and Culture of Human ES Cells from SCNT

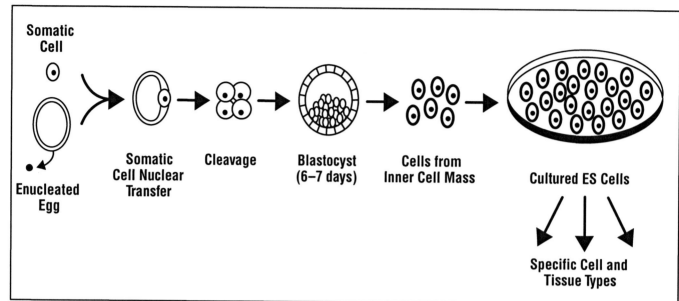

problematic to many citizens, because the cell created would not be an embryo with the potential to continue developing.

Stem Cells Found in the Postnatal and Adult Organism

In the adult mammal, cell division occurs in order to maintain a constant number of terminally differentiated cells in tissues in which cells have been lost due to injury, disease, or natural cell death. Cells with a high turnover rate are replaced through a highly regulated process of proliferation, differentiation, and apoptosis (programmed cell death) from relatively undifferentiated stem cells, or precursor cells (Thomson and Marshall 1998). The best known example of an AS cell is the hematopoietic stem cell, which is found in bone marrow and which is responsible for the production of all types of blood cells (Iscove 1990). Other examples of stem cells include the skin epithelium and the epithelium of the small intestine (Hall and Watt 1989). In the human small intestine, for example, approximately 100 billion cells are shed and must be replaced daily (Potten and Loeffler 1990). These tissues contain subpopulations of dividing stem cells that generate replacements for the relatively short-lived, terminally differentiated cells. Much of the debate in the stem cell field revolves around determining the breadth of the potential of these cells: Can they generate only the cells of that organ or are they capable of differentiating into several types of cells when given the proper stimuli?

The successful cloning of Dolly demonstrated that even somatic cells are capable of forming every cell of an organism after nuclear transfer into an oocyte (Wilmut et al. 1997). Preliminary studies of stem cells obtained from various systems of the adult organism suggest that in some cases the reactivation of dormant genetic programs may not require nuclear transfer or experimental modification of the genome. Although research in this area is preliminary, this particular class of stem cells (i.e., AS cells) might be able to differentiate along several cell lineages in response to an appropriate pattern of stimulation.

Neural Stem Cells

For a number of years, scientists have recognized that transplantation of fresh fetal neural tissue into the diseased adult brain may be a promising therapy for neurodegenerative disorders. This type of transplantation recently has been shown to be effective in younger patients with Parkinson's disease (Freed et al. 1999). This technique has several disadvantages, however, such as the need to time the surgery according to the availability of large amounts of fresh fetal tissue, the need to quickly screen for infectious diseases, and the limited amount of donor fetal tissue available (Bjorklund 1993; Cattaneo and McKay 1991). By developing techniques to culture and expand primary fetal neural cells before transplantation, some of the problems of using fresh tissue may be eliminated. In addition, it might be possible to direct cultured cells to develop along certain lineages or to express specific genes before they are transplanted, so that, for example, dopamine-producing cells could be selectively grown to treat Parkinson's disease (Cattaneo and McKay 1991; Snyder 1994).

Indeed, it has already been demonstrated that neural stem cells are capable of gene (Snyder, Taylor, and Wolfe 1995) and cellular (Rosario et al. 1997; Snyder et al. 1997; Yandava, Billinghurst, and Snyder 1999) replacement in models of neural disease. In many of these experiments, one stable clone of mouse neural stem cells could be used from individual to individual, strain to strain, and disease to disease, regardless of recipient age within the species, without immunorejection or the need for immunosuppression. This suggests that unique immune qualities may exist within stem cells that might allow them to be universal donors. Moreover, the possibility exists that many of the instructive cues for differentiation actually might originate from interaction with damaged central nervous system tissue itself.

The embryonic nervous system arises from the ectoderm. The first cell type to differentiate from the uncommitted precursor cells is the neuron, followed by the oligodendrocyte, and then the astrocyte (Frederiksen and McKay 1988). Recently, Angelo Vescovi, a neurobiologist at the National Neurological Institute Carlo Besta in Milan, Italy, and his colleagues reported that neural stem cells, which give rise to the three main types of brain cells, also can become blood cells when transplanted into mice in which the blood-forming tissue—the bone marrow—has been mostly destroyed (Bjornson et al. 1999). Although the study did not explain what caused the neural cells to turn into blood cells, the investigators speculate that the neural cells might be responding to the same signals that normally stimulate the few remaining blood stem cells to reproduce and mature after irradiation destroys most of the bone marrow (Strauss 1999). Although this research is preliminary and has not yet been conducted using human cells, it raises the possibility of using neural stem cell transplants to treat human blood cell disorders such as aplastic anemia and severe combined immunodeficiency. This is an appealing prospect, because bone marrow stem cells do not replenish themselves well in laboratory cultures. The problem of access to such cells in humans remains, as they must be obtained from the brain—an invasive and risky procedure. This research also opens up the possibility that other apparently restricted AS cells may retain the ability to differentiate into several different types of cells if exposed to a conducive external environment. It is clear that further research is required in this area.

Mesenchymal Stem Cells

Human mesenchymal stem cells, which are present in adult bone marrow, can replicate as undifferentiated cells and have the potential to differentiate into lineages of mesenchymal tissues, including bone, cartilage, fat, tendon, muscle, and marrow stroma (Pittenger et al. 1999). In a recent experiment, cells that have the characteristics of human mesenchymal stem cells were isolated from marrow aspirates of volunteer donors. Individual stem cells were identified that, when expanded to colonies, retained their multilineage potential. These results demonstrate that isolated expanded human mesenchymal stem cells in culture will differentiate, in a controlled manner, to multiple but limited lineages. One might speculate that these particular AS cells could be induced to differentiate exclusively into the adipocytic, chondrocytic, or osteocytic lineages, which then might be used to treat various bone diseases.

The specific environmental cues needed to initiate the proliferation and differentiation of these cells are not understood (Pittenger et al. 1999). The ability to isolate, expand, and direct the differentiation of such cells in culture to particular lineages, however, offers the opportunity to study events associated with cell commitment and differentiation. The human mesenchymal stem cells isolated by Pittenger and colleagues appear to have the ability to proliferate extensively and to maintain the ability to differentiate into certain cell types in culture. Their cultivation and selective differentiation should provide further information about this important progenitor of multiple tissue types and the potential of new therapeutic approaches for the restoration of damaged or diseased tissue (Pittenger et al. 1999).

Animal Models

ES cells were first derived from mouse embryos, and the mouse has become the principal model for the study of these cells (Evans and Kaufman 1981; Martin 1981). If mouse ES cells are injected into the developing blastocyst, they have the ability to contribute to all three germ layers of the mouse, including the germline, to form a chimeric animal. This is one of the unique properties of the mouse ES cell. More recently, cells with some properties of ES cells have been derived from cows, pigs, rats, sheep, hamsters, rabbits, and primates (Pedersen 1994). (See Table 2-1.) However, only in cows, pigs, and rats did these ES cells contribute to a chimeric animal, and in none of these cases was there contribution to the germline by ES cells, one of the most stringent criteria for defining ES cells.

Mouse ES Cells

ES cells were first isolated from mouse blastocysts in 1981 (Evans and Kaufman 1981; Martin 1981). These blastocysts were placed in culture and allowed to attach to the culture dish so that trophoblast cells spread out, while the undifferentiated inner cells (the inner cell mass) continued to grow as a tight but disorganized cluster. Before the inner cell mass developed into the equivalent of the embryonic disc, it was drawn up into a fine pipette, dissociated into single cells, and dispersed into another dish with a rich culture medium. Under these circumstances, the dissociated cells continued to grow rapidly for an extended period.

Mouse ES cells cannot become organized into an embryo by themselves or implant into the uterus if placed there. However, if the cells are injected back into a new blastocyst, they can intermingle with the host inner cell mass to make a chimera and participate in normal development, eventually contributing to all of the tissues of the adult mouse, including nerve, blood, skin, bone, and germ cells (Robertson and Bradley 1986). This

Table 2-1. Stem Cells Isolated from Mammals

Species	References
Mouse	Evans and Kaufman 1981
	Martin 1981
Rat	Iannaccone et al. 1994
Hamster	Doetschman, Williams, and Maeda 1988
Mink	Sukoyan et al. 1992
	Sukoyan et al. 1993
Rabbit	Moreadith and Graves 1992
	Giles et al. 1993
	Graves and Moreadith 1993
Sheep	Handyside et al. 1987
	Piedrahita, Anderson, and Bondurant 1990
	Notarianni et al. 1991
Pig	Piedrahita et al. 1988
	Evans et al. 1990
	Notarianni et al. 1990
	Piedrahita et al. 1990
	Hochereau-de Reiviers and Perreau 1993
	Talbot et al. 1993
	Wheeler 1994
	Shim et al. 1997
Cow	Evans et al. 1990
	Saito, Strelchenko, and Niemann 1992
	Strelchenko and Stice 1994
	Cibelli et al. 1998
Common Marmoset	Thomson et al. 1996
Rhesus Monkey	Thomson et al. 1995
Human	Bongso et al. 1994
	Shamblott et al. 1998
	Thomson et al. 1998

indicates that mouse ES cells have not lost the capacity to give rise to specialized tissues, but they will not do so unless placed in a conducive environment.

The ability of mouse ES cells to enter the germline in chimeras allows the introduction of specific genetic changes into the mouse genome and offers a direct approach to understanding gene function in the intact animal (Rossant, Bernelot-Moens, and Nagy 1993). Using the technique of homologous recombination in which a gene is either modified or disabled ("knocked out"), mouse ES cells that contain specific gene alterations may be derived. These genetically altered cells can then be used to form chimeras with normal embryos, subsequently generating a mouse lacking one specific gene or containing an extra or altered gene.

Mouse ES cells also have been extremely useful as models of the early differentiation events that occur during the development of mammalian embryos (Pedersen 1994), as shown in the following examples:

- When mouse ES cells were allowed to differentiate in culture, beating heart cells formed spontaneously, providing a model for cardiac-specific gene expression and the development of cardiac muscle and blood vessels (Chen and Kosco 1993; Doetschman et al. 1993; Miller-Hance et al. 1993; Muthuchamy et al. 1993; Robbins et al. 1990; Wobus, Wallukat, and Heschler 1991).

- Blood formation will occur spontaneously in ES cell-derived embryoid bodies and can be augmented by modifying the culture conditions (Snodgrass, Schmitt, and Bruyns 1992). Therefore, hematopoietic stem cells have been studied extensively in an effort to determine the conditions for differentiation, survival, and proliferation of blood cells.

- Several studies have highlighted the importance of growth and differentiation factors in the regulation of mammalian development. For example, the maintenance of mouse ES cells in an undifferentiated state was found to require the presence of leukemia inhibitory factor, a differentiation-inhibiting factor (Fry 1992). Other studies have found several growth and differentiation factors to be important in ES cell development and differentiation, including activins, colony-stimulating factor, erythropoietin, basic fibroblast growth factor, insulin-like growth factor 2, interleukins, parathryoid hormone-related peptide, platelet-derived growth factor, steel factor, and transforming growth factor ß (Pedersen 1994).

- In midgestation embryos and the adult mouse, only one parental allele of imprinted genes is expressed. However, studies have suggested that there is limited relaxation of imprinting in ES cells so that both maternal and paternal alleles are expressed (Pedersen 1994).

By understanding the mechanisms responsible for growth and differentiation in embryonic development, it may then be possible to attempt to regulate the differentiation of ES cells along specific pathways. The knowledge gained from these types of studies could someday lead to the effective treatment of certain important human diseases.

Historically, because of its well-defined genetics, short gestational time, ease of cultivation, and large litters, the mouse has been one of the primary models for the study of mammalian embryonic development. However, there are several differences between early mouse development and early human development, including

- the timing of embryonic genome expression (Braude, Bolton, and Moore 1988),

- the formation, structure, and function of the fetal membranes and placenta (Benirschke and Kaufmann 1990; Luckett 1975, 1978), and

- the formation of an egg cylinder (mouse) as opposed to an embryonic disc (human) (Kaufmann 1992; O'Rahilly 1987).

Thus, other animal models as well as new models that would allow the direct study of human embryonic development are crucial in order to comprehend early human development and to understand the growth requirements of human stem cells of specific lineages.

Bovine ES Cells

The first bovine ES-like cells were reported by Saito, Strelchenko, and Niemann in 1992. More recently, transgenic bovine ES-like cells were derived by using nuclear transfer of fetal fibroblasts to enucleated bovine oocytes (Cibelli et al. 1998). This technique involved introducing a marker gene into bovine fibroblasts from a 55-day-old fetus and then fusing the transgenic fibroblasts to enucleated oocytes to produce blastocyst-stage nuclear transplant embryos (Cibelli et al. 1998). ES-like cells then were derived from these embryos and were used to create chimeric embryos. When reintroduced into pre-implantation embryos, these transgenic ES-like cells differentiated into derivatives from the three EG layers—ectoderm, mesoderm, and endoderm (Cibelli et al. 1998). Bovine ES cells would be useful in agricultural production of transgenic cows and also may have the potential for generating tissues and organs for use in cross-species transplantation (xenotransplantation) in order to treat human diseases.

Primate ES Cells

Primate ES-like cells have been derived from both the rhesus monkey (Thomson et al. 1995) and the common marmoset (Thomson et al. 1996). When allowed to grow, both marmoset and rhesus ES cells spontaneously differentiate into more complex structures, including cardiac muscle, neurons, endoderm, trophoblast, and numerous unidentified cell types (Thomson and Marshall 1998).

Essential characteristics of these primate ES-like cells include 1) derivation from the pre-implantation or peri-implantation embryo, 2) prolonged undifferentiated proliferation, and 3) stable developmental potential to form derivatives of all three EG layers even after prolonged maintenance in culture (Thomson and Marshall 1998). In addition, although mouse ES cells rarely contribute to trophoblast in chimeras (Beddington and Robertson 1989), primate ES cells differentiate into all three germ layers and trophoblast-like cells (Thomson and Marshall 1998). Furthermore, some primate ES cell lines have maintained a normal karyotype through undifferentiated culture for at least two years, sustained a stable developmental potential throughout this culture period, and maintained the potential to form trophoblast *in vitro* (Thomson et al. 1995, 1996).

Although there is some variation between species, nonhuman primate ES cell lines appear to provide a useful *in vitro* model for understanding the differentiation of human tissues (Thomson and Marshall 1998), and primate ES cells provide a powerful model for understanding human development and disease. Furthermore, because of the similarities between human and primate ES cells, primate ES cells provide a model for developing strategies to prevent immune rejection of transplanted cells and for demonstrating the safety and efficacy of ES cell-based therapies (Thomson et al. 1995).

Human Models

Human ES Cell Lines Derived from Blastocysts

The first successful isolation of cells from the human inner cell mass of blastocysts and their culture *in vitro* for at least two series of cell divisions was reported by Bongso and colleagues in 1994. Starting with 21 spare embryos donated by nine patients in an IVF program,[3] this group isolated cells with typical stem cell characteristics from 17 five-day-old blastocysts (approximately 100 cells) (Bongso et al. 1994). These cells were like ES cells. They were small and round with high nuclear to cytoplasmic ratios, they stained positively for alkaline phosphatase (a biochemical marker for stem cells), and they maintained a normal diploid karyotype. However, after the second subculture, the cells differentiated into fibroblasts or died (Bongso et al. 1994).

In later work, Thomson and his colleagues were able to isolate human ES-like cell lines and grow them continuously in culture for at least five to six months. Although these cells have not passed the most stringent test—as have mouse ES cells—to determine whether they can contribute to the germline, we will continue to use the term *ES cell* throughout this report because both scientists and nonscientists alike have widely applied this term to refer to these cells. This renewable tissue culture source of human cells—capable of differentiating into a wide variety of cell types—is believed to have broad applications in basic research and transplantation therapies (Gearhart 1998).

In Thomson's work, human ES cells were isolated from embryos that were originally produced by IVF for clinical reproductive purposes. (See Exhibit 2-B.) Individuals donated the embryos, following an informed consent process. The consent forms and the entire research protocol were reviewed and approved by an appropriately constituted Institutional Review Board (IRB) (Thomson et al. 1998). Thirty-six embryos were cultured for approximately five days. The inner cell mass was isolated from 14 of the 20 blastocysts that developed,

> **Exhibit 2-B: *In Vitro* Fertilization (IVF)**
>
> The procedure of IVF today is widely available in many countries throughout the world, including the United States. Originally developed for the treatment of infertility due to blocked fallopian tubes, IVF has been extended to assist patients with premature depletion of oocytes, recurrent failure of embryos to implant, and low production of functional sperm. More recently, the technique has been used in conjunction with pre-implantation genetic diagnosis to enable fertile couples at risk for transmitting severe or fatal inherited diseases to have healthy children.
>
> Although details of the IVF procedures vary from center to center, the basic approach is to treat oocyte donors over several days with a regimen of hormones designed to stimulate the final maturation of several follicles within the ovary. This is known as hyperstimulation, a procedure that carries the risk of an adverse reaction of less than 1 percent. Following completion of the hormone treatment, mature follicles are detected by sonography and an average of ten are collected by transvaginal aspiration while the patient is sedated. The oocytes are then fertilized by sperm collected from a male donor and cultured in sterile fluid for about two days. When the zygote has reached the four- to eight-cell stage, between three and six zygotes are transferred to the uterus, and the untransferred embryos, if they are developing normally, are usually frozen. Nonviable embryos are discarded. (See also Figure 2-5.) More recently, IVF specialists have begun culturing embryos to the blastocyst stage before transfer to the uterus.
>
> The efficiency of the IVF procedure is relatively low, with approximately 20 percent of fertilized eggs resulting in successful pregnancies, depending on factors such as age of the recipient and the reason for infertility. In comparison, approximately 30 percent of normally conceived human embryos result in successful pregnancies. Embryos that are not transferred can be cryopreserved and stored indefinitely.
>
> **Sources:**
> National Institutes of Health (NIH). Human Embryo Research Panel. 1994. *Report of the Human Embryo Research Panel.* 2 vols. Bethesda, MD: NIH.
>
> New York State Task Force on Life and the Law. 1998. *Assisted Reproductive Technologies: Analysis and Recommendations for Public Policy.* New York: New York State Task Force on Life and the Law.

and five ES cell lines, originating from five separate embryos, were derived (Thomson et al. 1998). The technique used to derive these human ES cells is essentially the same as that used to isolate nonhuman primate ES cell lines (Thomson et al. 1995).

The resulting human ES cell lines had normal karyotypes (two male and three female) and were grown in culture continuously for at least five to six months (Thomson et al. 1998). In addition, the cell lines expressed cell surface markers that also are found on nonhuman primate ES cells (Thomson et al. 1998). Most important, the cell lines maintained the potential to form derivatives of all three EG layers—endoderm, mesoderm, and ectoderm (Thomson et al. 1998).

Many believe that research using human ES cells might offer insights into developmental events that cannot be studied directly in the intact human embryo but that have important consequences in clinical areas such as birth defects, infertility, and miscarriage. Some speculate that the origins of many human diseases (e.g., juvenile-onset diabetes) are due to events that occur early in embryonic development. Such cells also will be particularly valuable for the study of the development and function of tissues that differ between mice and humans. These cells allow for studies that focus on the differentiation of cells into specific tissues and the factors that bring about differentiation, so that cells can be manipulated to generate specific cell types for therapeutic transplantation. Moreover, it may be possible to identify gene targets for new drugs, to manipulate genes that could be used for tissue regeneration therapies, and to understand the teratogenic or toxic effects of certain compounds (Thomson et al. 1998).

Human EG Cells from Fetal Primordial Germ Cells

Primordial germ cells also can give rise to cells with characteristics of ES cells, and, as discussed previously,

Figure 2-5. 1996 Assisted Reproductive Technology (ART) Success Rates[a]

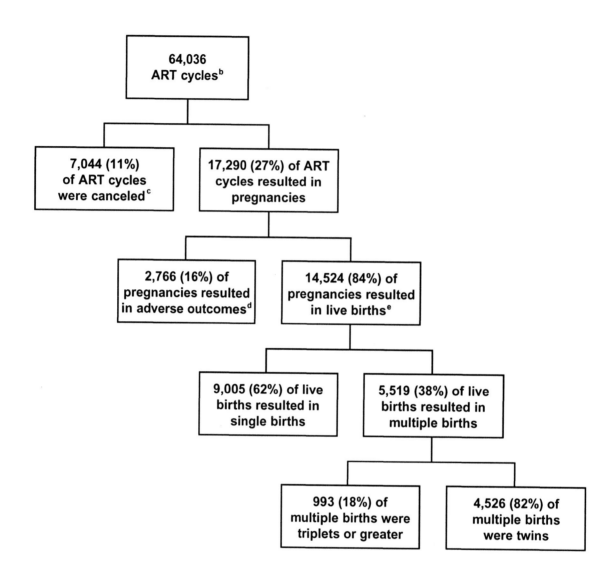

[a] Source: Centers for Disease Control and Prevention (CDC), American Society for Reproductive Medicine, Society for Assisted Reproductive Technology, and RESOLVE. 1998. *1996 Assisted Reproductive Technology Success Rates, National Summary and Fertility Clinic Reports*. Atlanta, GA: CDC.

[b] Data are from 300 U.S. fertility clinics that provided and verified information about the outcomes of all ART cycles started in their clinics in 1996.

[c] Fresh, nondonor cycles were canceled, most commonly because too few (egg) follicles developed. Illness unrelated to the ART procedure also may lead to cancelation. In general, cycles are canceled when chances of success are poor or risks are unacceptably high.

[d] Adverse outcomes included spontaneous abortion (83%), induced abortion (10%), stillbirth (4%), and ectopic pregnancy (3%)

[e] A total of 20,659 babies were born as a result of the 64,036 ART cycles carried out in 1996.

have been designated as EG cells in order to distinguish their tissue of origin (Gearhart 1998). A 1998 report from John D. Gearhart and his colleagues describes the establishment of human EG cell lines from human primordial germ cells (Shamblott et al. 1998). Using an IRB-approved protocol, the human EG cells were isolated from the developing gonads of five- to nine-week-old embryos and fetuses that were obtained following elective abortion (Shamblott et al. 1998). These human EG cell lines have morphological, immunohistochemical, and karyotypic features consistent with those of previously described ES cells and have a demonstrated ability to differentiate *in vitro* into derivatives of the three germ layers (Shamblott et al. 1998).

Fusion of Human Somatic Cells with Cow Eggs to Create Hybrid Embryonic Cells

Advanced Cell Technology of Worcester, Massachusetts, announced in November 1998 that its scientists had made human somatic cells revert to the primordial state by fusing them with cow eggs to create a hybrid embryo (Wade 1998). This work with human cells was performed in 1996 by Jose Cibelli. Using 52 of his own cells—some of them white blood cells and others scraped from the inside of his cheek—Cibelli used a pulse of electricity to fuse each cell with a cow egg from which the nucleus containing the DNA had been removed.[4] Out of these 52 attempts, only one embryo, derived from a cheek epithelial cell, developed into a blastocyst. Approximately 12 days after the fusion of cheek cell and cow egg, sufficient cells existed to allow harvesting of the inner cell mass to produce cells resembling human ES cells. The researchers observed that the hybrid cell quickly became more human-like as the human nucleus took control and displaced bovine proteins with human proteins. However, it is difficult to judge the validity of this work and the nature of the "embryo-like" material produced because the work is extremely preliminary and has not been submitted for peer review or for publication in a scientific journal.

The stated purpose of these experiments was to create an embryo solely for the purpose of establishing an ES cell line that might be used to treat any disease caused by the loss or malfunction of cells, such as Parkinson's disease, diabetes, and heart disease. The researchers emphasized that they had no intention of transferring the resulting hybrid embryos to a uterus, as they considered this to be both unethical and unsafe (Wade 1998).

Growth and Derivation of ES Cells

Human ES cells are different from many adult cells because they have the ability to divide extensively in culture. Although this property has been interpreted by nonscientists as an indication that investigators simply can use existing human ES and EG cell lines (which can be extensively reproduced for a limited time) to study their properties, this is not the case and is a reflection of a misunderstanding of the science that is involved. Evidence from mouse ES cell research suggests that it is essential to derive new ES cell lines repeatedly in order to further our understanding of how to differentiate these cells and grow them extensively in culture.

There are several reasons why it is necessary to repeatedly derive new ES cell lines. First, the properties of ES cells differ depending on the methods used to derive them.[5] Cells derived under some conditions may be limited in their potential to differentiate into a particular tissue type. Second, ES cells are not stable cell types that can simply be mass produced and supplied to an unlimited number of researchers. As these cells grow in culture they accumulate irreversible changes, and the conditions used to grow them can influence the speed at which these changes accumulate. Typically, researchers look only at the ability of ES cells to contribute to some tissues. In one study, however, the ability of ES cells to generate *all* tissue in a mouse was tested (Nagy et al. 1993). This research has shown dramatically that existing cell lines commonly in use by many researchers have lost the ability to generate all mouse tissues and thus to completely generate live mice. When new ES cells were derived and grown for only a short time in culture, they did allow all tissues to be generated. However, after about 14 doublings in culture, even these cells lost their ability to contribute to all tissues. The researchers conclude... "[P]rolonged passage in culture reduces the potential of the ES cell population as a whole. The proportion of cells that retain full potential diminishes with extended passage" (Nagy et al. 1993). Exactly what changes occur during culture are not yet clear. The chromosome complement remains normal, indicating that this criterion, although frequently used to characterize ES cells, is not a very stringent assay. It could be an accumulation of mutations, changes in gene expression, or epigenetic changes (Nagy et al. 1993). Thus, if one scientist were to obtain cells from a colleague's laboratory, the properties of the cells would depend greatly on the history of how those cells were grown. For this reason, many people who work with mouse ES cells re-derive the cells periodically to be sure the cells have the potential to differentiate into or contribute to many different tissues.

Finally, perhaps the most important reason for deriving new ES cell lines rather than simply working with existing cell lines is that a tremendous amount remains to be learned during the process of derivation itself. It took many laboratories more than ten years to ascertain appropriate conditions for the derivation and growth of mouse ES cells. Research on the growth and derivation of ES cells from other mammalian species is only in its early stages. In fact, only mouse ES cells have the property of contributing to the germline cell lineage—the most stringent criterion for ES cells. Thus, cells from other species are referred to as ES-like cells (Pedersen 1994). Further basic research into the proper conditions to maintain ES cells from many species is ongoing in an attempt to understand the factors necessary to generate stable ES cells. Given that only two successes have been reported on the derivation of human ES and EG cells, it is likely that significant basic research into the appropriate conditions to generate stable stem cells will be needed.

Potential Medical Applications of Human ES Cell and EG Cell Research

Although research into the use of ES and EG cells is still at an early stage, researchers hope to make a contribution to disease treatment in a variety of areas. The ability to elucidate the mechanisms that control cell differentiation is, at the most elemental level, the promise of human ES and EG cell research. This knowledge will facilitate the efficient, directed differentiation of stem cells to specific cell types. The standardized production of large, purified populations of human cells such as cardiomyocytes and neurons, for example, could provide a substantial source of cells for drug discovery and transplantation therapies (Thomson et al. 1998). Many diseases, such as Parkinson's disease and juvenile-onset diabetes mellitus, result from the death or dysfunction of just one or a few cell types, and the replacement of those cells could offer effective treatment and even cures.

Substantial advances in basic cell and developmental biology are required before it will be possible to direct human ES cells to lineages of human clinical importance. However, progress has already been made in the differentiation of mouse ES cells to neurons, hematopoietic cells, and cardiac muscle (Brustle et al. 1997; Deacon et al. 1998; Shamblott et al. 1998). Human ES and EG cells could be put to use in targeting neurodegenerative disorders, diabetes, spinal cord injury, and hematopoietic repopulation, the current treatments for which are either incomplete or create additional complications for those who suffer from them.

Use of Human ES Cells and EG Cells in Transplantation

One of the major causes of organ transplantation and graft failure is immune rejection, and a likely application of human ES and EG cell research is in the area of transplantation. Although much research remains to be done, ES cells derived through SCNT offer the possibility that therapies could be developed from a patient's own cells. In other words, a patient's somatic cells could be fused with an enucleated oocyte and developed to the blastocyst stage, at which point ES cells could be derived for the development of cell-based therapy. This essentially is an autologous transfer. Thus, issues of tissue rejection due to the recognition of foreign proteins by the immune system are avoided entirely. In addition, research to establish xenotransplantation (i.e., interspecies transplantation) as a safe and effective alternative to human organ transplantation is still in its infancy. Alternately, other techniques that would be immunologically compatible for transplantation purposes could be used to generate stem cells, such as

1) banking of multiple cell lines representing a spectrum of major histocompatibility complex (MHC) alleles to serve as a source for MHC matching, and/or

2) creating universal donor lines, in which the MHC genes could be genetically altered so rejection would not occur, an approach that has been tried in the mouse with moderate success (NIAID 1999).

Autologous transplants would obviate the need for immunosuppressive agents in transplantation as it would decrease a central danger to transplant patients—susceptibility to other diseases. Autologous transplants might address problems ranging from the supply of donor organs to the difficulty of finding matches between donors and recipients. Research on ES cells could lead to cures for diseases that require treatment through transplantation, including autoimmune diseases such as multiple sclerosis, rheumatoid arthritis, and systemic lupus erythematosus. These cells also might hold promise for treating type-I diabetes (Melton 1999; Varmus 1998), which would involve the transplantation of pancreatic islet cells or beta cells produced from autologous ES cells. These cells would enter the pancreas and provide normal insulin production by replacing the failing resident islet cells.

Studies of Human Reproduction and Developmental Biology

Research using human ES and EG cells could offer insights into developmental events that cannot be studied directly in the intact human embryo but that have important consequences in clinical areas, including birth defects, infertility, and pregnancy loss (Thomson et al. 1998). ES and EG cells provide large quantities of homogeneous material that can be used for biochemical analysis of the patterns of gene expression and the molecular mechanisms of embryonic differentiation.

Cancer Therapy

Human ES and EG cells may be used to reduce the tissue toxicity brought on by cancer therapy (NCI 1999). Already, bone marrow stem cells, representing a more committed stem cell, are used to treat patients after high-dose chemotherapy. However, the recovered blood cells appear limited in their ability to recognize abnormal cells, such as cancer cells. It is possible that injections of ES and EG cells would revive the complete immune response to patients undergoing bone marrow transplantation. Current approaches aimed at manipulating the immune system after high-dose chemotherapy so that it recognizes cancer cells specifically have not yet been successful.

Diseases of the Nervous System

Some believe that in no other area of medicine are the potential benefits of ES and EG cell research greater than in diseases of the nervous system (Gearhart 1998; Varmus 1998). The most obvious reason is that so many of these diseases result from the loss of nerve cells, and mature nerve cells cannot divide to replace those that are lost. For example, in Parkinson's disease, nerve cells that make the chemical dopamine die; in Alzheimer's disease, it is the cells that make acetylcholine that die; in Huntington's disease the cells that make gamma aminobutyric acid die; in multiple sclerosis, cells that make myelin die; and in amyotrophic lateral sclerosis, the motor nerve cells that activate muscles die. In stroke, brain trauma, spinal cord injury, and cerebral palsy and mental retardation, numerous types of cells are lost with no built-in mechanism for replacing them.

Preliminary results from fetal tissue transplantation trials for Parkinson's disease suggest that supplying new cells to a structure as intricate as the brain can slow or stop disease progression (Freed et al. 1999). Yet the difficulty of obtaining enough cells of the right type—that is, dopamine-producing nerve cells—limits the application of this therapy. In 1999, scientists developed methods in animal models to isolate dopamine precursor cells from the dopamine-producing region of the brain and coax them to proliferate for several generations in cell culture. When these cells were implanted into the brains of rodents with experimental Parkinson's disease, the animals showed improvements in their movement control (NINDS 1999). Scientists also have learned to instruct a stem cell from even a nondopamine region to make dopamine (Wagner et al. 1999). A large supply of "dopamine-competent" stem cells, such as ES cell lines, could remove the barrier of limited amounts of tissue. (See Exhibit 2-C.)

Another recent development eventually may provide treatments for multiple sclerosis and other diseases that attack the myelin coating of nerves. Scientists have successfully generated glial cells that produce myelin from

Chapter 2: Human Stem Cell Research and the Potential for Clinical Application

> **Exhibit 2-C: Potential Treatment for Parkinson's Disease**
>
> Parkinson's disease is a degenerative brain disease that affects 2 percent of the population over age 70. Symptoms include slow and stiff movements, problems with balance and walking, and tremor. In more advanced cases, the patient has a fixed, staring expression, walks with a stooped posture and short, shuffling pace, and has difficulty initiating voluntary movements. Falls, difficulty swallowing, incontinence, and dementia may occur in the late stages. Patients often lose the ability to care for themselves and may become bedridden.
>
> The cause of this illness is a deficiency of the neurotransmitter dopamine in specific areas of the brain. Treatment with drugs such as levodopa often is effective in relieving the symptoms. However, as the disease progresses, treatment often becomes more problematic, with irregular responses, difficulty adjusting doses, and the development of side effects such as involuntary writhing movements. Brain surgery with transplantation of human fetal tissue has shown promise as therapy.
>
> Stem cell transplantation also may be a promising therapy for Parkinson's disease. The injection of stem cells that can differentiate into brain cells may offer a means of replenishing neurons that are capable of synthesizing the deficient neurotransmitter. It is possible that stem cell transplantation may be simpler and more readily available than fetal tissue transplantation.

mouse ES cells (Brustle et al. 1999). When these ES cell-derived glial cells were transplanted in a rat model of human myelin disease, they were able to interact with host neurons and efficiently myelinate axons in the rat's brain and spinal cord (Brustle et al. 1999).

Other diseases that might benefit from similar types of approaches include spinal cord injury, epilepsy, stroke, Tay-Sachs disease, and pediatric causes of cerebral palsy and mental retardation. In mice, neural stem cells already have been shown to be effective in replacing cells throughout the brain and in some cases are capable of correcting neurological defects (Lacorazza et al. 1996; Rosario et al. 1997; Snyder et al. 1997; Snyder, Taylor, and Woolfe 1995; Yandava, Billinghurst, and Snyder 1999). Human neural stem cells also have recently been isolated and have been shown to be responsive to developmental signals and to be willing to replace neurons when transplanted into mice (Flax et al. 1998). These recent discoveries of ways to generate specific types of neural cells from ES cells hold much promise for the treatment of severe neurological disorders that today have no known cure.

Diseases of the Bone and Cartilage

Because ES and EG cells constitute a relatively self-renewing population of cells, they can be cultured to generate greater numbers of bone or cartilage cells than could be obtained from a tissue sample. If a self-renewing, but controlled, population of stem cells can be established in a transplant recipient, it could effect long-term correction of many diseases and degenerative conditions in which bone or cartilage cells are deficient in numbers or defective in function. This could be done either by transplanting ES and EG cells to a recipient or by genetically modifying a person's own stem cells and returning them to the marrow. Such approaches hold promise for the treatment of genetic disorders of bone and cartilage, such as osteogenesis imperfecta and the various chondrodysplasias. In a somewhat different potential application, stem cells perhaps could be stimulated in culture to develop into either bone- or cartilage-producing cells. These cells could then be introduced into the damaged areas of joint cartilage in cases of osteoarthritis or into large gaps in bone that can arise from fractures or surgery. This sort of repair would have a number of advantages over the current practice of tissue grafting (NIAMS 1999).

Blood Disorders

The globin proteins are essential for transport of oxygen in the blood, with different globins expressed at different developmental stages. The epsilon globin gene is expressed only in embryonic red blood cells. When this gene—which is not normally expressed in the adult—is artificially turned on in sickle cell patients, it blocks the sickling of the cells that contain sickle cell hemoglobin. Research involving ES cells could help answer questions about how to turn on the epsilon globin gene in adult blood cells and thereby halt the disease process. Stem cell research also may help produce transplantable cells that would not contain the sickle cell mutation.

Toxicity and Drug Testing

Human stem cell research offers promise for use in testing the beneficial and toxic effects of biologicals, chemicals, and drugs in the most relevant species for clinical validity—humans. Such studies could lead to fewer, less costly, and better designed human clinical trials yielding more specific diagnostic procedures and more effective systemic therapies. Beyond the drug development screening of pharmacological agents for toxicity and/or efficacy, human stem cell research could define new research approaches for clarifying the complex association of environmental agents with human disease processes (NIEHS 1999). It also makes possible a new means of conducting detailed investigations of the underlying mechanisms of the effects of environmental toxins or mixtures of toxins, including their subtle effects on the developing embryonic and fetal development tissue systems.

Transplantable Organs

Several researchers are investigating ways to isolate AS cells and create transplantable organs that may be used to treat a multitude of diseases that do not rely upon the use of embryonic or fetal tissue. Moreover, if it is found to be possible to differentiate ES cells into specific cell types, such stem cells could be an important source of cells for organ growth. For example, recent developments in animals have shown that it may be possible to create entire transplantable organs from a tissue base in a manner that would overcome such problems as the limited supply of organs and tissue rejection. Such a development—producing this tissue base by directing the growth of human embryonic cells—could be a major breakthrough in the field of whole organ transplantation.

For example, using tissue engineering methods, researchers have successfully grown bladders in the laboratory, implanted them into dogs, and shown them to be functional (Oberpenning et al. 1999). To create the bladders, small biopsies of tissue were taken from dog bladders. The biopsied tissue was then teased apart to isolate the urothelial tissue and muscle tissue, which were then grown separately in culture (Tanne 1999). The tissue was then applied to a mold of biodegradable material with the urothelial tissue on the inside and the muscle tissue on the outside. The new organs were transplanted within five weeks (Tanne 1999).

Dogs that received the tissue-engineered organs regained 95 percent of their original bladder capacity, were continent, and voided normally. When the new organs were examined 11 months later, they were completely covered with urothelial and muscle tissue and had both nerve and blood vessel growth. Dogs that did not undergo reconstructive procedures or only received implants of the biodegradable molds did not regain normal bladder function (Oberpenning et al. 1999). This accomplishment marks the first time a mammalian organ has been grown in a laboratory. The ability to create new organs by seeding molds with cells of specific tissue types would be extremely useful in treating children with congenital malformations of organs and people who have lost organs due to trauma or disease (Tanne 1999).

Summary

Currently, human ES cells can be derived from the inner cell mass of a blastocyst (those cells within the conceptus that form the embryo proper), and EG cells can be derived from the primordial germ cells of fetuses. These cells, present in the earliest stages of embryo and fetal development, can generate all of the human cell types and are capable, at least for some time, of self-renewal. A relatively renewable tissue culture source of human cells that can be used to generate a wide variety of cell types would have broad applications in basic research, transplantation, and other important therapies, and a major step in realizing this goal was taken in 1998 with the demonstration that human ES and EG cells can be grown in culture. The clinical potential for these stem cells is vast—they will be important for *in vitro* studies of normal human embryogenesis, human gene discovery, and drug and teratogen testing and as a renewable source of cells for tissue transplantation, cell replacement, and gene therapies.

Notes

1 For a summary of scientific progress in this field see Eiseman, E., "Human Stem Cell Research," RAND DRU-2171-NBAC, September 1999, a background paper prepared for the National Bioethics Advisory Commission.

2 Thomson, J.A., Testimony before NBAC. January 19, 1999. Washington, DC.

3 Consent to carry out this study was approved by the hospital ethical committee based on the guidelines on Assisted Reproductive Technology of the Ministry of Health, Singapore, that experimentation of human embryos up to day 14 of embryonic growth may be allowed (Bongso et al. 1994).

4 The details of this process are described in a European patent application (PCT/U397/12919 1997) and in testimony before the Commission by ACT President Michael West. November 17, 1998. Miami, FL.

5 Hogan, B., Testimony before NBAC. February 3, 1999. Princeton, NJ.

References

Beddingston, R.S., and E.J. Robertson. 1989. "An Assessment of the Developmental Potential of Embryonic Stem Cells in the Midgestation Mouse Embryo." *Development* 105:733–737.

Benirschke, K., and P. Kaufmann. 1990. *Pathology of the Human Placenta*. New York: Springer-Verlag.

Bjorklund, A. 1993. "Neurobiology: Better Cells for Brain Repair." *Nature* 362:414–415.

Bjornson C.R., R.L. Rietze, B.A. Reynolds, M.C. Magli, and A.L. Vescovi. 1999. "Turning Brain into Blood: A Hematopoietic Fate Adopted by Adult Neural Stem Cells *In Vivo*." *Science* 283:534–537.

Bongso, A., C.Y. Fong, S.C. Ng, and S. Ratnam. 1994. "Isolation and Culture of Inner Cell Mass Cells from Human Blastocysts." *Human Reproduction* 9(11):2110–2117.

Braude, P., V. Bolton, and S. Moore. 1988. "Human Gene Expression First Occurs Between the Four- and Eight-Cell Stages of Pre-Implantation Development." *Nature* 332:459–461.

Brustle, O., A.C. Spiro, K. Karram, K. Choudhary, S. Okabe, and R.D. McKay. 1997. "*In Vitro*-Generated Neural Precursors Participate in Mammalian Brain Development." *Proceedings of the National Academy of Sciences USA* 94:14809–14814.

Brustle, O., K.N. Jones, R.D. Learish, K. Karram, K. Choudhary, O.D. Weistler, I.D. Duncan, and R.D. McKay. 1999. "Embryonic Stem Cell-Derived Glial Precursors: A Source of Myelinating Transplants." *Science* 285:754–756.

Cattaneo, E., and R. McKay. 1991. "Identifying and Manipulating Neuronal Stem Cells." *Trends in Neurosciences* 14(8):338–340.

Chen, U., and M. Kosco. 1993. "Differentiation of Mouse Embryonic Stem Cells *In Vitro*: III. Morphological Evaluation of Tissues Developed After Implantation of Differentiated Mouse Embryoid Bodies." *Developmental Dynamics* 197:217–226.

Cibelli, J.B., S.L. Stice, P.J. Golueke, J.J. Kane, J. Jerry, C. Blackwell, F.A. Ponce de Leon, and J.M. Robl. 1998. "Transgenic Bovine Chimeric Offspring Produced from Somatic Cell-Derived Stem-Like Cells." *Nature Biotechnology* 16:642–646.

Corn, B., M. Reed, S. Dishong, L. Yunsheng, and T. Wagner. 1991. "Culture and Successful Transplantation of Embryonic Yolk-Sac Cells." *Clinical Biotechnology* 3:15.

Deacon, T., J. Dinsmore, L.C. Costantini, J. Ratliff, and O. Isacson. 1998. "Blastula-Stage Stem Cells Can Differentiate into Dopaminergic and Serotenergic Neurons After Transplantation." *Experimental Neurology* 149:28–41.

Diukman, R., and M.S. Golbus. 1992. "*In Utero* Stem Cell Therapy." *Journal of Reproductive Medicine* 37:515–520.

Doetschman, T., P. Williams, and N. Maeda. 1988. "Establishment of Hamster Blastocyst-Derived Embryonic Stem (ES) Cells." *Developmental Biology* 127:224–227.

Doetschman, T., M. Shull, A. Kier, and J.D. Coffin. 1993. "Embryonic Stem Cell Model Systems for Vascular Morphogenesis and Cardiac Disorders." *Hypertension* 22:618–629.

Donovan, P.J. 1994. "Growth Factor Regulation of Mouse Primordial Germ Cell Development." *Current Topics in Developmental Biology* 29:189–225.

———. 1998. "The Germ Cell: The Mother of All Stem Cells." *International Journal of Developmental Biology* 42:1043–1050.

Evans, M.J., and M.H. Kaufman. 1981. "Establishment in Culture of Pluripotential Cells from Mouse Embryos." *Nature* 292:154–156.

Evans, M.J., E. Notarianni, S. Laurie, and R.M. Moor. 1990. "Derivation and Preliminary Characterization of Pluripotent Cell Lines from Porcine and Bovine Blastocysts." *Theriogenology* 33:125–128.

Flax, J.D., S. Aurora, C. Yang, C. Simonin, A.M. Wills, L.L. Billinghurst, M. Jendoubi, R.L. Sidman, J.H. Wolfe, S.U. Kim, and E.Y. Snyder. 1998. "Engraftable Human Neural Stem Cells Respond to Developmental Cues, Replace Neurons and Express Foreign Genes." *Nature Biotechnology* 16:1033–1039.

Frederiksen, K., and R.D. McKay. 1988. "Proliferation and Differentiation of Rat Neuroepithelial Precursor Cells *In Vivo*." *Journal of Neuroscience* 8:1144–1151.

Freed, C.R., R.E. Breeze, C.O. Denver, P.E. Greene, W. Tsai, D. Eidelberg, J.Q. Trojanowski, J.M. Rosenstein, and S. Fahn. 1999. *Double-Blind Controlled Trial of Human Embryonic Dopamine Cell Transplants in Advanced Parkinson's Disease: Study Design, Surgical Strategy, Patient Demographics and Pathological Outcome.* Presentation to the American Academy of Neurology, 21 April.

Fry, R. 1992. "The Effect of Leukemia Inhibitory Factor (LIF) on Embryogenesis." *Reproduction, Fertility, and Development* 4:449–458.

Gardner, R.L. 1982. "Investigation of Cell Lineage and Differentiation in the Extra Embryonic Endoderm of the Mouse Embryo." *Journal of Embryological Experimental Morphology* 68:175–198.

Gearhart, J.D. 1998. "New Potential for Human Embryonic Stem Cells." *Science* 282:1061–1062.

Giles, J.R., X. Yang, W. Mark, and R.H. Foote. 1993. "Pluripotency of Cultured Rabbit Inner Cell Mass Cells Detected by Isozyme Analysis and Eye Pigmentation of Fetuses Following Injection into Blastocysts or Morulae." *Molecular Reproduction and Development* 36:130–138.

Graves, K.H., and R.W. Moreadith. 1993. "Derivation and Characterization of Putative Pluripotential Embryonic Stem Cells from Pre-implantation Rabbit Embryos." *Molecular Reproduction and Development* 36:424–433.

Hall, P. A., and F.M. Watt. 1989. "Stem Cells: The Generation and Maintenance of Cellular Diversity." *Development* 106:619–633.

Handyside, A., M.L. Hooper, M.H. Kaufman, and I. Wilmut. 1987. "Towards the Isolation of Embryonal Stem Cell Lines from the Sheep." *Roux's Archives of Developmental Biology* 196:185–190.

Hochereau-de Reviers, M.T., and C. Perreau. 1993. "In Vitro Culture of Embryonic Disc Cells from Porcine Blastocysts." *Reproduction, Nutrition, Development* 33:475–483.

Hollands, P. 1991. "Embryonic Stem Cell Grafting: The Therapy of the Future?" *Human Reproduction* 6:79–84.

Iannaccone, P.M., G.U. Taborn, R.L. Garton, M.D. Caplice, and D.R. Brenin. 1994. "Pluripotent Embryonic Stem Cells from the Rat are Capable of Producing Chimeras." *Developmental Biology* 163:288–292.

Iscove, N. 1990. "Haematopoiesis. Searching for Stem Cells." *Nature* 347:126–127.

Kaufmann, M.H. 1992. *The Atlas of Mouse Development*. London: Academic Press.

Lacorazza, H.D., J.D. Flax, E.Y. Snyder, and M. Jendoubi. 1996. "Expression of Human b-hexosaminidase a-subunit Gene (the Gene Defect of Tay-Sachs Disease) in Mouse Brains upon Engraftment of Transduced Progenitor Cells." *Nature Medicine* 2:424–429.

Luckett, W.P. 1975. "The Development of Primordial and Definitive Amniotic Cavities in Early Rhesus Monkey and Human Embryos." *American Journal of Anatomy* 144:149–167.

———. 1978. "Origin and Differentiation of the Yolk Sac Extra Embryonic Mesoderm in Presomite Human and Rhesus Monkey Embryos." *American Journal of Anatomy* 152:59–97.

Martin, G.R. 1981. "Isolation of a Pluripotent Cell Line from Early Mouse Embryos Cultured in Medium Conditioned by Teratocarcinoma Stem Cells." *Proceedings of the National Academy of Sciences USA* 78:7634–7638.

Matsui, Y. 1998. "Developmental Fates of Mouse Germ Cell Line." *International Journal of Developmental Biology* 42:1037–1042.

Melton, D. 1999. Testimony of the Juvenile Diabetes Foundation International Before the Senate Appropriations Subcommittee on Labor, Health and Human Services, and Education. January 12.

Miller-Hance, W.C., M. LaCorbiere, S.J. Fuller, S.M. Evans, G. Lyons, C. Schmidt, J. Robbins, and K.R. Chien. 1993. "In Vitro Chamber Specification During Embryonic Stem Cell Cardiogenesis. Expression of the Ventricular Myosin Light Chain-2 Gene Is Independent of Heart Tube Formation." *Journal of Biological Chemistry* 268:25244–25252.

Moreadith, R.W., and K.H. Graves. 1992. "Derivation of Pluripotential Embryonic Stem Cells from the Rabbit." *Transactions of the Association of American Physicians* 105:197–203.

Muthuchamy, M., L. Pajak, P. Howles, T. Doetschman, and D.F. Wieczorek. 1993. "Developmental Analysis of Tropomyosin Gene Expression in Embryonic Stem Cells and Mouse Embryos." *Molecular Cell Biology* 13:3311–3323.

Nagy, A., J. Rossant, R. Nagy, W. Abramow-Newerly, and J.C. Roder. 1993. "Derivation of Completely Cell Culture-Derived Mice from Early-Passage Embryonic Stem Cells." *Proceedings of the National Academy of Sciences USA* 90:8424–8428.

National Cancer Institute (NCI), NIH. 1999. Response to Senator Specter's Inquiry "What Would You Hope to Achieve from Stem Cell Research?" March 25.

National Institute of Allergy and Infectious Diseases (NIAID), NIH. 1999. Response to Senator Specter's Inquiry "What Would You Hope to Achieve from Stem Cell Research?" March 25.

National Institute of Arthritis and Musculoskeletal and Skin Diseases (NIAMS), NIH. 1999. Response to Senator Specter's Inquiry "What Would You Hope to Achieve from Stem Cell Research?" March 25.

National Institute of Environmental Health Sciences (NIEHS), NIH 1999. Response to Senator Specter's Inquiry "What Would You Hope to Achieve from Stem Cell Research?" March 25.

National Institute of Neurological Disorders and Stroke (NINDS), NIH. 1999. Response to Senator Specter's Inquiry "What Would You Hope to Achieve from Stem Cell Research?" March 25.

Notarianni, E., S. Laurie, R.M. Moor, and M.J. Evans. 1990. "Maintenance and Differentiation in Culture of Pluripotential Embryonic Cell Lines from Pig Blastocysts." *Journal of Reproduction and Fertility* (Supplement) 41:51–56.

Notarianni, E., L.C. Galli, S. Laurie, R.M. Moor, and M.J. Evans. 1991. "Derivation of Pluripotent, Embryonic Cell Lines from Pig and Sheep." *Journal of Reproduction and Fertility* (Supplement) 43:255–260.

Oberpenning, F., J. Meng, J.J. Yoo, and A. Atala. 1999. "De Novo Reconstruction of a Functional Mammalian Urinary Bladder by Tissue Engineering." *Nature Biotechnology* 17:149–155.

Chapter 2: Human Stem Cell Research and the Potential for Clinical Application

O'Rahilly, R. 1987. *Developmental Stages in Human Embryos.* Washington, DC: Carnegie Institution of Washington.

PCT/U397/12919. 1997. "Embryonic or Stem-Like Cell Lines Produced by Cross Species Nuclear Transplantation." *International Application Published Under the Patent Cooperation Treaty.* WO 98/07/841.

Pedersen, R.A. 1994. "Studies of In Vitro Differentiation With Embryonic Stem Cells." *Reproduction, Fertility, and Development* 6:543–552.

Piedrahita, J.A., G.B. Anderson, G.R. Martin, R.H. Bondurant, and R.L. Pashen. 1988. "Isolation of Embryonic Stem Cell-Like Colonies from Porcine Embryos." *Theriogenology* 29:286.

Piedrahita, J.A., G.B. Anderson, and R.H. Bondurant. 1990. "On the Isolation of Embryonic Stem Cells: Comparative Behavior of Murine, Porcine and Ovine Embryos." *Theriogenology* 34:879–891.

Pittenger, M.F., A.M. Mackay, S.C. Beck, R.K. Jaiswal, R.K. Douglas, J.D. Mosca, M.A. Moorman, D.W. Simonetti, S. Craig, and D.R. Marshak. 1999. "Multilineage Potential of Adult Human Mesenchymal Stem Cells." *Science* 284:143–147.

Potten, C.S., and M. Loeffler. 1990. "Stem Cells: Attributes, Cycles, Spirals, Pitfalls and Uncertainties: Lessons for and from the Crypt." *Development* 110:1001–1020.

Robbins, J., J. Gulick, A. Sanchez, P. Howles, and T. Doetschman. 1990. "Mouse Embryonic Stem Cells Express the Cardiac Myosin Heavy Chain Genes During Development In Vitro." *Journal of Biological Chemistry* 265:11905–11909.

Robertson, E., and A. Bradley. 1986. "Production of Permanent Cell Lines from Early Embryos and Their Use in Studying Developmental Problems." In *Experimental Approaches to Mammalian Embryonic Development*, eds. J. Rossant and R.A. Pedersen, 475–508. New York: Cambridge University Press.

Rosario, C.M., B.D. Yandava, B. Kosaras, D. Zurakowski, R.L. Sidman, and E.Y. Snyder. 1997. "Differentiation of Engrafted Multipotent Neural Progenitors Towards Replacement of Missing Granule Neurons in Meander Tail Cerebellum May Help Determine the Locus of Mutant Gene Action." *Development* 124:4213–4224.

Rossant J., C. Bernelot-Moens, and A. Nagy. 1993. "Genome Manipulation in Embryonic Stem Cells." *Philosophical Transactions of The Royal Society of London. Series B: Biological Science* 339(1288):207–215.

Saito, S., N.S. Strelchenko, and H. Niemann. 1992. "Bovine Embryonic Stem Cell-Like Cell Lines Cultured over Several Passages." *Roux's Archives of Developmental Biology* 201:134–141.

Shamblott, M.J., J. Axelman, S. Wang, E.M. Bugg, J.W. Littlefield, P.J. Donovan, P.D. Blumenthal, G.R. Higgins, and J.D. Gearhart. 1998. "Derivation of Pluripotent Stem Cells from Cultured Human Primordial Germ Cells." *Proceedings of the National Academy of Sciences USA* 95:13726–13731.

Shim, H., A. Gutierrez-Adan, L.R. Chen, R.H. BonDurant, E. Behboodi, and G.B. Anderson. 1997. "Isolation of Pluripotent Stem Cells from Cultured Porcine Primordial Germ Cells." *Biology of Reproduction* 57:1089–1095.

Snodgrass, H.R., R.M. Schmitt, and E. Bruyns. 1992. "Embryonic Stem Cells and In Vitro Hematopoiesis." *Journal of Cellular Biochemistry* 49:225–230.

Snyder, E.Y. 1994. "Grafting Immortalized Neurons to the CNS." *Current Opinion in Neurobiology* 4:742–751.

Snyder, E.Y., R.M. Taylor, and J.H. Wolfe. 1995. "Neural Progenitor Cell Engraftment Corrects Lysosomal Storage Throughout the MPS VII Mouse Brain." *Nature* 374:367–370.

Snyder, E.Y., C. Yoon, J.D. Flax, and J.D. Macklis. 1997. "Multipotent Neural Precursors Can Differentiate Toward Replacement of Neurons Undergoing Targeted Apoptotic Degeneration in Adult Mouse Neocortex." *Proceedings of the National Academy of Sciences USA* 94:11663–11668.

Strauss, E. 1999. "Brain Stem Cells Show Their Potential." *Science* 283:471.

Strelchenko, N., and S. Stice. 1994. "Bovine Embryonic Pluripotent Cell Lines Derived from Morula Stage Embryos." *Theriogenology* 41:304.

Sukoyan, M.A., A.N. Golubitsa, A.I. Zhelezova, A.G. Shilov, S.Y. Vatolin, L.P. Maximovsky, L.E. Andreeva, J. McWhir, S.D. Pack, S.I. Bayborodin, A.Y. Kerkis, H.I. Kizilova, and O.L. Serov. 1992. "Isolation and Cultivation of Blastocyst-Derived Stem Cell Lines from American Mink (Mustela vison)." *Molecular Reproduction and Development* 33:418–431.

Sukoyan, M.A., S.Y. Vatolin, A.N. Golubitsa, A.I. Zhelezova, L.A. Semenova, and O.L. Serov. 1993. "Embryonic Stem Cells Derived from Morulae, Inner Cell Mass, and Blastocysts of Mink: Comparisons of Their Pluripotencies." *Molecular Reproduction and Development* 36:148–158.

Talbot, N.C., C.E. Rexroad Jr., V.G. Pursel, and A.M. Powell. 1993. "Alkaline Phosphatase Staining of Pig and Sheep Epiblast Cells in Culture." *Molecular Reproduction and Development* 36:139–147.

Tanne, J.H. 1999. "Researchers Implant Tissue Engineered Bladders." *British Medical Journal* 318:350B.

Thomson, J.A., J. Itskovitz-Eldor, S.S. Shapiro, M.A. Waknitz, J.J. Swiergiel, V.S. Marshall, and J.M. Jones. 1998. "Embryonic Stem Cell Lines Derived from Human Blastocysts." *Science* 282:1145–1147.

Thomson, J.A., J. Kalishman, T.G. Golos, M. Durning, C.P. Harris, R.A. Becker, and J.P. Hearn. 1995. "Isolation of a Primate Embryonic Stem Cell Line." *Proceedings of the National Academy of Sciences USA* 92:7844–7848.

Thomson, J.A., J. Kalishman, T.G. Golos, M. Durning, C.P. Harris, and J.P. Hearn. 1996. "Pluripotent Cell Lines Derived from Common Marmoset (*Callithrix jacchus*) Blastocysts." *Biology and Reproduction* 55:254–259.

Thomson, J.A., and V.S. Marshall. 1998. "Primate Embryonic Stem Cells." *Current Topics in Developmental Biology* 38:133–165.

Van Blerkom, J. 1994. "The History, Current Status and Future Direction of Research Involving Human Embryos." In National Institutes of Health (NIH) Human Embryo Research Panel, *Report of the Human Embryo Research Panel, Vol. II: Papers Commissioned for the Human Embryo Research Panel*, 1–25. Bethesda, MD: NIH.

Varmus, H. 1998. Testimony Before the Senate Appropriations Subcommittee on Labor, Health and Human Services, and Education. December 2.

Wade, N. 1998. "Researchers Claim Embryonic Cell Mix of Human and Cow." *New York Times* 12 November, A-1.

Wagner, J., P. Åkerud, D.S. Castro, P.C. Holm, J.M. Canals, E.Y. Snyder, T. Perlmann, and E. Arenas. 1999. "Induction of a Midbrain Dopaminergic Phenotype in *Nurr1*-Overexpressing Neural Stem Cells by Type 1 Astrocytes." *Nature Biotechnology* 17:653–659.

Wheeler, M.B. 1994. "Development and Validation of Swine Embryonic Stem Cells: A Review." *Reproduction, Fertility, and Development* 6:563–568.

Wilmut I., A.E. Schnieke, J. McWhir, A.J. Kind, and K.H. Campbell. 1997. "Viable Offspring Derived from Fetal and Adult Mammalian Cells." *Nature* 385:810–813.

Winkel, G.K., and R.A. Pedersen. 1988. "Fate of the Inner Cell Mass in Mouse Embryos as Studied by Microinjection of Lineage Tracers." *Developmental Biology* 127(1):143–156.

Wobus, A.M., G. Wallukat, and J. Hescheler. 1991. "Pluripotent Mouse Embryonic Stem Cells are Able to Differentiate into Cardiomyocytes Expressing Chronotropic Responses to Adrenergic and Cholinergic Agents and Ca2+ Channel Blockers." *Differentiation* 48:173–182.

Yandava, B.D., L.L. Billinghurst, and E.Y. Snyder. 1999. "'Global' Cell Replacement is Feasible via Neural Stem Cell Transplantation: Evidence from the Dysmyelinated *Shiverer* Mouse Brain." *Proceedings of the National Academy of Sciences USA* 96:7029–7034.

Chapter Three

The Legal Framework for Federal Support of Research to Obtain and Use Human Stem Cells

Introduction

In the course of attempting to realize the promise of human *embryonic stem (ES) cell* and *embryonic germ (EG) cell* research to advance basic and applied science as well as to develop new, life-saving therapies, biomedical researchers encounter uncertainties in the law as well as explicit restrictions (including bans on federal research funding) that were created in response to earlier developments in biomedical science and public policy. At the same time, provisions also exist in state and federal law designed to facilitate this field of research and to establish—or offer models for establishing—appropriate safeguards to ensure that all efforts to obtain or use stem cells are carried out in an ethically acceptable way. To date, three sources of ES or EG cells—cadaveric fetal tissue, embryos remaining after infertility treatments, and embryos created solely for research purposes using either *in vitro* fertilization (IVF) or, potentially, somatic cell nuclear transfer (SCNT) techniques—have been identified. The goal of this chapter is to examine separately the legal issues raised by research involving each source of EG or ES cells, noting as appropriate when common issues arise.

The Law Relating to Aborted Fetuses as Sources of EG Cells

Federal law permits funding of some research with cells and tissues from the products of elective as well as spontaneous abortions, and state law facilitates the donation and use of fetal tissue for research. Both state and federal law set forth several requirements for the process of retrieving and using material from this source, although amendments may be needed to federal law in order to make existing safeguards applicable to stem cell research.

Federal Law Regarding Research Using Cells and Tissues from Aborted Fetuses

Since as early as the 1930s, American biomedical research has utilized *ex utero* fetal tissue both as a medium and, increasingly, as an object for experimentation (Gelfand and Levin 1993; Zion 1996). "For many years, the production and testing of vaccines, the study of viral reagents, the propagation of human viruses, and the testing of biological products have been dependent on the unique growth properties of fetal tissue" (Duke 1988, D112, D114). For example, the 1954 Nobel Prize for Medicine was awarded to American immunologists who used cell lines obtained from human fetal kidney cells to grow polio virus in cell cultures, a key advance in the development of polio vaccines (Driscoll 1985; Gelfand and Levin 1993).

In 1972, allegations (some of them quite shocking) about experiments with fetuses both *in* and *ex utero* created an air of controversy (fueled by the greater societal debate about elective abortion) over the use of fetal tissue in research.[1] When Congress established the National Commission for the Protection of Human Subjects of Biomedical and Behavioral Research in 1974, it placed the topic of research using the human fetus at the top of the commission's agenda. Within four months of assuming office, the commissioners were mandated to report on the subject, with the proviso that the presentation of their report to the Secretary of the Department of Health, Education, and Welfare (DHEW)—now the Department of Health and Human Services (DHHS)—would lift the moratorium that Congress had imposed on federal

Chapter 3: The Legal Framework for Federal Support of Research to Obtain and Use Human Stem Cells

funding of research using live fetuses.[2] On July 25, 1975, the National Commission submitted its conclusions and recommendations, which formed the basis for regulations that the Department issued later that year on research involving fetuses, pregnant women, and human IVF (1975).

General Regulation of Research with Human Beings Including Fetuses

The 1975 provisions remain as elements of the current federal regulations that aim to protect human subjects participating in research conducted with federal funds—rules that also are followed on a voluntary basis by many institutions in the case of research performed without federal support. The core regulations are set forth in the *Federal Policy for the Protection of Human Subjects*, known as the Common Rule, because the same regulatory provisions have been adopted by most federal agencies and departments that conduct or sponsor research in which human subjects are used. The DHHS regulations appear in Volume 45, Part 46 of the *Code of Federal Regulations*—45 CFR 46. The Common Rule makes up Subpart A of the DHHS regulations, and additional protections for special populations of research subjects appear in three further subparts of 45 CFR 46.

The special provisions applicable to fetal material appear in Subpart B, which covers research on "1) the fetus, 2) pregnant women, and 3) human *in vitro* fertilization" and applies to all DHHS "grants and contracts supporting research, development, and related activities" involving those subjects.[3] The regulations primarily address research that could affect living fetuses adversely. They provide for stringent Institutional Review Board (IRB) consideration, which is based upon the results of preliminary studies on animals and nonpregnant women and on assurances that living fetuses will be exposed only to minimal risk except when the research is intended to meet the health needs of the fetus or its mother.[4] Specific restrictions also are imposed on the inclusion of pregnant women in research activities.

Section 46.210 of Subpart B states that the sole explicit requirement for research involving "cells, tissues, or organs excised from a dead fetus" is that such research "shall be conducted only in accordance with any applicable State or local laws regarding such activities."[5] Some analysts have argued that this is the *only* component of Subpart B applicable to research in which cells or tissues from dead abortuses are used in research (Areen 1988). It appears, however, that even prior to the adoption in 1993 of legislation establishing special rules for using fetal tissue for transplantation, National Institutes of Health (NIH) officials had regarded other, general requirements of Subpart B as applicable to research with tissue from dead fetuses.[6] Specifically, these other provisions exclude researchers from any involvement in the decision to terminate a pregnancy or in an assessment of fetal viability and forbid the payment of any inducements to terminate a pregnancy.[7] This dispute over the scope of Subpart B produces one of the points of uncertainty that may need to be resolved either through legislation or official commentary from NIH's Office for Protection from Research Risks (OPRR), if investigators using cadaveric fetal tissue to generate human EG cells are to proceed with confidence and in an ethical fashion.

The Conditions for Federal Support of Fetal Tissue Transplantation

In the 1980s, medical scientists began experimenting with implanting brain tissue from aborted fetuses into patients with Parkinson's disease as well as patients with other neurological disorders. NIH investigators were among those working in this field, and their protocol to use fetal tissue for transplantation was approved by an internal NIH review body. Although the research complied with Subpart B, then-NIH Director James B. Wyngaarden decided to seek approval from Assistant Secretary for Health Robert E. Windom before proceeding.[8] In March 1988, Windom responded by declaring a temporary moratorium on federally funded transplantation research involving fetal tissue from induced abortions. He also asked NIH to establish an advisory body to consider whether such research should be conducted and under what conditions (Windom 1988). The Human Fetal Tissue Transplantation Research Panel—composed of biomedical investigators, lawyers, ethicists, clergy, and politicians—deliberated until the fall of 1988. Panel members then voted 19-2 to recommend continued funding for fetal tissue transplantation research under guidelines designed to ensure the ethical integrity of any experimental procedures (Adams 1988; Duguay 1992;

Silva-Ruiz 1998). In November 1989, after the transition had been made from the Reagan to the Bush administration, DHHS Secretary Louis Sullivan extended the moratorium indefinitely, based upon the position taken by the minority-voting panel members that fetal tissue transplantation research would increase the incidence of elective abortion (Goddard 1996; Robertson 1993).[9] Attempts by Congress to override the Secretary's decision were not enacted or were vetoed by President Bush.[10]

On January 22, 1993, immediately after President Clinton took office, he instructed the incoming Secretary of DHHS to lift the ban on federal funding for human fetal tissue transplantation research.[11] On February 5, 1993, DHHS Secretary Donna Shalala officially rescinded the moratorium, and, in March 1993, NIH published interim guidelines for research involving human fetal tissue transplantation (OPRR 1994). Provisions to legislate these safeguards were promptly proposed in Congress and included in the NIH Revitalization Act of 1993, which President Clinton signed into law on June 10, 1993.[12]

The 1993 act mirrors most prior statutory and regulatory provisions on research involving tissue from dead fetuses.[13] In general, the Revitalization Act states that any tissue from any type or category of abortion may be used for research on transplantation, but only for "therapeutic purposes." Most agree that this means that research on transplantation that has as its goal the treatment of disease is covered by the act, but that basic laboratory research—which only tangentially can be described as having a therapeutic purpose—would not be covered. Under all conditions, the investigator's research scope is not, however, unfettered. First, research activities in this area must be conducted in accordance with applicable state and local law. The investigator also must obtain a written statement from the donor verifying that a) she is donating fetal tissue for therapeutic purposes, b) no restrictions have been placed on who the recipient will be, and c) the donor has not been informed of the identity of the recipient. Further, the attending physician must sign a statement affirming five additional conditions of the abortion, aimed at insulating a woman's decision to abort from her decision to provide tissue for fetal research. Finally, the person principally responsible for the experiment must also affirm his or her own knowledge of the sources of tissue, that others involved in the research are aware of the tissue status, and that the researcher had no part in the abortion decision or its timing.

The statute provides significant criminal penalties for violation of four prohibited acts: 1) purchase or sale of fetal tissue "for valuable consideration" beyond "reasonable payments [for] transportation, implantation, processing, preservation, quality control, or storage…," 2) soliciting or acquiring fetal tissue through the promise that a donor can designate a recipient, 3) soliciting or acquiring fetal tissue through the promise that the recipient will be a relative of the donor, or 4) soliciting or acquiring fetal tissue after providing "valuable consideration" for the costs associated with the abortion itself.[14]

Research of the type conducted by Gearhart and his colleagues at The Johns Hopkins University, in which primordial germ cells were obtained from the gonadal ridge of human fetuses that had been aborted five to nine weeks after fertilization, arguably is not covered by the fetal tissue transplantation provisions of the 1993 NIH Revitalization Act, because these fetal cells are intended to be cultured and used in laboratory experiments, not transplanted. Nevertheless, if such research were federally supported, it could be subject to the requirements of Subpart B of 45 CFR 46—both the general limitations of § 46.206 (separating the investigators from the abortion process and forbidding payments for pregnancy termination) and the special requirements of § 46.210 for activities involving cells and tissues from dead fetuses. Someday, with the advancement of knowledge about cell differentiation and the like, EG cells derived from dead fetuses may be linked more directly or indirectly with transplantation, at which point the 1993 Act would arguably become applicable.[15] In anticipation of that day, and in order to achieve simplicity in the meantime by applying the same rules to all federally supported research with fetal remains, whether or not for transplantation, it would appear desirable to amend the law to clarify that the safeguards of the 1993 Act apply to research in which EG cells are obtained from dead fetuses after a spontaneous or elective abortion.

Chapter 3: The Legal Framework for Federal Support of Research to Obtain and Use Human Stem Cells

State Law Regarding Using Aborted Fetuses as Sources of Stem Cells

As recognized by federal statutes and regulations, state law governs the manner in which cells and tissues from dead fetuses become available for research, principally by statutes, regulations, and case law on organ transplantation. The most basic legal provisions lie in the Uniform Anatomical Gift Act (UAGA), which was first proposed in 1968 and rapidly became the most widely adopted uniform statute. While the UAGA is largely consistent with relevant federal statutes and regulations and should facilitate researchers obtaining cadaveric fetal tissue, a number of states have adopted other statutes that limit or prohibit certain types of research with fetal remains.

Laws Facilitating Donation of Fetal Material for EG Cell Research: The UAGA

The UAGA is relevant not only because federal statutes and regulations explicitly condition funding for research with fetal tissue on compliance with state and local laws, but also because the act applies when EG cell research using fetal tissue does not receive federal funding. The original version of the UAGA was approved by all 50 states and the District of Columbia; a 1987 revision has been enacted by 22 states (Zion 1996).[16] The act establishes a system of voluntary donation of "anatomical gifts" for transplantation, education, and research. It was intended to make it easier for people to authorize gifts of their own body (or parts thereof) through a simple "donor card" executed before the occasion arose, as well as to allow donations to be made with the permission of the next-of-kin, following an order established by the statute. The revised UAGA includes "a stillborn infant or fetus" in the definition of "decedents,"[17] for whom parental consent is determinative.[18] The UAGA also provides that "neither the physician or surgeon who attends the donor at death nor the physician or surgeon who determines the time of death" may be involved in the team that will use the organs removed from the decedent.[19] This section, although it may be waived, seems comparable to the separation that the 1993 NIH Revitalization Act and Subpart B of the DHHS regulations required between the research team and any physicians involved in terminating a pregnancy, determining fetal viability, or assisting in the clinical procedure during which fetal tissue is derived for research purposes.[20]

However, federal law restricts the procedures authorized by the UAGA in one area.[21] The UAGA permits donors to designate recipients—including individual patients—of anatomical gifts. The stricter provisions of the NIH Revitalization Act (which prohibits a donor from having knowledge of an individual transplant recipient) could override this state law in the case of federally supported fetal tissue transplantation, but the issue might not arise regarding stem cell research for two reasons. First, such research does not involve transplantation (and hence at this time is not relevant to the NIH Revitalization Act). Second, according to the Revitalization Act, the only recipient who may be designated by the parents of a dead fetus would be a stem cell researcher or research institution.

Laws Restricting Use of Donated Fetal Material for EG Cell Research

At present, 24 states do not have on their books any statutes "specifically addressing research on embryos or fetuses,"[22] and the restrictions in most of the remaining states principally involve embryos remaining after infertility treatments and limitations aimed at discouraging therapeutic abortions. For example, in 12 states, the law applies only to research with fetuses prior or subsequent to an elective abortion.[23] Six states ban research that involves aborted fetuses or their organs, tissues, or remains,[24] which could cause difficulties for researchers using stem cell lines derived from aborted fetuses "if cell lines are considered 'tissue.'"[25] Six other states permit fetal research when the fetus is deceased, but mandate that the donor must provide consent,[26] although none "specifically address[es] the type of information that must be provided to the progenitors before they are asked for consent."[27] In Pennsylvania, investigators using fetal tissue and recipients of the tissue are required to be informed if the tissue was procured as a result of stillbirth, miscarriage, ectopic pregnancy, abortion, or some other means.[28]

In order to diminish the impact that the potential use of a fetus in research might have on the decision to abort, states have enacted many restrictions on payment for fetal remains. The broadest prohibitions appear as part of state statutes regulating or prohibiting fetal research. Bans on sale vary in their terminology—an "aborted product of conception,"[29] an "aborted unborn child or the remains thereof,"[30] an "aborted fetus or any tissue or organ thereof,"[31] or an "unborn child"[32] —and exist both in states that permit research on a dead fetus with the mother's consent[33] and in those where it is illegal to conduct research upon any aborted product of conception.[34]

The bans on commercialization have a number of interesting twists. Rhode Island outlaws the selling of an embryo or fetus for purposes that violate the statute (such as research on living embryos or fetuses), but apparently allows payment to the mother for allowing a dead fetus to be used in research, because such research is permissible.[35] Minnesota prohibits the sale of living fetuses or nonrenewable organs but explicitly permits "the buying and selling of a cell culture line or lines taken from a [dead fetus]."[36]

The most widely adopted prohibitions on commercialization of fetal remains are those in Sections 10(a) and (b) of the 1987 revision of the UAGA, which prohibit the sale or purchase of any human body parts for any consideration beyond that necessary to pay for expenses incurred in the removal, processing, and transportation of the tissue.[37] On the federal level, what is in essence the same proscription is included both in the 1993 NIH Revitalization Act, which bars the acquisition or transfer of fetal tissue for "valuable consideration" with the same exceptions,[38] and in the National Organ Transplant Act of 1984 (NOTA), which prohibits the sale of any human organ for "valuable consideration for use in human transplantation"[39] if the sale involves interstate commerce.[40] (In 1988, Congress amended NOTA to include fetal organs within the definition of "human organ," in order to foreclose the sale of fetal tissue as well.[41]) Yet both federal statutes could be interpreted to apply only to sales for transplant or therapeutic purposes, not laboratory research. Moreover, the definition of reasonable processing fees in the federal law (and by extension, the UAGA)[42] is arguably too vague, "leav[ing]...room for unscrupulous tissue processors to abuse the law" (Goddard 1996, 394). If special provisions are adopted to govern federal support of research with fetal material to create human EG cell lines, it would seem advisable to ensure that the provisions lay out more clearly what payments may be made to whom and on what basis for fetal cells and tissues.

The state statutes regulating fetal research have been challenged in several court cases. Generally, limitations have been approved as they relate to live fetuses or to the disposal of aborted fetuses.[43] A few cases have dealt with restrictions on research with dead fetuses or fetal remains. In 1978, Louisiana adopted a statute forbidding virtually all experimentation involving a living fetus ("a live child or unborn child") that was not "therapeutic" to that child, a ban it expanded in 1981 to encompass research with aborted fetal tissue as well.[44] Plaintiffs who argued that the prohibition on research burdened their right of privacy challenged the law.[45] Agreeing, the federal district court concluded that the ban on research did not further the state's compelling interest in protecting the health of the woman, nor did the state's interest in the potential life of the unborn continue past the death of the fetus.[46] Finally, the district court addressed the statute's vagueness, noting that it was not possible, *ex utero*, to distinguish between fetal and maternal tissue and the products of spontaneous and induced abortions.[47] On appeal, the Fifth Circuit ignored the district court's analysis entirely, finding instead that the term "experiment" as used in the statute's prohibition against fetal experimentation was unconstitutionally vague.[48]

The Law Relating to Embryos as Sources of ES Cells

Turning to the second source of human ES cells—embryos created through IVF—one finds that in contrast to the regulatory complexity of the federal and state laws governing research using fetal tissue, the legal framework for research using human embryos is relatively straightforward. With the exception of a few state statutes, no viable regulatory system exists to guide or control the practice of human embryo research in the United States.[49] Regarding federally supported scientists, law prohibits such experimentation, while research conducted in the private sector takes place without any federal medical or bioethical oversight specific to the human embryo.[50] The central issue raised by existing law is whether the recent scientific developments are important enough to justify modifying, in part, the current blanket ban on federal support by creating a limited exception for certain types of human stem cell research.

Federal Law Regarding Research Using Cells and Tissues from Human Embryos

Federal law regarding research using human embryos by investigators employed or funded by the federal government may best be understood by reviewing Subpart B of the DHHS policy on the protection of human subjects and the rider that has been attached for several years to the DHHS appropriation, most recently in the Omnibus Consolidated and Emergency Supplemental Appropriations Act for Fiscal Year 1999 (OCESAA).[51]

The former, which continues to provide a basic framework for research, even though reasons exist to question its applicability, originated in concerns about research on the human fetus, but it also applies to "grants and contracts supporting research, development, and related activities involving...human in vitro fertilization."[52] At the time these provisions were first promulgated, IVF was still an experimental technique: The birth in England of Louise Brown, the first so-called test tube baby, did not occur until 1978. Recognizing that NIH scientists and others would wish to pursue research on IVF and the earliest stages of human development, the regulations provided that "no application or proposal involving human in vitro fertilization may be funded by the Department [until it] has been reviewed by the Ethical Advisory Board and the Board has rendered advice as to its acceptability from an ethical standpoint."[53] In 1977, NIH received an application from an academic researcher for support of a study involving IVF. After the application had undergone scientific review within NIH, it was forwarded to the Ethics Advisory Board (EAB) appointed by Joseph Califano, then Secretary of DHEW. At its May 1978 meeting, the EAB agreed to review the research proposal. With the increased public interest that followed the birth of Louise Brown that summer, Secretary Califano asked the EAB to study the broader social, legal, and ethical issues raised by human IVF. On May 4, 1979, in its report to the Secretary, the EAB concluded that federal support for IVF research was "acceptable from an ethical standpoint" provided that certain conditions were met, such as informed consent for the use of gametes, an important scientific goal "not reasonably attainable by other means," and not maintaining an embryo "in vitro beyond the stage normally associated with the completion of implantation (14 days after fertilization)" (DHEW EAB 1979, 106, 107). No action was ever taken by the Secretary with respect to the board's report; for other reasons, the Department dissolved the EAB in 1980. Because it failed to appoint another EAB to consider additional research proposals, DHEW effectively forestalled any attempts to support IVF, and no experimentation involving human embryos was ever funded pursuant to the conditions set forth in the May 1979 report or through any further EAB review.

Because the Revitalization Act of 1993 effectively ended the de facto moratorium on IVF and other types of research involving human embryos[54] by nullifying the regulatory provision that mandated EAB review,[55] NIH Director Harold Varmus convened the Human Embryo Research Panel to set forth standards for determining which projects could be funded ethically and which should be considered "unacceptable for federal funding."[56] The panel identified several areas of potential research activity that it considered ethically appropriate for federal support, including studies involving the development of ES cells, though only with embryos resulting from IVF or clinical research that have been donated with the consent of the progenitors. The most controversial aspect of the report was its conclusion that it might be ethical to allow researchers to create human embryos for certain research purposes.[57]

In September 1994, the panel submitted its report to the Advisory Committee to the Director (ACD) of NIH, which formally approved the recommendations and transmitted them to Varmus on December 1, 1994. The following day, pre-empting NIH's response, the President declared that federal funds should not be used to support the creation of human embryos for research purposes and directed that NIH not allocate any resources for such requests.[58] Thereafter, Varmus decided to implement the panel's recommendations not proscribed by the President's directive, concluding that NIH could begin to fund research activities involving "surplus" embryos

(Feiler 1998). Before any funding decisions could be made, however, Congress attached a rider to that year's DHHS appropriations bill that stipulated that none of the funds appropriated could be used to support any activity involving "1) the creation of a human embryo or embryos for research purposes; or 2) research in which a human embryo or embryos are destroyed, discarded, or knowingly subjected to risk of injury or death greater than that allowed for research on fetuses *in utero* under 45 CFR 46.208(a)(2) and section 498(b) of the Public Health Service Act (42 USC 289g(b))."[59]

When the question arose of whether to provide federal funding for human ES cell research using IVF embryos remaining from infertility treatments, Varmus sought the opinion of Harriet Rabb, DHHS General Counsel, regarding the effect of the prohibition in the current appropriations rider. Rabb reported to Varmus that the OCESAA does not prevent NIH from supporting research that uses ES cells derived from this source because the cells themselves do not meet the statutory, medical, or biological definition of a human embryo (NIH OD 1999).[60]

Having concluded that NIH may fund internal and external research that utilizes but does not create human ES cells, NIH has delayed actual funding until an Ad Hoc Working Group of the ACD develops guidelines for the ethical research in this area.[61] The working group began its deliberations in early 1999 and completed draft guidelines on April 8, 1999, which are still undergoing internal review and public comment.[62]

In addition to these guidelines, ES cell research that was supported by federal funds and directly involved human embryos might arguably be subject to the requirements of both Subpart A (the Common Rule) and Subpart B of 45 CFR 46—that is, the research would be required to meet general and specific substantive requirements, would have to be approved by the IRB of the investigator's institution, and might have to undergo further review at the national level. We use the word "arguably" because OPRR has provided no definitive guidance regarding such an interpretation. Indeed, in response to the Commission's inquiry of May 18, 1999, OPRR acknowledged that "although Subpart B does not apply to research involving a human embryo, per se, it does apply to research that involves the process of *in vitro* fertilization. An embryo formed by a means that does not involve *in vitro* fertilization would not be subject to Subpart B."[63] Because no other guidance is provided, we are left to interpret whether embryos (which are not defined in regulation) are human subjects and therefore protected by Subpart B.

Subpart A, which contains the basic requirements for IRB review, informed consent, privacy protection, and the like, aims to protect a "human subject," defined as "a living individual about whom an investigator...obtains (1) data through intervention or interaction...."[64] This definition creates uncertainties about whether the Common Rule applies to embryo research, the derivation of ES cells, and research involving successor stem cells from embryonic sources that require resolution. This is another point upon which a clearer, more accessible interpretation is needed from OPRR if investigators and IRBs are to proceed with confidence regarding a range of stem cell research activities involving human embryos. Assuming that the DHHS regulations apply, the special requirements of Subpart B also would be applicable, because (as previously described) NIH has long taken the position that human IVF research, which is clearly encompassed in Subpart B, encompasses *any* DHHS-funded research involving human embryos not *in utero*. This would mean not only that another EAB could be impaneled by the Secretary pursuant to 45 CFR § 46.204, but also that special responsibility would fall on investigators and IRBs under 45 CFR § 46.205.[65] In addition, special standards would have to be met under 45 CFR § 46.206, including mandates for prior studies involving animals and ensuring the least possible risk. The newly revised 45 CFR 46, Subpart B (not yet finalized) makes no substantive changes that would affect these requirements.[66]

State Law Regarding Research Using Cells and Tissues from Human Embryos

State legislatures have apparently been more concerned about regulating and restricting research using human fetuses or their remains instead of addressing research involving laboratory manipulation of human gametes and early stage embryos. Nonetheless, although

Chapter 3: The Legal Framework for Federal Support of Research to Obtain and Use Human Stem Cells

the statutes usually ignore issues (other than commercialization) specific to IVF (Robertson 1990), some could be construed broadly enough to encompass a range of experimental activities involving IVF, including cryopreservation, pre-implantation screening, gene therapy, twinning, cell line development, and basic research (Coleman 1996). The latter two are of obvious relevance to creating stem cell cultures from embryonic sources.

States that regulate cell line development from human embryos either prohibit the practice entirely or restrict it substantially (Coleman 1996). "All ten states that prohibit embryological research have vaguely worded statutes which could encompass cell line development if the statutes were interpreted broadly...[although] some [activity] could be characterized as non-experimental, thus removing it from the scope of experimentation bans" (Coleman 1996, 1358). Issues inherent in cell line development will include the potential for restrictions on downstream commercialization and uncertainty over the extent to which gamete donors must be informed about the nature of and potential commercial uses of the biological materials they donate (Coleman 1996).

Basic research typically involves precommercial scientific activity designed to explore biological processes or to understand genetic and cellular control mechanisms. As noted previously, 24 states and the District of Columbia do not restrict research involving fetuses or embryos.[67] Of the remaining 26 states that regulate embryo or fetal research in one form or another, basic embryological research is prohibited or restricted in 10 (Feiler 1998). Although the degree of regulation of experimental use of embryos under the New Hampshire statute is unlikely to impair ES cell research in that state,[68] the remaining nine states have legislated more broadly, effectively banning all research involving *in vitro* embryos, with penalties mandated in some states, including civil fines and imprisonment.[69]

The subject of commercialization is a potentially important one, affecting both researchers who must acquire embryos from for-profit IVF clinics or other sources and downstream users who may develop derivative, commercial applications from basic embryological and stem cell research. Currently, five states prohibit payment for IVF embryos for research purposes.[70] Eight additional states prohibit payment for human embryos for any purpose.[71] Five states apply ambiguous restrictions that may or may not prohibit sale of embryos, depending upon interpretation or, in some cases, action by state officials.[72] More troubling, some statutes could be interpreted to prevent payment for ES cell lines derived from human embryos (Coleman 1996), although "it is possible that because a cell line is new tissue produced from the genetic material of, but not originally a part of, the embryo, laws proscribing the sale of embryonic tissue may not apply."[73] In line with NOTA and the 1987 revisions of the UAGA, state statutes on organ transplantation now typically prohibit sale of human organs or parts, but none include language likely to impede research involving human embryos.

The Law Relating to Deriving Stem Cells from Organisms Created Through Cloning

The third potential source of human ES cells would involve the use of cloning—that is, SCNT. One possible use of SCNT would be to derive ES cells themselves, thus avoiding the need for embryos. If such a transfer directly into an enucleated stem cell were to be successful, the therapeutic potential of creating cells and tissues for autologous transplantation might be realized without any of the ethical and regulatory problems associated with the creation of embryos.

At present, however, the method for creating human ES cells through SCNT, which has been announced by one scientific team (although not yet published in a scientific journal), involves inserting a somatic cell nucleus into an enucleated oocyte, which, if it then developed, would become a blastocyst from which ES cells would be derived. This approach creates two problems. First, if the blastocyst were characterized as a human embryo (albeit one created asexually rather than by uniting egg and sperm *in vitro*), then the prohibition on federal funding (as well as the restrictions on embryo research in several states) would come into play. Second, the process of carrying out SCNT using human cells has been outlawed by at least two states and may or may not be eligible for federal funding. On March 4, 1997, shortly after the initial announcement that the Roslin Institute had succeeded in creating Dolly, the cloned sheep, the Office of the White House Press Secretary released a "Memorandum for the Heads of Executive Departments and Agencies," in which the President stated that

> Federal funds should not be used for cloning of human beings. The current restrictions on the use of Federal funds for research involving human embryos do not fully assure this result. In December 1994, I directed the National Institutes of Health not to fund the creation of human embryos for research purposes. The Congress extended this prohibition in FY 1996 and FY 1997 appropriations bills, barring the Department of Health and Human Services from supporting certain human embryo research. However, these restrictions do not explicitly cover human embryos created for implantation and do not cover all Federal agencies. I want to make it absolutely clear that no Federal funds will be used for human cloning. Therefore, I hereby direct that no Federal funds shall be allocated for cloning of human beings.[74]

On June 9, 1997, the President received NBAC's report entitled *Cloning Human Beings* and announced his acceptance of its recommendations, which included a moratorium on publicly or privately funded research to create a child through SCNT but not on laboratory research using the technique. A number of bills have been introduced in Congress to achieve this result—as have other bills that would enact a broader prohibition—but no federal legislation has been adopted. On February 9, 1998, responding to one of those bills (S. 1601, The Human Cloning Prohibition Act), the Executive Office of the President released a Statement of Administration Policy, which provides in part that

> the Administration supports amendments to S. 1601 that would...permit somatic cell nuclear transfer using human cells for the purpose of developing stem cell (unspecified cells capable of giving rise to specific cells and tissues) technology to prevent and treat serious and life-threatening diseases and other medical conditions, including the treatment of cancer, diabetes, genetic disorders, and spinal cord injuries and for basic research that could lead to such treatments.[75]

This statement does not, however, have the force or effect of a Presidential Directive or Executive Order and does not modify the March 1997 Presidential Directive prohibiting funding for human cloning by federal agencies. The resulting uncertainty must be resolved, taking into account the ethical analysis presented in the next chapter.

Summary

As described in Chapter 2, the development of human ES and EG cell lines represents an important advance in biomedicine that promises not only to expand basic scientific understanding but also to improve health and extend life for millions of patients. Even the greatest supporters of this new field recognize, however, that current methods of deriving EG and ES cells from cadaveric fetal tissue and embryos remaining after infertility treatments raise significant ethical issues. Further ethical analysis, which appears in the next chapter of this report, is needed before conclusions can be reached about the goals and principles that should guide policymaking in this field.

Federal law permits the funding of some research that uses tissue from dead fetuses following spontaneous or elective abortion, provided the researchers follow safeguards that aim to separate the decision to abort from the decision to donate material for research, to ensure appropriate consent, and to avoid commercialization of fetal material. The UAGA, which in every state facilitates the process of donating bodies and organs for research as well as transplantation, treats fetuses like other cadavers; the latest version of the statute imposes special conditions on the donation of fetal remains and reinforces the prohibition in federal law against paying for organ donation. The legal framework identified by these statutes is thus favorable to research in which EG cells would be derived from fetal tissue. Some questions remain, however, about the applicability of some of the statutes—for example, the principal set of federal safeguards appears in a statute dealing with fetal tissue transplantation, and EG cell research does not now, and may never, involve directly the transplantation of tissue or cells from a fetus to a patient. Therefore, to overcome the uncertainties and ensure that ethical safeguards are understood to be applicable to fetal stem cell research, statutory modification and regulatory clarification are desirable.

Confusion also is caused by restrictions and bans in several states on research use of the products of induced abortions; although these statutes seem aimed principally at research with living fetuses, some have—or may be read to have—broader reach. The common theme of these statutes—as in the law on federally funded research—is to erect a significant barrier between a woman's decision to abort a fetus and the separate question of whether fetal remains will be donated for research. To support that barrier, many states employ consent requirements and prohibit payment for fetal remains, so that such material does not become commercialized and thus inappropriately influence the abortion decision.

The picture is clearer but less favorable to research in the area of embryos remaining after infertility treatments. In addition to restrictions and even outright prohibitions in the law of a number of states, riders to DHHS appropriation statutes in recent years rule out the use of these funds in any process in which human embryos are created for research or are destroyed or subject to a risk of injury. Once it has developed special guidelines to ensure that investigators will safeguard the ethics of the process, NIH will fund suitable research projects using human ES cells derived from IVF embryos, although it will not fund the derivation process itself. This position has been denounced by many members of Congress who supported the ban on federal funding of research with embryos and who believe that however the statutory language may be read, its intent clearly is to prohibit research that depends upon the prohibited acts.

The questions raised by this disagreement go beyond interpretation of the language and intent of the DHHS appropriations rider. First, is the justification for research using human ES cells compelling enough to permit an exception to the ban on federal funding for embryo research? Second, can an exception be crafted in a way that continues to give appropriate weight to the values that underlie the ban in the first place? And third, is the justification for using ES cells strong enough to permit funding of the process of deriving these cells from IVF embryos remaining after infertility treatments? Answers to these questions will require evaluation of the scientific and medical aspects of human ES cell research that are described in Chapter 2 in the context of the ethical considerations that are discussed in Chapter 4.

Notes

1 Proposed guidelines for fetal tissue research were released by NIH and DHEW in 38 *Fed. Reg.* 31,738 (1973) (Gelfand and Levin 1993).

2 See National Research Act, Public Law 93-348, Section 201(a), 88 Stat. 348 (1974).

3 45 CFR § 46.201(a) (1997). "The purpose of this subpart [is] to...assure that [applicable research] conform[s] to appropriate ethical standards and relate[s] to important societal needs" (Ibid. at § 46.202).

4 The portions of Subpart B dealing with research on living fetuses were re-enforced by the Human Research Extension Act of 1985. The act directs that no federally supported research may be conducted on a nonviable living human fetus *ex utero* or on a living human fetus *ex utero* for whom viability has not been determined, unless a) the research or experimentation may enhance the health, well-being, or probability of survival of the fetus itself; or b) will pose no added risk of suffering, injury, or death to the fetus where the research or experimentation is for "the development of important biomedical knowledge which cannot be obtained by other means." In either instance, the degree of risk must be the same for fetuses carried to term as for those intended to be aborted (42 USC § 289g 1998).

5 On May 20, 1998, DHHS released for public comment proposed revisions of Subpart B, most of which relate to research with living fetuses. In these revisions, § 46.210 would become § 46.206, which would retain the requirement that research with material from a dead fetus would have to conform to state law. The revised regulation would add that any living individual who becomes personally identified as a result of research on dead fetal or placental material must be treated as a research subject and accorded the protections of the federal Common Rule.

6 During the period of 1987–92, the NIH Office of Science Policy repeatedly stated that NIH applies Subpart B broadly to a range of fetal research activities. For example, in a 1988 memorandum, NIH Director James B. Wyngaarden informed Assistant Secretary for Health Robert E. Windom that "[a]s you know, the NIH conducts all human fetal tissue research in accordance with Federal Guidelines (45 CFR 46)," and provided a 1987 summary of fetal tissue research at NIH that stated that "NIH-supported human fetal tissue research is conducted in compliance with all Federal... regulations regarding the use of human fetal tissue. These regulations include restrictions on tissue procurement [Subpart B] that are intended to prevent possible ethical abuses" (NIH 1987; Memorandum from James B. Wyngaarden to Robert E. Windom, February 2, 1988).

7 45 CFR §§ 46.206 (a)(3) and 46.206(b)(1997).

8 "Although such approval was not required, the Assistant Secretary was consulted because of the scientific and ethical implications of the study" (Ryan 1991, 687).

9 Letter from Louis Sullivan to William Raub, November 2, 1999.

10 See H.R. 2507, 102d Cong., 1st Sess. (1991) (amending Part G of Title IV of the Public Health Service Act). See also H.R. 5495, 102d Cong., 2nd Sess. (1992) (amending Part G of Title IV of the Public Health Service Act and incorporating the establishment of a federally operated national tissue bank as provided by Exec. Order No. 12,806 [1992]). During this period, in an apparent attempt to find an alternative to fetal tissue derived from elective abortion, the administration established (without success) a tissue bank to collect fetal tissue for research from ectopic pregnancies and miscarriages. Exec. Order No. 12,806, 57 *Fed. Reg.* 21,589 (1992). Because spontaneously aborted tissue may contain viral infections or pathological defects, the use of ectopic and miscarried abortuses is disfavored for transplantation and most other research. In October 1992, a consortium of disease advocacy organizations filed suit against DHHS Secretary Sullivan, alleging that the Hyde Amendment, which bars federal funding for abortions, Departments of Labor, Health, Education, and Welfare Appropriations Act of 1977, Public Law 94-439, did not apply to research on and transplantation of fetal tissue. The plaintiffs argued, moreover, that the fetal tissue transplantation research ban was beyond the Department's statutory authority under the law (Bell 1994).

11 See 58 *Fed. Reg.* 7457 (1993).

12 The administration's policies on fetal tissue transplantation did not entirely quell public controversy or congressional interest (GAO 1997).

13 The policy initiated by President Clinton in 1993 and formalized in the 1993 NIH Revitalization Act is in line with the position taken in many other countries that the use of fetal tissue from elective abortions in therapy for people with conditions such as Parkinson's disease is acceptable. As with U.S. laws and regulations, international guidelines emphasize the need to separate the decision to terminate pregnancy from the decision to donate fetal tissue and the need for informed consent for the donation. See Knowles, L.P., 1999, "International Perspectives on Human Embryo and Fetal Tissue Research." This background paper was prepared for NBAC and is available in Volume II of this report.

14 42 USC § 289g-2(a)-(c) (1997). But see Goddard (1996).

15 DHHS General Counsel Harriet Rabb apparently believes that research of the type conducted by Gearhart is already sufficiently connected to transplantation to be subject to the NIH Revitalization Act, though she does not explain how she reached that conclusion. In a January 15, 1999, memorandum to NIH Director Varmus, Rabb concluded that "[t]o the extent human pluripotent stem cells are considered human fetal tissue by law, they are subject to…the restrictions on fetal tissue transplantation research that is conducted or funded by DHHS, as well as to the federal criminal prohibition on the directed donation of fetal tissue." Rabb examined the definition of "fetal tissue" at 42 USC 289g-1(g) which defines it as "tissue or cells obtained from a dead human embryo or fetus after a spontaneous or induced abortion, or after a stillbirth" and observed that "some stem cells, for example those derived from the primordial germ cells of non-living fetuses, would be considered human fetal tissue for purposes of [federal law]." Having concluded that primordial germ cells extracted from nonliving fetuses are a type of fetal tissue, the General Counsel went on, without further explanation, to apply the prohibition on sale of fetal tissue, the firewall restrictions, and the donative limitations stipulated in the NIH Revitalization Act, as well as the requirements of 45 CFR § 46.210.

16 National Conference of Commissioners on Uniform State Laws (NCCUSL), *A Few Facts About the Revised Uniform Anatomical Gift Act*, 1987.

17 Uniform Anatomical Gift Act (UAGA) § 1(3). But see Zion (1996): "UAGA…does not differentiate between a fetus donated from a miscarriage or one given through an elective abortion. Presumably, either type of donation is included, but a certain determination is difficult" (1293).

18 Under § 3 of the UAGA, the first two categories of individuals who may consent to donate are a spouse or adult child of the decedent, which would be irrelevant in the case of a fetus, thus giving priority to the next class, the parents. Usually, permission from any member of a class is adequate, unless a majority of the class objects, though as revised, the "UAGA makes the mother's consent determinative unless the father objects, and…does not provide for notice to the father" (Gelfand and Levin 1993, 679). Gelfand and Levin contrast this UAGA provision with 45 CFR § 46.209(d), which requires the father's consent unless his identity or whereabouts "cannot reasonably be ascertained" or he is "unavailable" to consent; however, these provisions apply only "until it has been ascertained whether or not a fetus *ex utero* is viable," and do not apply to donation of a dead fetus or fetal remains.

19 UAGA § 8(b).

20 See, for example, 45 CFR § 46.206(a)(3) ("Individuals engaged in the activity [of research] will have no part in: (i) Any decisions as to the timing, method, and procedures used to terminate the pregnancy, and (ii) determining the viability of the fetus at the termination of the pregnancy"); see also Zion (1996): "These provisions create a 'Chinese Wall' between the individuals effecting the abortion and those conducting fetal tissue research and transplantation.…While this language standing alone would likely preclude most undue influence, the UAGA also provides for the waiver of the 'Chinese Wall'.…[R]evision may be necessary" (1294).

21 There are also state laws whose restrictions regarding choosing tissue recipients are broader, and may have implications for stem cell research. In Pennsylvania, for example, "No person who consents to the procurement or use of any fetal tissue or organ may designate the recipient of that tissue or organ, nor shall any other person or organization act to fulfill that designation" (18 Pa. Cons. Stat. Ann. § 3216(b)(5)). This law unintentionally would create the

situation where an IVF patient could donate her excess embryo for stem cell research, but she could specify that it be used by a particular medical center. She would have to blindly turn it over, and risk it going to a researcher or entity (such as a for-profit company) that she might not approve of. See Andrews, L.B., 1999, "State Regulation of Embryo Stem Cell Research." This background paper was prepared for NBAC and is available in Volume II of this report.

22 Andrews 1999.

23 See Ariz. Rev. Stat. Ann. § 36-2302(A) (subsequent); Ark. Stat. Ann. § 20-17-802 (subsequent); Cal. Health and Safety Code § 123440 (subsequent); Fla. Stat. Ann. § 390.0111(6) (prior or subsequent); Ind. Code Ann. § 16-34-2-6 (subsequent); Ky. Rev. Stat. § 436.026 (subsequent); Mo. Ann. Stat. § 188.037 (prior or subsequent); Neb. Rev. Stat. § 28-346 (subsequent); Ohio Rev. Code Ann. § 2919.14(A) (subsequent); Okla. Stat. Ann. tit. 63, § 1-735(A) (prior or subsequent); Tenn. Code Ann. § 39-15-208 (subsequent); Wyo. Stat. Ann. § 35-6-115 (subsequent).

24 Ariz. Rev. Stat. Ann. § 36-2302, -2303; Ind. Code Ann. § 1 6.34-2-6; N.D. Cent. Code § 14-02.2-01 to -02; Ohio Rev. Code Ann. § 2919.14; Okla. Stat. Ann. tit. 63, § 1-735; S.D. Codified Laws Ann. § 34-23A-17.

25 Andrews 1999. Similarly, Arizona's statute provides that a "person shall not knowingly use any human fetus or embryo, living or dead, or any parts, organs or fluids of any such fetus or embryo resulting from an induced abortion in any manner" (Ariz. Rev. Stat. § 36-2302(A)).

26 Ark. Stat. Ann. § 20-17-802(2); Mass. Ann. Laws ch. 112 § 12J(a)(II); Mich. Comp. Laws Ann. § 333.2687 (must also comply with state's version of the UAGA, Mich. Comp. Laws Ann. § 333.10101 et seq.); 18 Pa. Cons. Stat. Ann. § 3216(b)(1) (mother's consent valid only after decision to abort has been made; no compensation allowed); R.I. Gen. Laws § 11-54-1(d); Tenn. Code Ann. § 39-15-208(a).

27 Even in the context of research on live fetuses, only New Mexico's statute describes the information that must be provided before consent to research involving a fetus is valid. Under the New Mexico law, a woman who is asked to participate in research must be "fully informed regarding possible impact on the fetus" (Andrews 1999, citing N.M. Stat. Ann. § 24-9A-2(b)).

28 18 Pa. Cons. Stat. Ann. § 3216(b)(4).

29 Ohio Rev. Code Ann. § 2919.14.

30 Okla. Stat. Ann. § 1-735.

31 N.D. Cent. Code § 14-02.2-01(2); Mo. Stat. Ann. § 188.036(5).

32 Tenn. Code Ann. § 39-15-208 (also prohibits sale of an aborted fetus); Utah Code Ann. § 76-7-311.

33 Ark. Stat. Ann. § 20-17-802(c); also a crime to possess such material, § 20-17-802(d).

34 See, for example, Ind. Stat. § 35-46-5-1 (applies both to aborted and stillborn fetuses); Ohio Rev. Code Ann. § 2919.14(A); Okla. Stat. Ann. § 1-735(A).

35 R.I. Geb. Laws § 11-54-1(f).

36 Minn. Stat. Ann. § 145.422(3).

37 Of the 23 states in which organ transplant laws forbid payment, two appear inapplicable to using fetal remains in stem cell research: Arizona's statute defines a decedent to include a stillborn infant but not a fetus (Ariz. Rev. Stat. § 36-849(1)), and Kentucky excludes "fetal parts or…any products of the birth or conception" from its definition of "transplantable organs" that may not be sold (Ky. Rev. Stat. Ann. § 311.165(5)(b)).

38 42 USC § 289g-2(a) (1997).

39 National Organ Transplant Act (NOTA) 42 USC § 274e(a) (1997). "Valuable consideration" is defined at 42 USC § 274e(c)(2) (1997) negatively: "'valuable consideration' does not include the reasonable payments associated with the removal, transportation, implantation, processing, preservation, quality control, and storage of a human organ or the expenses of travel, housing, and lost wages incurred by the donor of a human organ in connection with the donation of the organ." A similar definition (excluding donor costs) is provided in the NIH Revitalization Act at 42 USC § 289g-2(d)(3) (1997).

40 Because the definition of "interstate commerce" in NOTA is based upon the Federal Food, Drug and Cosmetic Act, which defines it as "commerce between any State or Territory and any place outside thereof," 21 USC § 321(b), NOTA's prohibitions extend to purchasing organs abroad for importation into the United States. Most countries explicitly prohibit the commercialization of human fetal tissue. The Canadian Royal Commission on New Reproductive Technologies stated that the noncommercialization of reproduction should be considered a guiding principle. The commission recommended that no for-profit trade be permitted in fetal tissue and that the "prohibition on commercial exchange of fetuses and fetal tissue extend to tissues imported from other countries" (1993). This prohibition was intended to prevent the exploitation of poor women, especially in developing countries, who might be persuaded to begin and end pregnancies for compensation.

41 Organ Transplants Amendment Act of 1988, 42 USC § 274(e)(c)(1) (1997). The amendment was specifically intended to prevent the "sale or exchange for any valuable consideration" of fetal organs and tissue. 134 *Cong. Rec.* S10, 131 (27 July 1988).

42 As enacted in six states, the statutes prohibit the sale of human organs but fail to include a definition of "valuable consideration" that stipulates an exemption for miscellaneous overhead expenses; sixteen states provide such an exemption (Andrews 1999).

43 See for example, *Doe v. Rampton*, 366 F. Supp. 189, 194 (D. Utah 1973) (suggesting in dicta that statute provision prohibiting research on live fetus may not be otherwise unconstitutional), vacated and remanded, 410 U.S. 950 (1973) (directing further consideration in light of *Roe*); *Wolfe v. Schroering*, 388 F. Supp. 631, 638 (W.D. Ky. 1974), aff'd in part, rev'd in part on other grounds, 541 F.2d 523 (6th Cir. 1976) (upholding prohibition on experimentation on a viable fetus due to state's interest in the fetus after viability); *Planned Parenthood Association v. Fitzpatrick*, 401 F. Supp.

554 (E.D. Penn. 1975), aff'd without opin sub nom.; *Franklin v. Fitzpatrick*, 428 U.S. 901 (1976) (affirming legitimate state interest in disposal of fetal remains); *Wynn v. Scott*, 449 F. Supp. 1302, 1322 (N.D. Ill. 1978) (medical researchers have no fundamental rights under the Constitution to perform fetal experiments), aff'd on other grounds sub nom.; *Wynn v. Carey*, 599 F.2d 193 (7th Cir. 1979) (upholding state's rational interest in regulating medicine as to viable fetus); *Leigh v. Olson*, 497 F. Supp. 1340 (D.N.D. 1980) (striking fetal disposal statute as vague where it left "humane disposal" undefined and required mother to determine method of disposal); *Akron v. Akron Center for Reproductive Health, Inc.*, 462 U.S. 416 (1983) (struck down local ordinance that, inter alia, mandated humane and sanitary disposal of fetal remains, finding the provision impermissibly vague because it was unclear whether it mandated a decent burial of the embryo at the earliest stages of formation); *Planned Parenthood Association v. City of Cincinnati*, 822 F.2d 1390, 1391 (6th Cir. 1987) (struck down on other grounds, the court noted in dicta that the wording used by the municipal code regulating disposal of aborted fetal tissue might be precise enough to survive scrutiny); *Planned Parenthood of Minnesota v. Minnesota*, 910 F.2d 479 (8th Cir. 1990) (upholding Minnesota's fetal disposal statute against challenge of vagueness and infringement of privacy).

44 La. Rev. Stat. Ann. § 40:1299.35.13. See Clapp (1988): "The Louisiana statute effectively prohibits any research, experimentation, or even observational study on any embryo, fetus, or aborted fetal tissue. The ban encompasses a range of activities, including studies of the safety of ultrasound and pathological study of fetal tissues removed from a woman for the purpose of monitoring her health. Research on IVF is likewise barred. Since the aborted previable fetus is not living or cannot survive for long, no procedure performed upon it could be considered 'therapeutic,' and therefore use of this tissue is likewise prohibited. If performed on tissues from a miscarriage, such experimentation would be acceptable under the statutory scheme" [footnote omitted] (1076–1077).

45 *Margaret S. v. Treen*, 597 F. Supp. 636 (E.D. La. 1984), aff'd sub nom.; *Margaret S. v. Edwards*, 794 F.2d 994 (5th Cir. 1986). See Clapp (1988): The court "specifically note[d] that reproductive choice was 'not limited to abortion decisions…but extends to both childbirth and contraception.' Prohibiting experimentation on fetal tissues could deny a woman knowledge that would influence her own future pregnancies, as well as prohibit procedures of immediate medical benefit such as pathological examination of tissues. The court also found that the prohibition curtailed the development and use of prediagnostic techniques, including amniocentesis. This result constituted a 'denial of health care' and a 'significant burden' on choice made during the first trimester" [footnote omitted] (1078–1079).

46 *Margaret S. v. Treen*, 597 F. Supp. 636, 674-75 (E.D. La. 1984). See Clapp (1988): "The court further suggested the statute would fail even a rational relation test because it failed to serve its own stated purpose of treating the fetus like a human being, since it treated fetal tissue differently from other human tissue" (1079).

47 *Margaret S. v. Treen*, 597 F. Supp. 636, 675-76 (E.D. La. 1984).

48 *Margaret S. v. Edwards*, 794 F.2d 994, 999 (5th Cir. 1986). "The whole distinction between experimentation and testing, or between research and practice, is…almost meaningless, [such that] 'experiment' is not adequately distinguishable from 'test'…every medical test that is now 'standard' began as an 'experiment.'" But see Clapp (1988): "[T]he court hypothesized that the statute was intended 'to remove some of the incentives for research-minded physicians…to promote abortion' and was therefore 'rationally related to an important state interest.' This language suggests that if the statute had not been vague, the court would have applied less than strict scrutiny to a ban on fetal research. The court also implied, in dicta, that the rationale was based on the 'peculiar nature of abortion and the state's legitimate interest in discouraging' it, relying on *H.L. v. Matheson*, 450 U.S. 398, 411–413 (1981)" (1080). A concurring opinion "criticized the majority for avoiding the real constitutional issue raised—that any statutory ban on experimentation would inevitably limit the kinds of tests available to women and their physicians and thus could not help but infringe on fundamental rights" Ibid. at 999–1002 (Williams, J., concurring) (Clapp 1988, 1080). See also *Jane L. v. Bangerter*, 61 F.3d 1493 (10th Cir. 1995) (striking down as vague Utah's criminal prohibition on fetal research which permitted experimentation aimed at acquiring genetic information about the embryo or fetus).

49 Members of Congress who have opposed stem cell funding maintain that "current law…also specifically covers cells and tissue obtained from embryos," citing as applicable 42 USC § 289g-1(b)(2)(ii) ("no alternation of the timing, method, or procedures used to terminate the pregnancy…made solely for the purposes of obtaining the [fetal] tissue") (Members of the House of Representatives 1999). The apparent basis for this assertion is the definition of "human fetal tissue" at 42 USC § 289g-1(g) ("for purposes of this section, the term 'human fetal tissue' means tissue or cells obtained from a dead human embryo or fetus after a spontaneous or induced abortion"). Two elements render the congressional arguments unpersuasive: 1) neither 42 USC § 289g-1 nor 289g-2 is directed at embryo or IVF research; rather, both sections are exclusively centered in a conventional understanding of aborted fetal tissue and the issues arising from fetal tissue research; and 2) biological embryology, IVF, and ES cell research typically include only "live" embryos that are maintained in a living state for research purposes until they are either implanted, disaggregated for living unicellular components, or terminated upon the experiment's completion. A "dead human embryo" would, by definition, comprise a multicellular tissue mass in which all cellular functions associated with life activity had previously ceased (clinical cell death), and would be more in the nature of a stored pathology specimen. The draft guidelines of the NIH Ad Hoc Working Group of the Advisory Committee to the Director support this interpretation (NIH Ad Hoc Working Group 1999, 5).

50 Some private sector biotechnology companies have voluntarily undertaken to self-regulate their research activities using IVF embryos through the use of advisory boards and ethical protocols (Geron Ethics Advisory Board 1998).

51 Public Law No. 105-277, 112 Stat. 2681 (1998).

52 45 CFR § 46.201(a).

Chapter 3: The Legal Framework for Federal Support of Research to Obtain and Use Human Stem Cells

53 45 CFR § 46.204(d), nullified by section 121(c) of the NIH Revitalization Act of 1993, Public Law 103-43, June 10, 1993; see 59 Fed. Reg. 28276 (June 1, 1994).

54 DHHS has considered human embryo research only under the category of IVF research, as defined in Subpart B ("any fertilization of human ova which occurs outside the body of a female, either through admixture of donor human sperm and ova or by any other means," 45 CFR § 46.203(g)) and hence it had been subject to the requirement of EAB review prior to funding.

55 The 1993 Act deleted the requirement that IVF research be reviewed by an EAB before it could be funded, but it did not remove the remaining subsections of 45 CFR § 46.204, which prescribe the basic structure and functions of the "one or more Ethical Advisory Boards" that "shall be established by the Secretary" to provide advice as needed on individual applications or "general policies, guidelines, and procedures" covered by Subpart B, including the setting of "class of applications or proposals which: (1) must be submitted to the Board, or (2) need not be submitted to the Board" 45 CFR § 46.204 (a)-(c).

56 59 Fed. Reg. 28874, 28875 (June 3, 1994) (notice of meeting); (NIH 1994, vol. 1, ix).

57 "[It] would not be wise to prohibit altogether the fertilization and study of oocytes for research purposes....[H]owever, the embryo merits respect as a developing form of human life and should be used in research only for the most serious and compelling reasons....The Panel believes that the use of oocytes fertilized expressly for research should be allowed only under two conditions. The first condition is when the research by its very nature cannot otherwise be validly conducted. The second condition...is when a compelling case can be made that this is necessary for the validity of a study that is potentially of outstanding scientific and therapeutic value" (NIH 1994, vol. 1, xi–xii).

58 "The Director of the National Institutes of Health has received a recommendation regarding federal funding of research on human embryos. The subject raises profound ethical and moral questions as well as issues concerning the appropriate allocation of federal funds. I appreciate the work of the committees that have considered this complex issue and I understand that advances in *in vitro* fertilization research and other areas could derive from such work. However, I do not believe that federal funds should be used to support the creation of human embryos for research purposes, and I have directed that NIH not allocate any resources for such research. In order to ensure that advice on complex bioethical issues that affect our society can continue to be developed, we are planning to move forward with the establishment of a National Bioethics Advisory Commission over the next year" (Office of the White House Press Secretary, Statement by the President, December 2, 1994). Although technically superseded in its effect by the congressional appropriations rider governing DHHS, the Directive remains effective throughout other Executive agencies. This has not been formally inscribed as an Executive Order.

59 Public Law No. 104-99, Title I, § 128, 110 Stat. 26, 34 (1996). The rider defines "human embryo" as "any organism, not protected as a human subject under 45 CFR 46 as of the date of the enactment of this Act, that is derived by fertilization, parthenogenesis, cloning, or any other means from one or more human gametes or human diploid cells." NIH has described the effect of the ban as prohibiting "*in vitro* fertilization of a human egg for research purposes where there is no direct therapeutic intent...as well as...research with embryos resulting from clinical treatment and research on parthenogenesis." The rider has been attached to the subsequent DHHS appropriations, through the current Fiscal Year. See Public Law No. 104-208, Div. A, § 101(e), Title V, § 512, 110 Stat. 3009, 3009-270 (1996); Public Law No. 105-78, Title V, § 513, 111 Stat. 1467, 1517 (1997); Public Law No. 105-277, 112 Stat. 2461 (1998).

60 Memorandum from Harriet Rabb to Harold Varmus, January 15, 1999.

61 "NIH funds (including equipment, facilities, and supplies purchased on currently funded grants) should not be used to conduct research using human pluripotent stem cells derived from human fetal tissue or human embryos until further notice....While the NIH proposes to support research utilizing these human pluripotent stem cells, it will not do so until public consultation has occurred, guidelines are issued, and an oversight committee has ensured that each project is in accord with these guidelines. Research on human stem cells derived from sources other than human embryos or fetal tissue will not be subject to these guidelines and oversight: this research will continue to be funded under existing policies and procedures" (NIH 1999). The NIH Director's caution has not avoided public controversy, however (Members of the House of Representatives 1999; Lanza, Arrow, Axelrod, et al. 1999).

62 "Opening Statement of Co-Chair Ezra C. Davidson, Jr., M.D.," Meeting of the NIH Ad Hoc Working Group of the Advisory Committee to the Director, April 8, 1999 (NIH Ad Hoc Working Group 1999).

63 Letter from Gary B. Ellis, Director of the Office for Protection from Research Risks, to Eric M. Meslin, Executive Director of the National Bioethics Advisory Commission (NBAC), June 3, 1999.

64 45 CFR § 46.102(f).

65 In addition to their other duties, IRBs reviewing research subject to Subpart B must "1) Determine that all aspects of the activity meet the requirements of this subpart; 2) Determine that adequate consideration has been given to the manner in which potential subjects will be selected, and adequate provision has been made by the applicant or offeror for monitoring the actual informed consent process (e.g., through such mechanisms, when appropriate, as participation by the Institutional Review Board or subject advocates in: i) Overseeing the actual process by which individual consents required by this subpart are secured either by approving induction of each individual into the activity or verifying, perhaps through sampling, that approved procedures for induction of individuals into the activity are being followed, and ii) monitoring the progress of the activity and intervening as necessary through such steps as

visits to the activity site and continuing evaluation to determine if any unanticipated risks have arisen); 3) Carry out such other responsibilities as may be assigned by the Secretary" (45 CFR § 46.205(a) (1997)). See also 45 CFR § 46.205(c) (1997) ("Applicants or offerors seeking support for activities covered by this subpart must provide for the designation of an Institutional Review Board, subject to approval by the Secretary, where no such Board has been established under Subpart A of this part.").

66 See 45 CFR §§ 46.201-210, Subpart B, "Additional DHHS Protections for Pregnant Women, Human Fetuses, and Newborns Involved as Subjects in Research, and Pertaining to Human *In Vitro* Fertilization," *Fed. Reg.* 27794–27804 (May 20, 1998).

67 "In those states...embryo stem cell research is not banned," but see D.C. Code § 6-2601 (1998) prohibiting sale of any part of human body (even cells), a restriction that may extend to human embryos (Andrews 1999).

68 N.H. Rev. Stat. Ann. § 168-B:15 (limiting the maintenance of *ex utero* pre-implantation embryo in a noncryopreserved state to under 15 days and prohibiting the transfer of research embryo to the uterine cavity).

69 Louisiana broadly prohibits research involving IVF embryos. La. Rev. Stat. Ann. §§ 9:121–122 (West 1991). Eight other states restrict embryo research indirectly, banning all research on "live" embryos or fetuses. Fla. Stat. Ann. § 390.0111(6); Me. Rev. Stat. Ann. tit. 22, § 1593 (West 1992); Mass. Ann. Laws ch. 112, § 12j(a)(I) (Law. Co-op. 1996); Mich. Comp. Laws Ann. §§ 333.2685, 333.2686, 333.2692 (West 1992); Minn. Stat. Ann. § 145.422 Subd. 1,2 (West 1989); N.D. Cent. Code §§ 14-02.2-01, 14-02.2-02 (1991); 18 Pa. Cons. Stat. Ann. § 3216(a) (Supp. 1995); R.I. Gen. Laws § 11-54-1(a)-(c) (1994) (Andrews 1999).

70 Me. Rev. Stat. Ann. tit. 22 § 1593; Mass. Ann. Laws ch. 112 § 12(j)(A)(Iv); Mich. Comp. Laws § 333.2609; N.D. Cent. Code § 14-02.2-02(4); and R.I. Gen. Laws § 11-54-1(f).

71 Fla. Stat. Ann. § 873.05; Georgia Code Ann. § 16-12-160 (A) (Except for Health Services Education); Ill. Stat. Ann. Ch 110½ Para. 308.1; La. Rev. Stat. Ann. § 9:122; Minn. Stat. Ann. § 145.422(3) (Live); 18 Pa. Cons. Stat. Ann. § 3216(b)(3) (forbids payment for the procurement of fetal tissue or organs); Texas Penal Code § 48.02; Utah Code Ann. § 76-7-311. But see Feiler (1998): "Although some state laws prohibit the sale of fertilized embryos, they do nothing to prevent the sale of gametes (sperm and eggs), which can easily be converted into research embryos through deliberate fertilization. Payment for sperm and eggs is widespread among American infertility clinics" [citations omitted] (2455).

72 nn. 66; 75; 76; 80. Tenn. Code Ann. § 39-15-208 (199_) and Utah Code Ann. § 76-7-311 (199_) prohibit sale of an "unborn child"; D.C. Code § 6-2601 (199_) and Va. Code § 32.1-289.1 (199_) prohibit sale of all or a portion of the "human body" (D.C.) or a "natural body part" (Va.); two state statutes prohibit sale of specified organs (not including embryos), but permit state health officials to expand the list under prescribed conditions. N.Y. Public Health Law § 4307 (199_); W. Va. Code § 68.50.610(2) (199_) (Andrews 1999).

73 At least one state "prohibits the sale of living [embryos] or non-renewable organs but does allow 'the buying and selling of a cell culture line or lines taken from a non-living human [embryo],'" ibid., citing Minn. Stat. Ann. § 145.422(3) (Andrews 1999, citing Minn. Stat. Ann. § 145.422(3)).

74 Office of the White House Press Secretary, "Memorandum for the Heads of Executive Departments and Agencies," March 4, 1997.

75 Executive Office of the President of the United States, 1998, *Statement of Administration Policy [on] S.1601 (Human Cloning Prohibition Act)* (Washington, DC: Executive Office of the President).

References

Adams, A.M. 1988. "Background Leading to Meeting of the Human Fetal Tissue Transplantation Research Panel Consultants." *Report of the Human Fetal Tissue Transplantation Research Panel*. Vol. 2, A3–A5. Bethesda, MD: National Institutes of Health (NIH).

Areen, J.C. 1988. "Statement on Legal Regulation of Fetal Tissue Transplantation for the Human Fetal Tissue Transplantation Research Panel." *Report of the Human Fetal Tissue Transplantation Research Panel*. Vol. 2, D21–D26. Bethesda, MD: NIH.

Bell, N.M.C. 1994. "Regulating Transfer and Use of Fetal Tissue in Transplantation Procedures: The Ethical Dimensions." *American Journal of Law and Medicine* 20:277–294.

Canadian Royal Commission on New Reproductive Technologies. 1993. *Proceed with Care: Final Report of the Royal Commission on New Reproductive Technologies*. 2 vols. Ottawa: Minister of Government Services.

Clapp, M. 1988. "State Prohibition of Fetal Experimentation and the Fundamental Right of Privacy." *Columbia Law Review* 88:1073–1097.

Coleman, J. 1996. "Playing God or Playing Scientist: A Constitutional Analysis of State Laws Banning Embryological Procedures." *Pacific Law Journal* 27:1331–1399.

Department of Health, Education, and Welfare (DHEW). Ethics Advisory Board (EAB). 1979. *Report and Conclusions: HEW Support of Research Involving Human In Vitro Fertilization and Embryo Transfer*. Washington, DC: U.S. Government Printing Office.

Driscoll, Jr., E. 1985. "Nobel Recipient John Enders, 88; Virus Work Led to Polio Vaccine." Obituary. *Boston Globe*, 10 September, 85.

Duguay, K.F. 1992. "Fetal Tissue Transplantation: Ethical and Legal Considerations." *CIRCLES: Buffalo Women's Journal of Law and Social Policy* 1:36.

Duke, R.C. 1988. "Statement of the Population Crisis Committee." In *Report of the Human Fetal Tissue Transplantation Research Panel*. Vol. 2. D112–D121. Bethesda, MD: NIH.

Chapter 3: The Legal Framework for Federal Support of Research to Obtain and Use Human Stem Cells

Feiler, C.L. 1998. "Human Embryo Experimentation: Regulation and Relative Rights." *Fordham Law Review* 66:2435–2469.

General Accounting Office (GAO). 1997. *Report to the Chairmen and Ranking Minority Members, Committee on Labor and Human Resources, U.S. Senate, and Committee on Commerce, House of Representatives: National Institutes of Health-Funded Research: Therapeutic Human Fetal Tissue Transplantation Projects Meet Federal Requirements.* Washington, DC: U.S. Government Printing Office.

Gelfand, G., and T.R. Levin. 1993. "Fetal Tissue Research: Legal Regulation of Human Fetal Tissue Transplantation." *Washington and Lee Law Review* 50:647–694.

Geron Ethics Advisory Board. Geron Corporation. 1998. *A Statement on Human Embryonic Stem Cells.*

Goddard, J.E. 1996. "The National Institutes of Health Revitalization Act of 1993 Washed Away Many Legal Problems with Fetal Tissue Transplantation Research But a Stain Remains." *Southern Methodist University Law Review* 49:375–399.

Lanza, R.P., K.J. Arrow, J. Axelrod, et al. 1999. "Science over Politics." *Science* 284:1849–1850.

Members of the House of Representatives. 1999. *Statement of the Members of the House of Representatives to the Working Group of the Advisory Committee to the Director of the National Institutes of Health.* April 8.

National Commission for the Protection of Human Subjects of Biomedical and Behavioral Research. 1975. *Research on the Fetus: Report and Recommendations.* Washington, DC: U.S. Government Printing Office.

NIH. 1987. "Human Fetal Tissue Research Supported by the National Institutes of Health." *Summary Highlights of FY 1987.* Bethesda, MD: NIH.

———. 1999. "National Institutes of Health Position on Human Pluripotent Stem Cell Research." Bethesda, MD: NIH. April 19, www.nih.gov/grants/policy/stemcells.htm.

NIH. Ad Hoc Working Group of the Advisory Committee to the Director. 1999. "Draft Guidelines for Research Involving Pluripotent Stem Cell Research." Bethesda, MD: NIH.

NIH. Human Embryo Research Panel. 1994. *Report of the Human Embryo Research Panel.* 2 vols. Bethesda, MD: NIH.

NIH. Office of the Director (OD). 1999. "Fact Sheet on Stem Cell Research." Bethesda, MD: NIH, www.nih.gov/news/pr/apr99/od-21.htm.

Office for Protection from Research Risks (OPRR). 1994. *Human Subjects Protections: Fetal Tissue Transplantation—Ban on Research Replaced by New Statutory Requirements.* 29 April.

Robertson, J.A. 1990. "Reproductive Technology and Reproductive Rights. In the Beginning: The Legal Status of Early Embryos." *Virginia Law Review* 76:437–517.

———. 1993. "Abortion to Obtain Fetal Tissue for Transplant." *Suffolk University Law Review* 27:1359–1389.

Ryan, K.J. 1991. "Tissue Transplantation from Aborted Fetuses, Organ Transplantation from Anencephalic Infants and Keeping Brain-Dead Pregnant Women Alive Until Fetal Viability." *Southern California Law Review* 65:683–696.

Silva-Ruiz, P.F. 1998. "Section II: The Protection of Persons in Medical Research and Cloning of Human Beings." *American Journal of Comparative Law* 46:151–163.

Windom, R.E. 1988. "Memorandum to James B. Wyngaarden, March 22, 1988." *Report of the Human Fetal Tissue Transplantation Research Panel.* Vol. 2, A3. Bethesda, MD: NIH.

Zion, C. 1996. "The Legal and Ethical Issues of Fetal Tissue Research and Transplantation." *Oregon Law Review* 75:1281–1296.

Chapter Four

Ethical Issues in Human Stem Cell Research

Ethical Issues Relating to the Sources of Human Embryonic Stem or Embryonic Germ Cells

Research involving human embryonic stem (ES) cells and embryonic germ (EG) cells raises several important ethical issues, principally related to the current sources and/or methods of deriving these cells. If, for example, ES and EG cells could be derived from sources other than human embryos or cadaveric fetal material, fewer ethical concerns would be involved in determining a policy for their use for scientific research or clinical therapies. At present, however, the only methods available to isolate and culture human ES and EG cells involve the use of human embryos or cadaveric fetal tissue. Therefore, careful consideration of the ethical issues involved in the use of these sources is an unavoidable component of the advancement of this type of research.

This chapter first considers the ethical issues arising from research involving the derivation and/or use of ES or EG cells from three potential sources: cadaveric fetal tissue, embryos resulting from and remaining after infertility treatments, and embryos created solely for research purposes either by *in vitro* fertilization (IVF) or somatic cell nuclear transfer (SCNT) techniques. The chapter then reviews separately the specific arguments for and against federal funding of this research. Finally, the chapter discusses relevant ethical issues in federal oversight and review of research involving the derivation and/or use of ES or EG cells.[1]

Research with EG Cells Derived from Cadaveric Fetal Tissue

Many of the ethical questions regarding research involving the use of cadaveric fetal tissue were analyzed in depth by the 1988 National Institutes of Health (NIH) Human Fetal Tissue Transplantation Research Panel. What is new in the present context is that, in the near term at least, the materials derived from this tissue would not be transplanted; rather, gonadal tissue (both male and female) would be used as a source for human EG cells. Initially, these cell lines would be used in basic research to determine their nature, to understand their relationship to human development, and to identify differentiation factors that enable such cells to develop into particular tissue types. Later, such cell lines also might be used for the development of transplantation for particular tissue types. The value of cadaveric fetal tissue already has been demonstrated; a broad variety of research materials and reagents derived from cadaveric fetal tissue currently are used in federally funded research.[2]

The ethical acceptability of deriving EG cells from the tissue of aborted fetuses is, for some, closely connected to the ethical acceptability of abortion. Those who believe that elective abortions are morally acceptable are less likely to identify insurmountable ethical barriers to research that involves the derivation and use of EG cells derived from cadaveric fetal tissue. This group might agree that it is necessary to restrict such research by requiring that the decision to donate fetal tissue be separate from the decision to terminate the pregnancy. The purpose of such a requirement would be to protect the pregnant woman against coercion and exploitation rather than to protect the fetus. In addition, even those who find it acceptable to use cadaveric fetal tissue in research might hold that certain uses of such tissue—for example, uses that treat it as nothing more than any other bodily tissue—should be ruled out as disrespectful.

Chapter 4: Ethical Issues in Human Stem Cell Research

Those who view elective abortions as morally unjustified often—but not always—oppose the research use of tissue derived from aborted fetuses. They usually have no moral difficulty with the use of tissue from spontaneously aborted fetuses or—if they recognize exceptions to the moral prohibition on abortion—from fetuses in cases that they believe are morally justifiable abortions (e.g., to save the pregnant woman's life). However, in general they do not believe that it is possible to derive and use tissue from what they believe are unjustifiably aborted fetuses without inevitable and unacceptable association with those abortions. This association, they believe, usually taints the actions of all those involved in using these materials or in financing research protocols that rely on such tissue. Nevertheless, some opponents of elective abortions believe that it is still possible to support such research as long as effective safeguards are in place to separate abortion decisions from the procurement and use of fetal tissue in research. For them, when appropriate safeguards are in place, using cadaveric fetal tissue from elective abortions for research is relevantly similar to using nonfetal cadavers donated for scientific and medical purposes.

Association with Abortion

Opponents of the research use of fetal materials obtained from elective abortions dispute the claim that it is possible to separate the moral issues surrounding the abortion from those involved in obtaining and using fetal material. They argue that those who obtain and use fetal material from elective abortion inevitably become associated, in ethically unacceptable ways, with the abortions that are the source of the material.[3] They identify two major types of unacceptable association or cooperation with abortion: 1) causal responsibility for abortions and 2) symbolic association with abortions.

1. Causal Responsibility

Some believe that those who provide cadaveric fetal tissue in research are indirectly, if not directly, responsible for the choice of some women to have an abortion. Direct causal responsibility exists where, in this case, someone's actions directly lead a pregnant woman to have an abortion—for example, the researcher offers financial compensation for cadaveric fetal tissue and this compensation leads the pregnant woman to have an abortion she would not otherwise have had. In part because of concerns about direct causal responsibility, the Human Fetal Tissue Transplantation Research Panel (1988) recommended the following safeguards to separate the pregnant woman's decision to abort from her decision to donate fetal tissue:

- The consent of women for abortions must be obtained prior to requesting or obtaining consent for the donation of fetal tissue.
- Those who seek a woman's consent to donate should not discuss fetal tissue donation prior to her decision to abort, unless she specifically requests such information.
- Women should not be paid for providing fetal tissue.
- A separation must be maintained between abortion clinic personnel and those involved in using fetal tissue.
- There should be a prohibition against any alteration of the timing of or procedures used in an abortion solely for the purpose of obtaining tissue.
- Donors of cadaveric fetal tissue should not be allowed to designate a specific recipient of transplanted tissue.

As noted in Chapter 3, several of these safeguards were later adopted in federal legislation regarding the use of aborted fetal tissue in transplantation research, and they appear to be sufficient to avoid direct causal responsibility for abortions in human EG research as well as in transplantation research.

Those involved in research uses of EG cells derived from fetal tissue could be indirectly responsible for abortions if the perceived potential benefits of the research contributed to an increase in the number of abortions. Opponents of fetal tissue research argue that it is unrealistic to suppose that a woman's decision to abort can be kept separate from considerations of donating fetal tissue, as many women facing the abortion decision are likely to have gained knowledge about fetal tissue research through the media or other sources. The knowledge that having an elective abortion might have benefits for future patients through the donation of fetal tissue for research may tip the balance in favor of going through with an abortion for some women who are ambivalent about it. Some argue that the benefits achieved through the routine use of fetal tissue will further legitimize abortion and result in more permissive societal attitudes and policies concerning elective abortion.

It is impossible to eliminate the possibility completely, however slight it may be, that knowledge of the promise of research on EG cells derived from fetal tissue will play a role in some elective abortion decisions, even if only rarely. However, it is not clear how much moral weight ultimately attaches to this possibility. One might be justified in some instances in asserting that if it were not for the research use of fetal tissue following an abortion, a woman might not have chosen to terminate her pregnancy.

But one could assign this kind of causal responsibility to a number of factors that figure into abortion decisions without making ascriptions of indirect causal responsibility, or what is sometimes called moral complicity. For example, a woman might choose to have an abortion principally because she does not want to slow the advancement of her education and career. She might not have had an abortion in the absence of expectations that encourage women to develop their careers. Yet, we would not think it appropriate to charge those who promote such expectations and/or policies as complicit in her abortion. In both this case and that of research, the opportunity to choose abortion is a consequence of a legitimate social policy. The burden on those seeking to end such policies is to show that the risks of harm—both the probability and the magnitude of harm—resulting from the policies outweigh the expected benefits (Childress 1991). This criterion minimally requires evidence of a high probability of a large number of elective abortions that would not have occurred in the absence of those policies. There is, however, no such evidence at present. If compelling evidence did emerge that elective abortions did, or probably would, increase as a result of the research use of cadaveric fetal tissue, this would require a re-examination of the balance of benefits and harms as well as the safeguards that had been put into place to eliminate the potential for direct causal responsibility and reduce the likelihood of indirect causal responsibility for abortions.

2. Symbolic Association

People can become inappropriately associated with what they believe are wrongful acts for which they are not causally responsible. Particularly problematic for many is an association that appears to symbolize approval of the wrongdoing. For example, James Burtchaell maintains that those involved in research on fetal tissue enter a symbolic alliance with the practice of abortion in producing or deriving benefits from it (1988).

A common response is that persons can benefit from what they might consider immoral acts without tacitly approving of those acts. For example, transplant surgeons and transplant recipients may benefit (the latter more directly than the former) from donated organs from victims of murder or drunken driving but nevertheless condemn those wrongful acts (Robertson 1988; Vawter et al. 1991). A researcher who uses cadaveric fetal material in studies to answer important research questions or to study its potential therapeutic effects or the patient who receives the donated tissue need not sanction the act of abortion any more than the transplant surgeon who uses the organs of a murder victim approves of the homicidal act.

Some opponents of fetal tissue research maintain that it implicates those involved in a kind of wrongdoing that cannot be attributed to the transplant surgeon in the example above. Unlike drunken driving and murder, abortion is an institutionalized practice in which certain categories of human life (the members of which are considered by some to have the same moral status as human adults) are allowed to be killed. In this respect, some opponents of abortion go so far as to suggest that fetal tissue research is more analogous to research that benefits from experiments conducted by Nazi doctors during World War II (Bopp 1994).

But whatever one thinks of comparisons between the victims of Nazi crimes and aborted fetuses—and many are outraged by these comparisons—it is possible to concede the comparisons without concluding that human stem cell research involving cadaveric fetal tissue is morally problematic. Of course, some believe that those who use data derived from Nazi experiments are morally complicit with those crimes. For example, William Seidelman writes:

> By giving value to (Nazi) research we are, by implication, supporting Himmler's philosophy that the subjects' lives were 'useless.' This is to argue that, by accepting data derived from their misery we are, post mortem, deriving utility from otherwise 'useless' life. Science could thus stand accused of giving greater value to knowledge than to human life itself (1988, 232).

But one need not adopt this stance. Instead, one can reasonably believe that a scientist's actions must be understood and judged not by their consequences or uses but rather by several other factors, including the scientist's intentions, the social practices of which his or her actions are a part, and the social context in which those practices are embedded. As philosopher Benjamin Freedman wrote:

> A moral universe such as our own must, I think, rely on the authors of their own actions to be primarily responsible for attaching symbolic significance to those actions...[I]n using the Nazi data, physicians and scientists are acting pursuant to their own moral commitment to aid patients and to advance science in the interest of humankind. The use of data is predicated upon that duty, and it is in seeking to fulfill that duty that the symbolic significance of the action must be found (1992, 151).

It is likewise reasonable to maintain that the symbolic significance of support for research using EG cells derived from aborted fetal tissue lies in the commitment and desire to gain knowledge, promote health, and save lives. This research is allied with a worthy cause, and any taint that might attach from the source of the cells appears to be outweighed by the potential good that the research may yield.

Consent and Donation

In previous debates about the use of fetal tissue in research, questions have been raised about who has the moral authority to donate the material. Some assert that, from an ethical standpoint, a woman who chooses abortion forfeits her rights to determine the disposition of the dead fetus. Burtchaell, for instance, argues that "the decision to abort, made by the mother, is an act of such violent abandonment of the maternal trusteeship that no further exercise of such responsibility is admissible" (1988, 9). By contrast, John Robertson argues that this position mistakenly assumes that the persons disposing of cadaveric remains act only as the guardians or proxies of the deceased. Instead, "a more accurate account of their role is to guard their own feelings and interests in assuring that the remains of kin are treated respectfully" (1988, 6).

In our view, obtaining consent to donate fetal tissue is an ethical prerequisite for using such material to derive EG cells, even though the woman or couple are not research subjects per se, and even though the cadaveric fetus is not a human subject. This view is consistent with the conclusion of the Human Fetal Tissue Transplantation Research Panel, which held that "[e]xpress donation by the pregnant woman after the abortion decision is the most appropriate mode of transfer of fetal tissues because it is the most congruent with our society's traditions, laws, policies, and practices, including the Uniform Anatomical Gift Act and current Federal research regulations" (1988, 6). According to this panel, a woman's choice of a legal abortion does not disqualify her legally and should not disqualify her morally from serving "as the primary decisionmaker about the disposition of fetal remains, including the donation of fetal tissue for research." She "has a special connection with the fetus and she has a legitimate interest in its disposition and use." In addition, her decision to donate fetal tissue would not violate the dead fetus's interests. The panel concluded that "in the final analysis, any mode of transfer other than maternal donation appears to raise more serious ethical problems" (6). Fetal tissue should not be used without the woman's consent. Not only should her consent be necessary, it should also be sufficient to donate the tissue, except where the father's objection is known.

We concur with the Human Fetal Tissue Transplantation Research Panel that a woman undergoing an elective abortion should be authorized to donate fetal tissue, unless the father is known to object. We further agree with the panel and with subsequent federal legislation that it is important to establish safeguards to separate the pregnant woman's decision to abort from the decision to donate cadaveric fetal tissue. The guidelines already in place for fetal tissue transplantation research generally are appropriate and appear to be sufficient if they also apply to research involving human EG cells.

As already noted, some opponents of elective abortion can support fetal tissue research as long as there are safeguards to avoid direct causal responsibility and to reduce the likelihood of indirect causal responsibility. Many who view elective abortion as morally problematic, even if not always morally unjustified, also may endorse these safeguards as a way to avoid certain forms of association with morally problematic actions and at the same time as a way to prevent the exploitation and coercion of pregnant women. Even those who do not find elective abortions morally problematic may accept these safeguards in order to protect pregnant women from exploitation and coercion as well as to sustain social practices that reflect important social and cultural values and to respect the moral concerns of opponents of elective abortion. We believe, therefore, that there can be wide agreement on appropriate safeguards for the process of donating cadaveric fetal tissue.

At a minimum, these safeguards should separate the decision to have an abortion from the decision to donate by ensuring, as much as possible, that the former occurs before the latter by not providing before the abortion decision is made information about the possibility of using fetal materials in research and by prohibiting the provision of financial compensation for the fetal tissue to the woman (or to the couple) having the abortion. If these and other requirements that already have been adopted in regulations governing federally funded human fetal tissue transplantation research do not clearly extend to research to generate EG cells from cadaveric fetal tissue, the regulations should be modified to do so.

Research with ES Cells Derived from Embryos Remaining After Infertility Treatments

Ethical issues arising from research involving the use of human embryos have generated a sustained public policy discussion and a valuable body of literature that spans at least 20 years. Some of these issues were considered in depth by the Department of Health, Education and Welfare (DHEW) Ethics Advisory Board (EAB) in 1978 and in 1979 (DHEW Ethics Advisory Board 1979). The ethical debate was continued here and abroad by other national advisory bodies, including the British Warnock Committee (Committee of Inquiry 1984) and the Canadian Royal Commission on New Reproductive Technologies (1993). In 1994, the NIH Human Embryo Research Panel considered multiple types of present and future human embryo research and discussed both ethical and public policy issues (NIH 1994). In contrast, for example, SCNT has been seriously debated in the United States and elsewhere only for about two years, and the research use of SCNT has been debated for an even shorter period.

One source of embryos for ES cells is those remaining after infertility treatments. Couples who provide such embryos have decided that they no longer need them to achieve their reproductive goals. If the couple prefers to discontinue storing the remaining embryos and does not wish to donate them to other couples, the only alternatives are to direct that the embryos be discarded (that is, to destroy them through the thawing process) or to donate them for research. When only these latter two alternatives remain, the situation is somewhat similar to that in which a woman is deciding whether to donate fetal tissue for research following elective abortion and the situation in which families are deciding whether to donate the organs or tissues of a loved one who has recently died. However, whether this similarity is decisive depends upon one's perception of the moral status of embryos. Derivation of ES cells involves destroying the embryos, whereas abortion precedes the donation of the fetal tissue and death precedes the donation of whole organs for transplantation.

The Moral Status of Embryos

To say that an entity has "moral status" is to say something both about how one should act towards that thing or person and about whether that thing or person can expect certain treatment from others. The debate about the moral status of embryos traditionally has revolved around the question of whether the embryo has the same moral status as children and adult humans do—with a right to life that may not be sacrificed by others for the benefit of society. At one end of the spectrum of attitudes is the view that the embryo is a mere cluster of cells that has no more moral status than any other collection of human cells. From this perspective, one might conclude that there are few, if any, ethical limitations on the research uses of embryos.

At the other end of the spectrum is the view that embryos should be considered in the same moral category as children or adults. According to this view, research involving the destruction of embryos is absolutely prohibited. Edmund D. Pellegrino, a professor of bioethics at Georgetown University, described this perspective in testimony given before the Commission:

Chapter 4: Ethical Issues in Human Stem Cell Research

> The Roman Catholic perspective…rejects the idea that full moral status is conferred by degrees or at some arbitrary point in development. Such arbitrariness is liable to definition more in accord with experimental need than ontological or biological reality.[4]

In contrast, scholars representing other religious traditions testified that moral status varies according to the stage of development.[5] For example, Margaret Farley, a professor of Christian ethics at Yale University, pointed out that

> There are clear disagreements among Catholics—whether moral theologians, church leaders, ordinary members of the Catholic community—on particular issues of fetal and embryo research….A growing number of Catholic moral theologians, for example, do not consider the human embryo in its earliest stages…to constitute an individualized human entity.[6]

Other scholars from Protestant, Jewish, and Islamic traditions noted that major strands of those traditions support a view of fetal development that does not assign full moral status to the early embryo.[7] For example, Jewish scholars testified that the issue of the moral status of extra-corporeal embryos is not central to an assessment of the ethical acceptability of research involving ES cells. Rabbi Elliot Dorff noted that

> Genetic materials outside the uterus have no legal status in Jewish law, for they are not even a part of a human being until implanted in a woman's womb and even then, during the first 40 days of gestation, their status is 'as if they were water.' As a result, frozen embryos may be discarded or used for reasonable purposes, and so may stem cells be procured from them.[8]

As a result, for some Jewish thinkers, the derivation and use of ES cells from embryos remaining after infertility treatments may be less problematic than the use of aborted fetal tissue, at least following morally unjustified abortions.

On this issue, the Commission adopted what some have described as an intermediate position, one with which many likely would agree: that the embryo merits respect as a form of human life, but not the same level of respect accorded persons. We recognize that, on such a morally contested issue, there will be strong differences of opinion. Moreover, it is unlikely that, by sheer force of argument, those with particularly strong beliefs on either side will be persuaded to change their opinions (Murray 1996). However, there is, in our judgment, considerable value in describing some of these positions, not only to reveal some of the difficulties of resolving the issue, but to seek an appropriate set of recommendations that can reflect the many values we share as well as the moral views of those with diverse ethical commitments.

A standard approach taken by those who deny that embryos are persons with the same moral status as children and adults is to identify one or more psychological or cognitive capacities that are considered essential to personhood (and a concomitant right to life) but that embryos lack. Most commonly cited are consciousness, self-consciousness, and the ability to reason (Feinberg 1986; Tooley 1983; Warren 1973). The problem with such accounts is that they appear to be either under- or over-inclusive, depending on which capacities are invoked. For example, if one requires self-consciousness or the ability to reason as an essential condition for personhood, most very young infants will not be able to satisfy this condition. On the other hand, if sentience is regarded as the touchstone of the right to life, then nonhuman animals also possess this right.

Those who deny that embryos have the same moral status as persons might maintain that the embryo is simply too nascent a form of human life to merit the kind of respect accorded more developed humans. However, some would argue that, in the absence of an event that decisively (i.e., to everyone's satisfaction) identifies the first stage of human development—a stage at which destroying human life is morally wrong—it is not permissible to destroy embryos.

The fundamental argument of those who oppose the destruction of human embryos is that these embryos are human beings and, as such, have a right to life. The very humanness of the embryo is thus thought to confer the moral status of a person. The problem is that, for some, the premise that all human lives at any stage of their development are persons in the moral sense is not self-evident. Indeed, some believe that the premise conflates two categories of human beings: namely, beings that belong to the species *homo sapiens*, and beings that belong to a particular moral community (Warren 1973). According to this view, the fact that an individual is a member of the species *homo sapiens* is not sufficient to confer upon it membership in the moral community of persons. Although it is not clear that those who advance this view are able to establish the point at which, if ever, embryos first acquire the moral status of persons, those who oppose the destruction of embryos likewise fail to establish, in a convincing manner, why society should ascribe the status of persons to human embryos.

It is not surprising that these different views on the moral status of the embryo appear difficult to resolve, given their relationship to the issues surrounding the abortion debate, a debate the philosopher Alastair MacIntyre describes as interminable: "I do not mean by this just that such debates go on and on and on—although they do—but also that they can apparently find no terminus. There seems to be no rational way of securing moral agreement in our culture" (1984, 6). This difficulty has led most concerned observers to search for a position that respects the moral integrity of different perspectives, but to the extent possible, focuses public policy on ethical values that may be broadly shared.

The Importance of Shared Views

Once again, we are aware that the issue of the moral status of the embryo has occupied the thoughtful attention of previous bodies deliberating about fetal tissue and embryo research.[9] Further, as already noted, we do not presume to be in a position to settle this debate, but instead have aimed to develop public policy recommendations regarding research involving the derivation and use of ES cells that are formulated in terms that people who hold differing views on the status of the embryo can accept. As Thomas Nagel argues, "In a democracy, the aim of procedures of decision should be to secure results that can be acknowledged as legitimate by as wide a portion of the citizenry as possible" (1995, 212). In this vein, Amy Gutmann and Dennis Thompson argue that the construction of public policy on morally controversial matters should involve a "search for significant points of convergence between one's own understandings and those of citizens whose positions, taken in their more comprehensive forms, one must reject" (1996, 85).

R. Alta Charo suggests an approach for informing policy in this area that seeks to accommodate the interests of individuals who hold conflicting views on the status of the embryo. Charo argues that the issue of moral status can be avoided altogether by addressing the proper limits of embryo research in terms of political philosophy rather than moral philosophy:

> The political analysis entails a change in focus, away from the embryo and the research and toward an ethical balance between the interests of those who oppose destroying embryos in research and those who stand to benefit from the research findings. Thus, the deeper the degree of offense to opponents and the weaker the opportunity for resorting to the political system to impose their vision, the more compelling the benefits must be to justify the funding (1995, 20).

In Charo's view, once one recognizes that the substantive conflict among fundamental values surrounding embryo research cannot be resolved in a manner that will satisfy all sides, the most promising approach is to seek to balance all the relevant considerations in determining whether to proceed with the research. Thus, although it is clear that embryo research would offend some people deeply, she would argue that the potential health benefits for this and future generations outweigh the pain experienced by opponents of the research.

It is, however, questionable whether Charo's analysis successfully avoids the issue of moral status. It might be argued, for example, that placing the lives of embryos in this kind of utilitarian calculus will seem appropriate only to those who presuppose that embryos do not have the status of persons. Those who believe—or who genuinely allow for the possibility—that embryos have the status of persons will regard such consequentialist grounds for destroying embryos as extremely problematic.

In our view, an appropriate approach to public policy in this arena is to develop policies that demonstrate respect for all reasonable alternative points of view and that focus, when possible, on the shared fundamental values that these divergent opinions, in their own ways, seek to affirm. This particular perspective was recommended by Patricia King in her testimony before the Commission and elsewhere (1997).[10] As long as the

Chapter 4: Ethical Issues in Human Stem Cell Research

disagreement is cast strictly as one between those who think the embryo is a person with a right to life and those who think it has little or no moral status, the quest for convergence will be an elusive one. But there are grounds for supposing that this may be a misleading depiction of the conflict. Indeed, there may be a sufficiently broad consensus regarding the respect to be accorded to embryos to justify, under certain conditions, not only the research use of stem cells but also the use of embryos remaining after infertility treatments to generate ES cells.

The abortion debate offers an illustration of the complex middle ground that might be found in ethically and politically contentious areas of public policy. Philosopher Ronald Dworkin maintains that, despite their rhetoric, many who oppose abortion do not actually believe that the fetus is a person with a right to life. This is revealed, he claims, through a consideration of the exceptions that they often permit to their proposed prohibitions on abortion.

For example, some hold that abortion is morally permissible when a pregnancy is the result of rape or incest. Yet, as Dworkin comments, "[i]t would be contradictory to insist that a fetus has a right to live that is strong enough to justify prohibiting abortion even when childbirth would ruin a mother's or a family's life, but that ceases to exist when the pregnancy is the result of a sexual crime of which the fetus is, of course, wholly innocent" (1994, 32).

The importance of reflecting on the meaning of such exceptions in the context of the research uses of embryos is that they suggest that even in an area of great moral controversy it may be possible to identify some common ground. If it is possible to find common ground in the case of elective abortions, we might be able to identify when it would be permissible in the case of destroying embryos. For example, conservatives allow such exceptions implicitly hold with liberals that very early forms of human life may sometimes be sacrificed to promote the interests of other humans.[11] Although liberals and conservatives disagree about the range of ends for which embryonic or fetal life may ethically be sacrificed, they may be able to reach some consensus. Conservatives who accept that destroying a fetus is permissible when necessary to save a pregnant woman or spare a rape victim additional trauma might agree with liberals that it also is permissible to destroy embryos when it is necessary to save lives or prevent extreme suffering. We recognize, of course, that these cases are different, as the existence of the fetus may directly conflict with the pregnant woman's interests, while a particular *ex utero* embryo does not threaten anyone's interests. But this distinction obscures the fact that these two cases share an implicit attribution of greater value to the interests of children and adults.

We believe that the following would seem to be a reasonable statement of the kind of agreement that could be possible on this issue:

> Research that involves the destruction of embryos remaining after infertility treatments is permissible when there is good reason to believe that this destruction is necessary to develop cures for life-threatening or severely debilitating diseases and when appropriate protections and oversight are in place in order to prevent abuse.

Given the great promise of ES cell research for saving lives and alleviating suffering, such a statement would appear to be sufficient to permit, at least in certain cases, not only the use of ES cells in research, but also the use of certain embryos to generate ES cells. Some might object, however, that the benefits of the research are too uncertain to justify a comparison with the conditions under which one might make an exception to permit abortion. But the lower probability of benefits from research uses of embryos is balanced by a much higher ratio of potential lives saved relative to embryonic lives lost and by two other characteristics of the embryos used to derive ES cells: first, that they are at a much earlier stage of development than is usually true of aborted fetuses, and second, that they are about to be discarded after infertility treatment and thus have no prospect for survival even if they are not used in deriving ES cells. In our view, the potential benefits of the research outweigh the harms to embryos that are destroyed in the research process.

Another objection is that the availability of alternative means of obtaining (and sources of) stem cells makes it unnecessary to use embryos to obtain ES cells for research. Richard Doerflinger of the National Conference of Catholic Bishops testified before the Commission that "it is now clearer than ever that new research involving adult stem cells...offers the promise that embryonic stem cells may simply be irrelevant to future medical progress."[12] In our judgment, the derivation of stem cells from embryos remaining following infertility treatments is justifiable only if no less morally problematic alternatives are available for advancing the research. But as we have noted, ES cells from embryos appear to be different in scientifically important ways from AS cells and also appear to offer greater promise of therapeutic breakthroughs. The claim that there are alternatives to using stem cells derived from embryos is not, at the present time, supported scientifically. We recognize, however, that this is a matter that must be revisited continually as science advances.

Nevertheless, if research is to proceed with the derivation and use of ES cells from embryos that remain following infertility treatments, we must consider what kinds of conditions and constraints should apply to this work. Many of these conditions, discussed below, also are reflected in our recommendations that are provided in the next chapter.

First, ideally, those who have the authority to decide about the disposition of the remaining embryos should make the decision about whether to donate them to another couple, to continue to store them, or to destroy them before they are asked about donating them for research. This will reduce the likelihood that a desire to benefit research will lead to a choice to destroy the embryos. If the decision to destroy the embryos precedes the decision to donate them for research purposes, then the research use of such embryos affects only how, not whether, the destruction occurs. Obviously, this separation may not be possible, particularly because the couple may be given several options simultaneously, either at the outset of treatment for infertility or after its completion. Indeed, some infertility programs provide patients with multiple consent forms at the outset of treatment, forms that include options to donate to research, discard, or transfer any embryos that remain. But even then, it may be appropriate to view the options as consisting of donation of the embryos to another couple, their continued storage, or their destruction, with destruction of the embryos taking one of two forms—discarding them through thawing or through the process of research. If embryo destruction is permissible, then it certainly should be permissible to destroy them in a way that would generate stem cells for bona fide research.

Second, the couple's or the individual's decision to donate any remaining embryos for research should be a voluntary one, free from coercion and undue pressure. Third, donors of embryos for research should not be allowed to designate or restrict the recipients of derivative tissues or cell lines for research or therapy. Fourth, even though it is legal to sell sperm and ova, it should remain illegal to sell embryos; the demonstration of respect for embryos requires this prohibition. Fifth, only the minimum number of embryos that are needed to derive sufficient stem cells for important research should be used in this way.

Sixth, it is important to develop and widely disseminate additional professional standards of practice in reproductive medicine that will reduce the likelihood that infertility clinics will increase the numbers of embryos remaining after infertility treatments in order to increase the supply for possible research purposes. These standards could address issues such as the production of embryos, the number of embryos implanted and allowed to develop to term, and the care and handling of gametes and embryos.

Seventh, any research use of embryos or embryonic cell lines imported from outside of the country must satisfy all the regulations for the use of such materials when they are produced in the United States. Eighth, if possible, institutions, researchers, and potential recipients of therapies should be informed in some way about the source of the stem cells—perhaps by tagging the cells—so that all concerned can avoid using any cells that are believed to have been derived unethically. This last condition is intended to enable institutions, researchers, and patients to make their own conscientious choices about the acceptability of using stem cells that have been derived from ethically controversial sources.

Ethical Distinctions and Relationships Between the Derivation and Use of ES Cells Derived from Embryos Remaining After Infertility Treatments

There is significant debate regarding whether the *use* of cultures of ES cells should be regarded or treated differently from the *derivation* of such cells, given that the derivation arises from the destruction of an intact embryo. For purposes of this report, three questions will help frame this issue: First, are derivation and use ethically distinct? Second, is the use of ES cells, their derivation, or both ethically justifiable? Third, should use, derivation, or both be eligible for public funding? Here we discuss our views on the first two questions. Later in this chapter, we discuss the third question in more detail.

Even though many individuals would want to avoid the use of ES cells because of their source, the processes of derivation and use are sufficiently different to warrant being regarded as morally distinct from one another. The NIH Human Embryo Research Panel reached this conclusion as well (1994). Moreover, we heard testimony that would support this distinction.[13] However, there is vigorous debate regarding whether this distinction, even if morally relevant, is morally decisive or determinative for judgments about particular actions and public policies.

As previously discussed, most moral concerns about the derivation of ES cells from embryos that remain after infertility treatments focus on the fact that derivation involves destruction. If embryos could be destroyed by allowing them to thaw—the standard approach to discarding them—and researchers could then derive ES cells, the moral issues would be parallel to those that arise in the derivation of germ cells from aborted fetuses. Destruction and derivation could be separated in principle as well as through various practical measures. However, in practice, destruction and derivation cannot be separated; therefore, this option is not available. The question, then, is whether the use of ES cells derived in a process that destroys the embryos can be morally separated from that of derivation.

There are several possible responses. One position holds that such use is morally unacceptable because it necessarily involves association with the wrongful act of embryo destruction. Another position is that the problem of associating the use of the cells and the destruction of the embryo disappears if the destruction of the embryo is not viewed as problematic, as some traditions hold. There is no association with wrongdoing if the initial act is not on balance wrong. A third position holds that even if embryo destruction is viewed as morally wrong, there still may be ways to separate at least some uses from derivation. John Robertson suggests that there may be some circumstances in which researchers using ES cells would not be considered complicit with the destruction of embryos. He indicates, for example, that there would be no meaningful association where an investigator's "research plans or actions had no effect on whether the original immoral derivation occurred" (1999, 113).

Some commentators hold that it would be ethically justifiable, though regrettable, to use existing cell lines that were derived through unethical embryo destruction. A version of this position was suggested by Father Demetrios Demopulos, who explained his views from the perspective of Eastern Orthodoxy in testimony before the Commission:

> ...I cannot condone any procedure that threatens viability, dignity, and sanctity of that life. In my view the establishment of embryonic stem cell lines...was done at the cost of human lives. Even though not yet a human person, an embryo should not be used for or sacrificed in experimentation, no matter how noble the goal may seem.[14]

Yet, in response to a Commissioner's inquiry about whether it might still be permissible to use existing ES cell lines, Demopulos stated:

> In my opinion, yes, since the lines exist and they have some benefit. I wish they had not been derived in the way that they were but since they are there....I do not think it would be a good thing to not take advantage of [their availability].[15]

In our reflections on both the distinction and relationship between derivation and use, especially for purposes of determining ethically acceptable public policy, we were influenced by testimony that stressed how important it is for public policy to be clear and to be justified in terms that are widely understood. Individuals representing widely differing views about the moral status of the embryo and the moral justifiability of embryo destruction offered similar testimony. For example, Gilbert Meilaender called for the Commission to avoid misleading and even deceptive language in its statement and justification of public policies, whatever those policies turned out to be, on the grounds that misleading language would be a disservice to public discourse.[16] While affirming a different view regarding the moral status of embryos and embryo destruction, Dena Davis made a similar point by stressing that public policy and its rationale should pass the "straight-face test," a test failed, in her judgment, by an interpretation of federal law that permits federal funding of research using stem cells while denying federal funding of research that involves deriving the cells themselves. According to Davis, "it is disrespectful to suggest that those who believe that human embryos are persons look the other way when embryos are destroyed to obtain stem cells as long as public funding only kicks in once the stem cells are derived." Moreover, she argued that it is "more respectful, both of individuals opposed to the research and the public discourse generally, to be explicit about what is going on here and to acknowledge the ethical if not legal linkage between embryo destruction and the deriving of stem cells."[17]

The legal opinion rendered by the Department of Health and Human Services distinguishes the current legality of providing federal funds for the downstream use of ES cells from the legality of providing funds for the derivation of these cells. Indeed, as noted in Chapter 3, our own independent legal analysis reached a similar conclusion.[18] However, because our report focuses on the ethical issues involved in human ES and EG cell research, we find that there is no inconsistency between accepting this legal analysis and, at the same time, concluding that research involving both the derivation and use of these cells can under certain circumstances be justified ethically and that federal funds should be provided for both. We examine the ethical arguments for and against funding both derivation and use after we consider another possible source of stem cells–embryos that are created solely for research.

Research with ES Cells Derived from Embryos Created Solely for Research

Ever since the NIH Human Embryo Research Panel recommended that under certain conditions embryos could be created solely for research purposes (1994), there has been an ongoing discussion about the ethical and scientific merit of such a practice. Following is a discussion of this issue as it relates to two sources of ES cells derived from embryos that are created solely for research purposes.

Embryos Created Using IVF Procedures

There are two significant arguments in favor of creating human embryos using IVF technologies solely for stem cell research: The first is that there may be an inadequate supply of embryos remaining after infertility treatments. The second is that important research that could be of great medical benefit cannot be undertaken except with well-defined embryos that are created specifically for research and/or medical purposes. However, recommending federal funding for research using or deriving ES cells from embryos expressly created for research purposes presents two ethical problems. First, unlike in the case of embryos that remain following infertility treatments, there does not appear to be sufficient societal agreement on the moral acceptability of this practice at this time. Second, it is unclear whether an adequate supply of ES cells from embryos is available to meet scientific need or whether specialized cells are needed. We do not, at this time, support the federal sponsorship of research involving the creation of embryos solely for research purposes. However, we recognize that, in the future, scientific evidence and public support for this type of stem cell research may be sufficient in order to proceed. Therefore, to promote ongoing dialogue on this topic, we offer the following discussion.[19]

The "Discarded-Created" Distinction: On the Importance of Intentions

Various parties have discussed whether there is a moral difference between conducting research on embryos created with the intention of using them for reproduction and conducting research on embryos created with the intention of using them for research (Annas,

Caplan, and Elias 1996; Capron 1999; Davis 1995; Edwards 1990). Embryos created with the intention of using them for reproduction become available for research only when it is known that they are no longer intended to be used for infertility treatments; only then are they considered discarded, and only then do they become potentially available for research. The second group of embryos—research embryos—are those that are created without the intention that they will be used for procreative purposes. Rather, they are developed solely for research purposes or to generate research and medical materials such as stem cells or other cell lines, clones, DNA sequences, or proteins.

For some observers, it is difficult to defend an ethical distinction between what one can do with an embryo that has been created solely for research purposes and what one can do with an embryo remaining from infertility treatments (Davis 1995). For others, conducting research on embryos that were originally created for reproduction but which were then discarded is far easier to justify than is research conducted on embryos that were originally created for research (Harris 1998).

An ethical intuition that seems to motivate the "discarded-created" distinction is that the act of creating an embryo for reproduction is respectful in a way that is commensurate with the moral status of embryos, while the act of creating an embryo for research is not. Embryos that are discarded following the completion of IVF treatments were presumably created by individuals who had the primary intention of implanting them for reproductive purposes. These individuals did not consider the destruction of these embryos until it was determined that they were no longer needed. By contrast, research embryos are created for use in research and, in the case of stem cell research, their destruction in the process of research. Hence, one motivation that encourages serious consideration of the "discarded-created" distinction is a concern about instrumentalization—treating the embryo as a mere object—a practice that may increasingly lead us to think of embryos generally as means to our ends rather than as ends in themselves.

The Use of SCNT to Produce ES Cells

Somatic cell nuclear transfer of a diploid nucleus into an oocyte also has been suggested as a method to generate embryos from which ES cells could be derived. If successful, tissues derived from such cells could be useful in avoiding graft rejection if the donor nucleus were taken from the eventual transplant recipient. Although fertilization of an egg with sperm *in vitro* clearly results in a human zygote that will divide to become an embryo and has the potential to develop into a human if implanted, it is less clear whether the embryo created through SCNT has that potential. Nevertheless, the fact that this technique can produce living animals such as sheep and cows strongly suggests that it is likely that the cell that results from insertion of an adult nucleus into an oocyte is a zygote and can become an embryo.

Some have argued, however, that it is not clear that a zygote produced in this manner is similar to an embryo created by fertilization, because there are significant differences in the ability to generate different animals using these techniques, and we do not understand the potential of the human cell in this context. Because it is unclear whether SCNT works equally well in all species, we do not yet know whether this technique works in humans. Currently, therefore, we are uncertain whether cells created using SCNT have the full potential to become human. Because of previous work showing the potential of SCNT to create an animal in some situations, many would argue that similar concerns about the creation of embryos for research purposes apply to embryos created by SCNT. Thus, because of moral concerns outlined above regarding the creation of life only for research purposes, this category of research is disturbing to some. In the future, however, research may define conditions under which SCNT can be carried out while culturing the cells in such a manner that the resulting cell is directed to immediately differentiate into a specific tissue, precluding further development into an embryo. Perhaps in the future, then, it will be possible to use SCNT without the creation of an embryo.

One major distinction between IVF and SCNT embryos is that while creation of embryos by IVF would only generate more embryos, generation of embryos by SCNT would generate a specific kind of cell that might be useful in treating disease by allowing autologous transplant of a specific tissue type. Thus, in balancing the moral concern over the creation of an embryo and the value to society of the SCNT embryo, the potential therapeutic uses of the resulting ES cells from SCNT embryos must be evaluated carefully. At the present time, insufficient scientific evidence exists to evaluate this potential; however, within the next several years, such information should become more abundant. We recognize that if our recommendations are accepted, the most likely way that this information will accumulate is through research carried out in the private sector.

We are aware, however, that if the use of SCNT to create embryos for research purposes were deemed to be both scientifically and medically necessary, other ethical issues still would need to be addressed. For example, we would need to revisit the current prohibition on designating a recipient of fetal or embryonic tissue, in light of the likelihood that this would be an important motivator for producing such embryos.

The Arguments Relating to Federal Funding of Research Involving the Derivation and/or Use of ES and EG Cells

This chapter has described several issues that arise when considering the ethical acceptability of stem cell research, depending on the source of the ES or EG cells. These issues are not unique to the source of funding, however, as they could apply equally to stem cell research conducted in either the private or public sector. Because our main interest is in providing advice and guidance regarding the federal government's role in funding research that involves the derivation and/or use of ES and EG cells, we now turn to an examination of arguments both for and against such funding.

Arguments Against Federal Funding of Certain Types of Human Stem Cell Research

In our deliberations, we considered three major arguments against federal funding of certain types of stem cell research: its association with abortion and embryo destruction, objections by some citizens to having federal funds used for research they consider to be objectionable, and the possibility that federal funds could be used for research using AS cells rather than ES or EG cells. Each argument is briefly considered below.

Association with Abortion and Embryo Destruction

As discussed earlier, research in this area is controversial in part because of the belief, held by some, that there is a direct or indirect association with abortion. For those who hold this belief, federal funding of research that derives EG cells from cadaveric fetal tissue after elective abortion also would involve moral association with the act of abortion.[20] Similarly, federal funding for the use of embryos remaining after infertility treatments to obtain ES cells would involve the federal government in deliberately destroying biologically human entities.

Federal Funding for ES and/or EG Cell Research Violates the Deeply Held Moral Beliefs of Some Citizens

By funding research of this kind, opponents argue that the federal government is violating the beliefs of some citizens, including the belief that they should not be required to subsidize a practice they consider to be morally objectionable. If it is possible to achieve essentially the same legitimate public goals with a policy that does not offend some citizens' sincere moral sensibilities, it would be better to do so. Sometimes, the federal government decides not to support an activity because it would be offensive to many people and because the benefits lost from this support are minimal, either because the activity is of only marginal value or because other sponsors will ensure that a worthwhile activity receives the support it needs. Not infrequently, however, activities that produce valuable results and that are legitimate objects of government funding receive such support despite the objections raised by some taxpayers. Providing such support does not violate democratic principles or infringe on the rights of dissent of those in the minority. Of course, the existence of such strongly held dissenting views makes more necessary a careful assessment of the arguments in favor of government support of the activity.

Funding Alternative Sources of Stem Cell Research Is Morally Preferable

The Commission has considered the argument that a targeted and vigorous program that aims to develop alternative sources of human stem cells could discover ways to achieve the same therapeutic goals with the use of ethically less controversial means. As noted above and in Chapter 2, research on AS cells is still developing and should be encouraged, but on scientific grounds there is good reason to believe that ES cells will provide a more reliable source of cells that can differentiate into a variety of tissues. It also should be noted that the harvesting of AS cells is technically difficult and risky to human beings. For some types of adult cells, such as bone marrow cells, a certain amount of pain and discomfort is involved. For other types of stem cells, such as neuronal cells from the brain, there are significant risks to the donor from the brain biopsy procedure.

Although these objections to federal funding are important, they are not decisive. Regarding the objection based on association with wrongdoing, this report joins others in supporting various safeguards in the context of abortion in order to avoid any direct causal responsibility, to reduce the likelihood of any indirect causal responsibility, and to blunt symbolic association. Our report also proposes safeguards to prevent inappropriate and unnecessary use of embryos that remain following IVF. Regarding the second objection—avoiding offense to those who are morally opposed to using embryos for this purpose—we believe that public policy should avoid such offense in cases where the costs are not great, and we propose ways in which to reduce such offense. However, in this area of moral controversy, we believe that the arguments in favor of federal funding outweigh the offense that federal funding would create for some. Finally, we agree that alternative sources of stem cells should be sought when possible and that federal funds should be allocated to finding those sources. However, at the same time, we believe that on balance the ethical and scientific arguments support pursuing important research with EG cells obtained from cadaveric fetal tissue, with ES cells from embryos remaining after infertility treatments, and with other promising alternative sources. We now turn to additional arguments that lead us to support federal funding for certain types of ES and EG cell research.

Arguments in Favor of Federal Funding for Certain Types of Stem Cell Research

One of the principal ethical justifications for public sponsorship of research with human ES and EG cells is the same as for all biomedical and behavioral research in this country: Such research has the potential to produce health benefits for individuals suffering from disease. Many of the potential benefits of research using human ES or EG cells are discussed in Chapter 2.

The appeal to the potential benefits of stem cell research provides strong moral grounds for federal support of such research, but these potential benefits are not necessarily sufficient to justify this support. The pursuit of social benefit is always subject to moral constraint. Concerns for justice and respect for the rights of individuals can trump the morally laudable pursuit of potential benefits. Such concerns also may justify additional constraints on public funding of research.

The Enhancement of Scientific Progress Through Federal Support of the Derivation of ES Cells

Although ES cell lines already exist from studies conducted in the past year, relying upon these lines or upon the few other cell lines that might be derived by private companies for basic research on human stem cells could severely limit progress in this area of science. As discussed in Chapter 2, the potential to realize the possible medical benefits of ES cells depends on additional research into the nature and properties of ES cells. There are three main scientific reasons why it is beneficial for a broader segment of the scientific community to conduct research that involves both the derivation and use of ES cells.

First, there is great scientific value in understanding the process of ES cell derivation. Basic scientists who are interested in fundamental cellular processes are likely to make important discoveries about the nature of ES cells as they derive them in the laboratory. Moreover, by funding both derivation and use, under appropriate circumstances, federally funded researchers will be able to take advantage of the knowledge that arises from a detailed understanding of the source of the materials and the methods of derivation. Experience with animal studies indicates that research that involves both the derivation and use of particular cell lines has the greatest probability of generating promising new results.

Second, the properties of ES cells differ depending upon the conditions that were used to derive them. Moreover, the conditions for derivation of human ES cells that will differentiate into all tissue types are not yet fully understood by researchers. It is clear that the conditions used for mouse ES cells do not translate directly when using cells from other mammals. There is a significant amount of basic research that needs to be done regarding the process of ES cell derivation before the benefits from cell-based therapies can be realized.

Third, ES cells in culture are not stable indefinitely. As the cells are grown in culture, irreversible changes occur in their genetic makeup. Thus, especially in the first few years of human ES cell research, it is important to repeatedly derive ES cells to be sure that the properties of the cells that are being studied have not changed.

The Benefits of Encouraging Both Public and Private Support for ES and EG Cell Research

We anticipate that in order for stem cell research to proceed most effectively, it will require an environment in which both public and private funding will be available. Indeed, in his testimony before the Commission, David Blumenthal suggested that "since prohibition of federal funding of stem cell research will result in reliance on private companies to support almost all the investigation utilizing stem cells, the differences between industrially funded and publicly funded university investigation are pertinent to your [deliberations]."[21] Increasingly, research is being supported and conducted by industry. Support for biomedical research and development from private sector pharmaceutical and biotechnology companies now outstrips the funding from all federal sources for this research, and it is likely that the field will continue to develop even if no federal funding is forthcoming. The drug industry recently estimated that $24 billion will be spent on drug research and development in 1999, up from $2 billion in 1980 (PhRMA 1998). In light of this, some might question whether federal funding for the derivation and use of ES cells from embryos remaining from infertility treatments is necessary for future progress in this field.

We believe that a combination of federal and private sector funding is more likely to produce rapid progress in this field than would private sector funding alone. An entire cadre of researchers is likely to be drawn into this field of research through the establishment of a federal funding program. Perhaps an analogy with the field of higher education is useful. It would be possible for all college and university education in the United States to be offered solely by privately funded colleges or universities. However, the combination of publicly and privately funded schools allows the higher education system as a whole to capitalize on the unique strengths of each type of institution. Competing, yet often working together, the two types of institutions may be able to achieve levels of excellence that neither type could achieve by itself.

Synergy from a Combined Federal Effort for Research Involving Use and Derivation

Federal funding provides the opportunity for collaboration and coordination among a much larger group of researchers. Moreover, the availability of federal funding would likely increase greatly the number of scientists carrying out ES and EG cell research and thus increase the chance of important findings. Federal support for research will encourage basic research on the biology of stem cells, in addition to the product-oriented research typically supported by biotechnology firms that are focused on developing marketable products. However, in the long run, advances in the basic biology of stem cells—for example, increased understanding of the conditions and signals that lead stem cells to differentiate or of the detailed mechanisms of differentiation—are essential for therapeutic advances. Such basic research will require long-term efforts, which traditionally have been supported by NIH.*

---

*Commissioner Capron makes the following observations: "As described in Chapter 3 and mentioned earlier in this chapter, NIH, relying on the opinion of the General Counsel of DHHS, has concluded that the present rider to the Department's appropriation allows the funding of research using but not deriving ES cells from embryos because the latter would involve destroying embryos for research purposes. The alternative policy urged in this report would, in addition to its scientific benefits, also enable the federal government to play a stronger role in ensuring that ethically acceptable processes are used in deriving the ES cells that federally supported scientists use in their research. Specifically, adopting a limited exception to the funding ban solely to allow support of ES cell line derivation from embryos donated from fertility programs provides a stronger platform for the federal government to enforce the distinction between research using this group of embryos and that which would use embryos created solely for research purposes.

Of course, even if NIH funds only 'use' research, it could still try to require that the ES cells used not be derived from embryos created for research purposes. But its moral leverage is undermined by its own rationale: By insisting that federal funding of research using human ES cells does not implicate federal sponsors in the process by which the ES cells have been derived, it limits its ability to mandate that one process rather than another be used. Plainly, federal law could restrict federal support to activities that do not, for example, cause unlawful pollution; by extension, the limitation could extend to activities that do not purchase materials that were produced in processes that pollute. In the present case, however, the appropriations rider bans federal support for research that creates or destroys human embryos, which means that a federal agency cannot claim to be implementing federal policy were it to limit funding to research that uses only those ES cells that were derived from discarded embryos but not from embryos created for the purpose of deriving ES cells. Thus, NIH may be hard pressed to justify differentiation based on the type of embryos from which ES cells are derived, thereby losing an opportunity to oversee the derivation process directly and to enforce an important ethical distinction.

Adopting a limited exception to the embryo research ban solely for research to derive ES cells from embryos remaining from fertility programs would also avoid relying on the theoretical line between derivation and use research that underlies the NIH policy. Such a line is difficult to defend in practical terms when the question is not whether an activity is inherently licit or illicit but whether it ought to be paid for with federal research dollars. Any such line is merely theoretical because the funding provided for research using ES cells would of course flow directly to researchers deriving those cells, perhaps even in an adjacent laboratory. The only difference would be that the federal funds would not go directly as salary and laboratory expenses for the derivation process but indirectly in the form of funds to purchase the ES cells (which funds would then pay salaries, laboratory expenses, and so forth)."

Chapter 4: Ethical Issues in Human Stem Cell Research

Requiring That Recipients Conduct Their Research in Accordance with the Federal Regulations

As with all federally sponsored research, conditions attached to funding provide the federal government with the authority to require compliance with relevant regulations, policies, and guidelines. Among these regulations are those pertaining to human subjects research, tissue donation and transplantation, oversight, and review. In addition, federal funding agencies can stipulate that recipients of federal funding for human stem cell research must share both research results and research materials (including cell lines) with other recipients of federal funds or with all other researchers. Thus, federal funding may lead to more widespread dissemination of findings and sharing of materials, which ultimately may enhance scientific discoveries.

In contrast, many privately funded studies require that the scientists not distribute their findings until after a review by the company and that materials can be shared only after the institution receiving the materials has signed a material transfer agreement. Some of these agreements make it difficult for scientists to share or secure the reagents necessary for their research, even if they wish to do so. As the Institute of Medicine noted in its report, *Resource Sharing in Biomedical Research*, "The perception that scientific data and research materials (e.g., animals, reagents, etc.) have potential commercial value frequently causes universities to be even more reluctant than individual scientists with respect to sharing" (1996, 81).

Sustaining U.S. Leadership in Science and Technology

In supporting federal funding for certain types of stem cell research, we are not opposing research in the private sector. On the contrary, we recognize the value for the nation's investment in science and technology for research sponsored and conducted by both the public and the private sectors and the quality of private sector research. Indeed, stem cell research is receiving, and probably will continue to receive, increasing support from industry. There are, however, certain specific advantages that arise from the federal investment in science that should be acknowledged. An observation made by the Office of Technology Assessment, in its 1986 report, *Research Funding as an Investment*, is relevant in this context.

> The goal of federally funded research is not profitability, but a means of achieving social objectives, whether they are health, national security, or the enhancement of knowledge and education. The Federal research infrastructure is designed to provide a stable environment for these goals, despite a changing political environment....In addition, Federal research programs must be responsive to many more groups than industrial research efforts, and this affects the manner in which the research agenda is shaped. (1986, 61)

Federal funding is probably required in order for the United States to sustain a leadership position in this increasingly important area of research. By funding research, the federal government conveys the clear message that, under particular conditions and constraints, certain types of human stem cell research can be morally legitimate research that is worthy of public support.

Just Distribution of Potential Benefits from Stem Cell Research

Much of the testimony we heard indicated that the just distribution of the benefits of stem cell research, including both the knowledge gained and any potential therapeutic benefits, should be taken into account in any recommendation that would permit the federal government to support ES and EG cell research. For example, there was widespread agreement among the religious scholars who testified before us that in order for this research to be morally acceptable, several "background factors" must be in place, including equitable access to the benefits of the research and appropriate prioritization of this research relative to other social needs, both of which involve procedural and substantive justice. (See Appendix E.)

Issues of procedural and substantive justice are not unique to stem cell research but rather arise in various societal decisions about the use of funds for research, medical care, and other goods. Although we can note these issues here, we cannot resolve them. In addition, federal funding of stem cell research does not guarantee that greater numbers of the American public will have access to the fruits of basic or applied research or that this will occur more quickly than it would if federal funding were not available. However, by recommending federal funding for certain types of human stem cell research, we acknowledge that there is a basis for an argument for broader access to any therapies developed from that research.

Ethical Issues in Adopting Federal Oversight and Review Policies for ES and EG Cell Research

Concerns have been expressed regarding the likelihood of accountability depending on whether ES and EG cell research is sponsored and/or conducted by the public or private sector. Arthur Caplan, a bioethicist at the University of Pennsylvania, in testimony before the Senate Subcommittee on Labor, Health and Human Services, Education and Related Agencies, said that

> ...it is better to do things in this area that are accountable and public, than it is to ask them to become private and commercial. And if we continue the policies we have, we're not going to be able to bring the nuanced supervision and oversight that this area of stem cell research requires from us....That's why we need public funding, public accountability, to make the right tradeoffs.[22]

One of the principal benefits of federal funding of biomedical and behavioral research is that it is relatively easy to put in place an effective system of public oversight and review. By oversight, we are referring to the mechanism of monitoring categories of research or other activities to determine compliance with policies, procedures, rules, guidelines, and regulations and to prevent abuses. It is a policy strategy designed to provide the appropriate checks and balances and ensure ethically acceptable research protocols. The existing federal system of oversight has its origins both in the legislative and executive branches of the federal government: Congress, through its appropriations authority, may (and often does) direct that certain research be undertaken or avoided. Seen in this way, federal oversight can provide the public with two assurances: first, that stem cell research will receive national attention and scrutiny through the appropriations process undertaken by Congress; and second, that stem cell research would be conducted in accordance with relevant federal regulations. These oversight components are necessary but not sufficient for providing the public with confidence that research, especially research involving human subjects, is being undertaken appropriately. There also are mechanisms maintained by individual agencies such as the Food and Drug Administration.

In contrast, review usually refers to the evaluation of individual research protocols involving human subjects to assess their scientific merit and ethical acceptability—the activity usually carried out by Institutional Review Boards. As noted above, however, some research involving human stem cells may not be considered research involving human subjects, as defined by the Common Rule. In our view, the considerable sensitivity and public concern regarding stem cell research merits both national and local approaches to oversight and review, the details of which are described in the following chapter. We are persuaded that federal oversight and review of some types of stem cell research is required in order to make federal funding available to support such research. The types of questions about ES and EG cell research that we consider important for such an oversight and review body to ask are enumerated in Appendix F.

Summary

We were asked by the President to thoroughly review the issues associated with stem cell research, "balancing all ethical and medical considerations." In this chapter, we have endeavored to do just that. Specifically, we recognized that there are many different views on the ethical appropriateness of this type of research and also on the appropriateness of providing federal funding for such research. We believe that the ethical arguments that support the use of federal funds for stem cell research using cadaveric fetal tissue and for both deriving and using ES cells from embryos remaining after infertility treatments have considerable merit. However, such research should be conducted only within the context of a framework of national oversight and review. At the same time, we were not persuaded that we should recommend that federal funds be available at this time for the creation of embryos solely for research purposes. We arrived at these conclusions with full awareness of the strongly held views (from both religious and secular ethical perspectives) on all sides of the main issues regarding the morality of stem cell research.

Notes

1 The arguments presented here were helpfully informed by two papers prepared for the National Bioethics Advisory Commission (NBAC) by Fletcher, J.C., 1999, "Deliberating Incrementally on Human Pluripotential Stem Cell Research," and Siegel, A.W., 1999, "Locating Convergence: Ethics, Public Policy and Human Stem Cell Research." Both of these papers are available in Volume II of this report.

2 Eiseman, E., 1999, "Quick Response: Use of Human Fetal Tissue in Federally Funded Research." This paper was prepared for NBAC and is available in Volume II of this report.

3 Several terms have been used to refer to inappropriate connections between one agent's actions and another agent's wrongdoing. We have mainly used the term *association*, but other terms include *cooperation, collaboration,* and *complicity*. See, for example, Childress (1990). For a discussion of cooperation and complicity with evil in Roman Catholic moral theology, see Maguire (1986).

4 Pellegrino, E.D., Testimony before NBAC. May 7, 1999. Washington, DC. Meeting transcript, 10.

5 For a summary of these positions, see Appendix E.

6 Farley, M., Testimony before NBAC. May 7, 1999. Washington, DC. Meeting transcript, 18.

7 Dorff, E., M. Tendler, L. Zoloth, A. Sachedina. Testimony before NBAC. May 7, 1999. Washington, DC.

8 Dorff, E., Testimony before NBAC. May 7, 1999. Washington, DC. Meeting transcript, 48.

9 For a discussion of these issues see the paper prepared for NBAC by Knowles, L.P., 1999, "International Perspectives on Human Embryo and Fetal Tissue Research," available in Volume II of this report.

10 King, P.A., Testimony before NBAC. January 19, 1999. Washington, DC.

11 The terms *liberal* and *conservative* used here are used in the context intended by Dworkin (1994), *Life's Dominion*.

12 Doerflinger, R., Written testimony before NBAC. April 16, 1999. Charlottesville, VA. Meeting transcript, 1.

13 Pellegrino, E.D., Testimony before NBAC. May 7, 1999. Washington, DC.

14 Demopulos, D., Testimony before NBAC. May 7, 1999. Washington, DC. Meeting transcript, 89.

15 Ibid.

16 Meilaender, G., Testimony before NBAC. May 7, 1999, Washington, DC.

17 Davis, D., Testimony before NBAC. May 7, 1999. Washington, DC. Meeting transcript, 164.

18 This opinion was provided by Flannery, E., 1999, in "Analysis of Federal Laws Pertaining to Funding of Human Pluripotent Stem Cell Research," available in Volume II of this report.

19 For a discussion of these issues see the paper prepared for NBAC by Parens, E., 1999, "What Has the President Asked of NBAC? On the Ethics and Politics of Embryonic Stem Cell Research," available in Volume II of this report.

20 It is important to note, however, that the abortion exceptions, which serve as the basis for the type of shared views identified above, are exceptions to the law banning federal funding for abortions (Title V, Labor, HHS, and Education Appropriations, 112 Stat. 3681-385, Sec. 509 (a) (1)&(2)). Thus, federal funding for research use of cadaveric fetal tissue, within appropriate limits, might be viewed as consistent with current federal funding practices in the abortion context.

21 Blumenthal, D., Written testimony before NBAC. February 2, 1999. Princeton, NJ. Meeting transcript, 1.

22 Caplan, A.L., Testimony before the Senate Appropriations Subcommittee on Labor, Health, and Human Services, Education and Related Agencies. December 2, 1998.

References

Annas, G.J., A.L. Caplan, and S. Elias. 1996. "The Politics of Human-Embryo Research: Avoiding Ethical Gridlock." *New England Journal of Medicine* 334:1329–1332.

Bopp, J., 1994. "Fetal Tissue Transplantation and Moral Complicity with Induced Abortion." In *The Fetal Tissue Issue: Medical and Ethical Aspects*, eds. P.J. Cataldo and A. Moraczewski, 61–79. Braintree, MA: Pope John Center.

Burtchaell, J.T. 1988. "University Policy on Experimental Use of Aborted Fetal Tissue." *IRB* 10(4):7–11.

Canadian Royal Commission on New Reproductive Technologies. 1993. *Proceed with Care: Final Report of the Royal Commission on New Reproductive Technologies*. 2 vols. Ottawa: Minister of Government Services.

Capron, A.M. 1999. "Good Intentions." *Hastings Center Report* 29(2):26–27.

Charo, R.A. 1995. "The Hunting of the Snark: The Moral Status of Embryos, Right-to-Lifers, and Third World Women." *Stanford Law and Policy Review* 6:11–27.

Childress, J.F. 1990. "Disassociation from Evil: The Case of Human Fetal Tissue Transplantation Research." *Social Responsibility: Business, Journalism, Law, Medicine* 16:32–49.

———. 1991. "Ethics, Public Policy, and Human Fetal Tissue Transplantation Research." *Kennedy Institute of Ethics Journal* 1(2):93–121.

Committee of Inquiry into Human Fertilisation and Embryology. 1984. *Report of the Committee of Inquiry into Human Fertilisation and Embryology*. London: Her Majesty's Stationary Office.

Davis, D. 1995. "Embryos Created for Research Purposes." *Kennedy Institute of Ethics Journal* 5(4):343–354.

Department of Health, Education, and Welfare (DHEW). Ethics Advisory Board. 1979. *Report and Conclusions, HEW Support of Research Involving Human In Vitro Fertilization and Embryo Transfer*. Washington, DC: U.S. Government Printing Office.

Dworkin, R. 1994. *Life's Dominion: An Argument About Abortion, Euthanasia, and Individual Freedom*. New York: Vintage.

Edwards, R. 1990. "Ethics and Embryology: The Case for Experimentation." In *Experiments on Embryos*, eds. A. Dyson and J. Harris, 42–54. New York: Routledge.

Feinberg, J. 1986. "Abortion." In *Matters of Life and Death*, ed. T. Regan, 256–293. New York: Random House.

Freedman, B. 1992. "Moral Analysis and the Use of Nazi Experimental Results." In *When Medicine Went Mad: Bioethics and the Holocaust*, ed. A.L. Caplan, 141–154. Totowa, NJ: Humana Press.

Gutmann, A., and D. Thompson. 1996. *Democracy and Disagreement*. Cambridge, MA: Belknap Press.

Harris, J. 1998. *Clones, Genes and Immortality: Ethics and the Genetic Revolution*. New York: Oxford University Press.

Institute of Medicine (IOM). Committee on Resource Sharing in Biomedical Research. 1996. *Resource Sharing in Biomedical Research*. Washington, DC: National Academy Press.

King, P.A. 1997. "Embryo Research: The Challenge for Public Policy." *Journal of Medicine and Philosophy* 22(5):441–455.

MacIntyre, A. *After Virtue*. 1984. Notre Dame: University of Notre Dame Press.

Maguire, D.C. 1986. "Cooperation with Evil." In *Westminster Dictionary of Christian Ethics*, 2d ed., eds. J.F. Childress and J. Maquarrie. Philadelphia: Westminster Press.

Murray, T.H. 1996. *The Worth of a Child*. Berkeley: University of California Press.

Nagel, T. 1995. "Moral Epistemology." In *Society's Choices*, eds. R.E. Bulger, E.M. Bobby, and H.V. Fineberg, 201–214. Washington, DC: National Academy Press.

National Institutes of Health (NIH). 1988. *Report of the Human Fetal Tissue Transplantation Research Panel*. 2 vols. Bethesda, MD: NIH.

NIH. Human Embryo Research Panel. 1994. *Report of the Human Embryo Research Panel*. 2 vols. Bethesda, MD: NIH.

Office of Technology Assessment (OTA). 1986. *Research Funding as an Investment: Can We Measure the Returns? A Technical Memorandum*. Washington, DC: U.S. Government Printing Office.

Pharmaceutical Manufacturers and Research Association. (PhRMA). 1998. "Drug Companies Add 39 New Treatments to Nation's Medicine Chest, Increase R & D Investment to $24 Billion," www.phrma.org/charts/nda_i98.html.

Robertson, J.A. 1988. "Fetal Tissue Transplant Research Is Ethical: A Response to Burtchaell: II." *IRB* 10(6):5–8.

———. 1999. "Ethics and Policy in Embryonic Stem Cell Research." *Kennedy Institute of Ethics Journal* 9(2):109–136.

Seidelman, W.E. 1988. "Mengele Medicus: Medicine's Nazi Heritage." *The Milbank Quarterly* 66:221–239.

Tooley, M. 1983. *Abortion and Infanticide*. New York: Oxford University Press.

Vawter, D.E., W. Kearney, K.G. Gervais, A.L. Caplan, D. Garry, and C. Tauer. 1991. "The Use of Human Fetal Tissue: Scientific, Ethical and Policy Concerns." *International Journal of Bioethics* 2(3):189–196.

Warren, M.A. 1973. "On the Moral and Legal Status of Abortion." *The Monist* 57:43–61.

Chapter Five

Conclusions and Recommendations

Introduction

In November 1998, President Clinton charged the National Bioethics Advisory Commission with the task of conducting a thorough review of the issues associated with human stem cell research, balancing all ethical and medical considerations. The President's request was made in response to three separate reports that brought to the fore the exciting scientific and clinical prospects of stem cell research while also raising a series of ethical controversies regarding federal sponsorship of scientific inquiry in this area. Such research raises ethical issues because it involves the derivation of human *embryonic germ (EG) cells* from aborted fetuses or the derivation of human *embryonic stem (ES) cells* from early stage embryos remaining after infertility treatments. A number of these important ethical concerns previously have been identified in public debate, both here and abroad. The Commission reviewed these concerns in light of both the medical and scientific promise in this significant new field and the existing statutes and regulations that affect research in this area. Our task, however, was neither to engage in moral analysis for its own sake nor to address all the regulatory issues that might be raised, but rather to offer advice on how the balance of ethical, scientific, and medical considerations should shape policies on the use of federal funds to support research that involves deriving or using human ES or EG cells.

Scientific and Medical Considerations

The stem cell is a unique and essential cell type found in animals. Many kinds of stem cells are found in the human body, with some more differentiated, or committed, to a particular function than others. In other words, when stem cells divide, some of the progeny mature into cells of a specific type (heart, muscle, blood, or brain cells), while others remain stem cells, ready to repair some of the everyday wear and tear undergone by our bodies. These stem cells are capable of continually reproducing themselves and serve to renew tissue throughout an individual's life. For example, they continually regenerate the lining of the gut, revitalize skin, and produce a whole range of blood cells. Although the term *stem cell* commonly is used to refer to those cells within the adult organism that renew tissue (e.g., hematopoietic stem cells, a type of cell found in the blood), the most fundamental and extraordinary of the stem cells are found in the early stage embryo. These ES cells, unlike the more differentiated adult stem (AS) cells or other cell types, retain the special ability to develop into nearly any cell type. EG cells, which originate from the primordial reproductive cells of the developing fetus, have properties similar to ES cells.

It is the potentially unique versatility of the ES and EG cells derived, respectively, from the early stage embryo and cadaveric fetal tissue that presents such unusual scientific and therapeutic promise. Indeed, scientists have long recognized the possibility of using such cells to generate more specialized cells or tissue, which could allow the newly generated cells to be used to treat injuries or diseases such as Alzheimer's disease, Parkinson's disease, heart disease, and kidney failure. In addition, scientists regard these cells as important—perhaps essential—in understanding the earliest stages of human development and in developing life-saving drugs and cell-replacement therapies to treat disorders caused by early cell death or

Chapter 5: Conclusions and Recommendations

impairment. At the same time, the techniques for deriving these cells have not been fully developed as standardized and readily available research tools, and the development of any therapeutic application remains some years away.

Research also is under way to determine whether human stem cells could be obtained from the differentiated stem cells of fully developed organisms. Thus far, however, studies in animals indicate that this approach faces substantial scientific and technical limitations; indeed, the anatomic source of certain cells might preclude easy or safe access in human beings. In addition, important biological differences apparently exist between ES cells, EG cells, and AS cells. Furthermore, differences among species mean that for full scientific and clinical benefits to be realized, some research will need to be conducted with human ES and EG cells, even as the emphasis remains on laboratory and animal research. In summary, research using stem cells from animals or from human adults is not a substitute for human ES and EG cell research, and it is toward the latter that we direct our ethical and policy analyses.

Ethical and Policy Considerations

The longstanding controversy about the ethics of research involving human embryos and cadaveric fetal material arises from fundamental and sharply differing moral views regarding elective abortion or the use of embryos for research. Indeed, an earnest national and international debate continues over the ethical, legal, and medical issues that arise in this arena. This debate represents both a challenge and an opportunity: a challenge because it concerns important and morally contested questions regarding the beginning of life, and an opportunity because it provides another occasion for serious public discussion about important ethical issues. We are hopeful that this dialogue will foster public understanding about the relationships between the opportunities that biomedical science offers to improve human welfare and the limits set by important ethical obligations.

Although we believe most would agree that human embryos deserve respect as a form of human life, disagreements arise regarding both what form such respect should take and what level of protection is required at different stages of embryonic development. Therefore, embryo research that is not therapeutic to the embryo is bound to raise serious concerns for some about how to resolve the tensions between two important ethical commitments: to cure disease and to protect human life. For those who believe that from the moment of conception the embryo has the moral status of a person, research (or any other activity) that would destroy the embryo is considered wrong and should be prohibited. For those who believe otherwise, arriving at an ethically acceptable policy in this arena involves a complex balancing of many important ethical concerns. Although many of the issues remain contested on moral grounds, they can exist within a broad area of consensus upon which public policy can, at least in part, be constructed.

For most observers, the resolution of these ethical and scientific issues depends to some degree upon the source of the stem cells. The use of cadaveric fetal tissue to derive EG cell lines—like other uses of tissues or organs from dead bodies—is generally the most acceptable of these sources, provided that the research complies with the system of public safeguards and oversight already in place for such scientific inquiry. With respect to embryos and the ES cells from which they can be derived, some draw an ethical distinction among three potential types of embryos. One is referred to as the *research embryo*, an embryo created through *in vitro* fertilization (IVF), with gametes provided solely for research purposes. Many people, including the President, have expressed the view that the federal government should not fund research that involves creating such embryos. The second type of embryo is that which was created for treatment of infertility, but is now intended to be discarded because it is unsuitable or no longer needed for such treatment. The use of these embryos raises fewer ethical questions because it does not alter their final disposition. Finally, the recent demonstration of cloning techniques (somatic cell nuclear transfer [SCNT]) in nonhuman animals suggests that the transfer of a human somatic cell nucleus into an oocyte might create an embryo that could be used as a source of ES cells. The creation of a human organism using this technique raises questions similar to those raised by the creation of research embryos through IVF, and at this time federal funds may not be used for such research. In addition, if the enucleated oocyte that was to be combined with a human somatic cell nucleus came from a nonhuman animal, other issues would arise about the nature of the embryo produced. Thus, each source of material raises distinct ethical questions as well as scientific, medical, and legal ones.

Conscientious individuals have reached different conclusions regarding both public policy and private actions in the area of stem cell research. Their differing perspectives by their very nature cannot easily be bridged by any single public policy. But the development of such policy in a morally contested area is not a novel challenge for a pluralistic democracy such as that which exists in the United States. We are profoundly aware of the diverse and strongly held views on the subject of this report and has wrestled with the implications of these different views at each of our meetings devoted to this topic. Our aim throughout these deliberations has been to formulate a set of recommendations that fully reflects widely shared views and that, in our view, would serve the best interests of society.

Most states place no legal restrictions on any of the means of creating ES and EG cells that are described in this report. In addition, current Food and Drug Administration (FDA) regulations do not apply to this type of early stage research. (See Appendix D.) Therefore, because the public controversy surrounding such activities in the United States has revolved around whether it is appropriate for the federal government to sponsor such research, this report focuses on the question of whether the scientific merit and the substantive clinical promise of this research justify federal support, and, if so, with what restrictions and safeguards.

Views about the status of embryos and fetuses vary widely. Some believe that what matters is the potential for a new human life that arises at the moment of conception, while others identify the relevant concept as personhood, which they say begins only at some postembryonic stage. We heard from many members of the public, including those who are eager for this area of research to move forward as rapidly as possible, as well as those who oppose the research if it is built upon any activity that is connected to abortion or to the destruction of fertilized human eggs. In addition, our deliberations have been informed by testimony from scientists and physicians, lawyers and other experts in governmental regulation, philosophers, and Catholic, Protestant, Jewish, Islamic, and Eastern Orthodox theologians. As a result of these discussions, it has become clear that the question of whether federal policy should sponsor human ES or EG cell research is characterized by a tension between the desire to realize the great therapeutic benefits that may be derived from such work and the need to recognize that the materials involved must be treated with respect. We concluded that sufficient safeguards can be put in place in order to prevent abuse and to ensure that any use of embryos that remain after infertility treatments—like any use of fetal remains following elective abortion—is based upon and embodies the kind of respect for the embryos that most Americans would expect and demand of any activity that is carried out with the support of the federal government. Beyond the regulatory effects of the rules adopted to govern federal support for research in this area—with which we hope private sponsors of research involving ES and EG cells will comply voluntarily—the states also can influence research in this field through statutes and regulations on abortion, embryo research, and the donation of human body parts, embryos, and gametes.

Conclusions and Recommendations

The conclusions and recommendations presented in this chapter are grouped into several categories:

- the ethical acceptability of federal funding for research that either derives or uses ES or EG cells,
- the requirements for the donation of cadaveric fetal tissue and embryos for research,
- restrictions on the sale of these materials and designation of those who may benefit from their use,
- the need—and the means—for national oversight and institutional review,
- the need for local review of derivation protocols,
- the responsibilities of federal research agencies,
- the issues that must be considered regarding the private sector, and
- the need for ongoing review and assessment.

Chapter 5: Conclusions and Recommendations

The Ethical Acceptability of Federal Funding of ES Cell and EG Cell Research

Despite the enormous scientific and clinical potential offered by research use of ES or EG cells derived from various sources, many find that certain sources are more ethically problematic than others. Our recommendations reflect respect for these diverse views, which varied even among the Commissioners, regarding the ethical acceptability of the derivation and use of ES and EG cells from various sources.

As described in Chapter 2, human ES and EG cells can be derived from the following sources:

- human fetal tissue following elective abortion (EG cells),
- human embryos that are created by IVF and that are no longer needed by couples being treated for infertility (ES cells),
- human embryos that are created by IVF with gametes donated for the sole purpose of providing research material (ES cells), and
- potentially, human (or hybrid) embryos generated asexually by SCNT or similar cloning techniques in which the nucleus of an adult human cell is introduced into an enucleated human or animal ovum (ES cells).

A principal ethical justification for public sponsorship of research with human ES or EG cells is that this research has the potential to produce health benefits for those who are suffering from serious and often fatal diseases. We recognize that it is possible that all the various sources of human ES or EG cells eventually could be important to research and clinical application because of, for example, their differing proliferation potential, differing availability and accessibility, and differing ability to be manipulated, as well as possibly significant differences in their cell biology.

Although each source of stem cells poses its own scientific, ethical, and legal challenges and opportunities, much of the ethical analysis leading to public policy recommendations depends upon the scientific and/or clinical necessity of using a specific source of the cells. In our judgment, the immediate scientific uses of ES or EG cells can be satisfied by the derivation and use of cell lines derived from fetal tissues (i.e., EG cells) and from embryos (i.e., ES cells) remaining after infertility treatments have ended.

The potential use of matched tissue for autologous cell-replacement therapy from ES cells may in the future require the use of cell lines developed by SCNT techniques. In addition, embryos created through IVF specifically as a source of ES cells might be essential for creating banks of multiple cell lines representing a spectrum of alleles for the major histocompatibility complex. This goal might require that ova and sperm of persons with specific genotypes be selected to make embryos from which to derive particular classes of ES cells.

Finally, although much promising research currently is being conducted with stem cells obtained from adult organisms, studies in animals suggest that this approach will be scientifically and technically limited, and, in some cases, the anatomic source of the cells might preclude easy or safe access. Important research can and should go forward in this area, although important biological differences exist between ES and AS cells, and the use of AS cells should not be considered an alternative to ES and EG cell research.

Much research into the generation of specific tissue types from stem cells can be conducted using EG cells derived from fetal material and ES cells derived from embryos remaining after infertility treatments. In the future, adequate scientific evidence and increased prospect for medical benefits may be available to generate public support for using human ES cells derived from embryos produced through IVF for research purposes or by SCNT for autologous transplant. We note, however, that a responsible federal science policy does not necessarily require public funding for access to all sources of ES or EG cells at once. **At this time, therefore, the Commission believes that federal funding for the use and derivation of ES and EG cells should be limited to two sources of such material: cadaveric fetal tissue and embryos remaining after infertility treatments.** Specific recommendations and their justifications are provided below.

Recommendation 1:

Research involving the derivation and use of human EG cells from cadaveric fetal tissue should continue to be eligible for federal funding. Relevant statutes and regulations should be amended to make clear that the ethical safeguards that exist for fetal tissue transplantation also

apply to the derivation and use of human EG cells for research purposes.

Considerable agreement exists, both in the United States and throughout the world, that the use of fetal tissue in therapy for those with serious disorders, such as Parkinson's disease, is acceptable.[1] Research that uses cadaveric tissue from aborted fetuses is analogous to the use of fetal tissue in transplantation. The rationales for conducting EG research are equally strong, and the arguments against it are not persuasive. The removal of fetal germ cells does not occasion the destruction of a live fetus, nor is fetal tissue intentionally or purposefully created for human stem cell research. Although abortion itself doubtless will remain a contentious issue in our society, the procedures that have been developed to prevent fetal tissue donation for therapeutic transplantation from influencing the abortion decision offer a model for creating such separation in research to derive human EG cells. Because the existing statutes are written in terms of tissue transplantation, which is not a current feature of EG cell research, changes are needed to make explicit that the relevant safeguards will apply to research to derive EG cells from aborted fetuses.

Due to the contentious and polarizing nature of the abortion debate in the United States, restrictions were enacted over a decade ago to block the use of federal funding of fetal tissue transplantation therapy research. Until 1993, the only permissible source of tissue for such research was tissue from spontaneously aborted fetuses or ectopic pregnancies—sources that were largely unsuitable for research. In 1993, President Clinton lifted the ban on the use of fetal tissue from elective abortions for fetal tissue transplantation research.

Previous moral opposition to fetal tissue transplant research, because of its association with abortion, helped shape a system of safeguards to prevent the encouragement of the practice. These rules require that the consent process for women making abortion decisions must precede separately from the consent process for donation of fetal tissue for transplant research. Although some disagree, sufficient consensus exists that society should respect the autonomous choices of women who have chosen to have legal abortions to donate fetal tissue for research. If women have a right to choose to have an abortion, it follows that the self-determination or autonomy expressed in that right extends to the choice to donate fetal tissue for research purposes.

Research using fetal tissue obtained after legal elective abortions will greatly benefit biomedical science and also will provide enormous therapeutic benefits to those suffering from various diseases and other conditions. In our view, there is no overriding reason for society to discourage or prohibit this research and thus forgo an important opportunity to benefit science and those who are suffering from illness and disease—especially in light of the legality of elective abortions that provide access to fetal tissue and of the risks involved in losing these valuable opportunities. Indeed, the consequences of forgoing the benefits of the use of fetal tissue may well be harmful. Moreover, if not used in research, this tissue will be discarded.

The Acceptability of Federal Support for Research Using Embryos Remaining After Infertility Treatments to Derive ES Cells

The current congressional ban on embryo research prohibits federal support of any research "in which a human embryo...[is] destroyed, discarded, or knowingly subjected to risk of injury greater than that allowed for research on fetuses *in utero*."[2] The term *human embryo* in the statute is defined as "any organism, that is derived by fertilization, parthenogenesis, cloning, or any other means from one or more human gametes or human diploid cells."

The ban, which concerns only federally sponsored research, reflects a moral stance either that embryos deserve some measure of protection from society because of their moral status as persons, or that sufficient public controversy exists such that federal funds should not be used for this type of research. However, some effects of the embryo research ban raise serious moral and public policy concerns for those who hold differing views of the ethics of embryo research. In our view, the ban conflicts with several of the ethical goals of medicine, especially healing, prevention, and research—goals that are rightly characterized by the principles of beneficence and nonmaleficence, jointly encouraging the pursuit of each social benefit and avoiding or ameliorating potential harm.

In the United States, moral disputes—especially those concerning certain practices in the area of human reproduction—are sometimes resolved by denying federal funding for those practices (e.g., elective abortion), while not interfering with the practice in the private sector. In this case, investigative embryo research guided only by self-regulation is a widespread practice in the private sector, and the ban on embryo research has served to discourage the development of a coherent public policy, not only regarding embryo research but also regarding health research more generally. The ban also may have more profound effects on other areas of federally supported research that are dedicated to the relief of human suffering, raising concerns about the distribution and allocation of federal research resources. For example, by limiting the federal government's ability to fund promising areas of basic research, a complete ban on embryo research could prevent promising, collaborative studies in other areas, such as cancer and genetics. We recognize that many factors affect how federal research priorities are set in this country. However, in our view, the intentional withholding of federal funds for research that may lead to promising treatments may be considered unjust or unfair.

Although no consensus has been reached regarding the moral status of the embryo, there is agreement that if embryo research is permissible, some limitations and/or regulations are necessary and appropriate. Such regulation reflects an appreciation of the disparate views regarding the acceptability and unacceptability of this area of scientific investigation and serves as a way of providing accountability, allaying public anxiety, promoting beneficial research, and demonstrating respect for human embryos.

Recommendation 2:

Research involving the derivation and use of human ES cells from embryos remaining after infertility treatments should be eligible for federal funding. An exception should be made to the present statutory ban on federal funding of embryo research to permit federal agencies to fund research involving the derivation of human ES cells from this source under appropriate regulations that include public oversight and review. (See Recommendations 5 through 9.)

Based on advice from the Department of Health and Human Services (DHHS) General Counsel, the Director of the National Institutes of Health (NIH) announced in January 1999 that NIH will apply the ban only to research involving the derivation of ES cells from human embryos but not to research involving only the use of ES cells. NIH has indicated that research proposals that involve the use of ES cells will be considered for funding once NIH has established a set of special guidelines that are currently under development. The DHHS General Counsel concluded that ES cells are not, in themselves, organisms and hence cannot be embryos as defined by the statute. Thus, it could be surmised from this interpretation that the only activity that could amount to "research in which a human embryo or embryos are destroyed" would be an attempt to derive ES cells from living embryos. This, in fact, is the interpretation adopted by DHHS and NIH. More than 70 members of Congress have protested this interpretation, claiming that whatever the language of the statute, Congress clearly intended to prohibit not just research in which human embryos are destroyed but also research that depends on the prior destruction of a human embryo. Yet the plain meaning of the statutory wording differs from this interpretation, and nothing in its legislative history indicates that either proponents or opponents of the rider anticipated a situation in which research that destroyed the embryo would be conducted separately from research that used the cells derived from the embryo. Thus, in legal terms, the General Counsel's interpretation appears to be reasonable, even though it does not address any of the ethical concerns involved.

Although some may view the derivation and use of ES cells as ethically distinct activities, we believe that it is ethically acceptable for the federal government to finance research that both derives cell lines from embryos remaining after infertility treatments and that uses those cell lines. Although one might argue that some important research could proceed in the absence of federal funding for research that derives stem cells from embryos remaining after infertility treatments (i.e., federally funded scientists merely using cells derived with private funds), we believe that it is important that federal funding be made available for protocols that also derive such cells. Relying on cell lines that might be derived exclusively by a subset of privately funded researchers who are interested in this area could severely limit scientific and clinical progress.

An ethical problem is presented in trying to separate research in which human ES cells are used from the process of deriving those cells, because doing so diminishes the scientific value of the activities receiving federal support. This division—under which neither biomedical researchers at NIH nor scientists at universities and other research institutions who rely on federal support could participate in some aspects of this research—rests on the mistaken notion that derivation and use can be neatly separated without affecting the expansion of scientific knowledge. We believe that this misrepresentation of the new field of human stem cell research has several implications.

First, researchers using human ES cell lines will derive substantial scientific benefits from a detailed understanding of the process of ES cell derivation, because the properties of ES cells and the methods for sustaining the cell lines may differ depending upon the conditions and methods used to derive them. Thus, scientists who conduct basic research and who are interested in fundamental cellular processes are likely to make elemental discoveries about the nature of ES cells as they derive them in the laboratory. Second, significant basic research must be conducted regarding the process of ES cell derivation before cell-based therapies can be fully realized, and this work must be pursued in a wide variety of settings, including those exclusively devoted to basic academic research. Third, ES cells are not indefinitely stable in culture. As these cells are grown, irreversible changes occur in their genetic makeup. Thus, especially in the first few years of human ES cell research, it is important to be able to repeatedly derive ES cells in order to ensure that the properties of the cells that are being studied have not changed.

Thus, anyone who believes that federal support of this important new field of research should maximize its scientific and clinical value within a system of appropriate ethical oversight should be dissatisfied with a position that allows federal agencies to fund research using human ES cells but not research through which the cells are derived from embryos. Instead, recognizing the close connection in practical terms between the derivation and the use of these cells, it would be preferable to enact provisions that apply to funding by all federal agencies, provisions that would carve out a narrow exception for funding of research to use or to derive human ES cells from embryos that would otherwise be discarded by infertility treatment programs.

The Ethical Acceptability of Creating Embryos Through IVF Specifically as a Source of ES Cells

ES cells can be obtained from human research embryos created from donor gametes through IVF for the sole purpose of deriving such cells for research. The primary objection to creating embryos specifically for research is that many believe that there is a morally relevant difference between producing an embryo for the sole purpose of creating a child and producing an embryo with no such goal. Those who object to creating embryos for research often appeal to arguments that speak to respecting human dignity by avoiding the instrumental use of human embryos (i.e., using embryos merely as a means to some other goal does not treat them with appropriate respect or concern as a form of human life). Currently, we believe that cadaveric fetal tissue and embryos remaining after infertility treatments provide an adequate supply of research resources for federal research projects involving human embryos. Therefore, embryos created specifically for research purposes are not needed at the current time in order to conduct important research in this area.

Recommendation 3:

Federal agencies should not fund research involving the derivation or use of human ES cells from embryos made solely for research purposes using IVF.

In 1994, the NIH Human Embryo Research Panel argued in support of federal funding of the creation of embryos for research purposes in exceptional cases, such as the need to create banks of cell lines with different genetic make-ups that encoded various transplantation antigens—the better to respond, for example, to the transplant needs of groups with different genetic profiles. Such a project would require the recruitment of embryos from genetically diverse donors.

Chapter 5: Conclusions and Recommendations

A number of points are worth considering in determining how to deal with this issue. First, it is possible that the creation of research embryos will provide the only opportunity to conduct certain kinds of research—such as research into the process of human fertilization. Second, as IVF techniques improve, it is possible that the supply of embryos for research from this source will dwindle. Nevertheless, we have concluded that, either from a scientific or a clinical perspective, there is no compelling reason to provide federal funds for the creation of embryos for research at this time.

The Use of SCNT to Obtain ES Cells

The use of SCNT to transfer the nucleus of an adult somatic cell into an enucleated human egg likely has the potential of creating a human embryo. To date, although little is known about these embryos as potential sources of human ES cells, there is significant reason to believe that their use may have therapeutic potential. For example, the possible use of matched tissue for autologous cell-replacement therapy from ES cells may require the use of SCNT. Arguably, the use of this technique to create an embryo is different from the other cases we have considered—because of the asexual origin of the source of the ES cells—although oocyte donation is necessarily involved. We conclude that at this time, because other sources are likely to provide the cells needed for the preliminary stages of research, federal funding should not be provided to derive ES cells from SCNT. Nevertheless, the medical utility and scientific progress of this line of research should be monitored closely.

Recommendation 4:

Federal agencies should not fund research involving the derivation or use of human ES cells from embryos made using SCNT into oocytes.

Requirements for the Donation of Cadaveric Fetal Tissue and Embryos for Research

Potential donors of embryos for ES cell research must be able to make voluntary and informed choices about whether and how to dispose of their embryos. Because of concerns about coercion and exploitation of potential donors, as well as controversy regarding the moral status of embryos, it is important, whenever possible, to separate donors' decisions to dispose of their embryos from their decisions to donate them for research. Potential donors should be asked to provide embryos for research only if they have decided to have those embryos discarded instead of donating them to another couple or storing them. If the decision to discard the embryos precedes the decision to donate them for research purposes, then the research determines only how their destruction occurs, not whether it occurs.

Recommendation 5:

Prospective donors of embryos remaining after infertility treatments should receive timely, relevant, and appropriate information to make informed and voluntary choices regarding disposition of the embryos. Prior to considering the potential research use of the embryos, a prospective donor should have been presented with the option of storing the embryos, donating them to another woman, or discarding them. If a prospective donor chooses to discard embryos remaining after infertility treatment, the option of donating to research may then be presented. (At any point, the prospective donors' questions—including inquiries about possible research use of any embryos remaining after infertility treatment—should be answered truthfully, with all information that is relevant to the questions presented.)

During the presentation about potential research use of embryos that would otherwise be discarded, the person seeking the donation should

a) **disclose that the ES cell research is not intended to provide medical benefit to embryo donors,**

b) **make clear that consenting or refusing to donate embryos to research will not affect the quality of any future care provided to prospective donors,**

c) **describe the general area of the research to be carried out with the embryos and the specific research protocol, if known,**

d) **disclose the source of funding and expected commercial benefits of the research with the embryos, if known,**

e) **make clear that embryos used in research will not be transferred to any woman's uterus, and**

f) **make clear that the research will involve the destruction of the embryos.**

This proposal also stresses the separation that existing laws and policies seek between the pregnant woman's decision to abort and her decision to donate cadaveric fetal tissue for transplantation research. Recommendation 1 proposes to extend that separation to the donation of cadaveric fetal tissue for stem cell research. It may be difficult to achieve this separation in making decisions about embryos that remain after infertility treatments, in part because potential donors at the outset of treatment may have chosen to donate them to research. But, however difficult it may be to achieve, this separation will reduce the chance that potential donors could be pressured or coerced into donating their embryos for stem cell research.

The parts of this recommendation that deal with providing information to donors are designed to ensure that potential donors understand the range of available options and that their decisions are not influenced by anticipated personal medical benefits or by concerns about the quality of subsequent care; that they understand that the research will involve the destruction of the embryos; and that they understand the nature of the proposed research, its source of funding, and its anticipated commercial benefits, if known. Several additional suggested information items are proposed in a document entitled "Points to Consider in Evaluating Basic Research Involving Human Embryonic Stem Cells and Embryonic Germ Cells," presented in Appendix F.

Although the ethical considerations that support the prohibition of the designated donation of human fetal tissue are less acute for EG cell research than they are for transplantation, cause for concern remains. Potential donors of cadaveric fetal tissue for EG cell derivation would not have a direct therapeutic incentive to create or abort tissue for research purposes, as might occur in a transplant context. However, we agree that the prohibition remains a prudent and appropriate way to assure that no incentive—however remote—is introduced into a woman's decision to have an abortion. Any suggestion of personal benefit to the donor or to an individual known to the donor would be untenable and potentially coercive. Thus, the potential donor should be informed both before and after the decision to donate that there is no obligation to make such a gift, that no personal benefit will accrue as a result of the decision to donate, and that no penalty or sanction will result from a decision to refuse to donate.

Recommendation 6:

In federally funded research involving embryos remaining after infertility treatments, researchers may not promise donors that ES cells derived from their embryos will be used to treat patient-subjects specified by the donors.

Current provisions regulating fetal tissue research (42 USC § 289g-1 and g-2) have been narrowly interpreted by NIH and DHHS to apply only where fetal cellular material is intended for transfer into a living human recipient for therapeutic or clinical purposes. No comparable rules exist for human embryos. We believe that this statute should be applied more broadly to include *any* research involving human fetal or embryonic tissue, regardless of its immediate or eventual therapeutic benefit or intended method of intervention. Advances in EG cell research have demonstrated that bioethical concerns are not limited to fetal tissue transplantation.

As noted in Chapter 3, the Uniform Anatomical Gift Act (UAGA), currently enacted in some form in all 50 states and the District of Columbia, also may require clarification. Current versions of the UAGA explicitly permit donors to make an anatomical gift of the human body or body parts. Because a fetus is included within the UAGA's definition of *decedent*, either directly or by implication depending upon the statutory language enacted, the statute's anatomical gift provision undermines any federal prohibition of designated donation of human fetal tissue. What would otherwise qualify for federal statutory preemption is clouded by provisions of the NIH Revitalization Act of 1993 and the federal Common Rule, which direct that fetal tissue transplant researchers must abide by local and state laws, including (by implication) the UAGA.

Finally, if and when sufficient scientific evidence becomes available, clinical benefits are clearly anticipated, and agreement is reached among the various elements in society that the creation of embryos specifically for research or therapeutic purposes is justified (specifically through the use of SCNT), prohibitions on directed donation should be revisited. For obvious reasons, the use of SCNT to develop ES cells for autologous transplantation might require that the recipient be specified.

Chapter 5: Conclusions and Recommendations

Prohibitions Against the Sale of Embryonic and Fetal Material

Existing rules prohibit the practice of designated donation, the provision of monetary inducements to women undergoing abortion, and the purchase or sale of fetal tissue. We concur in these restrictions and in the recommendation of the 1988 Human Fetal Tissue Transplantation Research Panel that the sale of fetal tissue for research purposes should not be permitted under any circumstances. The potential for coercion is greatest when financial incentives are present, and the treatment of the developing human embryo or fetus as an entity deserving of respect may be greatly undermined by the introduction of any commercial motive into the donation or solicitation of fetal or embryonic tissue for research purposes.

Recommendation 7:

Embryos and cadaveric fetal tissue should not be bought or sold.

Policies already in place state that no for-profit trade in fetal tissue should be permitted, and some recommend that the "prohibition on commercial exchange of fetuses and fetal tissue extend to tissues imported from other countries" (Canadian Royal Commission 1993). This prohibition is intended to prevent the exploitation of poor women—especially those in developing countries—who might be persuaded to begin and end pregnancies for money. An important distinction must be made between the possible exploitation of persons that occurs when they are coerced or inappropriately induced to sell parts of their bodies and the exchanges that occur when companies, research institutions, or other groups provide reasonable compensation. This is a familiar issue in discussions about remuneration for participation in research and about which federal regulations defer to Institutional Review Boards (IRBs) for their judgment.

Current regulations (42 USC §§ 289g-2(a), 289g-2(b)(3), 274e, and 42 CFR § 46.206(b)) attempt to codify this recommendation. Further, depending upon whether a state has enacted the most recent revision of the UAGA (and not all states have enacted the UAGA restriction) and has included the fetus within its definition of *decedent*, the sale of fetal remains may or may not be prohibited by individual state statutes. Many states appear to rely on federal statutes and regulations to prohibit fetal tissue sale, and none address human embryos, except by implication.

We strongly encourage those who will draft modified legislation to frame their language in clear terms that are specifically defined. In particular, terms such as *valuable consideration, processing,* and *reasonable payments* require precise definitions.

We believe that, with respect to these regulations, different categories of research intermediaries should be treated differently. One approach would be to establish three intermediary categories: 1) entities responsible for tissue harvest or embryo collection, 2) entities responsible for stem cell derivation or other preresearch preparation *and* postderivation investigators; and 3) *de minimis* intermediaries (including courier or supply services, off-site specimen evaluation, pathological or chemical analysis for research suitability, and other insignificant non- or preresearch patient or specimen interactions). We believe that the first category warrants the greatest degree of regulation. An abortion provider, IVF clinic, or other third party responsible for obtaining consent and/or collecting biological materials should not be able to commercially solicit, pay for, or be paid for the fetal or embryonic material it obtains (permitting only a specifically defined, cost-based reimbursement exception for entities in that category). By placing such prohibitions against paying those who obtain the embryos, it is our intention to discourage the creation of excess embryos during routine infertility procedures, which would later be used for research purposes.

The National Organ Transplant Act (NOTA) prohibition on tissue sale (42 USC § 274e(a)) has been criticized for a statutory construction that focuses exclusively on the sale of human organs. Although we agree that fetal organ sale (as well as the sale of embryonic material) should be prohibited, we believe that NOTA's terms are unacceptably narrow and that pre-organ tissues characteristic of early fetal and embryonic development should be included in the tissue sale prohibition.

The Need for National Oversight and Review

The need for national oversight and review of ES and EG cell research is crucial. At present, no such system exists in the United States. A national mechanism to review protocols for deriving human ES and EG cells and to monitor research using such cells would ensure strict adherence to guidelines and standards across the country. Thus, federal oversight can provide the public with the assurance that research involving stem cells is being undertaken appropriately. Given the ethical issues involved in ES and EG cell research—an area in which heightened sensitivity about the very research itself led the President to request that the Commission study the issue—the public and the Congress must be assured that oversight can be accomplished efficiently, constructively, in a timely fashion, and with sufficient attention to the relevant ethical considerations.

Several countries, such as the United Kingdom, have recommended the establishment of regulatory boards or national commissions to license and regulate assisted reproductive treatments and embryo research. The use of a national oversight mechanism of this kind has certain advantages, particularly because the use of law to regulate (rather than to set limits) in this area would be burdensome, given the rapid development of biomedical science and technology. On the other hand, some kind of national commission or authority could provide the necessary flexibility and adaptability, and, in addition, such an entity could ensure more consistent ongoing application of safeguards as well as greater public accountability.[3]

In 1994, the NIH Human Embryo Research Panel considered and then explicitly rejected reconstituting the Ethics Advisory Board (EAB) for the purpose of reviewing proposals involving embryos or fertilized eggs:

> Although revisiting the EAB experience offers the potential for public consensus development and a consistent application of the new guidelines, it nonetheless has significant disadvantages. These include the creation of an additional standing government board, the likelihood of significant delay before embryo research could be funded in order to meet legal requirements for new rulemaking prior to the official creation of the government body, and possible further delay if all proposals for embryo research were required to be considered individually by an EAB-type board, despite appearing to be consistent with a developed consensus at NIH about acceptability for funding (1994, 72).

Instead, the NIH panel recommended that

> national review of all protocols by a diverse group of experts is warranted for a time. It is the hope of the Panel that this ad hoc group will develop additional guidance gained from experience with actual protocols that can be communicated to IRBs through existing mechanisms at NIH (1994, 73).

These recommendations envisioned a time when, following sufficient experience by the ad hoc panel, guidelines for embryo research review could be decentralized to the local IRBs. (It was recommended that the ad hoc panel function for at least three years.) We used similar reasoning in a previous report when recommending that the Secretary of Health and Human Services convene a Special Standing Panel to review certain categories of research involving persons with mental disorders (NBAC 1998). Like the NIH panel, we did not specify when such guidelines could be decentralized; but unlike the NIH panel, we did recommend that the panel be a standing rather than an ad hoc body.

The NIH panel's recommendations must be viewed in the context of its reporting relationship: the panel was charged with advising the NIH Director about research that could be sponsored or conducted by that agency. We note that NIH is not the only federal agency that might be interested in sponsoring or conducting research involving human stem cells; thus, some accommodation must be made for the review of proposals that are not funded by NIH.

Other elements of the NIH panel's recommendation also require additional assessment. For example, the panel recommended that "all such research proposals continue to be specially monitored by the councils and the NIH Office for Protection from Research Risks [OPRR]" (1994, 74). We are less sanguine than the NIH panel about the ability of OPRR to provide the needed oversight and monitoring for ES and EG cell research at this time, particularly given the recent decision to move this office from NIH to DHHS. Although OPRR's role in the oversight of human subjects research, like that of the FDA, remains central to the structure of human subjects protections in this country, we believe that at this time, an additional mechanism is needed for the review and oversight of federally sponsored research involving human ES and EG cells.

We do, however, share the concern of the 1994 NIH panel, investigators, and IRBs that the process of protocol review should not be viewed as simply a bureaucratic hurdle that researchers must successfully leap solely to satisfy a procedural or regulatory requirement. Done well, protocol review often improves the quality of studies by identifying concerns in the areas of study design, selection of subjects, recruitment, informed consent, and dissemination of results.

Recommendation 8:

DHHS should establish a National Stem Cell Oversight and Review Panel to ensure that all federally funded research involving the derivation and/or use of human ES or EG cells is conducted in conformance with the ethical principles and recommendations contained in this report. The panel should have a broad, multidisciplinary membership, including members of the general public, and should

a) **review protocols for the derivation of ES and EG cells and approve those that meet the requirements described in this report,**

b) **certify ES and EG cells lines that result from approved protocols,**

c) **maintain a public registry of approved protocols and certified ES and EG cell lines,**

d) **establish a database—linked to the public registry—consisting of information submitted by federal research sponsors (and, on a voluntary basis, by private sponsors, whose proprietary information shall be appropriately protected) that includes all protocols that derive or use ES or EG cells (including any available data on research outcomes, including published papers),**

e) **use the database and other appropriate sources to track the history and ultimate use of certified cell lines as an aid to policy assessment and formulation,**

f) **establish requirements for and provide guidance to sponsoring agencies on the social and ethical issues that should be considered in the review of research protocols that derive or use ES or EG cells, and**

g) **report at least annually to the DHHS Secretary with an assessment of the current state of the science for both the derivation and use of human ES and EG cells, a review of recent developments in the broad category of stem cell research, a summary of any emerging ethical or social concerns associated with this research, and an analysis of the adequacy and continued appropriateness of the recommendations contained in this report.**

We recommend several functions that the panel should carry out. In order to accomplish its purposes, the panel should maintain a public registry of federally funded protocols that employ or derive human ES and EG cells and, to the degree possible, a comprehensive listing of privately funded protocols. The purpose of the registry is to make it possible to track not only the protocols themselves and their adherence to the principles described above, but also their outcomes and the outcomes of all research based on their results. The panel should be able to describe the history and trajectory of research that uses these cells and to guard against the promiscuous use of the cells. As they are submitted, new federally funded protocols involving the derivation of ES cells must include a statement that only certified cell lines will be used.

Knowledge about the history and ultimate outcome and use of research using human ES and EG cells should be open to the public. Thus, the information accumulated by the panel through the registry should be used not only for ethical review, but also for public education. This is an important educational and informational function that may encourage the active participation of the private sector in the registry—even in the absence of any federal regulatory requirement to do so. In addition, within five years, the panel and the registry should be independently reviewed. This review, which should include an evaluation of the processes of the oversight and review mechanisms, will help to determine whether the level of limitations on this area of research is appropriate as well as to determine whether case-by-case review of derivation protocols is still warranted.

There are several benefits to a national review process for all federally funded research on ES and EG cells. These include preventing ethically problematic research, assuring the public that the research is scientifically meritorious and ethically acceptable; providing information by which to evaluate issues of social justice in the use of the knowledge or other products of the research; providing public oversight of controversial research practices; assuring consistency in the review of protocols; evaluating this type of research; and educating the public.

Although we are aware that NIH will likely conduct and/or fund the majority of federally sponsored stem cell research in the country and will be developing its own set of guidelines for the conduct of ES and EG cell research, we are persuaded that it is important to distance to some degree the review and oversight of stem cell research from what is the principal source of funding in this country. The proximity of NIH within DHHS (our recommended location for the panel) makes it possible for a number of beneficial arrangements to develop. These include developing requirements for data sharing as a condition for receiving grants; developing guidelines for sharing cell lines; and providing a common review mechanism for other federal agencies that are conducting/funding research involving ES and EG cells (e.g., through a Memorandum of Understanding).

The Need for Local Review of Derivation Protocols

For more than two decades, prospective review by an IRB has been the principal method for assuring that federally sponsored research involving human subjects will be conducted in compliance with guidelines, policies, and regulations designed to protect human beings from harm. This system of local review has been subject to criticism, and, indeed, in previous analyses we have identified a number of concerns regarding this system of review. In preparing this report, we considered a number of proposals that would allow for the local review of research protocols involving human ES and EG cell research, bearing in mind that a decision by the Commission to recommend a role for IRBs might be incorrectly interpreted as endorsing the view that human ES or EG cells or human embryos are human subjects.

We adopted the principle, reflected in these recommendations, that for research involving the derivation of ES and EG cells, a system of national oversight and review would be needed to ensure that important federal sponsorship of stem cell research could proceed—but only under specific conditions. We recognized that for such research proposals, many of the ethical issues could be considered at the local level—that is, at the institutions where the research would be conducted. In general, the IRB is an appropriate body for reviewing protocols that aim to derive ES or EG cells. Although few review bodies (including IRBs) have extensive experience in the review of such protocols, IRBs remain the most visible and expert entities available. It is for this reason, for example, that a number of recommendations presented in this report (8, 9, 10, 11, and 12) discuss the importance of developing additional guidance for the review of protocols that involve human stem cell research.

Recommendation 9:

Protocols involving the *derivation* of human ES and EG cells should be reviewed and approved by an IRB or by another appropriately constituted and convened institutional review body prior to consideration by the National Stem Cell Oversight and Review Panel. (See Recommendation 8.) This review should ensure compliance with any requirements established by the panel, including confirming that individuals or organizations (in the United States or abroad) that supply embryos or cadaveric fetal tissue have obtained them in accordance with the requirements established by the panel.

As noted earlier, for research proposals that involve the derivation of human ES or EG cells, particular ethical issues require attention through a national review process. However, this process should begin at the local level, because institutions that intend to conduct research involving the derivation of human ES cells or EG cells should continue to take responsibility for ensuring the ethical conduct of that research. More important, however, IRBs can play an important role—particularly by reviewing consent documents and by assuring that collaborative research undertaken by investigators at foreign institutions has satisfied any regulatory requirements for the sharing of research materials.

We noted in Chapter 3 that currently there is no definitive answer to the question of whether the

Common Rule, 45 CFR 46, and/or Subpart B apply to research involving fetal tissue transplantation, to human embryo research, and by extension to EG and ES cell research. If the regulations do apply, then IRBs would be expected to review protocols, consistent with the regulatory requirements. We have indicated, however, that even if these regulations do not apply, we believe that IRBs should be expected to review derivation protocols to assess their ethical acceptability without having to commit to a position that the activities are human subjects research as defined by the regulations. If, as a matter of public policy, ES and EG cell research were found to be human subjects research, certain clarifying changes in the regulations might be needed. For example, Subpart B would need to provide clearly that *any* living donor of human biological material constitutes a human subject for purposes of research protection, and IRB review and informed consent under all subparts of the DHHS version of the Common Rule would need to apply. Similarly, we have made clear that the authorization a woman may give to donate fetal tissue following an elective abortion may better be understood as consent to donate—analogous to donating organs—rather than as providing informed consent for research participation. Even if these models differ, the principle we adopt remains the same: opportunity for consent should rest exclusively with the individual or individuals legally empowered to assume a donative role.[4]

Responsibilities of Federal Research Agencies

We have recommended that protocols involving the *derivation* of ES or EG cells should be reviewed by both a local review group and the national panel described in Recommendation 8. For protocols that involve only the use but not the derivation of ES or EG cells, oversight and review are still necessary, but these protocols do not require reliance on such a system. In our judgment, these protocols can be appropriately reviewed using the existing system for the submission, review, and approval of research proposals that is in place at federal research agencies, which includes the use of a peer review group—sometimes called a study section or initial review group—that is established to assess the scientific merit of the proposals. In addition, in some agencies, such as NIH, staff members review protocols before they are transmitted to a national advisory council for final approval. These levels of review all provide an opportunity to consider ethical issues that arise in the proposals. When research proposals involve human subjects, in order to assure that it is ethically acceptable, federal agencies rely on local IRBs for review and approval. (See Recommendation 9.) At every point in this continuum—from the first discussions that a prospective applicant may have with program staff within a particular institute to the final decision by the relevant national advisory council—ethical and scientific issues can be addressed by the sponsoring agency. But even if—based on a particular interpretation of the federal regulation—these research proposals do not involve human subjects, we believe the system of oversight and review can adequately address the relevant ethical issues.

Recommendation 10:

All federal agencies should ensure that their review processes for protocols using human ES or EG cells comply with any requirements established by the National Stem Cell Oversight and Review Panel (see Recommendation 8), paying particular attention to the adequacy of the justification for using such cell lines.

Research involving human ES and EG cells raises critical ethical issues, particularly when the proposals involve the derivation of ES cells from embryos that remain after infertility treatments. We recognize that these research proposals may not follow the paradigm that is usually associated with human subjects research. Nevertheless, research proposals that are being considered for funding by federal agencies must, in our view, meet the highest standards of scientific merit and ethical acceptability. To that end, the recommendations made in this report, including a proposed set of points to consider in evaluating basic research involving human ES cells and EG cells (see Appendix F), constitute a set of ethical and policy considerations that should be reflected in the respective policies of federal agencies conducting or sponsoring human ES or EG cell research.

Attention to Issues for the Private Sector

Although this report primarily addresses the ethical issues associated with the use of federal funds for research involving the derivation and/or use of ES and

EG cells, we recognize that considerable work in both of these areas will be conducted under private sponsorship. Thus, our recommendations may have implications for those working in the private sector. First, for cell lines to be eligible for use in federally funded research, they must be certified by the National Stem Cell Oversight and Review Panel described in Recommendation 8. Therefore, if a private company aims to make its cell lines available to publicly funded researchers, it must submit its derivation protocol(s) to the same oversight and review process recommended for the public sector, (i.e., local review; see Recommendation 9) and for certification by the proposed national panel that the cells have been derived from embryos remaining after infertility treatments or from cadaveric fetal tissue.

Second, we hope that nonproprietary aspects of protocols developed under private sponsorship will be made available in the public registry, as described in Recommendation 8. The greater the participation of the private sector in providing information on human ES and EG cell research, the more comprehensive the development of the science and related public policies in this area.

Third, and perhaps most relevant in an ethically sensitive area of emerging biomedical research, it is important that all members of the research community, whether in the public or private sector, conduct the research in a manner that is open to appropriate public scrutiny. During the last two decades, we have witnessed an unprecedented level of cooperation between the public and private sectors in biomedical research, which has resulted in the international leadership position of the United States in this area. Public bodies and other authorities, such as the Recombinant DNA Advisory Committee, have played a crucial role in enabling important medical advances in fields such as gene therapy by providing oversight of both publicly and privately funded research efforts. We believe that voluntary participation by the private sector in the review and certification procedures of the proposed national panel, as well as in its deliberations, can contribute equally to the socially responsible development of ES and EG cell technologies and accelerate their translation into biomedically important therapies that will benefit patients.

Recommendation 11:

For privately funded research projects that involve ES or EG cells that would be eligible for federal funding, private sponsors and researchers are encouraged to adopt voluntarily the applicable recommendations of this report. This includes submitting protocols for the derivation of ES or EG cells to the National Stem Cell Oversight and Review Panel for review and cell line certification. (See Recommendations 8 and 9.)

In this report, we recommend that federally funded research that involves the derivation of ES cells should be limited to those efforts that use embryos that remain after infertility treatments. Some of the recommendations made in this context—such as the requirement for separating the decision by a woman to cease infertility treatment when embryos still remain from her decision to donate those embryos to research—simply do not apply to efforts to derive ES cells from embryos created (whether by IVF or by SCNT) solely for research purposes—activities that might be pursued in the private sector. Nevertheless, other ethical standards and safeguards embodied in the recommendations, such as provisions to prevent the coercion of women and the promotion of commerce in human reproduction, remain vitally important, even when embryos are created solely for research purposes.

Recommendation 12:

For privately funded research projects that involve deriving ES cells from embryos created solely for research purposes and that are therefore not eligible for federal funding (see Recommendations 3 and 4)

a) **professional societies and trade associations should develop and promulgate ethical safeguards and standards consistent with the principles underlying this report, and**

b) **private sponsors and researchers involved in such research should voluntarily comply with these safeguards and standards.**

Professional societies and trade associations dedicated to reproductive medicine and technology play a central role in establishing policy and standards for clinical care, research, and education. We believe that these organizations

Chapter 5: Conclusions and Recommendations

can and should play a salutary role in ensuring that all embryo research conducted in the United States, including that which is privately funded, conforms to the ethical principles underlying this report. Many of these organizations already have developed policy statements, ethics guidelines, or other directives addressing issues in this report, and we have benefited from a careful review of these materials. These organizations are encouraged to review their professional standards to ensure not only that they keep pace with the evolving science of human ES and EG cell research, but also that their members are knowledgeable about and in compliance with them. For those organizations that conduct research in this area but that lack statements or guidelines addressing the topics of this report, we recommend strongly that they develop such statements or guidelines. No single institution or organization, whether in the public or the private sector, can provide all the necessary protections and safeguards.

The Need for Ongoing Review and Assessment

No system of federal oversight and review of such a sensitive and important area of investigation should be established without at the same time providing an evaluation of its effectiveness, value, and ongoing need. The pace of scientific development in human ES and EG cell research likely will increase. Although one cannot predict the direction of the science of human stem cell research, in order for the American public to realize the promise of this research and to be assured that it is being conducted responsibly, close attention to and monitoring of all the mechanisms established for oversight and review are required.

Recommendation 13:

The National Stem Cell Oversight and Review Panel described in Recommendation 8 should be chartered for a fixed period of time, not to exceed five years. Prior to the expiration of this period, DHHS should commission an independent evaluation of the panel's activities to determine whether it has adequately fulfilled its functions and whether it should be continued.

There are several reasons for allowing the national panel to function for a fixed period of time and for evaluating its activities before it continues its work. First, some of the hoped-for results will be available from research projects that are using the two sources we consider to be ethically acceptable for federal funding. Five years is a reasonable period in which to allow some of this information to amass, offering the panel, researchers, members of Congress, and the public sufficient time to determine whether any of the knowledge or potential health benefits are being realized. The growing body of information in the public registry and database described above (particularly if privately funded researchers and sponsors voluntarily participate) will aid these considerations.

Second, within this period the panel may be able to determine whether additional sources of ES cells are necessary in order for important research to continue. Two arguments have been offered for supporting research using embryos created specifically for research purposes: One is the concern that not enough embryos remain for this purpose from infertility treatments, and the other is the recognition that some research requires embryos that are generated for specific research and/or medical purposes. The panel should assess whether additional sources of ES cells that we have judged to be ineligible for federal funding at this time (i.e., embryos created solely for research purposes) are legitimately needed.

Third, an opportunity to assess the relationship between local review of protocols using human ES and EG cells and the panel's review of protocols for the derivation of ES cells will be offered. It will, of course, take time for this national oversight and review mechanism to develop experience with the processes of review, certification, and approval described in this report.

Fourth, we hope that the panel will contribute to the broad and ongoing national dialogue on the ethical issues regarding research involving human embryos. A recurring theme of our deliberations, and in the testimony we heard, was the importance of encouraging this national conversation.

The criteria for determining whether the panel has adequately fulfilled its functions should be set forth by an independent body established by DHHS. However, it would be reasonable to expect that the evaluation would rely generally on the seven functions described above in Recommendation 8 and that this evaluation would be conducted by a group with the requisite expertise. In addition, some of the following questions might be considered when conducting this evaluation: Is there reason to believe that the private sector is voluntarily submitting descriptions of protocols involving the derivation of human ES cells to the panel for review? Is the panel reviewing projects in a timely manner? Do researchers find that the review process is substantively helpful? Is the public being provided with the assurance that social and ethical issues are being considered?

Summary

Recent developments in human ES and EG cell research have raised the prospect that new therapies will become available that will serve to relieve human suffering. These developments also have served to remind society of the deep moral concerns that are related to research involving human embryos and cadaveric fetal tissue. Serious ethical discussion will (and should) continue on these issues. However, in light of public testimony, expert advice, and published writings, we have found substantial agreement among individuals with diverse perspectives that although the human embryo and fetus deserve respect as forms of human life, the scientific and clinical benefits of stem cell research should not be foregone. We were persuaded that carrying out human ES cell research under federal sponsorship is important, but only if it is conducted in an ethically responsible manner. After extensive deliberation, the Commission believes that acceptable public policy can be forged, in part, based upon these widely shared views. Through this report, we not only offer recommendations regarding federal funding and oversight of stem cell research, but also hope to further stimulate the important public debate about the profound ethical issues regarding this potentially beneficial research.

Notes

1 Use of fetal tissue in research is also permitted in Canada, the United Kingdom, Australia, and in most countries in the European Union. Germany, for example, does not permit embryo research but does permit the use of fetal tissue for the derivation of EG cells. The German statement concerning human ES cells upholds the ban on destructive embryo research, effectively banning the derivation of ES cells, because the option of deriving EG cells exists in that country. See the German statement concerning the question of human ES cells, March 1999, 8–10 (DFG 1999).

2 Public Law No. 105-78, 513(a) (1997).

3 EGE *Opinion* (1998) at Art. 2.11. See also the Australian NHMRC *Guidelines* (1996) advocating that complementary national assisted reproductive technology standards or legislation be adopted in the Australian States.

4 See *Fed. Reg.* 27804, proposed rule 45 CFR § 46.204(d)-(e) and Table 1, "Current and Proposed 45 CFR 46, Subpart B," 27798, explanatory text ("consent of the father is not required"; rather, "consent of the mother *or* her legally authorized representative is required" [after she is]..."informed of the reasonably foreseeable impact of the research on the fetus").

References

Canadian Royal Commission on New Reproductive Technologies. 1993. *Proceed with Care: Final Report of the Royal Commission on New Reproductive Technologies.* 2 vols. Ottawa: Minister of Government Services.

Deutsche Forschungsgemeinschaf (DFG). 1999. "Statement Concerning the Question of Human Embryonic Stem Cells," www.dfg-bonn.de/english/press/eszell_e.html.

European Commission. European Group on Ethics (EGE). 1998. "Opinion of the European Group on Ethics in Science and New Technologies."

National Bioethics Advisory Commission (NBAC). 1998. *Research Involving Persons with Mental Disorders That May Affect Decisiomaking Capacity.* 2 vols. Rockville, MD: U.S. Government Printing Office.

National Health and Medical Research Council (NHMRC). 1996. *Draft Statement on Ethical Guidelines on Assisted Reproductive Technology.* Canberra, Australia: Australian Government Publishing Service.

National Institutes of Health (NIH). Human Embryo Research Panel. 1994. *Report of the Human Embryo Research Panel.* 2 vols. Bethesda, MD: NIH.

Appendix A

Acknowledgments

This report benefited from the input of many individuals and groups. Several organizations responded to a February 1999 request from the National Bioethics Advisory Commission for input on the scientific, medical, and ethical issues involved in human stem cell research, and nearly 40 scientific, medical, professional, religious, and health organizations were asked to provide their perspectives on these complex issues. The Commission gratefully acknowledges the thoughtful comments provided by the following groups:

- American Bioethics Advisory Commission (Stafford, Virginia)
- The American College of Obstetricians and Gynecologists (Washington, DC)
- The American Society for Cell Biology (Bethesda, Maryland)
- American Society for Reproductive Medicine (Birmingham, Alabama)
- Association of American Medical Colleges (Washington, DC)
- Biotechnology Industry Association (Washington, DC)
- College of American Pathologists (Northfield, Illinois)
- Pharmaceutical Research and Manufacturers of America (Washington, DC)
- RESOLVE (Somerville, Massachusetts)

In addition, the Commission asked the following individuals to review portions of the draft report for scientific, legal, and ethical accuracy. The comments provided by these individuals improved the quality and outcome of the report and are greatly appreciated:

- Brigid L.M. Hogan (Vanderbilt University School of Medicine; Nashville, Tennessee)
- Anna Mastroianni (University of Washington; Seattle, Washington)
- John A. Robertson (The University of Texas; Austin, Texas)
- Janet Rossant (Samuel Lunenfeld Research Institute, Mt. Sinai Hospital; Toronto, Ontario)
- Lee Silver (Princeton University; Princeton, New Jersey)
- Evan Y. Snyder (Harvard Medical School; Cambridge, Massachusetts)
- James A. Thomson (University of Wisconsin; Madison, Wisconsin)

We are also grateful to Michelle Myer, a graduate student at the University of Virginia, for preparing a summary of the presentations that were provided to the Commission on May 7, 1999, on religious perspectives relating to research involving human stem cells. The summary appears as Appendix E of this report.

Appendix B

Glossary

adult stem (AS) cells – stem cells found in the adult organism (e.g., in bone marrow, skin, and intestine) that replenish tissues in which cells often have limited life spans. They are more differentiated than embryonic stem (ES) cells or embryonic germ (EG) cells.

ART (assisted reproductive technology) – all treatments or procedures that involve the handling of human eggs and sperm for the purpose of helping a woman become pregnant. Types of ART include *in vitro* fertilization, gamete intrafallopian transfer, zygote intrafallopian transfer, embryo cryopreservation, egg or embryo donation, and surrogate birth.

blastocyst – a mammalian embryo in the stage of development that follows the morula. It consists of an outer layer of trophoblast to which is attached an inner cell mass.

blastomere – one of the cells into which the egg divides after its fertilization; one of the cells resulting from the division of a fertilized ovum.

chimera – an organism composed of two genetically distinct types of cells.

cloning – the production of a precise genetic copy of a molecule (including DNA), cell, tissue, plant, or animal.

differentiation – the specialization of characteristics or functions of cell types.

diploid cell – the cell containing two complete sets of genes derived from the father and the mother respectively; the normal chromosome complement of somatic cells (in humans, 46 chromosomes).

ectoderm – the outer layer of cells in the embryo; the origin of skin, the pituitary gland, mammary glands, and all parts of the nervous system.

embryo – 1) the beginning of any organism in the early stages of development, 2) a stage (between the ovum and the fetus) in the prenatal development of a mammal, 3) in humans, the stage of development between the second and eighth weeks following fertilization, inclusive.

embryonic stem (ES) cells – cells that are derived from the inner cell mass of a blastocyst embryo.

embryonic germ (EG) cells – cells that are derived from precursors of germ cells from a fetus.

endoderm – the innermost of the three primary layers of the embryo; the origin of the digestive tract, the liver, the pancreas, and the lining of the lungs.

ex utero – outside of the uterus.

fibroblast – a cell present in connective tissue, capable of forming collagen fibers.

gamete – 1) any germ cell, whether ovum or spermatozoon, 2) a mature male or female reproductive cell.

gastrulation – the process of transformation of the blastula into the gastrula, at which point the embryonic germ layers or structures begin to be laid out.

germ cells – gametes (ova and sperm) or the cells that give rise directly to gametes.

haploid cell – a cell with half the number of chromosomes as the somatic diploid cell, such as the ova or sperm. In humans, the haploid cell contains 23 chromosomes.

Appendix B: Glossary

in vivo – in the natural environment (i.e., within the body).

in vitro – in an artificial environment, such as a test tube or culture medium.

in vitro **fertilization (IVF)** – a process by which a woman's eggs are extracted and fertilized in the laboratory and then transferred after they reach the embryonic stage into the woman's uterus through the cervix. Roughly 70 percent of assisted reproduction attempts involve IVF, using fresh embryos developed from a woman's own eggs.

karyotype – the chromosome characteristics of an individual cell or of a cell line, usually presented as a systematic array of metaphase chromosomes from a photograph of a single cell nucleus arranged in pairs in descending order of size.

mesoderm – the middle of the three primary germ layers of the embryo; the origin of all connective tissues, all body musculature, blood, cardiovascular and lymphatic systems, most of the urogenital system, and the lining of the pericardial, pleural, and peritoneal cavities.

morula – 1) the mass of blastomeres resulting from the early cleavage divisions of the zygote, 2) solid mass of cells resembling a mulberry, resulting from the cleavage of an ovum.

oocyte – 1) a diploid cell that will undergo meiosis (a type of cell division of germ cells) to form an egg, 2) an immature ovum.

ovum – female reproductive or germ cell.

pluripotent cells – cells, present in the early stages of embryo development, that can generate all of the cell types in a fetus and in the adult and that are capable of self-renewal. Pluripotent cells are not capable of developing into an entire organism.

pre-implantation embryo – 1) the embryo before it has implanted in the uterus, 2) commonly used to refer to *in vitro* fertilized embryos before they are transferred to a woman's uterus.

somatic cells – [from *soma* - the body] 1) cells of the body which in mammals and flowering plants normally are made up of two sets of chromosomes, one derived from each parent, 2) all cells of an organism with the exception of germ cells.

stem cells – cells that have the ability to divide indefinitely and to give rise to specialized cells as well as to new stem cells with identical potential.

totipotent – having unlimited capacity. Totipotent cells have the capacity to differentiate into the embryo and into extra-embryonic membranes and tissues. Totipotent cells contribute to every cell type of the adult organism.

trophoblast – the outermost layer of the developing blastocyst of a mammal. It differentiates into two layers, the cytotrophoblast and syntrophoblast, the latter coming into intimate relationship with the uterine endometrium with which it establishes nutrient relationships.

zygote – 1) the cell resulting from the fusion of two gametes in sexual reproduction, 2) a fertilized egg (ovum), 3) the diploid cell resulting from the union of a sperm and an ovum, 4) the developing organism during the first week after fertilization.

Appendix C

Letters of Request and Response

THE WHITE HOUSE

WASHINGTON

November 14, 1998

Dr. Harold Shapiro
Chair
National Bioethics Advisory Commission
Suite 3C01
6100 Executive Boulevard
Bethesda, Maryland 20892-7508

Dear Dr. Shapiro:

This week's report of the creation of an embryonic stem cell that is part human and part cow raises the most serious of ethical, medical, and legal concerns. I am deeply troubled by this news of experiments involving the mingling of human and non-human species. I am therefore requesting that the National Bioethics Advisory Commission consider the implications of such research at your meeting next week, and to report back to me as soon as possible.

I recognize, however, that other kinds of stem cell research raise different ethical issues, while promising significant medical benefits. Four years ago, I issued a ban on the use of federal funds to create human embryos solely for research purposes; the ban was later broadened by Congress to prohibit any embryo research in the public sector. At that time, the benefits of human stem cell research were hypothetical, while the ethical concerns were immediate. Although the ethical issues have not diminished, it now appears that this research may have real potential for treating such devastating illnesses as cancer, heart disease, diabetes, and Parkinson's disease. With this in mind, I am also requesting that the Commission undertake a thorough review of the issues associated with such human stem cell research, balancing all ethical and medical considerations.

I look forward to receiving your reports on these important issues.

Sincerely,

Bill Clinton

NATIONAL BIOETHICS ADVISORY COMMISSION

6100 Executive Blvd
Suite 5B01
Rockville, MD
20892-7508

Telephone
301-402-4242
Facsimile
301-480-6900
Website
www.bioethics.gov

Harold T. Shapiro, Ph.D.
Chair

Patricia Backlar

Arturo Brito, M.D.

Alexander M. Capron, LL.B.

Eric J. Cassell, M.D.

R. Alta Charo, J.D.

James F. Childress, Ph.D.

David R. Cox, M.D., Ph.D.

Rhetaugh G. Dumas, Ph.D., R.N.

Laurie M. Flynn

Carol W. Greider, Ph.D.

Steven H. Holtzman

Bette O. Kramer

Bernard Lo, M.D.

Lawrence H. Miike, M.D., J.D.

Thomas H. Murray, Ph.D.

Diane Scott-Jones, Ph.D.

Eric M. Meslin, Ph.D.
Executive Director

Henrietta Hyatt-Knorr, M.A.
Deputy Executive Director

November 20, 1998

The President
The White House
Washington, DC 20500

Dear Mr. President:

I am responding to your letter of November 14, 1998 requesting that the National Bioethics Advisory Commission discuss at its meeting in Miami this week the ethical, medical, and legal concerns arising from the fusion of a human cell with a cow egg.

The Commission shares your view that this development raises important ethical and potentially controversial issues that need to be considered, including concerns about crossing species boundaries and exercising excessive control over nature, which need further careful discussion. This is especially the case if the product resulting from the fusion of a human cell and the egg from a non-human animal is transferred into a woman's uterus and, in a different manner, if the fusion products are embryos even if no attempt is made to bring them to term. In particular, we believe that any attempt to create a child through the fusion of a human cell and a non-human egg would raise profound ethical concerns and should not be permitted.

We devoted time at our meeting to discussing various aspects of this issue, benefiting not only from the expertise of the Commissioners, but from our consultation (via telephone) with Dr. Ralph Brinster, a recognized expert in the field of embryology, from the University of Pennsylvania. Also in attendance at our meeting was Dr. Michael West, of Advanced Cell Technology, who was given an opportunity to answer questions from Commission members. As you know, however, the design and results of this experiment are not yet publicly available, and as a consequence the Commission was unable to evaluate fully its implications.

As a framework for our initial discussion, we found it helpful to consider three questions:

1. ***Can the product of fusing a human cell with the egg of a non-human animal, if transferred into a woman's uterus, develop into a child?***

At this time, there is insufficient scientific evidence to answer this question. What little evidence exists, based on other fusions of non-human eggs with non-human cells from a different species, suggests that a pregnancy cannot be maintained. If it were possible, however, for a child to develop from these fused cells, then profound ethical issues would be raised. An attempt to develop a child from these fused cells should not be permitted.

This objection is consistent with our views expressed in *Cloning Human Beings*, in which we concluded that:

> "...at this time it is morally unacceptable for anyone in the public or private sector, whether in a research or clinical setting, to attempt to create a child using somatic cell nuclear transfer cloning."

2. ***Does the fusion of a human cell and an egg from a non-human animal result in a human embryo?***

The common understanding of a human embryo includes, at least, the concept of an organism at its earliest stage of development, which has the potential, if transferred to a uterus, to develop in the normal course of events into a living human being. At this time, however, there is insufficient scientific evidence to be able to say whether the combining of a human cell and the egg of a non-human animal results in an embryo in this sense. In our opinion, if this combination does result in an embryo, important ethical concerns arise, as is the case with all research involving human embryos. These concerns will be made more complex and controversial by the fact that these hybrid cells will contain both human and non-human biological material.

It is worth noting that these hybrid cells should not be confused with human embryonic stem cells. Human embryonic stem cells, while derived from embryos, are not themselves capable of developing into children. The use of human embryonic stem cells, for example to generate cells for transplantation, does not directly raise the same type of moral concerns.

3. ***If the fusion of a human cell and the egg of a non-human animal does not result in an embryo with the potential to develop into a child, what ethical issues remain?***

If this line of research does not give rise to human embryos, we do not believe that totally new ethical issues arise. We note that scientists routinely conduct non-controversial and highly beneficial research that involves combining material from human and other species. This research has led to such useful therapies as: blood clotting factor for hemophilia, insulin for diabetes, erythropoietin for anemia, and heart valves for transplants. Combining human cells with non-human eggs might possibly lead some day to methods to overcome transplant rejections without the need to create human embryos, or to subject women to invasive, risky medical procedures to obtain human eggs.

We recognize that some of the issues raised by this type of research may also be pertinent to stem cell research in general. We intend to address these and other issues in the report that you requested regarding human stem cell research.

Sincerely,

Harold T. Shapiro
Chair

Appendix D

The Food and Drug Administration's Statutory and Regulatory Authority to Regulate Human Stem Cells[1]

An Overview of Food and Drug Administration Regulations Pertinent to Human Cellular Materials and Tissues

The Food and Drug Administration (FDA) has had in place a regulatory framework for cellular and tissue materials that has evolved over time as the development and use of such biological materials for therapeutic purposes has increased. The Public Health Service Act (PHS Act), 42 USC 262 and 264, the Federal Food, Drug, and Cosmetic Act (FD&C Act), 21 USC 201 et seq., and implementing regulations of the FDA provide the agency with broad authority to regulate both the research into and the use of human stem cells that are *intended to be used as* biological products, drugs, or medical devices in order to prevent, treat, cure, or diagnose a disease or condition.[2] Scientific research not intended for use in the development of any FDA-regulated product is not under the oversight and control of the FDA.

In order for the FDA to assert its regulatory authority over stem cell-related research and products, such research *and* products must fall within one of the product categories over which the FDA exercises jurisdiction and must move in interstate commerce. To the extent that the FDA determines that a particular product falls within the definition of a biological product, a drug, or a medical device, it will assert its jurisdiction. Whether a particular product falls within the definition of any of the FDA-regulated product categories will depend, in part, upon the intended use of the product.

The manufacturer's objective intent—as evidenced by labeling, promotional, and other relevant materials for the product—has long been regarded as the primary source for establishing a product's intended use and thus its status for purposes of FDA regulation.[3] Although this approach would seem to grant manufacturers unlimited control over the regulatory status of their products, courts in fact have recognized the FDA's right to look beyond the express claims of manufacturers in order to consider more subjective indicia of intent—such as the foreseeable and actual use of a product—to prove that its intended use subjects it to agency jurisdiction.[4]

Regardless of whether the FDA or the manufacturer is characterizing the intended use of a product for purposes of evaluating FDA jurisdiction, it is clear that FDA regulatory authority will not extend automatically to all scientific research on stem cells. Indeed, to the extent that such nonhuman research is preliminary in nature and/or is undertaken without intent to develop a therapeutic product, stem cell research is not subject to FDA jurisdiction. Thus, for example, basic research to develop stem cell models to evaluate the safety and efficacy of therapeutic products would not be regulated directly. Instead, the FDA would review any scientific data generated from such a model and submitted as part of a marketing application. It is only at the juncture when the science of stem cell research has progressed to the point that development of a particular therapeutic product and its use in humans is envisioned that FDA regulatory authority will apply, and further research then must be conducted in compliance with FDA requirements.

Even if a product falls within one of the defined categories over which the FDA asserts its jurisdiction, no statutory authority over the product exists unless it moves in interstate commerce. The FDA takes an expansive view of what constitutes interstate commerce; in regard to biological products, the FDA has been particularly aggressive. For example, in its 1993 policy statement regarding somatic cell therapy products, the FDA concluded that

> [t]he interstate commerce nexus needed to require premarket approval under the statutory provisions governing biological products and drugs may be created in various ways in addition to shipment of the finished product by the manufacturer. For example, even if a biological drug product is manufactured entirely with materials that have not crossed State lines, transport of the product into another State by an individual patient creates the interstate commerce nexus. If a component used in the manufacture of the product moves interstate, the interstate commerce prerequisite for the prohibition against drug misbranding is also satisfied even when the finished product stays within the State. Products that do not carry labeling approved in a PLA (or NDA) are misbranded under section 502(f)(1) of the [FD&C] Act....Moreover, falsely labeling a biological product is prohibited under section 351(b) of the PHS Act without regard to any interstate commerce nexus (42 U.S.C. 262(b)) (58 *Fed. Reg.* at 53250).

It can be expected that the FDA would apply the same logic to all cellular and tissue materials that are used in the prevention, treatment, cure, or diagnosis of a disease or condition.

Application to Stem Cells

In recent congressional testimony, National Institutes of Health Director Harold Varmus described three potential applications of research using human "pluripotent stem cells" that illustrate the inconsistencies of FDA regulation. He noted that the FDA does not regulate two of the examples, but will regulate one. First, stem cell research could include basic research such as "the identification of the factors involved in the cellular decision-making process that determines cell specialization."[5] Second, "[h]uman pluripotent stem cell research could also dramatically change the way we develop drugs and test them for safety and efficacy. Rather than evaluating safety and efficacy of a candidate drug in an animal model of a human disease, these drugs could be tested against a human cell line that had been developed to mimic the disease process."[6] It is unlikely that the FDA would regulate either of these potential applications directly. Varmus also made the following comments:

> Perhaps the most far-reaching potential application of human pluripotent stem cells is the generation of cells and tissue that could be used for transplantation, so-called cell therapies. Pluripotent stem cells stimulated to develop into specialized cells offer the possibility of a renewable source of replacement cells and tissue to treat a myriad of diseases, conditions and disabilities including Parkinson's and Alzheimer's disease, spinal cord injury, stroke, burn, heart disease, diabetes, osteoarthritis and rheumatoid arthritis.[7]

These stem cell products, based on their *intended use*, would be subject to FDA regulation.

Case-by-Case Regulation

The FDA has been cautious in exercising its regulatory discretion regarding cellular and tissue materials and in fact never has overseen a single regulatory program for human cellular and tissue-based products. Instead, the FDA has regulated these products on a case-by-case basis, responding as it deemed appropriate to the particular characteristics of and concerns raised by each type of product.[8]

One example has been the FDA's approach to regulating bone marrow. Although for years the FDA has licensed blood and blood components pursuant to section 351 of the PHS Act (42 USC 262), it voluntarily has refrained from regulating minimally manipulated bone marrow, the earliest source of stem cells used for transplantation, despite its status as a blood component. Indeed, not until the early 1990s did the FDA announce that to the extent that bone marrow was subject to extensive manipulation prior to transplantation, it would be treated the same as somatic cell therapy and gene therapy products subject to the investigational new drug (IND) regulations and would require PHS Act licensure (58 *Fed. Reg.* 53248, 53249 (Oct. 14, 1993)).

Also in 1993, in response to concerns regarding the transmission of the human immunodeficiency virus (HIV) and other infectious diseases, the FDA published an emergency final rule that mandated certain processing, testing, and recordkeeping procedures for specific types of tissue products.[9] This rule, however, did *not* mandate premarket approval or notification for all tissues, but rather provided, among other things, for donor screening, documentation of testing, and FDA inspection of tissue facilities.[10]

Another example of the FDA's case-by-case approach is the publication in 1996 of a guidance that stated that manipulated autologous structural cells (autologous cells manipulated and then returned to the body for structural repair or reconstruction) would be subject to PHS licensure.[11] In addition, until recently, the FDA carefully chose not to regulate reproductive tissues. Then, in 1997, it proposed that, in the future, certain reproductive tissues (i.e., semen, ova, and embryos) should be regulated in some form.

Traditional tissue products (including but not limited to bone, skin, corneas, and tendons) also have been subject to the FDA's piecemeal regulatory approach. Historically, the FDA regulated these products on an ad hoc basis as medical devices under section 201 of the FD&C Act. However, with the advent of HIV and the potential for its transmission, the FDA concluded in the early 1990s that a more comprehensive program for regulating the use of traditional tissues was necessary. In 1991, the FDA concluded that human heart valves were medical devices subject to premarket approval requirements.[12] Following litigation, the FDA decided that while these products were indeed medical devices, they would not be subject to premarket approval requirements.[13] In defining tissue subject to this rule, the FDA exempted a number of products, including vascularized organs, *dura mater*, allografts, and umbilical cord vein grafts.

A New Approach to Regulating Human Cellular and Tissue-Based Products

In February 1997 the FDA proposed a new approach to the regulation of human cellular and tissue-based products. This framework is intended to "protect the public health without imposing unnecessary government oversight" ("Reinventing the Regulation of Human Tissue," *National Performance Review*, February 1997). Although it is still considered a proposed approach, the 1997 document utilizes FDA's existing statutory authority under the PHS and FD&C Acts to regulate a broad array of cellular and tissue materials. The framework proposed is a tiered approach to regulation (FDA, "A Proposed Approach to the Regulation of Cellular and Tissue-Based Products," February 28, 1997). Products that pose increased risks to health or safety would be subject to increased levels of regulation (i.e., either licensure under the PHS Act or premarket approval under the FD&C Act), while products that pose little or no risk of transmitting infectious disease would be subject to minimal regulation (e.g., facility registration and product listing). However, products that are 1) highly processed (more-than-minimally manipulated); 2) are used for other than their usual purpose; 3) are combined with nontissue components (e.g., devices or other therapeutic products); or 4) are used for metabolic purposes (e.g., systemic, therapeutic purposes) will be subject to clinical investigation as INDs, must be documented with investigational device exemption applications (IDEs), and will be subject to premarket approval as biological products, medical devices, or new drugs.

This proposed approach addresses the FDA's regulation of stem cell products. In the case of a minimally manipulated product for autologous use and allogeneic use of cord blood stem cells by a close blood relative, the FDA has proposed requiring compliance with standards consistent with section 361 of the PHS Act, rather than an IND and licensure pursuant to section 351 of the act. However, minimally manipulated products that will be used by an unrelated party will be regulated under section 351 of the Act. The FDA also intends to develop standards—including disease screening requirements, establishment controls, processing controls, and product standards: "If sufficient data are not available to develop processing and product standards after a specified period of time, the stem cell products would be subject to IND and marketing application requirements."[14] Stem cell products that are more-than-minimally manipulated will require INDs and licensing under section 351 of the PHS Act. For example, stem cell products that are to be used for a nonhomologous function or are more-than-minimally manipulated will be required to be licensed under section 351. The FDA also has articulated "increased safety and effectiveness concerns for cellular and tissue-based products that are used for nonhomologous function, because there is less basis on which to predict the product's behavior."[15]

Implementation of the Proposed Approach

The FDA has begun to implement the proposed approach with the publication on January 20, 1998, of a *Request for Proposed Standards for Unrelated Allogeneic Peripheral and Placental/Umbilical Cord Blood Hematopoietic Stem/Progenitor Cell Products*" (63 Fed. Reg. 2985), utilizing its standards-setting authority under section 361 of the PHS Act.[16] In this notice, the FDA requests product standards to ensure the safety and effectiveness of stem cell products, which should be supported by clinical and nonclinical laboratory data. The FDA also announced its intention to phase in over a three-year period implementation of IND application and license application requirements for minimally manipulated unrelated allogeneic hematopoietic stem/progenitor cell products. The notice states that "[i]f adequate information can be developed, the agency intends to issue guidance for establishment controls, processing controls, and product standards....FDA intends to propose that, in lieu of individual applications containing clinical data, licensure may be granted for products certified as meeting issued standards." If, however, the FDA determines that adequate standards cannot be developed, the agency has expressed its intention to enforce IND and licensing requirements at the end of three years. Proposals are due on or before January 20, 2000.

On May 14, 1998, the FDA proposed *Establishment Registration and Listing for Manufacturers of Human Cellular and Tissue-Based Products* (63 Fed. Reg. 26744). The agency describes the proposed registration and listing requirements as a first step towards accomplishing its goal of putting into place a comprehensive new system of regulation for human cellular and tissue-based products. Registration and listing is intended to allow the FDA to assess the state of the cell and tissue industry, "to accrue basic knowledge about the industry that is necessary for its effective regulation," and to facilitate communication between the agency and industry (Ibid. at 26746). As proposed, the registration and listing requirements would apply to human cellular and tissue-based products that the FDA will regulate under section 361 of the PHS Act.[17] Among the products designated for regulation under that section and consequently subject to registration and listing are bone, tendons, skin, corneas, as well as peripheral and cord blood stem cells under certain conditions, and sperm, oocytes, and embryos for reproductive use (Ibid. at 26746).

FDA Discretion Entitled to Great Deference

Today there is a vast array of biological products that have been approved by the FDA and many others that are awaiting FDA action.[18] These products are scientifically complex and rarely lend themselves to categorization. As a result, the FDA invariably is required to determine on a case-by-case basis whether its existing statutory authority applies to a new product, which particular authority to apply, and, if so, what evidence will adequately demonstrate proof of safety, purity, and potency (efficacy). The decision of whether and how to regulate a product is made based upon the FDA's expert determination and upon the particular facts and circumstances, the historical application of the law to similar products, the applicable statutory and regulatory criteria, and the state of the FDA's scientific understanding at the time of the approval.

The FDA's exercise of the significant discretion provided to the agency by Congress is entitled to great deference by the courts.[19] In a recent challenge to the FDA's approval of a biological product under the PHS Act, the District Court for the District of Columbia held that "FDA's policies and its interpretation of its own regulations will be paid special deference *because of the breadth of the Congress' delegation of authority to FDA and because of FDA's scientific expertise.*"[20]

Moreover, even if the FDA has not asserted jurisdiction previously with regard to reproductive tissue, for example, it is within the agency's statutory authority that its policies are evolutionary. The Supreme Court has recognized that expert administrative agency interpretations are not "carved in stone. On the contrary, the agency...must consider varying interpretations and the wisdom of its policy *on a continuing basis*" (emphasis added).[21] Furthermore, the Court has acknowledged that "regulatory agencies do not establish rules of conduct to last forever....[A]n agency must be given ample latitude to 'adapt their rules and policies to the demands of changing circumstances.'"[22]

Conclusion

The FDA has developed a comprehensive approach to the regulation of cellular and tissue-based therapeutic products under its jurisdiction, including human stem cells. Nonclinical and clinical stem cell research undertaken to develop a therapeutic product intended to treat human disease will continue to be regulated by the FDA, while basic scientific research and other nonhuman research will remain outside of the agency's purview.

Notes

1 The content of this appendix is based upon a paper commissioned by the National Bioethics Advisory Commission and prepared by Brady, R.P., M.S. Newberry, and V.W. Girard, "The Food and Drug Administration's Statutory and Regulatory Authority to Regulate Human Pluripotent Stem Cells," available in Volume II of this report.

2 The scope of this appendix is limited to human stem cells. The FDA has a similar regulatory structure to regulate animal stem cell products used as animal drugs (21 USC 360b). The U.S. Department of Agriculture has the authority to regulate animal stem cell products used in animal vaccines (21 USC 151).

3 See *United States v. An Article...Sudden Change*, 409 F.2d 734, 739 (2d Cir. 1969).

4 See *National Nutritional Foods Ass'n. v. Mathews*, 557 F.2d. 325, 334 (2d Cir. 1977); *Action on Smoking and Health v. Harris*, 655 F.2d 236, 240–41 (D.C. Cir. 1980).

5 Statement of Harold Varmus, M.D., Director, National Institutes of Health, before the Senate Appropriations Subcommittee on Labor, Health and Human Services, Education and Related Agencies. December 2, 1998. Meeting transcript, 3.

6 Ibid.

7 Ibid. 3–4.

8 63 *Fed. Reg.* 26744 (May 14, 1998) (FDA Proposed Rule "Establishment and Listing for Manufacturers of Human Cellular and Tissue-Based Products").

9 "Human Tissue Intended for Transplantation" 58 *Fed. Reg.* 65514 (Dec. 14, 1993).

10 In 1997, FDA finalized its 1993 emergency rule establishing processing, testing, and recordkeeping requirements for all tissue products. "Human Tissue Intended for Transplantation" 62 *Fed. Reg.* 40429 (July 29, 1997).

11 CBER, Guidance on Applications for Products Comprised of Living Autologous Cells Manipulated *Ex Vivo* and Intended for Structural Repair or Reconstruction (May 1996).

12 "Cardiovascular Devices; Effective Date of Requirement for Premarket Approval; Replacement Heart Valve Allograft" 56 *Fed. Reg.* 29177 (June 26, 1991).

13 FDA Rescission Notice, 59 *Fed. Reg.* 52078 (October 14, 1994).

14 Proposed Approach, 25.

15 Proposed Approach, 16.

16 While FDA may choose to implement this policy through regulation, FDA also may implement it on a case-by-case basis. See infra, Section VI.

17 Consistent with the discussion supra, Section III. A., the preamble to the proposed rule states that "use of human cellular or tissue-based products solely for nonclinical scientific or educational purposes does not trigger the registration or listing requirements. Any use for implantation, transplantation, infusion, or transfer into humans is considered clinical use and would be subject to part 1271 [the registration and listing requirements]" Ibid. 26748.

18 Today, biological products are available or under development to treat, diagnose, or prevent virtually every serious or life-threatening disease. Available products include, but are not limited to, vaccines (manufactured both in traditional ways and through the use of biotechnology); human blood and blood-derived products; monoclonal or polyclonal immunoglobulin products; human cellular (i.e., gene therapy) products; protein, peptide, and carbohydrate products; protein products produced in animal body fluids by genetic alteration of the animal (i.e., transgenic animals); animal venoms; and allergenic products.

19 *U.S. v. Rutherford*, 442 U.S. 544, 553 (1979); *Bristol-Myers Squibb Co. v. Shalala*, 923 F. Supp. 212, 216 (D.D.C. 1996).

20 *Berlex Laboratories, Inc. v. FDA et al.*, 942 F. Supp. 19 (D.D.C. 1996) (emphasis added). See also *Lyng v. Payne*, 476 U.S. 926 (1986).

21 *Chevron, U.S.A., Inc. v. Natural Resources Defense Council, Inc.*, 467 U.S. 837, 863–64 (1984).

22 *Motor Vehicle Mfrs. Ass'n. of the U.S. v. State Farm Mut. Auto. Ins. Co.*, 463 U.S. 29, 42 (1983) (citations omitted).

Appendix E

Summary of Presentations on Religious Perspectives Relating to Research Involving Human Stem Cells, May 7, 1999

Introduction

As part of the National Bioethics Advisory Commission's deliberations for this report, a meeting was convened on May 7, 1999, at Georgetown University in order for the Commission to hear testimony from prominent scholars of religious ethics on their traditions' views of human stem cell research. Although it would be inappropriate for religious views to determine public policy in our country, such views are the products of long traditions of ethical reflection, and they often overlap with secular views. Thus, the Commission believed that testimony from scholars of religious ethics was crucial to its goal of informing itself about the range, content, and rationale of various ethical positions regarding research in this area.

The Commission heard testimony from scholars who work within the Roman Catholic, Protestant, Eastern Orthodox, Jewish, and Islamic faiths. Although the presenters were able to reach consensus on several significant issues related to embryonic stem (ES) and embryonic germ (EG) cell research, disagreement emerged among the religious traditions represented and often within each tradition itself, particularly between restrictive and permissive positions on several issues.

Roman Catholic Perspectives

The restrictive, "official" position within Roman Catholicism opposes EG and ES cell research, primarily because obtaining stem cells from either aborted fetal tissue or embryos that remain following clinical *in vitro* fertilization (IVF) procedures involves the intentional destruction of a genetically unique, living member of the human species. According to this view, it is impermissible to obtain stem cells from *in vitro* fertilized blastocysts, because doing so results in the destruction of the blastocyst—a human life worthy of full moral protection from the moment of conception. No amount of benefit to others can justify the destruction of the blastocyst, an act that would be equivalent to murder.

Similarly, from this perspective, it is impermissible to obtain EG cells from the gonadal tissue of aborted fetuses, because although such harvesting is not directly responsible for the death of the fetus, it nevertheless involves complicity with the evil of abortion. Moreover, to make use of any therapy derived from research on either human embryonic or fetal tissue and to contribute to the development or application of such research through general taxation would involve complicity in the destruction of human life. Federal funding, which in a sense would make all citizens complicit in this research, thus would greatly impose upon the consciences of Catholics.

However, even the restrictive position of the Roman Catholic Church does not oppose stem cell research per se. The central moral impediment to such research concerns the sources from which stem cells are derived. The act of harvesting stem cells from other sources—miscarried fetuses, placental blood, or adult tissues—would not be intrinsically immoral. In fact, this perspective, recognizing the potential benefits to human health of stem cell research, encourages investigation into the feasibility of such alternative sources. In practice, however, stem cell research, even with alternative stem cell sources, would remain morally problematic for two reasons. First, some are concerned that any safeguards will be ineffective because, in the face of potentially promising and lucrative research, the temptation to transgress such safeguards might be irresistible. Second, many fear that the benefits of this research might not be distributed equitably and are concerned that stem cell research perhaps may not be the best use of national resources, given the preponderance of so many other unmet human needs.

Although all Roman Catholics share a variety of important basic convictions, individual Catholics often differ in how to interpret them in practice. According to a less restrictive Catholic perspective, this disagreement is due, at least in part, to a commitment to the theory of natural law—a commitment that, while a fundamental part of the Catholic tradition, also involves reliance upon an "imperfect science." A commitment to natural law involves belief in a moral order that can be "seen" by all human beings in the reality of creation itself. But because the act of "looking" entails "a complex process of discernment and deliberation, and a structuring of insights, a determination of meaning, from the fullest vantage point available, given a particular history—one that includes the illumination of Scripture and the accumulated wisdom of the tradition"—what any two human beings see will not always be the same.[1]

With respect to stem cell research, the major areas of disagreement among Catholics are also those upon which the restrictive voice within Catholicism most strongly bases its opposition: the moral status of the embryo and the moral permissibility of using aborted fetuses as sources of stem cells. In contrast to this restrictive view of the embryo, another Catholic might, with the aid of science, look to the reality of the early human embryo and see that which is not yet an "individualized human entity with the settled inherent potential to become a human person."[2] Because the early embryo, according to this less restrictive view, is not a person, it is sometimes permissible to use it in research, though as human life it must always be accorded some respect. Similarly, one might decide that adequate barriers—such as a prohibition against the directed donation of cadaveric fetal tissue, and the distinction between somatic cell nuclear transfer (SCNT) for research or therapy and SCNT for reproduction—can be erected between the use of aborted fetal tissue in research and the act of abortion itself so that engaging in the former does not amount to complicity in the latter. From this perspective, then, a Catholic may be able to support ES cell research without sacrificing a commitment to the fundamental principles that define Catholicism, including the duties to protect human life, honor the sacred, and promote distributive justice in health care. Finally, because of the diversity within and among ethical traditions, this perspective is congruent with the restrictive Catholic view that individuals who oppose this research should not be forced to contribute to it but, contrary to the restrictive view, favors an approach that would allow federal funding, but with accommodations made to permit conscientious objection.

To summarize the testimony of the Roman Catholic panel, all agree that in light of certain agreed-upon principles, major Catholic concerns with regard to both embryonic and nonembryonic stem cell research include the following issues: 1) the moral status of the early embryo, 2) complicity with abortion in using fetal tissue as a source of stem cells, 3) the need for safeguards, distributive justice, and just allocation of national resources, and 4) the difficulty in federally funding research to which many are opposed on moral and religious grounds. The major disagreements arise from conflicting interpretations of the broad principles, which in turn lead to different responses to these four major concerns.

Jewish Perspectives

The two main sources of Jewish ethics—theology and law—yield several principles relevant to a Jewish ethical analysis of stem cell research. First, human beings are merely the stewards of their bodies, which belong to God. Moreover, God has placed conditions on the use of the human body, including the command that health and life must be preserved. Second, human beings are God's partners in healing, and in order to fulfill God's command, they have a duty to use any means available to heal themselves, whether these means are natural or artificial. Third, because all human beings, regardless of ability, are created in the image of God, they are valuable. Fourth, human beings, unlike God, lack perfect knowledge of the consequences of their actions and in the process of trying to improve themselves or the world must, therefore, be careful to avoid causing harm to them.

Four potential moral impediments to EG and ES cell research arise from these Jewish principles: 1) the moral status of the fetus and of the act of abortion, 2) potential complicity with evil, 3) the commandments to respect the dead, and 4) the moral status of the embryo.

According to Conservative Judaism, the fetus until the 40th day after conception is "like water." Although the fetus becomes a potential and partial person after the 40th day, and is thus entitled to a certain amount of respect and protection, it remains primarily a part of the pregnant woman's body, and does not become an independent person with full moral rights until the greater part of its body emerges from the womb during birth. Because of the command to preserve human health and life, if either the health or the life of the woman is clearly threatened by the fetus, abortion is not only permissible but obligatory, as she is a full person while the fetus remains only a part of her and a potential person. When the woman's health is at some increased risk but is not clearly compromised by the pregnancy, abortion is permissible but not obligatory. More recently, some Jewish authorities also permit abortion in cases in which the fetus has a terminal disease or serious malformations.

According to Orthodox Judaism, on the other hand, after 40 days of gestation, the fetus becomes a person with full moral rights and may not be aborted except to protect the pregnant woman's health. Yet, even though abortion after 40 days is viewed by the Orthodox Jews as homicide, it does not follow from this perspective that life-saving use of stem cells procured from illegitimately aborted fetuses is impermissible (although the question of who can legitimately give consent to such procurement is problematic from this perspective). Although this perspective recognizes the possibility that therapeutic use of aborted fetuses may make abortion appear less heinous, the strength of the commandment to preserve life, for which all other laws must be suspended except those prohibiting murder, idolatry, and sexual transgressions, overrides this concern. Thus, despite the disagreement within Judaism regarding the moral status of the fetus and the permissibility of abortion after 40 days, all agree that neither source of stem cells is illegitimate. One caveat to this consensus is that some within Conservative Judaism who accept the permissibility of abortion to preserve the life or health of the woman nevertheless require that stem cells be procured only from fetuses that have been legitimately aborted; Orthodox Judaism, by contrast, appears to hold that although abortion after 40 days postconception is generally impermissible, there is no complicity involved in using these aborted fetuses as sources of stem cells.

Jewish thinkers agree that commandments to respect the dead, which require that corpses not be mutilated or left unburied longer than necessary, can be suspended in order to save lives. Because of the strong commandment to preserve life and health, for example, Jewish law permits both autopsies and organ procurement when they will benefit the living. Reasoning by analogy, if tissue procurement from the cadavers of full persons in order to benefit human health and life is permitted, then tissue procurement from dead fetuses—which according to some Jewish perspectives are less than full persons—must also be permitted for the same purpose provided that (for some interpreters) the abortion itself was permissible according to Jewish law.

There is also wide consensus within Judaism that no serious moral impediments exist to using IVF embryos as sources of stem cells because extra-corporeal embryos have no status under Jewish law. These entities lack status because all embryos prior to 40 days postconception are "like water" and because as extra-corporeal entities, they lack the status of potential and partial person that is accorded to fetuses, which develop from embryos implanted in a uterus. Although extra-corporeal embryos merit a certain respect as human life, they are closest in moral status to gametes and thus may be discarded, frozen, or used as life-saving sources of stem cells. In fact, so long as they are never implanted, there is no clear legal prohibition against creating embryos for research purposes, although extra-legal norms may raise ethical questions about this practice.

Because stem cells can be permissibly procured either from extra-corporeal embryos or from legitimately aborted fetuses, stem cell research is not considered intrinsically immoral. Rather, stem cell research becomes morally problematic when applied in a variety of contexts. First, Judaism views the provision of health care as a communal duty. Thus, a context in which the benefits of stem cell research are not accessible to all persons who are in need would be problematic. Similarly, it may be problematic to focus national resources in this area of research rather than in other areas of need. In addition, although obtaining consent to procure stem cells is necessary, it may be challenging. Finally, there is widespread agreement that stem cell research should not be used to enhance human beings, although some disagreement exists over whether it may be used to improve health or whether it must be reserved only for life-saving purposes.

Eastern Orthodox Perspectives

According to Eastern Orthodoxy, all human beings are created in the image of God and grow continuously toward the likeness of God. Although the embryo, fetus, and adult are each at different stages of this process, all share the same potential for attaining authentic personhood, and each, with God's grace, will attain such personhood. According to this belief, God has given us medicine in order to heal, and any misuse of this gift that results in the destruction of potentially authentic persons is considered illegitimate. Thus, although miscarried fetuses may be used as sources of EG cells, neither electively aborted fetuses nor blastocysts may be so used. However, despite the impermissibility of procuring ES cells from blastocysts, because cell lines from this source already exist and have the potential to save lives, it is considered wasteful to discard these lines, and it is in fact permissible to use them. No complicity is thought to arise from such use. On the other hand, it is not permissible to procure EG cells from aborted fetuses, as such procurement *would* involve complicity.

Even assuming that stem cells could be permissibly procured, Eastern Orthodoxy shares with other religious traditions a variety of concerns about the context in which stem cell research might be applied, including addressing the problems of equitable access to the benefits of the research and other problems that can occur when market forces control the research; using the research for eugenic or cosmetic purposes, rather than for healing; and obtaining the informed, voluntary consent of the woman or couple.

Islamic Perspectives

Islam consists of two major schools of thought—Sunni and Shi'i—both of which refer to the same historical sources. Although these two schools differ somewhat in their views of abortion, in general, Islam regards the life of the fetus as developing over several stages, and personhood is considered a process. Although from the moment of conception the embryo is a human life meriting some protection, it is not commonly thought to attain personhood until it is ensouled, some time around the fourth month of gestation. Thus, because of the enormous potential to improve human health through this type of research, the vast majority of followers of Islam would agree that it is permissible to use early human embryonic life for this purpose. Moreover, it is permissible to use the tissue from illegitimately aborted fetuses to save lives, just as it is permissible to use cadaveric organs to save lives, even when the cadaveric organ source has been wrongfully killed. Finally, with caution, it can be deduced that creating embryos for research purposes is also permissible from an Islamic perspective, as long as those embryos are not implanted.

Protestant Perspectives

Protestant positions range dramatically from the highly restrictive to the nonrestrictive in this area. For example, according to restrictive Protestant view, a person is not defined by his or her capacities; rather, a person is a human being with a personal history, regardless of whether he or she is aware of that history. From this perspective, embryos are simply the weakest and least advantaged people among us. Because procuring stem cells from embryos requires the destruction of the embryo, such procurement thus raises serious moral issues, despite the ease with which it might be used to attain undeniably positive consequences for others, and rather than accepting the use of illicit means to achieve a good end, we should search for alternative, permissible means. Similarly, using aborted fetuses as sources of EG cells amounts to complicity with evil, and procurement of EG cells even from permissibly aborted fetuses (however that category is defined) would involve using a human life twice for another's benefit—first, to benefit the woman who aborted and then to benefit society through EG cell research. Therefore, from this perspective, it is impermissible to derive stem cells from embryos, whether spare or created for this purpose, and from aborted fetuses, whether permissibly aborted or not. The use of alternative sources of stem cells—for example, from bone marrow or umbilical cord blood—would, however, be permissible.

For Protestants whose views are less restrictive on this issue, the moral status of the embryo is more ambiguous. Although even nascent human life—which retains the potential for full human life—deserves respect and protection from callous disregard, the early embryo and the late fetus are viewed in moral terms as significantly different. Because the potential benefits of ES and EG cell research are so substantial, the moral difference between the early embryo and the developed fetus becomes compelling in this case, and it is thus permissible to use human life at the blastocyst stage to benefit other lives. No embryos should be created solely for this purpose, however, unless no other sources are available, and attempts should be made to locate alternative sources of stem cells that do not involve the destruction of embryos. It is permissible to procure EG cells from aborted fetuses, as long as safeguards are erected to prevent the therapeutic use of aborted fetal tissue from either increasing the frequency of abortion or encouraging a callous view of early human life. Moreover, although less restrictive Protestant views permit the procurement of stem cells from both proposed sources, this procurement must occur within a context of respect for nascent human life, only when significant benefit can be derived from it, and only after broad public discussion and acceptance of such research. If the general public is excluded from a discussion of this research, then public support of this and future beneficial research may be compromised.

Furthermore, the requirement that all members of society have the opportunity to participate in open, sustained dialogue about these decisions is critical from this perspective, and if federal funds are to be allocated toward this research, conscientious objectors should be accommodated. Finally, most Protestants share previously articulated contextual concerns regarding 1) ensuring global access to the benefits of this research, 2) avoiding the negative consequences that might come with market-controlled research, and 3) assessing the priority of these research efforts relative to other current and pending health-related research projects.

Summary of Broad Areas of Agreement and Disagreement

Not surprisingly, the panelists did not reach unanimity on all aspects of human ES and EG cell research. Although some differences exist among the various religious traditions, these mostly concern the appropriate sources and methods of religious-ethical reasoning. On substantive issues, less restrictive individuals across most religious traditions appear to have more in common with each other than with restrictive members of their own faiths. (The same is true for commonalities among restrictive members of all faiths.) The substantive issues relevant to stem cell research on which there is internal disagreement include the following:

1) *The moral status of the embryo.* The perceived status of the embryo ranges from full moral personhood with correlative inviolable rights to life to an early, extracorporeal biological entity lacking any significant moral status. Between these poles, although the embryo tends to be viewed as valuable because of its current status as a form of human life and its potential status as a person, it is ultimately, if tragically, subordinate to the health needs of actual persons.

2) *Whether the use of EG cells derived from aborted fetuses involves complicity with the perceived evil of abortion.* On one end of the spectrum is the view that many abortions are permissible. Thus, complicity with evil is either never or rarely a consideration. On the other end of the spectrum is the view that all deliberate abortions are immoral, and that any use of EG cells derived from aborted fetuses involves complicity.

Those who take more moderate positions argue that even when abortion is wrong, it is not wrong to use tissue that would otherwise be discarded, or that complicity can be avoided by erecting barriers between abortion and stem cell procurement, such as a prohibition of directed donation.

3) *Whether stem cell research should, ideally, be federally funded.* Some, based on their belief in the duty to heal, hold that stem cell research should proceed as quickly as possible (given certain conditions; see below), while others hold that any federal funding that enables immoral research is itself immoral and would involve conscientious citizens in complicity against their will. The moderate view holds that in the absence of agreement on such issues as the moral status of the embryo, conscientious objectors should be allowed to opt out of federal support for the research and that without any federal support, privatized human ES and EG cell research will make contextual goals such as distributive justice even more difficult to realize.

Despite these areas of disagreement, widespread consensus was reached both within and among the various religious traditions on several important issues in ES and EG cell research:

1) Stem cell research is not inherently immoral, and in fact has the potential to contribute important knowledge that can lead to therapies for certain diseases, provided that morally legitimate sources of cells are used (although this is defined differently), and provided that important contextual factors of justice and regulation are addressed. (See #3 below.)

2) If society chooses to embark upon federally funded ES and EG cell research, it must do so under conditions of respect for the humanity of the embryo. It would be preferable if there existed alternative sources of stem cells that did not involve the direct or indirect destruction of human life, and efforts should be made to identify such sources.

3) In order for the research to be morally permissible, several "background factors" must be in place, including

- assurance of equitable access to the benefits of the research,
- appropriate prioritization of this research relative to other social needs,
- assurance that the research will be used to treat disease, not enhance humans,
- public education, discussion, and acceptance of human stem cell research, and
- public scrutiny, oversight, and regulation of the research.

4) Assuming that privately funded research will continue in this area, it is preferable that a public body—even one that is funded with tax dollars—be required by law to review all private sector research and to make this review part of the public record, despite the possibility that the connection between the government and ES and EG cell research may be perceived as legitimating research that some citizens will continue to consider immoral.

Meeting Participants

Catholicism
Kevin W. Wildes, S.J., Ph.D., Georgetown University
Edmund D. Pellegrino, M.D., Georgetown University
Margaret Farley, Ph.D., Yale University

Judaism
Rabbi Elliot N. Dorff, Ph.D., University of Judaism
Rabbi Moshe Tendler, Ph.D., Yeshiva University
Laurie Zoloth, Ph.D., San Francisco State University

Eastern Orthodoxy
Father Demetrios Demopulos, Ph.D., Holy Trinity Greek Orthodox Church

Islam
Abdulaziz Sachedina, Ph.D., University of Virginia

Protestantism
Gilbert C. Meilander, Jr., Ph.D., Valparaiso University
Nancy J. Duff, Ph.D., Princeton University Theological Seminary
Ronald Cole-Turner, M.Div., Ph.D., Pittsburgh Theological Seminary

Notes

1 Farley, M., "Roman Catholic Views on Research Involving Human Embryonic Stem Cells." Testimony before NBAC. May 7, 1999. Washington DC. Meeting transcript, 3.

2 Ibid. 5.

Appendix F

Points to Consider in Evaluating Basic Research Involving Human Embryonic Stem Cells and Embryonic Germ Cells

This document describes some of the ethical, scientific, and legal issues that could be considered when designing and/or reviewing studies that involve access to and use of human stem cells. These *Points to Consider* are relevant only for designing and evaluating studies in which the role of the individual(s) who provide gametes, cadaveric fetal tissue, or embryos is limited to providing these materials for research intended to develop generalizable new knowledge. This document results from the recommendations described in this report and therefore is intended for use by those who design, conduct, and review research involving human embryonic stem (ES) and embryonic germ (EG) cells using federal funds. Private researchers and sponsors also may find this document to be of use. These *Points to Consider* do not apply to situations in which an individual would be the recipient of a stem cell-based therapy, nor do they apply to studies involving human/animal hybrids.

I. Scientific and Research Design Considerations

The ethical acceptability of any research protocol depends, in part, on its scientific merit, the qualifications of investigators, the protocol's overall design characteristics, and the precise nature of the materials and operations employed. In these respects, several issues arise when designing research involving human ES and EG cells, consideration of which would help ensure not only that the research is well designed, important, feasible, and timely, but also that a number of important ethical matters are considered. These issues are of particular significance given the nature of the materials to be used in research.

A. What are the sources from which human ES and EG cells will be obtained?
   1. From existing cell lines
   2. From cadaveric fetal tissue (following elective abortion or surgical termination of ectopic pregnancy)
   3. From embryos remaining after infertility treatments
   4. From embryos created solely for research purposes[1]

B. Has previous and requisite research been conducted using nonhuman animal models?

C. Are there valid alternatives to using human ES and EG cells in the proposed research?

D. What are the future plans for conservation of gametes, cadaveric fetal tissue, and embryos?
   1. Will ES or EG cells be produced and stored for later use?
   2. If a particular protocol is being proposed that uses embryos remaining after infertility treatments, does it propose to use only the minimum number of embryos necessary?
   3. What plans exist in the event that additional ES or EG stem cells are needed?

E. In what setting will the research be conducted?
   1. Are the investigators scientifically qualified to carry out the proposed research?
   2. Is the research environment (including facilities) appropriate for the conduct of research involving stem cells?

II. Identification of Providers and Donors and Recruitment Practices and Compensation

Several issues should be considered when identifying individuals (or couples) who may be asked to consider providing gametes, fetal tissue, or embryos for research; consideration of these issues could help to ensure that no inappropriate burden, inducement, or exploitation would occur.

A. Identification and recruitment practices
   1. Are potential donors or providers identified through advertisements to the general public? Are they identified through direct solicitation? Do they self-select?
   2. Is the selection of such individuals equitable and fair?
   3. Are these individuals vulnerable to undue influence, coercion, or exploitation? Does the recruitment method raise concerns about undue influence or coercion of the prospective donors?
   4. Are the potential donors capable of consenting?
   5. In which circumstances is it appropriate to identify and recruit an individual as well as his or her partner?

B. Compensation and reimbursement
   1. Will any financial compensation be paid to individuals (or couples) who donate materials; and if so, will the details of this compensation be disclosed?
   2. Does the compensation reimburse the individual (or couple) solely for the additional expenses that relate to this particular project?
   3. When is the offer of compensation made relative to an individual's (or couple's) decision to make available the materials from which stem cells will be derived?

III. Consent to Donate

Several issues arise in the process of providing information to individuals and couples who may be donating cadaveric fetal tissue or embryos remaining after infertility treatments. Considering these issues would help to ensure that prospective donors or providers of source materials would receive timely, relevant, and appropriate information to make informed and voluntary choices. In some cases, these issues are unique to the provision of gametes, embryos, or fetal tissue; in other cases, the items are important in other situations as well.

A. General considerations for individuals (or couples) who donate cadaveric fetal tissue or embryos remaining after infertility treatments
   1. Who will seek the consent? Will a clinician and/or researcher be available to answer questions?
   2. Is it appropriate for others to participate in the consent process (e.g., partner or family member)?
   3. Will psychological support mechanisms be in place if needed?
   4. Are the purposes of ES or EG cell research (in general) described fully?
   5. Will the consent form clearly disclose that stem cell research is not intended to benefit the donor directly?
   6. Is it clear that decisions to consent to or refuse the procedures to obtain stem cells will not affect the quality of care the patient will receive?
   7. Will individuals be informed that no medical or genetic information about the fetal tissue, embryos, or stem cells derived from these sources will be available to any outside individual or entity?
   8. What measures will be taken to protect the privacy and confidentiality of individuals who provide cadaveric fetal tissue or embryos?
   9. Is the source of funding for the research (public, private, public/private, philanthropic) disclosed?
   10. What known commercial benefits, if any, are expected to arise for the investigators seeking to obtain human ES or EG cells?

B. Additional considerations specific to consent to donate cadaveric fetal tissue
   1. Is there a description of what usually is done with fetal tissue at the institution at which a pregnancy will be terminated? Is this information available in written form and provided to individuals?
   2. Is permission to conduct research immediately available?

C. Additional considerations specific to consent to donate embryos remaining after infertility treatments
   1. Are the methods of disposal of embryos remaining after infertility treatments described? Is this information available in written form and provided to patients?
   2. Will information be made available about whether the embryos were viable and normal or not?
   3. Is there a description of the options available (e.g., permit material to be used in research, cryopreserve, discard, or donate to another couple for infertility treatment)?
   4. Is it clear that the embryos used in research will not, under any circumstances, be transferred to any woman's uterus?
   5. Is it clear that the research will result in the destruction of the embryo? Is the method described?

IV. Review Issues

Because of the special nature of human ES and EG cells, several issues arise in the review and oversight of research involving their use. The Commission has recommended a system of national oversight and review, combined with local monitoring. Careful and thoughtful consideration of these issues will provide assurance that, regardless of the source of funding, appropriate compliance with applicable regulations, guidelines, and other standards will occur. These considerations would supplement, not replace, applicable federal and state regulations.

A. Applicability of relevant regulations
   1. What current guidelines, regulations, rules, or policies apply to the conduct of this research? If ambiguity exists, how will it be resolved?
   2. What mechanisms are in place to assure compliance with these regulations?
   3. What regulations apply for collaborating with international researchers (e.g., importing fetal tissue or embryos from other countries)?

B. Applicability of professional practice standards
C. Submission of research findings for publication
D. Other responsibilities of investigators and collaborating clinicians

Note

1 The National Bioethics Advisory Commission has recommended that federal agencies should not fund research involving the derivation or use of human ES cells from embryos created solely for research purposes. (See Recommendations 3 and 4.)

Appendix G

Public and Expert Testimony

January 19, 1999 (Washington, DC)

Public:
E.J. Suh, Collegians Activated to Liberate Life
Kneale Ewing, Collegians Activated to Liberate Life
Olga Fairfax
Will Goodman

Expert:
Harold Varmus, National Institutes of Health
John Gearhart, The Johns Hopkins University
James Thomson, University of Wisconsin
Austin Smith, University of Edinburgh
Daniel Perry, Alliance for Aging Research
Patricia King, Georgetown University School of Law
John Robertson, University of Texas School of Law
Erik Parens, The Hastings Center
Françoise Baylis, Dalhousie University
Ted Peters, Center for Theology and the Natural Sciences
Karen Lebacqz, Pacific School of Religion

February 2–3, 1999 (Princeton, New Jersey)

Expert:
David Blumenthal, Massachusetts General Hospital
Brigid Hogan, Vanderbilt University
Barbara Mishkin, Hogan & Hartson L.L.P.
Robert Brady, Hogan & Hartson L.L.P.

March 2–3, 1999 (Vienna, Virginia)

Expert:
John Fletcher, University of Virginia
Lori Knowles, The Hastings Center
LeRoy Walters, Georgetown University

April 16, 1999 (Charlottesville, Virginia)

Public:
Richard Doerflinger, National Conference of Catholic Bishops
Edward Furton, National Catholic Bioethics Center
Karen Poehailos
Sidney Gunst, Jr.
Ida Chow, American Society of Cell Biology
Ethics and Religious Liberty Commission of the Southern
  Baptist Convention (submitted written testimony)

May 7, 1999 (Washington, DC)

Public:
Dena Davis, Cleveland-Marshall College of Law
Richard Doerflinger, National Conference of Catholic Bishops

Expert:
Kevin Wildes, Georgetown University
Edmund Pellegrino, Georgetown University
Margaret Farley, Yale University
Demetrios Demopulos, Holy Trinity Greek Orthodox Church
Elliot Dorff, University of Judaism
Moshe Tendler, Yeshiva University
Laurie Zoloth, San Francisco State University
Abdulaziz Sachedina, University of Virginia
Gilbert Meilander, Jr., Valparaiso University
Nancy Duff, Princeton University Theological Seminary

May 11–12, 1999 (Northbrook, Illinois)

Public:
Daniel McConchie, Center of Bioethics and Human Dignity

Expert:
Lori Andrews, Chicago–Kent College of Law
Sander Shapiro, University of Wisconsin–Madison

June 28, 1999 (Washington, DC)

Public:
Phil Noguchi, Food and Drug Administration

Appendix H

Commissioned Papers

The following papers, prepared for the National Bioethics Advisory Commission, are available in Volume II of this report

State Regulation of Embryo Stem Cell Research
Lori B. Andrews
Chicago-Kent College of Law

The Food and Drug Administration's Statutory and Regulatory Authority to Regulate Human Pluripotent Stem Cells
Robert P. Brady, Molly S. Newberry, and Vicki W. Girard
Hogan & Hartson L.L.P.

Quick Response: Use of Human Fetal Tissue in Federally Funded Research
Elisa Eiseman
RAND Science and Technology Policy Institute

Analysis of Federal Laws Pertaining to Funding of Human Pluripotent Stem Cell Research
Ellen J. Flannery and Gail H. Javitt
Covington & Burling

Deliberating Incrementally on Human Pluripotential Stem Cell Research
John C. Fletcher
University of Virginia

Bioethical Regulation of Human Fetal Tissue and Embryonic Germ Cellular Material: Legal Survey and Analysis
J. Kyle Kinner, Presidential Management Intern
National Bioethics Advisory Commission

Regulating Embryonic Stem Cell Research: Biomedical Investigation of Human Embryos
J. Kyle Kinner, Presidential Management Intern
National Bioethics Advisory Commission

International Perspectives on Human Embryo and Fetal Tissue Research
Lori P. Knowles
The Hastings Center

What Has the President Asked of NBAC? On the Ethics and Politics of Embryonic Stem Cell Research
Erik Parens
The Hastings Center

Locating Convergence: Ethics, Public Policy, and Human Stem Cell Research
Andrew W. Siegel
The Johns Hopkins University

Ethical Issues in Human Stem Cell Research

VOLUME II
COMMISSIONED PAPERS

Rockville, Maryland
January 2000

National Bioethics Advisory Commission

Harold T. Shapiro, Ph.D., Chair
President
Princeton University
Princeton, New Jersey

Patricia Backlar
Research Associate Professor of Bioethics
Department of Philosophy
Portland State University
Assistant Director
Center for Ethics in Health Care
Oregon Health Sciences University
Portland, Oregon

Arturo Brito, M.D.
Assistant Professor of Clinical Pediatrics
University of Miami School of Medicine
Miami, Florida

Alexander Morgan Capron, LL.B.
Henry W. Bruce Professor of Law
University Professor of Law and Medicine
Co-Director, Pacific Center for Health Policy and Ethics
University of Southern California
Los Angeles, California

Eric J. Cassell, M.D., M.A.C.P.
Clinical Professor of Public Health
Cornell University Medical College
New York, New York

R. Alta Charo, J.D.
Professor of Law and Medical Ethics
Schools of Law and Medicine
The University of Wisconsin
Madison, Wisconsin

James F. Childress, Ph.D.
Kyle Professor of Religious Studies
Professor of Medical Education
Co-Director, Virginia Health Policy Center
Department of Religious Studies
The University of Virginia
Charlottesville, Virginia

David R. Cox, M.D., Ph.D.
Professor of Genetics and Pediatrics
Stanford University School of Medicine
Stanford, California

Rhetaugh Graves Dumas, Ph.D., R.N.
Vice Provost Emerita, Dean Emerita, and
 Lucille Cole Professor of Nursing
The University of Michigan
Ann Arbor, Michigan

Laurie M. Flynn
Executive Director
National Alliance for the Mentally Ill
Arlington, Virginia

Carol W. Greider, Ph.D.
Professor of Molecular Biology and Genetics
Department of Molecular Biology and Genetics
The Johns Hopkins University School of Medicine
Baltimore, Maryland

Steven H. Holtzman
Chief Business Officer
Millennium Pharmaceuticals Inc.
Cambridge, Massachusetts

Bette O. Kramer
Founding President
Richmond Bioethics Consortium
Richmond, Virginia

Bernard Lo, M.D.
Director
Program in Medical Ethics
The University of California, San Francisco
San Francisco, California

Lawrence H. Miike, M.D., J.D.
Kaneohe, Hawaii

Thomas H. Murray, Ph.D.
President
The Hastings Center
Garrison, New York

Diane Scott-Jones, Ph.D.
Professor
Department of Psychology
Temple University
Philadelphia, Pennsylvania

CONTENTS

State Regulation of Embryo Stem Cell Research A-1
Lori B. Andrews
Chicago-Kent College of Law

The Food and Drug Administration's Statutory and Regulatory
Authority to Regulate Human Pluripotent Stem Cells B-1
Robert P. Brady, Molly S. Newberry, and Vicki W. Girard
Hogan & Hartson L.L.P.

Quick Response: Use of Human Fetal Tissue
in Federally Funded Research C-1
Elisa Eiseman
RAND Science and Technology Policy Institute

Analysis of Federal Laws Pertaining to Funding
of Human Pluripotent Stem Cell Research D-1
Ellen J. Flannery and Gail H. Javitt
Covington & Burling

Deliberating Incrementally on Human Pluripotential Stem Cell Research E-1
John C. Fletcher
University of Virginia

Bioethical Regulation of Human Fetal Tissue
and Embryonic Germ Cellular Material: Legal Survey and Analysis F-1
J. Kyle Kinner, Presidential Management Intern
National Bioethics Advisory Commission

Regulating Embryonic Stem Cell Research:
Biomedical Investigation of Human Embryos G-1
J. Kyle Kinner, Presidential Management Intern
National Bioethics Advisory Commission

International Perspectives on Human Embryo
and Fetal Tissue Research H-1
Lori P. Knowles
The Hastings Center

What Has the President Asked of NBAC?
On the Ethics and Politics of Embryonic Stem Cell Research I-1
Erik Parens
The Hastings Center

Locating Convergence:
Ethics, Public Policy, and Human Stem Cell Research J-1
Andrew W. Siegel
The Johns Hopkins University

STATE REGULATION OF EMBRYO STEM CELL RESEARCH

Commissioned Paper
Lori B. Andrews
Chicago-Kent College of Law

Introduction

In a society deeply divided over the moral and legal status of embryos, any scientific or medical project using tissue or cells from the unborn is bound to raise serious philosophical and social concerns. Because varying viewpoints regarding the legal protection of embryos and fetuses have been enacted into law in different states, it is often difficult for researchers and physicians to determine which laws cover their work. Indeed, statutory and court precedents dealing with embryo and fetal research, abortion, organ transplant, and payment for body tissue all have ramifications for work involving embryo stem cells. Yet, no two states have identical laws covering these procedures.

Some type of embryo stem cell research is permissible in virtually every state.[1] Yet, because of differences in state laws, certain states would ban the collection of stem cells from embryos that were created through *in vitro* fertilization (IVF).[2] In other states, a prohibition would only apply to the isolation of stem cells from *aborted* embryos and fetuses.[3] (See Appendix A for a table showing bans under the embryo and fetal research laws and abortion laws by state. Appendix B presents bans on payment under the Uniform Anatomical Gift Act [UAGA] by state.)

The breadth of potential regulation affecting embryo stem cell research can be assessed by asking some of the following questions:

Source of Cells

Is an embryo (up to eight weeks post-conception) or fetus (after eight weeks post-conception) the source of the cells?

If an embryo is the source, was it an excess embryo from an IVF patient? Or was the embryo created especially to serve as a source of stem cells?

If the source is a fetus, was it the subject of a planned abortion, or was it stillborn?

If the source of the cells was an aborted fetus, was the treatment of the woman prior to the abortion varied in any way to aid in the research or the later collection of cells?

Is the research being conducted only on cell lines themselves?

Informed Consent

Did the male or female progenitor, or both, or neither, consent to the use of the embryo or fetus?

Was the consenting individual told specifically that the embryo or fetus would be used as a source of stem cells?

Who asked for the consent, and in what circumstances?

Was the female or male progenitor told of any plans to develop a commercial cell line or other commercial product out of the embryo's or fetus' tissue?

Privacy

Was the male or female progenitor, or both, or neither, told what infectious disease or genetic tests would be undertaken on the tissue?

Will they be told of the outcome of such tests?

Will it be possible to identify the progenitors based on information available with the cell line, the health institution, the research center, or by combining information held in various locations?

Commercialization

Was the male or female progenitor paid for access to the tissue?

Did the doctor or researcher pay someone else for the conceptus, the tissue, or the resulting cells?

Will the ultimate products—the cell lines or specialized cells or tissues or organs—be sold?

All of the factors implicit in the questions above are relevant to the legality of the collection and use of embryo stem cells. Researchers need to be aware that the social sensitivity about fetal tissue leads to it being treated differently than other tissue. For example, a Colorado law provides funding to disseminate information to promote organ and tissue donation; however, that funding cannot be used to encourage fetal tissue donation.[4]

Source of Cells

Varying social perspectives on the fetus underlie different states' approaches to regulating research using tissue from conceptuses. In 24 states, there are no laws specifically addressing research on embryos or fetuses.[5] In those states, then, embryo stem cell research is not banned. However, other legal precedents covering informed consent, privacy, and commercialization come into play.

The other 26 states do have laws regulating research on conceptuses.[6] The laws vary enormously, with differing implications. For example, in 12 of those states, there is greater freedom to use a fetus for research if the fetus was not the subject of an abortion.[7] In addition, an Illinois law specifically states that it is permissible to use tissues or cells from a dead fetus whose death did not result from an induced abortion so long as there is consent from one of the parents.[8] But, since there are many practical difficulties to obtaining tissue from a miscarried fetus—and there are medical reasons to believe such tissue may be less than optimal—most stem cell research will not use miscarried conceptuses, but rather will use IVF embryos or will use aborted embryos or fetuses.

A North Dakota law would ban embryo stem cell research using either IVF embryos or aborted conceptuses. In other states, though, the statutes could apply to either one type of research or the other. Some states forbid research on aborted conceptuses; such laws would not apply to the technique used by James Thomson and his colleagues involving *in vitro* fertilized (rather than aborted) embryos. Other state laws, however, apply only to "live" conceptuses.[9] These laws would not have an impact on the technique used by John Gearhart and his colleagues, which involved dead aborted embryos and fetuses. However, *in vitro* embryos would be considered "live," not dead, because they could be used to create a live child, and thus the laws banning research on live conceptuses would prohibit the Thomson technique.

Ten states' laws apply to *in vitro* embryos. In New Hampshire, the regulation of research on embryos prior to implantation is minimal.[10] The research must take place before day 14 post-conception, and the subject embryo must not be implanted in a woman. These stipulations could readily be met by researchers wanting to use IVF embryos as a source of stem cells. Thomson and his colleagues cultured the embryos for just five days before removing the cells.

In contrast, nine states ban research on *in vitro* embryos altogether.[11] In Louisiana, an *in vitro* fertilized ovum may not be farmed or cultured for research purposes;[12] the sole purpose for which an IVF embryo may be used is for human *in utero* implantations.[13] In the other eight states, embryo research is banned as part of the broader ban on all research involving live conceptuses. These laws could create problems for researchers who want to use the technique described by Thomson and his colleagues to culture stem cells from embryos.[14] The penalties are high—in some states, the punishment includes imprisonment.[15]

Other states have restrictions on research involving fetuses which would affect researchers who, like Gearhart and his colleagues, culture stem cells by using gonadal tissue from dead aborted embryos or fetuses.[16] In six states, if the aborted embryo or fetus is dead, the focus is primarily on assuring that the mother has given her consent.[17] But in six other states, research on dead aborted conceptuses is severely restricted.[18]

The six states' bans on research on dead aborted conceptuses vary slightly in how they refer to the prohibited practices, but all would cover the currently experimental technique of removing gonadal tissue from aborted conceptuses to create embryo stem cells. In North Dakota and South Dakota, for example, a researcher may not use an aborted embryo or fetus—or any tissue or organ thereof—for research, experimentation, study, or for animal or human transplantation.[19] Similarly, in Indiana, no experiments may be conducted on an aborted fetus (nor may an aborted fetus be transported out of Indiana for experimental purposes).[20] In Ohio, no person shall experiment upon an aborted product of human conception.[21] In Oklahoma, research is forbidden on an aborted unborn child or its remains.[22]

Some of the laws specifically apply not only to research on embryos and fetuses, but also to research involving their organs or tissues.[23] This could potentially cause problems not just for the scientists who actually retrieve the cells from the fetus, but—if cell lines are considered "tissue"—to other scientists who do research involving the cell lines. Similar problems are raised by an Arizona statute which provides, "A person shall not knowingly use any human fetus or embryo, living or dead, or any parts, organs or fluids of any such fetus or embryo resulting from an induced abortion in any manner...."[24]

The bans on research on aborted fetuses cannot be circumvented by using what would medically be called embryos (because they are less than eight weeks post-conception). The legal definition of a fetus is different than the medical one. The term "fetus" in the state laws usually refers to any product of conception beginning at fertilization.

The analysis of the reach of the embryo research bans is further complicated when the cloning procedure used by José Cibelli is employed because some of the statutes forbidding research on a conceptus or unborn child define those entities as the products of "fertilization." Whether Minnesota's and Pennsylvania's statutes would apply to cloning turns on whether the term "fertilization" includes cloning. Minnesota's statute bans research on a "living conceptus," created *in utero* or *ex utero*, "from fertilization through 265 days thereafter."[25] Pennsylvania prohibits experimentation on an unborn child, which is defined as a *homo sapiens* from fertilization to birth.[26] Since fertilization is not defined, a court might turn to a dictionary definition: "the process of union of two germ cells whereby the somatic chromosome number is restored and the development of a new individual is initiated...."[27] Cloning is not the union of two germ cells, but this process *does* result in the somatic chromosome number. It could lead to the development of a new individual, as was the case with the cloning of Dolly the sheep. Arguably, then, the two most important elements of fertilization are satisfied, and the third merely explains the only way previously known to accomplish the other two. Thus, fertilization could be interpreted to include cloning. The 265-day period of coverage in the Minnesota statute potentially creates a loophole, though. If an embryo is created through cloning, it could be argued that if it is cryopreserved for 265 days after "fertilization," it could be experimented upon thereafter. If the use of the term "fertilization" is interpreted to preclude the application of the Minnesota and Pennsylvania laws to the Cibelli technique, then only seven states would ban embryo stem cell research on cloned embryos.[28]

In some instances, the woman or couple who donate an embryo or fetus for research purposes also face liability. While many state laws focus only on the *use* of the unborn in research, some—like those in Maine, Michigan, North Dakota, and Rhode Island—additionally prohibit the transfer, distribution, or giving away of any live embryo for research purposes.[29] In Maine, a person who does so is subject to a fine of up to $5,000 and up to five years' imprisonment.[30]

Some states address how the conceptus should be treated *after* the research. In California, for example, fetal remains may be used for "any type of scientific or laboratory research or for any other kind of experimental study,"[31] but, after the study, the remains must be promptly interred or disposed of by incineration.[32]

Given the higher level of scrutiny that embryo and fetal research receives, it is ironic that in California, where every tissue bank must be licensed, there is an exception for the "collection, processing, storage, or distribution of tissue derived from a human embryo or fetus."[33] In addition, in Hawaii, a coroner's physician or medical examiner has a right to retain fetal tissue from an autopsy "to be used for necessary or advisable scientific investigation, including research, teaching, and therapeutic purposes."[34]

The Constitutionality of Bans on Research on Conceptuses

Not all regulations affecting research are constitutional. Laws restricting research on conceptuses may be struck down as too vague or as violating the right to privacy to make reproductive decisions. Such a challenge was successful in a federal district court case, *Lifchez v. Hartigan*,[35] which held that a ban on research on conceptuses was unconstitutional because it was too vague in that it failed to define the terms "experimentation" and "therapeutic."[36] The court pointed out that there are multiple meanings of the term "experimentation."[37] It could mean pure research, with no direct benefit to the subject. It could mean a procedure that is not sufficiently tested so that the outcome is predictable or a procedure that departs from present-day practice. It could mean a procedure performed by a practitioner or clinic for the first time. Or it could mean routine treatment on a new patient. Since the statute did not define the term, it was unconstitutionally vague. It violated researchers' and clinicians' due process rights under the fifth amendment since it forced them to guess whether their conduct was unlawful.[38]

A similar result was reached by a federal appellate court assessing the constitutionality of a Louisiana law prohibiting nontherapeutic experimentation on fetuses in *Margaret S. v. Edwards*. The appeals court declared the law unconstitutional because the term "experimentation" was so vague it did not give researchers adequate notice about what type of conduct was banned.[39] The court said that the term "experimentation" was impermissibly vague[40] since physicians do not and cannot distinguish clearly between medical experimentation and medical tests.[41] The court noted that "even medical treatment can be reasonably described as both a test and an experiment."[42] This is the case, for example, "whenever the results of the treatment are observed, recorded, and introduced into the database that one or more physicians use in seeking better therapeutic methods."[43]

A third case struck down as vague the Utah statute that provided that "live unborn children may not be used for experimentation, but when advisable, in the best medical judgment of the physician, may be tested for genetic defects."[44] The Tenth Circuit held "[b]ecause there are several competing and equally viable definitions, the term 'experimentation' does not place health care providers on adequate notice of the legality of their conduct."[45]

If an embryo stem cell researcher were to challenge the state statutory bans, it is unlikely that he or she would be as successful as the plaintiffs in Illinois, Louisiana, or Utah. Researchers presumably know when they are engaging in research. Part of the constitutional deficiency in the Illinois, Louisiana, and Utah cases was that physicians offering health care services for their patients (such as embryo donation from an infertile woman, preimplantation genetic screening on an embryo, or treatment of a pregnant woman for diabetes) would not know if the activity would be considered by prosecutors to be experimental with respect to the embryo or fetus. Moreover, the female patients' reproductive freedom was implicated in the previous cases, providing another reason for overturning the statute. Stem cell researchers do not have those potential legal arguments.

Consent

A cornerstone of research on human subjects is that consent is necessary before research may be undertaken. With respect to research involving a conceptus, the researched-upon subject is not capable of giving consent, so a policy decision must be made about who should be asked for permission to undertake research on the conceptus.

There is a consensus that at least the woman who provided the egg for fertilization should be asked for permission before research is undertaken involving her embryo. There is some evidence that, in earlier studies, the woman's consent was not always sought. Embryos removed during the course of medical procedures, such as hysterectomies, were apparently used without the knowledge or consent of the women undergoing the procedures.[46] Experimentation on a woman's embryo without her consent can create psychological harm to women.[47] It can also pose potential physical harm to particular women since, if they do not realize research is intended for the fetus, they may not question potentially risky medical procedures undertaken on them to facilitate the research.[48]

In the context of fetal research, it was originally argued that the woman who has decided to abort the fetus is not an appropriate person to give a proxy consent for research on it.[49] Her decision to abort was taken as a sign of abandonment or inability to act in the fetus' best interest. However, since women have a constitutional right to abort, taking away her right to consent to or refuse research because she exercised that right may be an unconstitutional penalty.[50] Similarly, the fact that a couple does not wish to create a pregnancy with an excess embryo should not be taken to signify that they have somehow "abandoned" the embryo and waived all decisionmaking rights.[51]

In most infertility clinics, the consent form for IVF asks couples whether they are willing to donate excess embryos for research. With respect to aborted fetuses, it is unclear whether consent to use fetal remains for research is currently always sought—and whether sufficient information is given to the parent about the nature of the research. Precedents such as the John Moore case would require that the parent of the fetus be told whether research was going to be undertaken and whether there were plans to commercialize the tissue or its products.[52]

With respect to research on IVF embryos, couples are not routinely given information about the *type* of research that will be undertaken. It is common for women undergoing IVF to be given fertility drugs, which often lead to the creation of more embryos that can safely be implanted into the woman at one time. Consequently, many couples freeze embryos to use later if their initial attempts at IVF with fresh embryos do not work. Currently there are at least 100,000 frozen embryos, increasing at a rate of nearly 19,000 per year.[53]

Prior to freezing, couples are generally given informed consent forms to fill out. They are asked what they want done with the frozen embryos if they divorce, die, or decide not to use the embryos for their own procreative purposes. Generally, they are given three choices: termination, donation to another couple, or donation for research purposes.

Yet, it seems unreasonable to think that couples who merely checked "research" on their forms have given an adequate informed consent to embryo stem cell research. Couples should be told specifically that their embryos will be used for such research. Some couples, who might be comfortable with allowing their embryos to be used in research to improve fertilization techniques, may be troubled by an experiment that turns what could have been their potential child into a cell line that is for sale. Clearly, any future consent forms should include more details about the specific categories of research that might be conducted on the embryos so that the couples can make an informed choice about whether to submit their embryos to researchers. The embryos already in storage under a general research authorization should not be used for embryo stem cell research without making an attempt to recontact the couples for specific consent. The new possibility of embryo stem cell research is sufficiently different from the types of research that couples assumed would be done on the embryos at the time the couples donated them that recontact seems prudent. One large IVF clinic already has a policy of recontacting all couples with more specific information before *any* embryo research is actually undertaken.[54]

The recipients of the embryo stem cell products, too, should be told of their origin. Some people may not want treatment that uses cells derived from human embryos, just as some Jehovah's Witnesses will turn down treatment involving blood transfusions. Moreover, the use of human tissue may pose infectious disease risks to the recipient which need to be factored into a decision about whether to employ a treatment based on embryo stem cells.

None of the state statutes affecting embryo and fetus research specifically address the type of information that must be provided to the progenitors before they are asked for consent. In contrast, other guidelines for human research emphasize providing information about the nature of the research, its purpose, and the effect on the subject. Of the state statutes governing research on live fetuses, only New Mexico's statute describes the information that must be given before consent to research involving a fetus is valid. Under the New Mexico law, a woman who is asked to participate in research must be "fully informed regarding possible impact on the fetus."[55] Although the New Mexico law does not cover research on embryos or apply to dead fetuses, it does underscore the importance of informed consent.

Intriguingly, one state requires disclosure about the nature of the research to the researchers and their staffs and the recipients of the tissue. The Pennsylvania statute contains unique and broad disclosure requirements. Anyone involved in the procurement, use, or transplantation of fetal tissue or organs must be told whether the tissue was procured as a result of stillbirth, miscarriage, ectopic pregnancy, abortion, or some other means.[56]

The laws adopted for fetal tissue transplantation may have unexpected implications here. Normally, under the UAGA, the next-of-kin, including the parents of a fetus, may designate the recipient of tissue from the deceased. This can be an institution, like a particular medical school, or an individual, such as a relative. Once fetal tissue transplantation for diseases such as Parkinson's disease became a possibility, however, legislators did not want women to conceive and abort just to provide tissue needed by a relative. This led to the adoption of a federal law saying that in National Institutes of Health-funded fetal tissue research, the woman donating the fetal tissue may not choose the recipient of the tissue for transplantation.[57] There are also state laws whose restrictions regarding choosing tissue recipients are broader and may have implications for stem cell research. In Pennsylvania, for example, "No person who consents to the procurement or use of any fetal tissue or organ may designate the recipient of that tissue or organ, nor shall any other person or organization act to fulfill that designation."[58] This law unintentionally would create the situation where an IVF patient could donate her excess embryo for stem cell research, but she could not specify that it be used by a particular medical center. She would have to blindly turn it over and risk it going to a researcher or entity (such as a for-profit company) of which she might not approve.

Privacy

Legal and ethical concerns are also raised about the privacy of the couples whose conceptuses are used for embryo stem cell research. Since abortion is controversial, protecting the identity of the couples is paramount. But there are also subtler privacy issues. In order to assure a safe "product" involving embryo stem cells, the tissue from a conceptus might be tested for infectious or genetic diseases. Yet, testing of tissue from the embryo or the fetus may provide information about the parents. The mother of the fetus should be told as part of the informed consent process what tests will be undertaken and whether she will be informed of the outcome of the tests. In some states, statutes provide further protection from testing without consent under laws that prohibit unauthorized genetic disease testing or unauthorized HIV testing.

Commercialization

There is considerable state legislation affecting whether embryos may be sold for research purposes or other purposes. Certain states include within their abortion laws or conceptus research laws bans on payments. Other states have adopted versions of the UAGA, the law governing donation of tissue and organs from a deceased individual for transplantation or treatment.

Five states would ban payment for IVF embryos for research purposes.[59] Eight states would ban payment for embryos or fetuses for any purpose.[60] While some state laws might prevent payment for embryonic cell lines, it is possible that because a cell line is new tissue produced from the genetic material of, but not originally a part of, the embryo, laws proscribing the sale of embryonic tissue may not apply. In fact, a Minnesota law prohibits the sale of living conceptuses or nonrenewable organs but does allow "the buying and selling of a cell culture line or lines taken from a non-living human conceptus...."[61] In contrast, Nevada's broadly worded statute making it a crime for anyone to use or "make available...the remains of an aborted embryo or fetus for any commercial purpose" could conceivably outlaw the sale of cell lines from fetal tissue or even products made from those cell lines.[62] Similarly, in Pennsylvania, no compensation or other consideration may be paid to any person or organization in connection with the procurement of fetal tissue or organs.[63]

There are also laws that affect payment in connection with the use of a dead aborted fetus. Some laws prohibit the sale of an aborted product of conception.[64] Some prohibit sale of an aborted unborn child or the remains thereof.[65] Others prohibit the sale of an aborted fetus or any tissue or organ thereof.[66] Still others prohibit the sale of an unborn child.[67]

In Arkansas, even though research on a dead aborted fetus is permissible with the mother's consent, a person may not "buy, sell, give, exchange, or barter or offer to buy, sell, give exchange, or barter" such a fetus or any organ, member, or tissue of such fetal material.[68] It is also a crime to possess such material.[69] In Tennessee, research upon an aborted fetus is permitted with the prior knowledge and consent of the mother, but "no person, agency, corporation, partnership or association shall offer money or anything of value for an aborted fetus."[70]

In Indiana, Ohio, and Oklahoma, where research upon an aborted product of conception is forbidden, there is also a ban on its sale.[71] In Georgia, it is illegal to sell a human fetus or part thereof[72] except for health services education.[73] In addition, payment in Georgia is permissible for the reasonable costs associated with the removal, storage, or transportation of a human body or part donated for scientific research.[74]

In the District of Columbia, the sale of any portion of the human body (including cells) for any purpose is illegal;[75] however, there is a question about whether an IVF embryo or aborted fetus will be considered a "human body." There is a similar statute in Virginia prohibiting the sale of any "natural body part" for any reason.[76]

The anti-commercialization laws have nuances. In Rhode Island, for example, it is illegal to sell an embryo or fetus for purposes that violate the statute.[77] This law would ban the sale of IVF embryos for research purposes due to the ban on research on live embryos and fetuses. However, since research on dead embryos and fetuses is permissible with the permission of the mother, she apparently could be paid.

In addition, in 21 states, the organ transplant laws use language which is broad enough to forbid payment for fetal tissue which will be used in transplant or therapy.[78] Furthermore, in two additional states, the law prohibits sale of certain organs (which do not include fetal tissue), but the list can be expanded by a state health official to include fetal tissue.[79]

The bans on payment for embryonic or fetal tissue are broad. They preclude any compensation, remuneration, or exchange of value for the tissue. They could also easily be interpreted to ban payment for any of the costs associated with removal, preparation, or preservation of the tissue. This obviously could create a problem if researchers at the National Institutes of Health purchased embryo stem cells from researchers in states with such bans.[80]

The laws that ban payment for organs and tissue in general (rather than just those from fetuses) are not as restrictive. Sixteen of the 21 states' bans on payments for organs do allow reasonable payments for the removal, processing, disposal, preservation, quality control, storage, transplantation, or implantation of a part.

Concerns about the money motive can also be seen in laws that address the treatment of the donor of embryonic or fetal tissue. A Massachusetts law provides that "no person shall perform or offer to perform an abortion where part or all of the consideration for said performance is that the fetal remains may be used for experimentation or other kind of research or study."[81] Five other states have similar statutes.[82]

Conclusion

State lawmakers have expressed their concern for the sanctity of life—and its inherent value—by the laws they have adopted regarding research on conceptuses and commercialization of body tissue. There is considerable variation in the approaches states have taken, and embryo stem cell researchers (as well as policymakers) need to be aware of these laws. The reach of the state statutes is broad—they affect any researcher, no matter what the source of funds or the type of institution with which he or she is affiliated.

Notes

1 However, North Dakota has statutes that could be used to prohibit both the Thomson and the Gearhart techniques.

2 See discussion infra at notes 10–15.

3 See discussion infra at notes 16–18.

4 Colo. Rev. Stat. Ann. § 42-2-107(4)(b)(III)(A).

5 These states are Alabama, Alaska, Colorado, Connecticut, Delaware, Georgia, Hawaii, Idaho, Iowa, Kansas, Maryland, Mississippi, Nevada, New Jersey, New York, North Carolina, Oregon, South Carolina, Texas, Vermont, Virginia, Washington, West Virginia, Wisconsin, and the District of Columbia. A March 29, 1999, database search of LEXIS statutory entries was conducted to assess whether any of these states have applicable laws. No applicable laws were located using the search terms *zygote, pre-embryo, morula, blastocyst, embryo, fetus, conceptus, products of conception,* or *unborn child.*

6 Ariz. Rev. Stat. Ann. § 36-2302, -2303; Ark. Stat. Ann. § 20-17-801, -802; Cal. Health & Safety Code § 25956, 25957; Fla. Stat. Ann. § 390.001(6), (7); 720 Ill. Comp. Stat. 510/12.1; Ind. Code Ann. § 35-1-58.5-6; Ky. Rev. Stat. § 436.026; La. Rev. Stat. Ann § 9:121 et seq.; Me. Rev. Stat. Ann. tit. 22, § 1593; Mass. Ann. Laws ch. 112, § 12J; Mich. Comp. Laws Ann. §§ 333.2685, -2692; Minn. Stat. Ann. § 145.421 to -.422; Mo. Ann. Stat. § 188.037; Mont. Code Ann. § 50-20-108(3); Neb. Rev. Stat. §§ 28-342 to -346; N.H. Rev. Stat. Ann. sec 168-b:15; N.M. Stat. Ann. § 24-9A-1 et seq.; N.D. Cent. Code § 14-02.02-01, -02; Ohio Rev. Code Ann. § 2919.14; Okla. Stat. Ann. tit. 63, § 1-735; 18 Pa. Cons. Stat. Ann. § 3216; R.I. Gen. Laws § 11-54-1; S.D. Codified Laws Ann. § 34-23A-17; Tenn. Code Ann. § 39-15-208; Utah Code Ann. §§ 76-7-310; and Wyo. Stat. § 35-6-115.

7 This is because the other 12 of these laws apply only to research with fetuses prior or subsequent to an elective abortion. Ariz. Rev. Stat. Ann. § 36-2302(A) (subsequent); Ark. Stat. Ann. § 20-17-802 (subsequent); Cal. Health & Safety Code § 123440 (subsequent); Fla. Stat. Ann. § 390.0111(6) (prior or subsequent); Ind. Code Ann. § 16-34-2-6 (subsequent); Ky. Rev. Stat. § 436.026 (subsequent); Mo. Ann. Stat. § 188.037 (prior or subsequent); Neb. Rev. Stat. § 28-346 (subsequent); Ohio Rev. Code Ann. § 2919.14(A) (subsequent); Okla. Stat. Ann. tit. 63, § 1-735(A) (prior or subsequent); Tenn. Code Ann. § 39-15-208 (subsequent); Wyo. Stat. Ann. § 35-6-115 (subsequent).

8 720 Ill. Comp. Stat. 510/12.1.

9 See, for example, Utah Code § 76-7-310: "Live unborn children may not be used for experimentation."

10 N.H. Rev. Stat. Ann. § 168-B:15.

11 Fla. Stat. Ann. § 390.0111(6); La. Rev. Stat. Ann. § 9:121 et. seq.; Me. Rev. Stat. tit. 22 § 1593; Mass. Ann. Laws. ch. 112 § 12J; Mich. Comp. Laws. §§ 333.2685 to 2692; Minn. Stat. Ann. § 145.421 (applies only until 265 days after fertilization); N.D. Cent. Code §§ 14-02.2-01 to -02; 18 Pa. Cons. Stat. Ann. § 3216; R.I. Gen. Laws. § 11-54-1.

12 La. Rev. Stat. Ann. § 9:122.

13 La. Rev. Stat. Ann. § 9:122.

14 James A. Thomson, Joseph Itskovitz-Eldor, Sander S. Shapiro, Michelle A. Waknitz, Jennifer J. Swiergiel, Vivienne S. Marshall, and Jeffrey M. Jones, "Embryonic Stem Cell Lines Derived from Human Blastocysts" 282 *Science* 1145 (1998).

15 The Maine law, which applies both to research on embryos and research on fetuses, carries a maximum five-year prison term. Me. Rev. Stat. tit. 22 § 1593. The Massachusetts and Michigan laws also carry with them a potential prison sentence of up to five years. Mass Ann. Laws 112 § 12J(a)(V); Mich. Comp. Laws § 333.2691.

16 Michael J. Shamblott, Joyce Axelman, Shunping Wang, Elizabeth M. Bugg, John W. Littlefield, Peter J. Donovan, Paul D. Blumenthal, George R. Huggins, and John D. Gearhart, "Derivation of Pluripotent Stem Cells from Cultured Human Primordial Germ Cells" 95 *Proc. Nat'l Acad. Sci.* 13726 (1998).

17 See, for example, Ark. Stat. Ann. § 20-17-802(2); Mass. Ann. Laws ch. 112 § 12J(a)(II); R.I. Gen. Laws § 11-54-1(d); Tenn. Code Ann. § 39-15-208(a). In Pennsylvania, the consent of the mother is required, and she must not receive any compensation (whether monetary or otherwise). Her consent is only valid once the decision to abort has been made. 18 Pa. Cons. Stat. Ann. § 3216(b)(1). In Michigan, the mother's consent is required (Mich. Comp. Laws Ann. § 333.2687), but the state's version of the UAGA must also be complied with. Mich. Comp. Laws Ann. § 333.10101 et seq.

18 Ariz. Rev. Stat. Ann. § 36-2302, -2303; Ind. Code Ann. § 16.34-2-6; N.D. Cent. Code § 14-02.2-01 to -02; Ohio Rev. Code Ann. § 2919.14; Okla. Stat. Ann. tit. 63, § 1-735; S.D. Codified Laws Ann. § 34-23A-17.

19 N.D. Cent. Code § 14-02.2-01(2); S.D. Codified Laws § 34-23A-17. In North Dakota, if the embryo or fetus has died for some reason other than an induced abortion, research may be performed with the consent of the mother. N.D. Cent. Code § 14-02.2-02. In South Dakota, the unborn child or fetus (or its parts) may be used if it was removed in the course of an ectopic or molar pregnancy. S.D. Codified Laws § 34-23A-17.

20 Ind. Code Ann. § 16-34-2-6.

21 Ohio Rev. Code Ann. § 2919.14(A).

22 Okla. Stat. Ann. § 1-735.

23 See, for example, N.D. Cent. Code § 14-02.2-02(4). (This particular provision would not apply if the tissue were obtained from a stillborn fetus and the mother authorized the research.)

24 Ariz. Rev. Stat. § 36-2302(A).

25 Minn. Stat. Ann. § 145.421.

26 18 Pa. Cons. Stat. § 3216.

27 *Webster's Third New International Dictionary, Unabridged* (Miriam Webster, Inc., 1986).

28 Fla. Stat. Ann. § 390.0111(6); La. Rev. Stat. Ann. § 9:121 et seq.; Me. Rev. Stat. tit. 22 § 1593; Mass. Ann. Laws. ch. 112 § 12J; Mich. Comp. Laws. §§ 333.2685 to 2692; N.D. Cent. Code §§ 14-02.2-01 to -02; R.I. Gen. Laws. § 11-54-1.

29 Me. Rev. Stat. Ann. tit. 22 § 1593; Mich. Comp. Laws Ann. § 333.2690; N.D. Cent. Code § 14-02.2-2(4); R.I. Gen. Laws § 11-54-1(f).

30 Me. Rev. Stat. Ann. tit. 22 § 1593. Michigan has a maximum five-year prison term. Mich. Comp. Laws Ann. § 333.2691.

31 Cal. Health & Safety Code § 123440.

32 Cal. Health & Safety Code § 123445.

33 Cal. Health & Safety Code § 1635.1.

34 Hawaii Rev. Stat. § 841-14.

35 735 F. Supp. 1361 (N.D. Ill. 1990).

36 Id. at 1364. The court also held that the statute violated couples' right to privacy to make reproductive decisions to undertake preimplantation genetic screening or procreate with a donated embryo. Id. at 1377. With embryo stem cell research, the progenitors' reproductive freedom is not an issue, however.

37 Id. at 1364–65.

38 Id. at 1364.

39 *Margaret S. v. Edwards*, 794 F.2d 994, 999 (5th Cir. 1986).

40 Id.

41 Id. A concurring judge found this analysis to be contrived and opined that the provision was not unconstitutionally vague. Id. at 1000 (Williams, J., concurring). Instead, he suggested that the prohibition was unconstitutional because "under the guise of police regulation the state has actually undertaken to discourage constitutionally privileged induced abortions." Id. at 1002, citing *Thornburgh v. American College of Obstetricians and Gynecologists*, 106 S. Ct. 2169, 2178 (1986). The concurring judge pointed out that the state had "failed to establish that tissue derived from an induced abortion presents a greater threat to public health or other public concerns than the tissue of human corpses [upon which experimentation is allowed]." Id. Moreover, the state had not shown a rational justification for prohibiting experimentation on fetal tissue from an induced abortion, rather than a spontaneous one. Id.

42 *Margaret S. v. Edwards*, 794 F.2d 994, 999 (5th Cir. 1986).

43 Id.

44 Utah Code Ann. § 76-7.3-310.

45 *Jane L. v. Bangerter*, 61 F.3d 1493, 1501 (10th Cir.), rev'd on other grounds sub nom., *Leavitt v. Jane L.*, 518 U.S. 137 (1996).

46 Gena Corea, *The Mother Machine* 102, 135 n.2 (1984).

47 Stephen Toulmin, "Fetal Experimentation: Moral Issues and Institutional Controls," in National Commission for the Protection on Human Subjects of Biomedical and Behavioral Research, Appendix: Research on the Fetus 10–11 (1975).

48 It has been charged that some abortionists performed abortion procedures that were more risky to the pregnant woman in order to obtain a live fetus for research purposes. H. Schulman, "Editorial: Major Surgery for Abortion and Sterilization" 40 *Obstet. Gynecol.* 738, 739 (1972).

49 See, for example, P. Ramsey, *The Ethics of Fetal Research* 88–99 (1975).

50 National Commission for the Protection of Human Subjects of Biomedical and Behavioral Research, *Report and Recommendations: Research on the Fetus* (1975) (hereinafter, *Research on the Fetus*). This document is reprinted in 40 Fed. Reg. 33,530 (1975). See also Alexander Capron, "The Law Relating to Experimentation with the Fetus," 13-1 in Appendix: Research on the Fetus, supra note 47. "Such attempts to take away parental custody and control on the grounds that the mother has abandoned the fetus or is unable to take account of its interests seem unwise (because of the burden placed on state officials which they are ill-equipped to handle), misguided (because it is based on misapprehension of the significance of the decision to abort), unnecessary (because the interests of such fetuses are already protected by the law from parental abuse to the same degree as those of other children), and perhaps unconstitutional (because it chills exercise of the right to have an abortion and operates arbitrarily through presumptions rather than actual facts about parental choices)."

51 See, for example, the American Fertility Society guidelines, which provide that the couple should have primary decisionmaking authority with respect to the fate of their embryos. Ethics Committee of the American Fertility Society, "Ethical Considerations of the New Reproductive Technologies" 53 *Fertility and Sterility* 60S (Supplement 2, June 1990).

52 *Moore v. Regents of the University of California* 793 P.2d 479 (Cal. 1990).

53 Lori Andrews, *The Clone Age: Adventures in the New World of Reproductive Technology* (New York: Henry Holt, 1999).

54 Statement of Richard Marrs, M.D., at Whittier College of Law Conference on Reproductive Technologies, April 23, 1999.

55 N.M. Stat. Ann. § 24-9A-2(b).

56 18 Pa. Cons. Stat. Ann. § 3216(b)(4).

57 42 U.S. Code Ann. § 289g-1(b)(1)(B).

58 18 Pa. Cons. Stat. Ann. § 3216(b)(5).

59 Me. Rev. Stat. Ann. tit. 22 § 1593; Mass. Ann. Laws. ch. 112 § 12(J)(a)(IV); Mich. Comp. Laws § 333.2609; N.D. Cent. Code § 14-02.2-02(4); and R.I. Gen. Laws § 11-54-1(f).

60 Fla. Stat. Ann. § 873.05; Georgia Code Ann. § 16-12-160(a) (except for health services education, id. at (b)(5)); 755 Ill. Comp. Stat. 50/8.1; La. Rev. Stat. Ann. § 9:122; Minn. Stat. Ann. § 145.422(3) (live); 18 Pa. Cons. Stat. Ann. § 3216(b)(3) (forbids payment for the procurement of fetal tissue or organs); Texas Penal Code § 48.02; and Utah Code Ann. § 76-7-311.

61 Minn. Stat. Ann. § 145.422(3).

62 Nev. Rev. Stat. Ann. § 451.015.

63 18 Pa. Cons. Stat. Ann. § 3216(b)(3).

64 See, for example, Ohio Rev. Code Ann. § 2919.14.

65 See, for example, Okla Stat. Ann. § 1-735.

66 See, for example, N.D. Cent. Code § 14-02.2-01(2); Mo. Stat. Ann. § 188.036(5).

67 See, for example, Tenn. Code Ann. § 39-15-208 (similarly prohibits the sale of an aborted fetus); Utah Code Ann. § 76-7-311.

68 Ark. Stat. Ann. § 20-17-802(c).

69 Ark. Stat. Ann. § 20-17-802(d).

70 Tenn. Code Ann. § 39-15-208.

71 Ind. Stat. § 35-46-5-1 (applies to both aborted and stillborn fetuses); Ohio Rev. Code Ann. § 2919.14(A); Okla. Stat. Ann. § 1-735(A).

72 Ga. Code Ann. § 16-12-160(a).

73 Ga. Code Ann. § 16-12-160(b)(5).

74 Ga. Code Ann. § 16-12-160(b)(6).

75 D.C. Code § 6-2601. (This law does *not* cover hair or blood.)

76 Va. Code § 32.1-289.1. (This does not apply to hair, ova, blood, and other self-replicating body fluids.)

77 R.I. Gen. Laws § 11-54-1(f).

78 Ark. Code Ann. § 20-17-610(a); Cal. Health & S § 7155(a); Conn. Gen. Stat.§ 19a-280a; Haw. Rev. Stat. § 327-10(a); Idaho Code § 39-3411(1); 755 Ill. Comp. Stat. 50/8.1(a); Iowa Code § 142C.10(1); Minn. Stat. § 525.9219(a); Nev. Rev. Stat. § 451.590(1); N.H. Rev. Stat. Ann. § 291-A:11(I); N.M. Stat. Ann. § 24-6A-10(A); N.Y. Pub Health § 4307; N.D. Cent. Code § 23-06.2-10(1); Ohio Rev. Code Ann. § 2108.12(A); R.I. Gen. Laws § 23-18.6-10(a); S.D. Codified Laws § 34-26-55; Utah Code Ann. § 26-28-10(1); Vt. Stat. Ann. Health 18 § 5246(a); Va. Code Ann. § 32,1-289.1; Wash. Rev. Code § 68.50.610(1); W. Va. Code § 16-19-7a. Arizona's laws apparently would not apply since they define a decedent to include a stillborn infant, but not a fetus. Ariz. Rev. Stat. § 36-849(1). Kentucky, though prohibiting the sale of "transplantable organs" (Ky. Rev. Stat. Ann. §311.171(1)), excludes "fetal parts or other tissues, hair, bones, blood, arteries, any products of the birth or conception...bodily fluids including sperm, ovum, ovaries, fetus [and] placenta" from its definition of "transplantable organ" Ky. Rev. Stat. Ann. § 311.165(5)(b).

79 N.Y. Pub Health § 4307; W. Va. Code § 16-19-7a.

80 Ariz. Rev. Stat. § 36-849(2); Ark. Code Ann. § 20-17-610(b); Cal. Health & S § 7155(b); Haw. Rev. Stat. § 327-10(b); Idaho Code § 39-3411(2); Iowa Code § 142C.10(2); Minn. Stat. § 525.9219(b); Nev. Rev. Stat. § 451.90(2); N.H. Rev. Stat. Ann. § 291-A:11(II); N.M. Stat. Ann. § 24-6A-10(B); N.D. Cent. Code § 23-06.2-10(2); Ohio Rev. Code Ann. § 2108.12(B); R.I. Gen. Laws § 23-18.6-10(b); Utah Code Ann. § 26-28-10(2); Vt. Stat. Ann. Health 18 § 5246(b); Wash. Rev. Code § 68.50.610(2).

81 Mass. Ann. Laws ch. 112 §12J(a)(III). The Massachusetts law specifically uses the term fetus to apply to embryos as well. Id. at 12J(a)(IV).

82 Mich. Comp. Laws Ann. § 333.2689; N.D. Cent. Code § 14-02.2-02(3); N. Mex. Stat. § 24-9A-5(B); 18 Pa. Cons. Stat. Ann. § 3216; R.I. Gen. Laws § 11-54-1(e).

Appendix A: Bans Under The Embryo and Fetal Research Laws and Abortion Laws

State	IVF Embryo Research Banned	Research on Dead Aborted Conceptus Banned	Research on Dead Aborted Conceptus Okay with Woman's Permission	Bans Payment for IVF Embryos for Research	Bans Payment for IVF Embryos for Any Purpose	Bans Payment for Aborted Conceptus for Any Purpose
Alabama						
Alaska						
Arizona		●				
Arkansas			●			●
California						
Colorado						
Connecticut						
Delaware						
District of Columbia						
Florida	●				●	
Georgia					●	●
Hawaii						
Idaho						
Illinois					●	
Indiana		●				●
Iowa						
Kansas						
Kentucky						
Louisiana	●				●	
Maine	●			●		
Maryland						
Massachusetts	●		●	●		
Michigan	●		●	●		
Minnesota	●				●	
Mississippi						
Missouri						●
Montana						
Nebraska						
Nevada						●
New Hampshire						
New Jersey						
New Mexico						
New York						
North Carolina						
North Dakota	●	●		●		●
Ohio		●				●
Oklahoma		●				●
Oregon						
Pennsylvania	●		●		●	
Rhode Island	●		●	●		
South Carolina						
South Dakota		●				
Tennessee			●			●
Texas					●	
Utah					●	●
Vermont						
Virginia						
Washington						
West Virginia						
Wisconsin						
Wyoming						

Appendix B: Bans On Payment Under State Uniform Anatomical Gift Acts

	"Decedent" includes stillborn infant or fetus	"Decedent" includes stillborn infant	"Decedent" defined other	"Part" not defined	"Part" defined as: organs, tissues, eyes, bones, arteries, blood or other fluid and other portions of human body	Sale or purchase prohibited: "a person shall not knowingly, for valuable consideration purchase or sell a part for transplantation or therapy if removal of the part is intended to occur after the death of the decedent"	"Valuable consideration does not include reasonable payment for the removal, processing, disposal, preservation, quality control, storage, transplantation or implantation of a part"	Other Provisions
Alabama	§ 22-19-41 (2)				§ 22-19-41(5)			
Alaska	§ 13.50.070(2)				§ 13.50.070(6)			
Arizona		§ 36-841(2)			§ 36-841(8)	§ 36-849(1)	§ 36-849(2)	
Arkansas	§ 20-17-601(2)				§ 20-17-601(7)	§ 20-17-610(a)	§ 20-17-610(b)	
California	§ 7150.1(b)				§ 7150.1(g) (+ pacemaker)	§ 7155(a) (or reconditioning)	§ 7155(b)	
Colorado	§ 12-34-102(2)				§ 12-34-102(5)			
Connecticut	§ 19a-279a(2)							§ 19a-280a *Prohibition against transfer of organs for valuable consideration* "Organ" defined as "...eye, bone, skin, fetal tissue or any other human organ or tissue"
Delaware	§ 2710(4)				§ 2710(10)			
District of Columbia	§ 2-1501(a)(2)				§ 2-1501(a)(5)			
Florida	(not defined)			§ 732.911				
Georgia	§ 44-5-142(3)				§ 44-5-142(10) (+ pacemaker)			
Hawaii	§ 327-1				§ 327-1	§ 327-10(a)	§ 327-10(b)	
Idaho	§ 39-3401(2)				§ 30-3401(8)	§ 39-3411(1)	§ 39-3411(2)	
Illinois	755 ILCS 50/2(c)				§ 755 ILCS 50/2(f)			755 ILCS 50/8.1(a) *Prohibition of payment for gift* "any person who knowingly pays or offers to pay...[for] an anatomical gift"

Appendix B: Bans On Payment Under State Uniform Anatomical Gift Acts *continued*

	"Decedent" includes stillborn infant or fetus	"Decedent" includes stillborn infant	"Decedent" defined other	"Part" not defined	"Part" defined as: organs, tissues, eyes, bones, arteries, blood or other fluid and other portions of human body	Sale or purchase prohibited: "a person shall not knowingly, for valuable consideration purchase or sell a part for transplantation or therapy if removal of the part is intended to occur after the death of the decedent"	"Valuable consideration does not include reasonable payment for the removal, processing, disposal, preservation, quality control, storage, transplantation or implantation of a part"	Other Provisions
Indiana	§ 29-2-16-1(b)				§ 29-2-16-1(e)			§ 29-2-16-116 *Prohibition of payment for gift* "a person who knowingly or intentionally purchases or sells a part for transplantation or therapy, if removal of the body part is to occur after the death of an individual..."
Iowa	§ 142C.2(3)				§ 142C.2 (10) (+ blood products)	§ 142C.10(1)	§ 142.C10(2)	
Kansas	§ 65-3209(b)				§ 65-3209(e)			
Kentucky	§ 311.165(2)							§ 311.156(5)(b) "transplantable organ" does not include fetal part or other tissues, hair, bones, blood arteries, any of the products of birth or conception, teeth, skin, bodily fluids including spinal fluid, plasma, sperm, ovum, ovaries, fetus, placenta... § 311.171(1) (1) no person shall sell or make a charge for any transplantable organ (2) no person shall offer remuneration for a transplantable organ (3) no person shall broker...

Louisiana	§ 2351(2)	§ 2351(4)			
Maine	§ 2901(2)	§ 2901(5)			
Maryland	(not defined)	§ 4-501(b)			
Massachusetts	§ 7	113 § 7			
Michigan	§ 333.10101(b)	§ 333.10101(f)			
Minnesota	See footnote[1]	§ 525.921(Subd. 6)	§ 525.9219(a)	§ 525.9219(b)	
Mississippi	(not defined)	§ 41-39-33(b)			
Missouri	§ 194.210(2)	§ 194.210			
Montana	§ 72-17-102(2)	§ 72-17-102(9)			
Nebraska	§ 71-4801(2)	§ 71-4801(5)			
Nevada	§ 451.520	§ 451.535	§ 451.590(1)	§ 451.590(2)	
New Hampshire	§ 291-A:2 (II)	§ 291-A:2 (VIII)	See footnote[2]	§ 291-A:11(II)	
New Jersey	§ 26:6-57(b)	§ 26:6-57			
New Mexico	§ 24-6A-1(B)	§ 24-6A-1(G)	§ 24-6A-10(A)	§ 24-6A-10(B)	
New York	§ 4300(2)	§ 4300			§ 4307 Prohibition of sales and purchases of human organs: "It shall be unlawful for any person to knowingly acquire, receive or otherwise transfer for valuable consideration any human organ for use in human transplantation. The term human organ means the human kidney, liver, heart, lung, bone marrow, and any other human organ or tissue as may be designated by the commissioner but shall exclude blood."
North Carolina	§ 103A-403(2)	§ 130A-403			§130A-410 Use of tissue declared a service
North Dakota	§ 23-06.2-2(2)	§ 23-06.2-01(7)	§ 23-06.2-10(1)	§ 23-06.2-10(2)	

[1] Minnesota § 525.921 defines "decedent" as "a deceased individual and includes a stillborn infant or an embryo or fetus that has died of natural causes *in utero*."

[2] § 291-A:11(I) does not qualify the limitation of the purchase or sale for transplantation or therapy. Instead it says "a person shall not knowingly, for valuable consideration, purchase or sell a part, if removal of the part occurs or is intended to occur after the death of the decedent."

Appendix B: Bans On Payment Under State Uniform Anatomical Gift Acts *continued*

	"Decedent" includes stillborn infant or fetus	"Decedent" includes stillborn infant	"Decedent" defined other	"Part" not defined	"Part" defined as: organs, tissues, eyes, bones, arteries, blood or other fluid and other portions of human body	Sale or purchase prohibited: "a person shall not knowingly, for valuable consideration purchase or sell a part for transplantation or therapy if removal of the part is intended to occur after the death of the decedent"	"Valuable consideration does not include reasonable payment for the removal, processing, disposal, preservation, quality control, storage, transplantation or implantation of a part"	Other Provisions
Ohio	§ 2108.01(B)					§ 2108.12(A)	§ 2108.12(B)	§ 2108.01 *Definitions* (F) "Part" means any portion of a human body (G) "Tissue" means any body part other than an organ or eye § 2108.12 *Sale of Human Body Parts Prohibited* (A) No person, for valuable consideration shall knowingly acquire, receive or otherwise transfer a human organ, tissue, or eye for transplantation. § 2108.11 Transactions involving human fluids or body parts are not a sale
Oklahoma	§ 2202(2)				§ 2202(5)			
Oregon	§ 97.950(3)				§ 97.950(9)			
Pennsylvania	§ 8601				§ 8601			
Rhode Island	§ 23-18.6-1				§ 23-18.6-1	§ 23-18.6-10(a)	§ 23-18.6-10(b)	
South Carolina	§ 44-43-320(b)				§ 44-43-320(e)			

South Dakota	§ 34-26-20(2)	§ 34-26-20(5)			§ 34-26-42 *Human organ defined* Human organ means the human kidney... or any other human organ nonrenewable or nonregenerative tissue except plasma and sperm § 34-26-44 *Sale or purchase of human organs prohibited*
Tennessee	§ 68-30-102(2)	§ 68-30-102			
Texas	§ 692.002(2)	§ 692.002(7)			
Utah	§ 26-28-2(2)	§ 26-28-2(7)	§ 26-28-10(1)³	§ 26-28-10(2)	
Vermont	§ 5238(2)	§ 18 VSA § 5238	§ 18 VSA § 5246(a)	§ 18 VSA § 5246(b)	
Virginia	§ 32.1-289	§ 32.1-289			§ 32.1-289.1 *Sale of body parts prohibited* "With the exception of hair, ova, blood and other self-replicating body fluids, it shall be unlawful for any person to sell, to offer to sell, to buy, or to procure through purchase any natural body part for any reason including but not limited to medical and scientific uses such as transplantation, implantation, infusion, or injection."
		"Decedent means deceased individual" § 68.50.530(2)			
Washington		68.50.530	§ 68.50.610(1)	§ 68.50.610(2)	

³ § 26-28-10 says: "a person may not knowingly, for valuable consideration, purchase or sell a part of his or another's body if the removal of the part is intended to occur after the death of the body from which the part would be removed." There is no qualification for transplant or therapy.

Appendix B: Bans On Payment Under State Uniform Anatomical Gift Acts *continued*

	"Decedent" includes stillborn infant or fetus	"Decedent" includes stillborn infant	"Decedent" defined other	"Part" not defined	"Part" defined as: organs, tissues, eyes, bones, arteries, blood or other fluid and other portions of human body	Sale or purchase prohibited: "a person shall not knowingly, for valuable consideration purchase or sell a part for transplantation or therapy if removal of the part is intended to occur after the death of the decedent"	"Valuable consideration does not include reasonable payment for the removal, processing, disposal, preservation, quality control, storage, transplantation or implantation of a part"	Other Provisions
West Virginia	§ 16-19-1(d)				§ 16-19-1(g)		§ 16-19-7a	*Prohibition of sales and purchases of human organs* "Human organ means the human kidney...and any other human organ or tissue as may be designated by the director of health but shall exclude blood."
Wisconsin			"Decedent means deceased individual" § 157.06(b)		§ 157.06(g)			
Wyoming	§ 35-5-101(iii)				§ 35-5-101			

THE FOOD AND DRUG ADMINISTRATION'S STATUTORY AND REGULATORY AUTHORITY TO REGULATE HUMAN PLURIPOTENT STEM CELLS

Commissioned Paper
Robert P. Brady, Molly S. Newberry, Vicki W. Girard
Hogan & Hartson L.L.P.

I. Executive Summary

The Public Health Service Act (PHS Act), 42 USC § 262 and 264, the Federal Food, Drug, and Cosmetic Act (FD&C Act), 21 USC § 201 et seq., and the Food and Drug Administration's (FDA's) implementing regulations thereof provide the agency with broad authority to regulate both the research into and the use of human pluripotent stem cells (stem cells) intended to be used as biological products, drugs or medical devices to prevent, treat, cure, or diagnose a disease or condition.[1] Scientific research not intended to be used to develop any FDA-regulated product is not under the oversight and control of FDA.

As described in detail below, FDA has utilized its existing statutory authority to develop a regulatory framework for cellular and tissue materials that has evolved over time as the development and use of such biological materials for therapeutic purposes has increased. This paper briefly reviews these statutory provisions and FDA's evolving regulatory framework.

II. Background on the Science of Human Pluripotent Stem Cells

After an egg is fertilized, it forms a single cell that has the potential to develop into a human being (National Institutes of Health, "Pluripotent Stem Cells: A Primer," January 15, 1999, at 1–2 [NIH Primer]). Because it can develop into an entire human being, this cell is called a "totipotent" cell. This cell then divides into two identical totipotent cells. After several days, the totipotent cell forms a blastocyst, which consists of an outer layer of cells and an inner cell mass. The "inner cell mass cells can form virtually every type of cell found in the human body" except the placenta and supporting tissues (id. at 2). Because these cells can develop into most but not all cells they are called "pluripotent" cells (id.). Pluripotent cells go on to specialize into "stem cells," which give rise to cells that have a particular function, such as blood stem cells.

Pluripotent stem cells have been developed in two different ways. First, they have been "isolated from the inner cell mass at the blastocyst stage" (id. at 3). "These cells were grown in culture and found to divide indefinitely and have the ability to form cells of the three major tissue types—endoderm (which goes on to form the lining of the gut), mesoderm (which gives rise to muscle, bone and blood) and ectoderm (which gives rise to epidermal tissues and the nervous system)" (Statement of Harold Varmus, M.D., Director, National Institutes of Health, before the Senate Appropriations Subcommittee on Labor, Health and Human Services, Education and Related Agencies, December 2, 1998, at 1). Second, they have been isolated from fetal tissue. It is also thought that "somatic cell nuclear transfer (SCNT) may be another way that pluripotent stem cells could be isolated" (id. at 4). In SCNT, the nucleus from a somatic cell is extracted and transferred to a reproductive cell (from a different person) whose own nucleus has been removed or inactivated. Insertion of the donor cell into the recipient cell may be accomplished directly by injection or by placing the donor nucleus and recipient cell side by side and applying a small burst of electricity to induce fusion of the two. The electrical burst also initiates cell division of the fused cell, which will result in the formation of a blastocyst, from which stem cells may be isolated.

The medical potential for human pluripotent stem cells is unknown at this time, but is thought to be extraordinary. "It is not too unrealistic to say that this research has the potential to revolutionize the practice of medicine and improve the quality and length of life" (NIH Primer at 7). In his recent congressional testimony, Dr. Varmus described three potential applications of pluripotent stem cells, two of which are not regulated by FDA and one of which will be regulated by FDA. First, stem cell research could include basic research such as "the identification of the factors involved in the cellular decision-making process that determines cell specialization" (Statement of H. Varmus at 3). Second, "[h]uman pluripotent stem cell research could also dramatically change the way we develop drugs and test them for safety and efficacy. Rather than evaluating safety and efficacy of a candidate drug in an animal model of a human disease, these drugs could be tested against a human cell line that had been developed to mimic the disease process" (id.). Neither of these potential applications likely would be directly regulated by FDA.

> Perhaps the most far-reaching potential application of human pluripotent stem cells is the generation of cells and tissue that could be used for transplantation, so-called cell therapies. Pluripotent stem cells stimulated to develop into specialized cells offer the possibility of a renewable source of replacement cells and tissue to treat a myriad of diseases, conditions and disabilities including Parkinson's and Alzheimer's disease, spinal cord injury, stroke, burn, heart disease, diabetes, osteoarthritis and rheumatoid arthritis (id. at 3–4).

These stem cell products, based on their intended use, would be the subject of FDA's regulation as set forth below.

III. FDA's Statutory and Regulatory Authority to Regulate Human Pluripotent Stem Cells

A. The Cornerstones of FDA Jurisdiction: Product Definition and Interstate Nexus

In order for FDA to assert regulatory authority over stem cell related research and products, they must fall within one of the product categories over which FDA exercises jurisdiction *and* must move in interstate commerce.

To the extent FDA determines that a particular product falls within the definition of a biological product, a drug, or a medical device, jurisdiction will be asserted. Whether a particular product falls within the definition for any of the FDA-regulated product categories will turn, in part, on the intended use of the product. The manufacturer's objective intent, as evidenced by labeling, promotional, and other relevant materials for the product, has long been regarded as the primary source for establishing a product's intended use and thus its status for purposes of FDA regulation (See *United States v. An Article...Sudden Change*, 409 F.2d 734, 739 (2d Cir. 1969)). While that approach would seem to grant manufacturers unbridled control over the regulatory status of their products, in fact, courts have recognized FDA's right to look beyond the express claims of manufacturers to consider more subjective indicia of intent, such as the foreseeable and actual use of a product, to prove that its intended use subjects it to agency jurisdiction. (See *National Nutritional Foods Ass'n. v. Mathews*, 557 F.2d. 325, 334 (2d Cir. 1977); *Action on Smoking and Health v. Harris*, 655 F.2d 236, 240–41 (D.C. Cir. 1980).)

Regardless of whether FDA or the manufacturer is characterizing the intended use of a product for purposes of evaluating FDA jurisdiction, it is clear that FDA regulatory authority will not automatically extend to all scientific research on stem cells. Indeed, to the extent such nonhuman research is preliminary in nature and/or is undertaken without an intent to develop a therapeutic product, stem cell research is not subject to FDA jurisdiction. Thus, for example, the basic research to develop stem cell models to evaluate the safety and efficacy of therapeutic products would not be directly regulated. In contrast, any scientific data generated from such a model and submitted to FDA as part of a marketing application would be reviewed by FDA. It is only when the science regarding stem cell use has progressed to the point that development of a particular therapeutic product and its use in humans is envisioned that FDA regulatory authority applies, and further research must be conducted in compliance with FDA's requirements.

Even if a product falls within one of the defined categories over which FDA asserts jurisdiction, no statutory authority over the product exists unless it moves in interstate commerce. FDA takes an expansive view of what constitutes interstate commerce in order to assure that its regulatory controls reach as many products and related research as possible. In regard to biological products FDA has been particularly aggressive. For example, in its 1993 policy statement regarding somatic cell therapy products, FDA concluded that

> The interstate commerce nexus needed to require premarket approval under the statutory provisions governing biological products and drugs may be created in various ways in addition to shipment of the finished product by the manufacturer. For example, even if a biological

drug product is manufactured entirely with materials that have not crossed State lines, transport of the product into another State by an individual patient creates the interstate commerce nexus. If a component used in the manufacture of the product moves interstate, the interstate commerce prerequisite for the prohibition against drug misbranding is also satisfied even when the finished product stays within the State. Products that do not carry labeling approved in a PLA (or NDA) are misbranded under section 502(f)(1) of the [FD&C] Act....Moreover, falsely labeling a biological product is prohibited under section 351(b) of the PHS Act without regard to any interstate commerce nexus (42 USC 262(b)) (58 *Fed. Reg.* at 53250).

In all likelihood, FDA would apply the same logic to all cellular and tissue materials that are used in the prevention, treatment, cure, or diagnosis of a disease or condition of human beings.

B. FDA Has Jurisdiction to Regulate Stem Cells Under Section 351 of the PHS Act

Under section 351 of the PHS Act, FDA is authorized to regulate biological products introduced into interstate commerce (42 USC § 262(a)). The PHS Act defines a "biological product" to mean "a virus, therapeutic serum, toxin, antitoxin, vaccine, *blood, blood component or derivative, allergenic product, or analogous product*, or arsphenamine or derivative of arsphenamine (or any other trivalent organic arsenic compound), *applicable to the prevention, treatment, or cure of a disease or condition of human beings*" (PHS Act § 351(i), 42 USC § 262(i) [emphasis added]). This definition includes stem cell products, which are considered by FDA to be analogous to blood or blood components or derivatives if they are used for the prevention, treatment, or cure of a disease or condition of human beings.

Cellular products that currently are regulated by FDA as biological products include: 1) autologous or allogeneic lymphocytes activated and expanded *ex vivo* (e.g., lymphokine-activated killer cells, tumor infiltrating lymphocytes, antigen specific clones); 2) encapsulated autologous, allogeneic, or xenogeneic cells or cultured cell lines intended to secrete a bioactive factor or factors (e.g., insulin, growth hormone, a neurotransmitter); 3) autologous or allogeneic somatic cells (e.g., hepatocytes, myocytes, fibro-blasts, lymphocytes) that have been genetically modified; 4) cultured cell lines; and 5) autologous or allogeneic bone marrow transplants using expanded or activated bone marrow cells when such products are used for the prevention, treatment, or cure of a disease or condition of human beings (58 *Fed. Reg.* 53248, 53250, Oct. 14, 1993, "Application of Current Statutory Authorities to Human Somatic Cell Therapy Products and Gene Therapy Products" [the "Somatic Cell Therapy Policy"]). In addition, peripheral and umbilical cord blood stem cells that have been more-than-minimally processed and are intended to prevent, treat, or cure disease also are regulated as biological products (FDA, "A Proposed Approach to the Regulation of Cellular and Tissue-Based Products," February 28, 1997).

Biologics license applications (BLAs) (historically referred to as Establishment License Applications [ELAs]) and Product License Applications (PLAs) are issued by FDA upon a showing that the establishment and product meet "standards, designed to insure the continued safety, purity, and potency of such products..." (Id. at § 351(d); 42 USC § 262(d)). These standards were first adopted in the law in 1902 (the Biologics Control Act of 1902, Chap. 1378, 32 Stat. 738 (1902)). In the early 1970s, FDA incorporated by regulation the requirement of efficacy into the approval standards for biological products (21 CFR § 601.25). Data to support licensure of a biological product usually must be developed through nonclinical and clinical research.

While a biological product is under clinical investigation, it must meet FDA's investigational new drug (IND) requirements set forth in 21 CFR Part 312 (21 CFR § 601.2). FDA defines the universe of clinical research subject to the agency's jurisdiction as "...all clinical investigations of products that are subject to section 505...of the Federal Food, Drug, and Cosmetic Act or to the licensing provisions of the Public Health Service Act..." (21 CFR § 312.2). FDA regulates "pre-clinical research" which it defines as "...non-clinical laboratory studies that support or are intended to support applications for research or marketing permits for products regulated by FDA, including...human and animal drugs...(and) biological products..." (21 CFR § 58.1). FDA regulates this area of research through the enforcement of Good Laboratory Practices on persons and entities carrying out such nonhuman research (21 CFR Part 58).

FDA's IND regulations require that prior to conducting clinical trials a company submit an IND application to FDA, setting forth its protocols for the study and the scientific basis for believing the product would be safe and effective for particular use(s) in humans. The study may begin within 30 days following submission of an IND application, unless FDA advises otherwise or requests additional time to review the application (21 CFR § 312.20). Clinical trials generally are conducted in three phases. Once trials have commenced, FDA may stop the trials by placing a "clinical hold" on them because of concerns about, for example, the safety of the product being tested (21 CFR § 312.42). In addition, all clinical studies must be approved and conducted under the supervision of the Institutional Review Board (IRB) responsible for the study (21 CFR Part 56). Lastly, all patients involved in such clinical research must be provided with informed consent in full compliance with FDA requirements (21 CFR Part 50).

The key elements of the PHS Act framework have been largely unchanged since its original enactment in 1902. As described in greater detail below, this framework has been able to remain in place because FDA has always retained the flexibility to address regulatory issues created by new technologies. Keeping up with continuing advances in the field of modern biotechnology, FDA issues regulations, guidance documents, or policy statements to describe whether and how its current statutory and regulatory authority applies to a new technology. As new technology and therapeutic products develop, FDA has carefully exercised its inherent discretion of how to apply the law to ensure that science can advance while the public health is protected. As will be discussed in detail below, FDA currently is exercising its authority over stem cells under section 351 of the PHS Act (see infra, Section V).

C. FDA Has the Statutory Authority to Regulate Stem Cells Under Section 361 of the PHS Act

In addition to having authority under section 351 of the PHS Act to regulate stem cell products meeting the applicable statutory definition, FDA also has authority to regulate stem cell products under section 361 of the PHS Act. Section 361 authorizes the Department of Health and Human Services (HHS) to "make and enforce such regulations as in [its] judgment are necessary to prevent the introduction, transmission, or spread of communicable diseases from foreign countries into the States or possessions, or from one State or possession into any other State or possession" (42 USC § 264). This provision provides the agency with broad discretion to enact regulations necessary to prevent the spread of communicable diseases.

Section 361 serves, in part, as the basis on which FDA currently regulates human tissue intended for transplantation (i.e., minimally manipulated tissue such as corneal tissue, bones, skin, or tendons) (21 CFR Parts 16 and 1270. See 62 Fed. Reg. 40429, July 29, 1997, "Human Tissue Intended for Transplantation"). Section 361 also has served as the basis under which FDA has regulated the source and use of potable water, milk pasteurization, and the transmission of communicable disease through shellfish, turtles, certain birds, and bristle brushes (id. at 40431. See *State of Louisiana v. Mathews*, 427 F. Supp. 174 (E.D. La. 1977) [FDA regulation issued pursuant to section 361 of PHS Act banning the sale and distribution of small turtles was permissible as necessary to prevent spread of communicable disease]). Section 361 also serves as part of the statutory basis on which FDA has imposed requirements to protect the nation's blood supply (id.).

As evidenced by this discussion, section 361 of the PHS Act provides FDA with broad authority to enact regulations necessary to protect the public health by preventing the spread of communicable disease. However, while FDA has utilized this provision to ban certain products in interstate commerce and to establish infectious disease testing and related processing standards for tissue, it has not been used by FDA to adopt premarket approval requirements or otherwise regulate biomedical research. Thus, FDA does regulate cellular products, in part, under section 361 of the PHS Act because the transfer of such cellular components could convey communicable diseases such as AIDS, hepatitis, and herpes simplex. Indeed, the agency currently uses this statutory authority in conjunction with its other premarket approval authorities to provide a comprehensive regulatory structure for cellular and tissue products, including stem cells (see id.).

D. FDA Has the Statutory Authority to Regulate Stem Cell Products as Drugs Under Section 505 of the FD&C Act

In addition to having authority to regulate stem cell products as biological products under the PHS Act, FDA also has concluded that it has the authority under section 505 of the FD&C Act (21 USC § 355) to regulate as a drug any stem cell product that meets the applicable statutory definition. The FD&C Act defines drugs as "articles intended for use in the diagnosis, cure, mitigation, treatment, or prevention of disease in man or other animals" and "articles (other than food) intended to affect the structure or any function of the body of man or other animals" (Section 201(g) of the FD&C Act, 21 USC § 321(g)). The vast majority of "new drugs" regulated under the FD&C Act are various dosage forms of synthetic chemicals or plant derivatives. In contrast, the majority of biological products licensed under the PHS Act are products derived from human cellular or tissue materials. FDA exercises its discretion, based roughly on the product categories described above, in approving products either as new drugs *or* biological products. The PHS Act makes it clear that if a biological product is licensed under section 351, it shall not be required to also have approval under the FD&C Act (42 USC § 262(j)).

FDA approves new drugs for marketing, based upon proof of efficacy and safety, under section 505 of the FD&C Act (21 USC § 355). Manufacturers submit their preclinical and clinical data to establish the safety and efficacy of a new drug pursuant to a New Drug Application (NDA). During the investigational stage, investigational drugs are regulated under the same authority and in the same manner as investigational biologics (see supra, Section III. B). In order to receive marketing approval, FDA requires properly conducted, adequate, and well-controlled studies demonstrating efficacy with sufficient levels of statistical assurance to support product approval. Reports of these clinical trials as well as preclinical data must be submitted along with information pertaining to the preparation of the drug, analytical methods, drug product formulation, details on the manufacture of finished products, and proposed packaging and labeling (21 CFR § 314.50). Once a drug product is approved it is subject to continuing regulation by FDA, such as compliance with Good Manufacturing Practices (GMPs) and marketing and advertising restrictions (21 CFR §§ 202, 210, 211). In addition, FDA may require additional clinical tests following approval to confirm safety and efficacy (phase IV clinical trials).

During the pre- and post-approval periods, drugs and biological products are subject to the adulteration and misbranding provisions of the FD&C Act (21 USC §§ 351, 352). Section 501 of the FD&C Act provides in part that a product is adulterated if it is a drug that was not manufactured in conformance with GMPs or was prepared, packed, or held under unsanitary conditions (21 USC § 351(a)). A product is misbranded if, among other things, its labeling is false or misleading in any particular, or if any word, statement, or other information requested to appear on the label or labeling is not prominently placed thereon (id. at § 352). Also during the pre- and post-approval periods, drug products are subject to FDA's general prohibitions against promoting products for unapproved or "off-label" uses.

In bringing enforcement actions against biological products licensed under the PHS Act, FDA routinely utilizes various provisions of the FD&C Act drug adulteration and misbranding authorities as part of any such action. In addition, FDA utilizes some of the enforcement authorities of the FD&C Act, such as seizures or injunctions, to enforce both laws against biological products they deem violative (21 USC §§ 322; 334).

E. FDA Has the Statutory Authority to Regulate Stem Cell Products as Devices Under the FD&C Act

Section 201(h) of the FD&C Act defines a medical device, in pertinent part, as "an instrument, apparatus, implement, machine, contrivance, implant, *in vitro* reagent, or other similar or related article which is (1) intended for use in the diagnosis of disease or other conditions, or in the cure, mitigation, treatment, or prevention of disease, in man or other animals, or (2) intended to affect the structure or any function of the body of man or other animals," and which is not dependent upon being metabolized for the achievement of its primary intended purposes (21 USC § 321(h)). To the extent FDA concludes that stem cell products meet the definition of a device and operate in a manner similar to human tissue products used for transplantation (e.g., heart valve allografts and human lenticules—corrective lenses derived from human corneal tissue), they may be subject to regulation as devices.

Under section 513 of the FD&C Act, all medical devices are classified into one of three classes—Class I, Class II, or Class III (21 USC § 360(c)). A device's class determines the types of regulatory controls it is subject to and the process it goes through to receive marketing approval from FDA. Most medical devices in the United States fall within Classes I or II and are marketed pursuant to a simplified approval process set forth in section 510(k) of the FD&C Act known as "Premarket Notification" (or "510(k) clearance") (21 USC § 360(k)). A medical device that does not qualify for 510(k) clearance is placed in Class III, which is reserved for devices classified by the FDA as posing the greatest risk (e.g., life-sustaining, life-supporting, implantable, or devices presenting a potentially unreasonable risk of injury). Stem cell products, to the extent FDA considers them to be devices, would most likely be placed in Class III. A Class III device generally must undergo the premarket approval (PMA) process, prior to marketing which requires the manufacturer to prove the safety and effectiveness of the device to the FDA's satisfaction. A PMA application must provide extensive preclinical and clinical trial data and also information about the device and its components regarding, among other things, manufacturing, labeling, and promotion (21 CFR § 814.20). As in the case of drugs and biologics, the data standards applied to devices in a PMA submission require the manufacturer to demonstrate that the device is safe and effective under the conditions of use recommended in the labeling (FD&C Act § 515(d); 21 USC § 360e(d)).

A clinical study in support of a PMA application requires an Investigational Device Exemption (IDE) application approved in advance by the FDA for a limited number of patients (FD&C Act § 520(g); 21 USC § 360j(g)). The IDE application must be supported by appropriate data, such as animal and laboratory testing results (21 CFR part 812). The clinical study may begin if the IDE application is approved by the FDA and the appropriate IRB at each clinical study site.[2] In all cases, the clinical study must be conducted under the auspices of an IRB pursuant to FDA's regulatory requirements intended for the protection of subjects, including execution of informed consent, and to assure the integrity and validity of the data (21 CFR Part 56).

As with drugs and biologics, devices manufactured or distributed pursuant to FDA clearance or approval are subject to pervasive and continuing regulation by the FDA and certain state agencies. They are also subject to the same rules regarding adulteration and misbranding (see supra, Section III. D).

IV. Historical Application of Statutory Authority to Regulate Cellular and Tissue Materials

A brief historical review of FDA's application of the statutes described above to cellular and tissue materials shows that FDA has been cautious in exercising its regulatory discretion. FDA has never had a single regulatory program for human cellular and tissue-based products. Instead, it has regulated these products on a case-by-case basis responding as it determined appropriate to the particular characteristics of and concerns raised by each type of product (63 *Fed. Reg.* 26744, May 14, 1998, FDA Proposed Rule "Establishment and Listing for Manufacturers of Human Cellular and Tissue-Based Products").

One example has been FDA's approach to regulating bone marrow. While for years FDA has licensed blood and blood components pursuant to section 351 of the PHS Act (42 USC § 262), FDA has voluntarily refrained from regulating minimally manipulated bone marrow, the earliest source of stem cells used for transplantation, despite its status as a blood component. Indeed, not until the early 1990s did FDA announce that to the extent bone marrow was subject to extensive manipulation prior to transplantation it would be treated the same as somatic cell therapy and gene therapy products subject to the IND regulations and requiring PHS Act licensure (58 *Fed. Reg.* 53248, 53249 (Oct. 14, 1993)).

Also in 1993, in response to concerns about the transmission of the human immunodeficiency virus (HIV) and other infectious diseases, FDA published an emergency final rule which established certain processing, testing, and recordkeeping requirements for certain types of tissue products ("Human Tissue Intended for Transplantation" (58 *Fed. Reg.* 65514, Dec. 14, 1993)). This rule, however, did not mandate premarket approval or notification for all tissues, but rather provided, among other things, for donor screening, documentation of testing, and FDA inspection of tissue facilities.[3]

As another example of FDA's case-by-case approach, in 1996 FDA published a guidance stating that manipulated autologous structural (MAS) cells, which are autologous cells manipulated and then returned to the body for structural repair or reconstruction, are subject to PHS licensure (CBER, Guidance on Applications for Products Comprised of Living Autologous Cells Manipulated *Ex Vivo* and Intended for Structural Repair or Reconstruction, May 1996). Similarly, until very recently, FDA carefully chose not to regulate reproductive tissues. Then, as will be discussed below, in 1997 FDA proposed that in the future certain reproductive tissues, such as semen, ova, and embryos, should come under some form of regulation.

Traditional tissue products, including but not limited to bone, skin, corneas, and tendons, also have been subject to FDA's piecemeal regulatory approach. Historically, FDA regulated these products on an ad hoc basis as medical devices under section 201 of the FD&C Act (see, e.g., 63 *Fed. Reg.* 26744, citing as examples, dura mater, corneal lenticules, and umbilical cord vein grafts). However, with the advent of HIV and the potential for its transmission, FDA concluded in the early 1990s that a more comprehensive program for traditional tissues was necessary. In 1991, FDA concluded that human heart valves were medical devices subject to premarket approval requirements ("Cardiovascular Devices; Effective Date of Requirement for Premarket Approval; Replacement Heart Valve Allograft" 56 *Fed. Reg.* 29177, June 26, 1991). After a period of litigation, FDA relented somewhat and concluded that while these products remained medical devices, they would not be subject to premarket approval requirements (FDA Rescission Notice, 59 *Fed. Reg.* 52078, October 14, 1994). In defining tissue subject to this rule, FDA exempted a number of products, including vascularized organs, dura mater, allografts, and umbilical cord vein grafts (id. at 40434).

This very brief review of the regulatory landscape shows a regulatory framework that, in FDA's own words, has been "fragmented." FDA has regulated most of these products on an ad hoc basis either as medical devices or biological products or, in certain instances, chose not to regulate certain of these products at all. As a result of the agency's own reevaluation and congressional concerns and pressure, in the mid-1990s, FDA concluded that a comprehensive approach to the regulation of these products was an important step forward in public health protection.

V. Comprehensive FDA Policy to Regulate Cellular or Tissue-Based Products, Including Pluripotent Stem Cells

A. The Proposed Approach

In February 1997, FDA proposed, consistent with the existing statutory framework set forth above, a new approach to the regulation of human cellular and tissue-based products. This framework is intended to "protect the public health without imposing unnecessary government oversight" ("Reinventing the Regulation of Human Tissue," *National Performance Review*, February 1997, at 1).

The 1997 document establishes the further evolution of FDA's application of the PHS Act and FD&C Act to cellular and tissue products. While still a proposed approach, it utilizes FDA's existing statutory authority under the PHS Act and FD&C Act to regulate a broad array of cellular and tissue materials.

The framework proposes a tiered approach to the regulation of cellular and tissue-based products (FDA, "A Proposed Approach to the Regulation of Cellular and Tissue-Based Products," February 28, 1997, the "Proposed Approach"). Products that pose increased risks to health or safety would be subject to increased levels of regulation (i.e., either licensure under the PHS Act or premarket approval under the FD&C Act). For example, products that pose little risk of transmitting infectious disease would be subject to minimal regulation (i.e., facility registration and product listing). However, products that are 1) highly processed (more-than-minimally manipulated), 2) used for other than their normal purpose, 3) combined with nontissue components (i.e., devices or other therapeutic products), or 4) used for metabolic purposes (i.e., systemic, therapeutic purposes) will be required to be clinically investigated under INDs and IDEs and subject to premarket approval as biological products, medical devices, or new drugs.

The Proposed Approach addresses FDA's regulation of stem cell products. In the case of a minimally manipulated product for autologous use and allogeneic use of cord blood stem cells by a close blood relative, FDA proposed requiring compliance with standards consistent with section 361 of the PHS Act rather than an IND and licensure pursuant to section 351 of the PHS Act. However, minimally manipulated products that will be used by an unrelated party will be regulated under section 351 of the PHS Act. The agency intends to develop standards, including disease screening requirements, establishment controls, processing controls, and product standards. "If sufficient data are not available to develop processing and product standards after a specified period of time, the stem cell products would be subject to IND and marketing application requirements" (Proposed Approach at 25). Stem cell products that are more-than-minimally manipulated will require INDs and licensing under section 351 of the PHS Act. For example, stem cell products that are used for a nonhomologous function or are more-than-minimally manipulated will be required to be licensed under section 351. FDA has "increased safety and effectiveness concerns for cellular and tissue-based products that are used for non-homologous function, because there is less basis on which to predict the product's behavior" (Proposed Approach at 16).

B. FDA Implementation of the Proposed Approach

FDA has begun to implement the Proposed Approach.[4] On January 20, 1998, FDA published a "Request for Proposed Standards for Unrelated Allogeneic Peripheral and Placental/Umbilical Cord Blood Hematopoietic Stem/Progenitor Cell Products" (63 *Fed. Reg.* 2985, Jan. 20, 1998), utilizing its standards-setting authority under section 361 of the PHS Act. In this notice, the agency requests product standards to ensure the safety and effectiveness of stem cell products, which should be supported by clinical and nonclinical laboratory data. FDA also announced its intention to phase in after three years implementation of IND application and license application requirements for minimally manipulated unrelated allogeneic hematopoietic stem/progenitor cell products (id.). The notice states that "[i]f adequate information can be developed, the agency intends to issue guidance for establishment controls, processing controls, and product standards....FDA intends to propose that, in lieu of individual applications containing clinical data, licensure may be granted for products certified as meeting issued standards." If, however, FDA determines that adequate standards cannot be developed, the agency has expressed its intention to enforce IND and licensing requirements at the end of three years. Proposals are due on or before January 20, 2000.

On May 14, 1998, FDA proposed "Establishment Registration and Listing for Manufacturers of Human Cellular and Tissue-Based Products" (63 *Fed. Reg.* 26744, May 14, 1998). The agency describes the proposed registration and listing requirements as a first step towards accomplishing the agency's goal of putting in place a comprehensive new system of regulation for human cellular and tissue-based products. Registration and listing is intended to allow FDA to assess the state of the cell and tissue industry, "to accrue basic knowledge about the industry that is necessary for its effective regulation," and to facilitate communication between the agency and industry (id. at 26746). As proposed, the registration and listing requirements would apply to human cellular and tissue-based products that FDA will regulate under section 361 of the PHS Act.[5] Among the products cited by FDA as regulated under that section and consequently subject to registration and listing are bone, tendons, skin, corneas, as well as peripheral and cord blood stem cells under certain conditions and sperm, oocytes, and embryos for reproductive use (id. at 26746).

VI. FDA Has the Legal Discretion to Regulate Stem Cell Products in a Variety of Ways. Moreover, FDA's Discretion Is Entitled to Great Deference

Although, as described above, the PHS Act sets forth the basic framework for the regulation of biological products and that law is complemented by the FD&C Act, Congress gave FDA significant discretion regarding the manner in which FDA approves and regulates these products. This is entirely appropriate given the rapidly changing nature of biotechnology.

Today there is a vast array of biological products that have been approved by FDA and many others that are awaiting FDA action.[6] These products are scientifically complex and rarely lend themselves to categorization. As a result, FDA invariably is required to determine on a case-by-case basis whether its existing statutory authority applies to a new product, which particular authority to apply and, if so, what evidence will adequately demonstrate proof of safety, purity, and potency (efficacy). The decision whether and how to regulate a product is made based on FDA's expert determination and is based on the particular facts and circumstances, the historical application of the law to similar products, the applicable statutory and regulatory criteria, and the state of FDA's scientific understanding at the time of the approval.

The decision to leave this determination to FDA's discretion and expertise was a wise policy decision by Congress. The success of the approval process for cellular and tissue products is in many ways dependent upon FDA's appropriate application of discretion to respond as it sees fit to any particular product within the basic statutory and regulatory framework discussed above. Otherwise, FDA would be unable to respond to the almost daily developments in biotechnology and the complex scientific issues presented by each particular product. Absent this discretion, the cellular and tissue product approval process would grind to a halt.

When FDA exercises the significant discretion provided to the agency by Congress, FDA's exercise of this discretion is entitled to great deference (*U.S. v. Rutherford*, 442 U.S. 544, 553 (1979); *Bristol-Myers Squibb Co. v. Shalala*, 923 F. Supp. 212, 216 (D.D.C. 1996)). In a recent challenge to FDA's approval of a biological product under the PHS Act, the District Court for the District of Columbia held that "FDA's policies and its interpretation of its own regulations will be paid special deference *because of the breadth of the Congress' delegation of authority to FDA and because of FDA's scientific expertise*" (*Berlex Laboratories, Inc. v. FDA et al.*, 942 F. Supp. 19 (D.D.C. 1996) [emphasis added]. See also *Lyng v. Payne*, 476 U.S. 926 (1986)).

Moreover, even if FDA has not asserted jurisdiction previously with regard to reproductive tissue, for example, it is appropriate that FDA's policies in this area are evolutionary. The Supreme Court has recognized that expert administrative agency interpretations are not "carved in stone. On the contrary, the agency...must consider varying interpretations and the wisdom of its policy *on a continuing basis*" (*Chevron, U.S.A., Inc. v. Natural Resources Defense Council, Inc.*, 467 U.S. 837, 863-64 (1984) [emphasis added]). Furthermore, the Court has acknowledged that "regulatory agencies do not establish rules of conduct to last forever....[A]n agency must be given ample latitude to 'adapt their rules and policies to the demands of changing circumstances'" (*Motor Vehicle Mfrs. Ass'n. of the U.S. v. State Farm Mut. Auto. Ins. Co.*, 463 U.S. 29, 42 (1983) [citations omitted]).

VII. Conclusion

The extensive statutory and regulatory authority available to FDA will ensure that the agency's regulatory approach can continue to evolve to keep up with the rapidly changing world of biotechnology. Despite the patchwork quilt of regulation applied through the mid-1990s, FDA has now developed a comprehensive regulatory approach to the regulation of cellular and tissue-based therapeutic products under its jurisdiction, including pluripotent stem cells. Nonclinical and clinical stem cell research undertaken to develop a therapeutic product intended to treat human disease will continue to be regulated by FDA, while basic scientific research and other nonhuman research will remain outside the agency's purview.

Notes

1 The scope of this paper is limited to human pluripotent stem cells. FDA has a similar regulatory structure to regulate animal stem cell products used as animal drugs. 21 USC § 360b. The U.S. Department of Agriculture would regulate animal stem cell products used in animal vaccines. 21 USC § 151.

2 While it is true that if the device presents a "nonsignificant risk" to the patient a sponsor may begin the clinical study after obtaining IRB approval without the need for FDA approval, this would not likely apply to stem cell research.

3 In 1997, FDA finalized its 1993 emergency rule establishing processing, testing, and recordkeeping requirements for all tissue products. "Human Tissue Intended for Transplantation" 62 *Fed. Reg.* 40429 (July 29, 1997).

4 While FDA may choose to implement this policy through regulation, FDA also may implement it on a case-by-case basis. See infra, Section VI.

5 Consistent with the discussion supra, Section III. A, the preamble to the proposed rule states that "use of human cellular or tissue-based products solely for nonclinical scientific or educational purposes does not trigger the registration or listing requirements. Any use for implantation, transplantation, infusion, or transfer into humans is considered clinical use and would be subject to part 1271 [the registration and listing requirements]" id. at 26748.

6 Today, biological products are available or under development to treat, diagnose, or prevent virtually every serious or life-threatening disease. Available products include, but are not limited to, vaccines (manufactured both in traditional ways and through the use of biotechnology); human blood and blood-derived products; monoclonal or polyclonal immunoglobulin products; human cellular (i.e., gene therapy) products; protein, peptide, and carbohydrate products; protein products produced in animal body fluids by genetic alteration of the animal (i.e., transgenic animals); animal venoms; and allergenic products.

QUICK RESPONSE: USE OF FETAL TISSUE IN FEDERALLY FUNDED RESEARCH

Commissioned Paper
Elisa Eiseman
RAND Science and Technology Policy Institute

Preface

This Project Memorandum was prepared in response to a request from the National Bioethics Advisory Commission (NBAC) for timely information on technical issues regarding the use of human fetal tissue in federally funded research. It is not a project deliverable. This document and ones like it are designed to meet specific needs of clients for information on science and technology issues on a short-term basis.

Originally created by Congress in 1991 as the Critical Technologies Institute and renamed in 1998, the Science and Technology Policy Institute is a federally funded research and development center sponsored by the National Science Foundation and managed by RAND. The Institute's mission is to help improve public policy by conducting objective, independent research and analysis on policy issues that involve science and technology. To this end, the Institute supports the Office of Science and Technology Policy and other Executive Branch agencies, offices, and councils; helps science and technology decisionmakers understand the likely consequences of their decisions and choose among alternative policies; and helps improve understanding in both the public and private sectors of the ways in which science and technology can better serve national objectives.

Science and Technology Policy Institute research focuses on problems of science and technology policy that involve multiple agencies. In carrying out its mission, the Institute consults broadly with representatives from private industry, institutions of higher education, and other nonprofit organizations.

Introduction

As part of its report, *Ethical Issues in Human Stem Cell Research*, NBAC had questions about whether the federal government is currently funding research that uses human embryonic and fetal tissue. These questions included the following:

- Is the federal government currently funding research that uses human embryonic tissue, or material and reagents derived from this tissue? If so, what research is being funded?

- Is the federal government currently funding research that uses human fetal tissue, or material and reagents derived from this tissue? If so, what research is being funded?

- What are the sources of the human embryonic and fetal tissue that are being used in federally funded research?

This Project Memorandum attempts to answer these questions and provides background information for NBAC's report.

Definitions

Because NBAC has adopted a scheme in its draft report for defining certain terms, such as *zygote, embryo,* and *fetus,* that scheme has been adopted here. To be accurate when referring to the details of the stages of human development, the following definitions are used: 1) for the first week of development, from fertilization until implantation into the uterus, the organism is a zygote; 2) after implantation, from the beginning of the second week through the end of the eighth week of development, the organism is an embryo; and 3) from the ninth week of development until the time of birth, the organism is a fetus. Therefore, commonly used terms, such as *embryo research, embryo donation,* and *embryo transfer,* are inaccurate, since these procedures all occur with zygotes, not embryos. In addition, fertility specialists refer to the products of *in vitro* fertilization (IVF) as embryos, even though for the first week of development they are actually zygotes. Because these terms are commonly used by fertility clinics and the public, and NBAC uses these terms in this way, products of IVF will be called embryos in this document.

NBAC also defined the terms *fetal tissue* and *embryonic tissue*. Fetal tissue is defined as tissue obtained from an organism between the ninth week of development and the time of birth. Fetal tissue comes from cadaveric fetal tissue that is obtained after either spontaneous or elective abortions. Embryonic tissue is defined as tissue obtained from an organism between the second and eighth weeks of development. In addition to spontaneous and elective abortions, embryonic tissue can also be obtained from embryos created by IVF. Therefore, for embryonic tissue, it is important to differentiate between tissue obtained from embryos created by IVF and tissue obtained from abortuses.

Methods

The following sources were consulted to determine the types of research involving human embryonic and fetal tissue that the federal government is currently funding:

- The American Type Culture Collection (ATCC), www.atcc.org, was queried for cell lines derived from human embryos or fetuses. The cell lines identified were then used as search terms in MEDLINE searches.

- A MEDLINE search of the literature using PUBMED was performed to identify papers about research using human embryonic or fetal tissue. These MEDLINE searches identified publications that had authors affiliated with several agencies of the federal government, including the National Institutes of Health (NIH), the Centers for Disease Control and Prevention (CDC), the Environmental Protection Agency (EPA), and the Department of Veterans Affairs. These agencies and departments were selected only to provide examples of those agencies that might be supporting or conducting this research, not to provide a comprehensive assessment thereof.

- RAND's RaDiUS (Research and Development in the United States) database was used to identify federally funded research with human embryonic or fetal tissue. RaDiUS is a comprehensive, real-time accounting of federal research and development (R&D) activities and spending. RaDiUS allows users to see the total R&D investment by all federal agencies, to compare the level of R&D investment in specific areas of science and technology across all federal agencies, or to examine the details of research investments within a specific agency. RaDiUS was searched using proximity and wildcard searches for combinations of seven different search terms: human, embryo(s), fetus, *in vitro* fertilization, IVF, assisted reproductive technology(ies), and fertility. Proximity searches allow the user to search the database using two or more terms that are within a certain number of words from each other. For example, one could search for the terms *human* and *embryo* within two words of each other and would find abstracts that contain the term *human embryo*, as well as abstracts that contain the phrase *the embryo of the human species*. Wildcard searches allow the user to search the database for several forms of a word, such as plurals, suffixes, and prefixes. For example, using the wildcard character "%" one could search for abstracts containing the words *embryo, embryos, embryonic*, etc., by searching with "embryo%."

Findings

Currently, the federal government funds research that uses cultured cell lines derived from abortuses, as well as DNA, cells, and tissues taken from abortuses. For example, the human embryonic kidney 293 cells, originally derived by researchers at McMaster University in Hamilton, Ontario, Canada (Graham et al., 1977), have been used by researchers at several institutes at NIH for basic research including the following:

1) gene therapy research at the National Institutes of Neurological Disorders and Stroke (Mochizuki et al., 1998);

2) research on insulin-like growth factor-I at the National Cancer Institute (Dey et al., 1998) and the National Institute of Diabetes and Digestive and Kidney Diseases (Okubo et al., 1998);

3) signal transduction research at the National Institute of Dental Research (NIDR) (Lin et al., 1997); and

4) research on follicle-stimulating hormone at the National Institute of Child Health and Human Development (NICHD) (Flack et al., 1994).

Cell lines derived from human embryonic or fetal tissue, such as the human embryonic kidney 293 cells, are commonly used by researchers in federally funded research. Many of these cell lines are commercially available from ATCC. A search of the ATCC's catalog for human embryonic cell lines revealed at least 32 cell lines derived from human embryos. However, in ATCC's catalog, embryonic cell lines include cells derived from embryos ranging from five weeks through the first trimester. Since the developing organism is only an embryo through the eighth week of development, some of these embryonic cell lines may actually be derived from cadaveric fetal tissue. The ATCC catalog also lists at least 28 cell lines derived from human fetuses. Several of these embryonic and fetal cell lines maintained by ATCC were originally developed and characterized by the Naval Biosciences Laboratory in Oakland, California, and then transferred to ATCC in 1982.

Researchers at NIDR have used tissues from human embryos and fetuses to study skeletal development (Hoang et al., 1996; Chang et al., 1994). In these studies, cadaveric embryos and fetuses ranging from 6–13 weeks gestation were fixed, embedded in paraffin, serially cut into 5–7 μm sections, and mounted on microscope slides. The tissue from the human embryos and fetuses was obtained from legally sanctioned procedures performed at the University of Zagreb Medical School in Zagreb, Croatia (Hoang et al., 1996). The procedure for obtaining autopsy materials was approved by the Internal Review Board of the Ethical Committee at the University of Zagreb School of Medicine and the Office of Human Subjects Research of NIH (Hoang et al., 1996).

In addition to researchers in the intramural program at NIH, there are many researchers at academic institutions that are funded by NIH's extramural program who perform research using human embryonic and fetal tissue. For example, researchers at the University of Iowa are studying the effects of maternal diabetes in human fetal lung development using cultured human fetal lung (Grant Number: R01HL50050). In addition, a group of researchers in the Department of Surgery at the University of California in San Francisco are using fetal skin to study wound healing in fetuses that occurs without scarring (Grant Number: R01GM55814; Stelnicki et al., 1998).

Furthermore, researchers funded by NIH are not the only federally funded researchers using human embryonic and fetal tissue. Researchers at the EPA and the Department of Veterans Affairs are also conducting research with these tissues. For example, the Reproductive Toxicology Division and the Developmental Toxicology Division at the EPA have used human embryonic palates to study the effects of teratogenic chemicals in development of cleft palate (Abbott et al., 1999; Abbott et al., 1998; Abbott et al., 1994; Abbott and Buckalew, 1992).

Researchers at the Neurophysiology and Spinal Cord Pharmacology Laboratories at the Veterans Affairs Medical Center in Miami, Florida, have studied neural transmitters in cultured embryonic human dorsal root ganglion neurons (Valeyev et al., 1999). The neurons were obtained with written informed consent and human subjects committee approval in accordance with published guidelines from 5–8 week old cadaveric human embryos through a collaborative agreement with the Central Laboratory for Human Embryology (see below) (Valeyev et al., 1999). In addition, researchers from the Medical Service at the Veterans Administration Medical Center in Palo Alto, California, have studied gene expression in Wilm's tumor, the most common abdominal tumor in children, and compared it to the expression of genes in normal human fetal tissue (Vu and Hoffman, 1996). Normal fetal tissue from 6–12 weeks gestation abortuses was obtained from the Central Laboratory for Human Embryology (Vu and Hoffman, 1996).

In addition to funding research that uses human embryonic and fetal tissue, the federal government funds tissue banks that collect, study, describe, and distribute tissues from abortuses. In 1997, the NICHD awarded ATCC a grant for a feasibility study to explore the utility of various tissues from human abortuses for transplantation and related research (Grant Number: R01HD32179). The Central Laboratory of Human Embryology has been funded by NIH for 30 years (Grant Number: 5R24HD00836-35). The Central Laboratory of Human Embryology collects, characterizes stages, identifies, processes, and distributes human embryonic and fetal tissues within minutes of delivery. Annually, it collects and analyzes 3,500 embryos and fetuses. The Central Laboratory of Human Embryology is also participating in a project directed by the Puget Sound Blood Center of Seattle to determine the availability and suitability of embryonic tissue for transplantation research and therapy. Two other institutions, the Northwest Tissue Center and the Fred Hutchinson Cancer Research Center, are participating in this project, and together with the Central Laboratory of Human Embryology have formed the Northwest Fetal Tissue Program.

In contrast, there is no evidence of federally funded research that used embryonic tissue derived from IVF embryos identified in searches of MEDLINE or RaDiUS. While the searches of these databases were broad, they were not designed to be comprehensive, so it was not possible to identify all federally funded research projects that use human embryonic tissue or materials and reagents derived from this tissue. In addition, even though the majority of materials and reagents from human embryonic tissue is derived from cadaveric embryonic tissue obtained after spontaneous and elective abortions, it is not possible to definitively determine the origins of all cultured cell lines and DNA routinely used in research by federally funded investigators.

Conclusion

The federal government is currently funding research that uses a broad variety of research materials and reagents derived from human embryonic and fetal tissue. The human embryonic and fetal tissue used in this research comes from cadaveric tissue that is obtained after either spontaneous or elective abortions. There is no evidence revealed through the search conducted for this Project Memorandum that the federal government is funding research that uses cells or tissue derived from embryos remaining after IVF procedures or, more relevantly, cells or tissues derived from embryos created solely for research purposes. But as noted above, lack of evidence of such funding is not equivalent to evidence of lack of funding.

References

Abbott B.D., and Buckalew A.R. 1992. Embryonic palatal responses to teratogens in serum-free organ culture. *Teratology* 45(4):369–382.

Abbott B.D., Held G.A., Wood C.R., Buckalew A.R., Brown J.G., and Schmid J. 1999. AhR, ARNT, and CYP1A1 mRNA quantitation in cultured human embryonic palates exposed to TCDD and comparison with mouse palate *in vivo* and in culture. *Toxicol Sci* 47(1):62–75.

Abbott B.D., Probst M.R., and Perdew G.H. 1994. Immunohistochemical double-staining for Ah receptor and ARNT in human embryonic palatal shelves. *Teratology* 50(5):361–366.

Abbott B.D., Probst M.R., Perdew G.H., and Buckalew A.R. 1998. AH receptor, ARNT, glucocorticoid receptor, EGF receptor, EGF, TGF alpha, TGF beta 1, TGF beta 2, and TGF beta 3 expression in human embryonic palate, and effects of 2,3,7, 8-tetrachlorodibenzo-p-dioxin (TCDD). *Teratology* 58(2):30–43.

Chang S.C., Hoang B., Thomas J.T., Vukicevic S., Luyten F.P., Ryba N.J., Kozak C.A., Reddi A.H., and Moos M. Jr. 1994. Cartilage-derived morphogenetic proteins. New members of the transforming growth factor-beta superfamily predominantly expressed in long bones during human embryonic development. *J Biol Chem* 269(45):28227–28234.

Dey B.R., Spence S.L., Nissley P., and Furlanetto R.W. 1998. Interaction of human suppressor of cytokine signaling (SOCS)-2 with the insulin-like growth factor-I receptor. *J Biol Chem* 273(37):24095–24101.

Flack M.R., Bennet A.P., Froehlich J., Anasti J.N., and Nisula B.C. 1994. Increased biological activity due to basic isoforms in recombinant human follicle-stimulating hormone produced in a human cell line. *J Clin Endocrinol Metab* 79(3):756–760.

Graham F.L., Smiley J., Russell W.C., and Nairn R. 1977. Characteristics of a human cell line transformed by DNA from human adenovirus type 5. *J Gen Virol* 36(1):59–74.

Hoang B., Moos M. Jr., Vukicevic S., and Luyten F.P. 1996. Primary structure and tissue distribution of FRZB, a novel protein related to Drosophila fizzled, suggest a role in skeletal morphogenesis. *J Biol Chem* 271(42):26131–26137.

Lin K., Wang S., Julius M.A., Kitajewski J., Moos M. Jr., and Luyten F.P. 1997. The cysteine-rich fizzled domain of Frzb-1 is required and sufficient for modulation of Wnt signaling. *Proc Natl Acad Sci USA* 94(21):11196–11200.

Mochizuki H., Schwartz J.P., Tanaka K., Brady R.O., and Reiser J. 1998. High-titer human immunodeficiency virus type 1-based vector systems for gene delivery into nondividing cells. *J Virol* 72(11):8873–8883.

Okubo Y., Blakesley V.A., Stannard B., Gutkind S., and Le Roith D. 1998. Insulin-like growth factor-I inhibits the stress-activated protein kinase/c-Jun N-terminal kinase. *J Biol Chem* 273(40):25961–25966.

Stelnicki E.J., Arbeit J., Cass D.L., Saner C., Harrison M., and Largman C. 1998. Modulation of the human homeobox genes PRX-2 and HOXB13 in scarless fetal wounds. *J Invest Dermatol* 111(1):57–63.

Valeyev A.Y., Hackman J.C., Holohean A.M., Wood P.M., Katz J.L., and Davidoff R.A. 1999. GABA-Induced Cl- current in cultured embryonic human dorsal root ganglian neurons. *J Neurophysiol* 82(1):1–9.

Vu T.H., and Hoffman A.R. 1996. Alterations in the promoter-specific imprinting of the insulin-like growth factor-II gene in Wilm's tumor. *J Biol Chem* 271(15):9014–9023.

Analysis of Federal Laws Pertaining to Funding of Human Pluripotent Stem Cell Research

Commissioned Paper
Ellen J. Flannery and Gail H. Javitt

I. Introduction

The National Bioethics Advisory Commission (NBAC) was established by President Clinton in 1995 to provide advice and make recommendations concerning bioethical issues arising in the context of government research programs.[1] The General Counsel for the Department of Health and Human Services (DHHS), Harriet Rabb, recently provided a legal opinion to Harold Varmus, Director of the National Institutes of Health (NIH), concluding that current law does not prohibit federal funding of research using human pluripotent stem cells derived from either human embryos or nonliving fetuses.[2] The NBAC has asked us to provide an independent analysis of whether the DHHS General Counsel's legal conclusions are reasonable.

This memorandum first reviews federal laws and regulations pertaining to research with human fetal tissue. It concludes that DHHS reasonably determined that these provisions would not prohibit federal funding of research with pluripotent stem cells derived from nonliving fetuses. Next, the memorandum reviews the statute prohibiting federal funding of human embryo research and concludes that DHHS reasonably determined that the statutory language does not prohibit federal funding of research using pluripotent stem cells that were derived from a human embryo. In addition, although the legislative history is ambiguous, DHHS could reasonably determine that it does not demonstrate that Congress intended to prohibit federal funding of research using pluripotent stem cells that were derived from an embryo without reliance on federal funds. Thus, DHHS reasonably concluded that federal funding of research using pluripotent stem cells derived from a human embryo is lawful, provided that federal funds are not used to derive the stem cells.

II. Federal Funding of Research on Stem Cells Derived from Nonliving Fetuses

Stem cells derived from human fetal tissue are subject to federal laws and regulations pertaining to human fetal tissue. It was reasonable for DHHS to conclude that these laws and regulations, discussed below, do not prohibit federal funding of research using pluripotent stem cells derived from nonliving fetuses,[3] but do impose certain requirements depending on the type of research performed.

A. Background on DHHS Regulation of Research Involving Fetuses

When DHHS provides federal funding, through grants or contracts, for research involving human subjects, such research is subject to conditions imposed by DHHS regulations governing the protection of human subjects.[4] DHHS promulgated regulations in May 1974 establishing basic protections for human subjects. Later that year, the Department proposed to amend those regulations to provide additional protections for certain subjects who might have diminished capacity for informed consent, including pregnant women, fetuses, prisoners, and the institutionalized mentally disabled.[5]

Shortly before the proposed rule was published, Congress enacted a prohibition on federally funded research on a living human fetus, to be in effect until recommendations were made to DHHS by the National Commission for the Protection of Human Subjects of Biomedical and Behavioral Research.[6] After considering the report of the Commission and public comments on the proposed rule, DHHS promulgated regulations in 1975 governing federally funded research involving pregnant women, fetuses, and "cells, tissue, or organs excised from a dead fetus."[7] DHHS also lifted the moratorium on fetal research.[8]

DHHS regulations provided that "[a]ctivities involving the dead fetus, macerated fetal material, or cells, tissue, or organs excised from a dead fetus shall be conducted only in accordance with any applicable State or local laws regarding such activities."[9] "Dead fetus" was defined as "a fetus *ex utero* which exhibits neither a heartbeat, spontaneous respiratory activity, spontaneous movement of voluntary muscles, nor pulsation of the umbilical cord (if still attached)."[10] The Commission report referred to organs, tissue, and cells from a dead fetus collectively as "fetal tissue," although the DHHS regulations included them under the general heading of dead fetus and fetal material.[11] In the preamble to the regulation, DHHS stated:

> The Department notes...that research involving the dead fetus and fetal material is governed in part by the Uniform Anatomical Gift Act which has been adopted by 49 States, the District of Columbia and Puerto Rico....Any applicable State or local laws regarding such activities are, of course, controlling.[12]

Thus, DHHS could fund research activities involving cells or tissue from a nonliving fetus, but that research would be subject to applicable state and local laws. Neither the regulations or the preamble to the regulations, nor the report of the Commission which was considered by DHHS in promulgating the regulations, limited those "activities" involving a dead fetus to any particular type of research, such as basic research or therapeutic research.[13]

B. DHHS Moratorium on Fetal Tissue Transplantation Research

In 1987, the Director of NIH requested approval from the DHHS Assistant Secretary for Health for a research protocol involving the experimental implantation of human fetal cells into the brain of a patient with Parkinson's disease.[14] The Assistant Secretary withheld his approval for that project and all other research involving transplantation of human fetal tissue and convened an expert panel to provide a report and recommendations.[15] In November 1989, the Secretary of DHHS continued indefinitely the moratorium on fetal tissue transplantation.[16] As described by the Secretary, the moratorium was limited in scope:

> I am continuing indefinitely the limited moratorium on federal funding of research in which human fetal tissue from induced abortions is transplanted into human recipients. This action does not affect funding of other research involving fetal tissue.[17]

The moratorium thus precluded federal funding of research where human fetal tissue from induced abortions was transplanted into human recipients.[18] It did not prohibit the "funding of such research in the private sector...[or] Federal support of therapeutic transplantation research that uses fetal tissue from spontaneous abortions or ectopic pregnancies."[19]

By its terms, the moratorium also did not prohibit federal funding of fetal cell and tissue research that did not involve transplantation into human recipients.[20] During the moratorium, NIH continued to fund research using fetal tissues and cells for a variety of purposes, including use in testing vaccines and virus research.[21]

C. Nullification of the Moratorium on Fetal Tissue Transplantation Research

In January 1993, President Clinton directed the Secretary of DHHS to lift the moratorium on fetal tissue transplantation research.[22] Thereafter, Congress enacted the National Institutes of Health Revitalization Act of 1993.[23] The Act authorizes the Secretary of DHHS to allocate federal funds for "research on the transplantation of human fetal tissue for therapeutic purposes," regardless of whether the tissue was obtained pursuant to an induced abortion.[24] The Act defines "human fetal tissue" as "tissue or *cells* obtained from a dead human embryo or fetus after a spontaneous or induced abortion, or after a stillbirth."[25]

The Act specifies that fetal tissue may be used for transplantation research only if informed consent is obtained from the woman donating the tissue[26] and from the researcher.[27] The Act also prohibits the solicitation or acceptance of fetal tissue for transplantation into a specific individual (i.e., directed donation) under certain circumstances.[28]

The Act also codifies the Clinton Administration's nullification of the 1988 moratorium.[29] The report accompanying the legislation stated:

> In taking this legislative approach, the Committee has nullified the 1988 moratorium on human fetal tissue transplantation research. The Committee intends that such research be allowed to proceed as do all other research efforts at NIH for therapeutic drugs and treatments for neurological, metabolic, hematologic, immunological, genetic, and other disorders. The Committee does not intend that fetal tissue transplantation research be treated with special scientific, peer review, or regulatory favor, but rather that it not be treated with administrative disfavor that prevents its progress.[30]

The 1993 Act did not alter the regulations that had previously been in effect with respect to federally funded fetal tissue research. Rather, it imposed additional protections and requirements for the subset of fetal tissue research involving transplantation of the tissue. It also established limitations on the sale of human fetal tissue.

D. Federal Funding of Research Using Pluripotent Stem Cells Derived from a Nonliving Fetus

The DHHS General Counsel's memorandum addresses research in which pluripotent stem cells were derived from "primordial germ cells isolated from the tissue of nonliving fetuses"[31] and maintained in culture.[32] The cells were obtained following a "therapeutic termination of pregnancy."[33] These stem cells thus would constitute "cells obtained from a dead human embryo or fetus after...induced abortion" and would meet the definition of "human fetal tissue" in 42 USC §§ 289g-1(g) and 289g-2(d)(1).

The DHHS General Counsel reasonably concluded that the statutory restrictions applicable to the transplantation of human fetal tissue—i.e., the restriction on directed donation and the informed consent requirements—would apply to research involving the transplantation for therapeutic purposes of human pluripotent stem cells derived from a nonliving fetus. In contrast, although this further point was not directly addressed by the DHHS General Counsel, to the extent that research using pluripotent stem cells derived from a nonliving fetus is basic research that does not involve transplantation into a human donee, the statutory restrictions should not apply to such research.[34]

The DHHS General Counsel apparently also concluded that the statutory prohibition in section 289g-2(a) on the transfer of "any human fetal tissue for valuable consideration" applies to the exchange of "research materials" involving pluripotent stem cells derived from nonliving fetuses.[35] This is a reasonable conclusion because such stem cells fall within the definition of "human fetal tissue" in section 289g-2(d)(1) and section 289g-2(a) and is not by its terms limited to human fetal tissue for transplantation. However, it is also possible to interpret this prohibition as applicable only to human fetal tissue intended for transplantation, because the provision was enacted in legislation addressing *transplantation* research.[36] In addition, the legislative history of section 289g-2(a) is focused on tissue for transplantation:

> The Committee adopts the prohibition on the sale of human fetal tissue to make the treatment of such tissue parallel to the treatment of other human organs intended for transplantation....Indeed, the Committee has dealt with fetal tissue more restrictively than other transplantation, for although current organ transplant law prohibits the sale of organs, it allows for payment for the removal of the organ.[37]

Given the statutory language and the ambiguous legislative history, DHHS reasonably concluded that the statutory prohibition on sale for valuable consideration applies to human pluripotent stem cells "to the extent [they] are considered human fetal tissue by law."[38]

III. Federal Funding of Research on Stem Cells Derived from Human Embryos

A. Current Law

The Omnibus Consolidated and Emergency Supplemental Appropriations Act for Fiscal Year 1999[39] provides that none of the funds appropriated for activities of DHHS may be used for:

(1) the creation of a human embryo or embryos for research purposes; or

(2) research in which a human embryo or embryos are destroyed, discarded, or knowingly subjected to risk of injury or death greater than that allowed for research on fetuses *in utero* under 45 CFR 46.208(a)(2) and section 498(b) of the Public Health Service Act (42 USC 289g(b)).[40]

The term "human embryo" is defined as "any organism, not protected as a human subject under 45 CFR 46 as of the date of the enactment of this Act, that is derived by fertilization, parthenogenesis, cloning, or any other means from one or more human gametes or human diploid cells."[41] This provision has been included in Congressional appropriations for DHHS activities since 1996, without alteration.[42]

B. Statutory Analysis

1. Statutory Language

In order to determine whether Congress intended the ban on federal funding of human embryo research to include research with pluripotent stem cells derived from human embryos, it is necessary first to examine the statutory language.[43] The statute provides that federal funds may not be used for "research in which a human embryo or embryos are destroyed." The statute clearly prohibits federal funding for research in which a human embryo is actually destroyed. The critical question here is whether it also prohibits DHHS funding of research that depends upon the prior destruction of a human embryo to enable its performance. If so, federal funding of research *using* stem cells derived from an embryo would be prohibited even though the researcher did not himself destroy the embryo in order to obtain the cells.[44]

The DHHS General Counsel's opinion concluded that federal funding for research using pluripotent stem cells derived from an embryo would not violate federal law, because the statute defines embryos as "organisms" and stem cells are not organisms and therefore are not embryos.[45] The conclusion that stem cells are not "embryos" appears to be consistent with the scientific evidence presented to Congress and the NBAC.[46] Accordingly, DHHS reasonably concluded that federal funding of the stem cell research is not prohibited on this basis.

However, the DHHS General Counsel did not address the question of whether the statute prohibits federal funding of research that is dependent on the prior destruction of a human embryo. This question is analyzed below.

The statute states that the ban on federal funding applies to research "in which…embryos are destroyed." The most straightforward interpretation of this text is that it prohibits federal funding only of the specific research during which the embryos are actually destroyed. Had Congress intended to extend the funding ban beyond such direct destruction, it could have used different statutory language; for example, this statutory provision could have tracked the language of the fetal tissue statute, which expressly applies to "cells obtained from a dead human embryo or fetus."[47] Since the provision does not contain such additional language, DHHS could reasonably conclude that the embryo research funding ban does not extend to research on pluripotent stem cells that were previously obtained from an embryo in research that was not supported by federal funds.

2. Legislative History

Since it might be argued that the statute should be interpreted to prohibit federal funding of research that is dependent on the prior destruction of a human embryo, it is instructive to review the legislative history of the provision as a means to discern congressional intent. This legislative history is discussed below.

a. Pre-Enactment Legislative History

Prior to 1993, DHHS imposed a de facto moratorium on the funding of human embryo research.[48] This moratorium was lifted by the NIH Revitalization Act of 1993.[49] The NIH thereafter convened the Human Embryo Research Panel "to consider various areas of research involving the *ex utero* human embryo and to provide advice as to those areas" that (1) "are acceptable for Federal funding," (2) "warrant additional review," and (3) "are unacceptable for Federal support."[50] The Panel was asked to recommend specific guidelines for the review and conduct of research that the Panel considered to be acceptable for federal funding.[51]

The Panel concluded that several areas of human embryo research would be acceptable for federal funding, including research "involving the development of embryonic stem cells...with embryos resulting from IVF treatment for infertility or clinical research that have been donated with the consent of the progenitors."[52] The Panel noted that the "therapeutic utility of embryonic stem cells lies in their potential as a source of differentiated cells for transplantation and tissue repair,"[53] for example, "to repair regions of the nervous system that have undergone damage (e.g., motor neurons in spinal cord injury) or degeneration (e.g., as in Parkinson's disease) or for bone marrow transplantation."[54]

NIH Director Varmus was questioned about the Panel Report during March 1995 hearings to discuss fiscal year 1996 appropriations for DHHS. In response to a question regarding why embryo research was important for NIH to pursue, Dr. Varmus specifically addressed research with embryonic stem cells:

> The second issue that's of great interest concerns the nature of so-called embryonic stem cells. These cells have the capacity to differentiate into virtually any kind of tissue. All work of this sort proceeds with animal models with considerable vigor throughout the biomedical research enterprise. We know from studies of embryonic stem cells of the mouse that we now have the potential to consider using embryonic stem cells as potential universal donors, for example, for bone marrow transplantation and other kinds of therapeutic maneuvers. But we can't begin to pursue such studies with human materials until we're able to use early stage embryos.[55]

The NIH Panel Report was opposed by a broad coalition of religious and anti-abortion organizations.[56] In addition, 28 members of the House of Representatives signed a letter to Director Varmus urging him to reject the Panel's recommendations.[57]

In July 1995, Representatives Jay Dickey and Roger Wicker introduced an amendment to H.R. 2127, a fiscal year 1996 appropriations bill for DHHS, that prohibited the use of federal funds to support human embryo research.[58] The amendment contained language identical to the statutory provision now in effect. The Dickey-Wicker amendment was approved by the House Committee on Appropriations on July 20, 1995, and by the full House of Representatives on August 4, 1995. In an editorial published shortly thereafter, Representative Dickey stated that the amendment "prevents federal funding of destructive experiments on live human embryos," and would "prevent taxpayer funding of experiments such as those proposed by a Human Embryo Research Panel advising the...NIH."[59] Twelve members of the House Committee on Appropriations dissented from the provision and raised the concern that it could prevent progress in many areas of medical research.[60] The Dickey-Wicker amendment was ultimately included in the Balanced Budget Downpayment Act, enacted January 26, 1996.[61]

The provision was again included in appropriations legislation for fiscal year 1997. One member of Congress who had dissented from the ban during the 1996 appropriations introduced an amendment to modify the ban to permit federal funding of research using "spare" embryos, i.e., embryos created for use in IVF that were then donated by the progenitors for use in research.[62] This amendment was defeated in the House. Supporters of the amendment argued that continuing the ban on federal funding for embryo research would prevent the advancement of medical research for a variety of diseases.[63] One House member argued that federal funding of embryo research "could also lead to breakthroughs in the use of embryonic stem cells."[64] Opponents of the amendment focused on the ban's effect of preventing federal funds from being used for "the destruction of" embryos or to destroy human life.[65] They also emphasized that the funding ban would not prohibit research: "Private embryo research is legal now, and it would continue to be legal."[66] The provision was again included in DHHS appropriations legislation for fiscal years 1998[67] and 1999,[68] but the legislative history reflects no further debate.

The Report of the Senate Committee on Appropriations on the DHHS funding bill for 1999 contains an interesting reference to embryonic stem cell research. The Committee recommended retention of the appropriations legislation "language on human embryo research."[69] At the same time, however, the Committee "strongly urge[d]" NIH to give "full consideration" to a grant proposal for a stem cell biology center at the University of Nebraska Medical Center, which would further research in "embryonic stem cell biology" as well as "solid organ and hematopoietic stem cell biology."[70] It cannot be determined from the report whether or not this center is intended to include human embryonic stem cell research.

The legislative history discussed above is ambiguous with respect to the question whether Congress intended to prohibit federal funding for research using stem cells derived from a human embryo. On the one hand, the sponsors of the provision prohibiting funding focused their remarks on "destructive experiments" and "destruction" of the embryo. They acknowledged that destruction of the embryo could occur in privately-funded research, but emphasized that such destruction should not be funded by taxpayer dollars.[71] On the other hand, opponents of the provision argued that a ban on federal funding for research on human embryos could restrict or prevent advances in various areas of medical research, perhaps even including embryonic stem cell research.[72] There is no indication that either proponents or opponents contemplated the situation under consideration here, in which research that destroyed the embryo was separately conducted from research using the cells derived from the embryo.[73]

Given that the legislative history of the provision is ambiguous, DHHS could reasonably conclude that Congress did not intend the funding ban to extend beyond the immediate research that destroyed the embryo. The most straightforward meaning of the statutory language "*in* which...embryos *are* destroyed" is that federal funds cannot be provided for the research in which the embryos are actually destroyed. Since the legislative history does not provide "strong evidence that Congress actually intended another meaning,"[74] it was reasonable for DHHS to give effect to "the ordinary meaning of the words used."[75]

b. Post-Enactment Application of the Ban to Specific Research

Since 1996, Congress has reviewed NIH's application of the ban to specific research on one occasion (of which we are aware). In June 1997, the House Commerce Committee's Subcommittee on Oversight and Investigations held hearings to review the events surrounding NIH's firing of one of its scientists for conducting human embryo research with federal funds.[76] According to testimony presented by Dr. Varmus, the only witness at the hearing, the researcher was fired for using NIH laboratory equipment to analyze DNA that was extracted from human embryos for the purpose of detecting genetic defects. NIH took the position that the ban prohibited federal support for preimplantation genetic analysis from DNA derived from a human embryo.[77] No evidence was presented that the researcher extracted the DNA from the embryos himself.[78] Dr. Varmus testified that the ban precluded all research "involving human embryos," including preimplantation genetic diagnosis, a position with which the Chairman of the Subcommittee concurred.[79]

The precedential value of this case for the issue under consideration here is unclear. NIH's position that it could not allocate federal funds for research on preimplantation genetic diagnosis might be viewed as inconsistent with the DHHS General Counsel's opinion that the agency is not prohibited from allocating funds for research using stem cells derived from an embryo, since both types of research involve derivatives of human embryos.[80] Furthermore, the DNA analysis could be conducted without destruction of the embryo. However, there may be grounds for distinguishing the two situations. For example, the researcher was fired not only because of the use of NIH resources for the DNA analysis, but also on the grounds that he was "diverting both…NIH-supported personnel…and equipment" to a private IVF clinic at Suburban Hospital.[81] In addition, preimplantation genetic diagnosis is arguably distinguishable from research using pluripotent stem cells, in that the information obtained from the embryonic DNA is used to make a decision about whether to implant or discard an embryo, and thus the DNA analysis might be deemed an integral part of the research enterprise in which an embryo may be actually discarded or destroyed.

IV. Conclusion

We conclude that DHHS reasonably determined that federal funding is not prohibited for research using pluripotent stem cells derived from nonliving fetuses, provided applicable legal requirements are satisfied. DHHS also reasonably determined that the prohibition on federal funding of human embryo research does not prohibit federal funding of research using pluripotent stem cells derived from an embryo, provided that those cells are derived without support from federal funds.

Notes

1 Exec. Order No. 12975, 3 CFR 409 (1996), 42 USC § 6601 (1996), as amended by Exec. Order No. 13018, 3 CFR 216 (1997), 42 USC § 6601 (1996), and Exec. Order No. 13046, 62 *Fed. Reg.* 27685 and 28109 (May 20 and 22, 1997).

2 Memorandum from Harriet S. Rabb, General Counsel, DHHS, to Harold Varmus, M.D., Director, NIH, at 1 (Jan. 15, 1999) (hereinafter *Rabb Memorandum*).

3 The DHHS General Counsel's opinion addressed only stem cells derived from nonliving fetuses. Our opinion is similarly limited. We note, however, that federal law imposes restrictions on the federal funding of research on nonviable or viable human fetuses. See 42 USC § 289(g) (Supp. 1997); see also 45 CFR § 46.209 (1998).

4 39 *Fed. Reg.* 18914 (May 30, 1974) (establishing 45 CFR Part 46); 40 *Fed. Reg.* 11854 (Mar. 13, 1975) (amending and readopting Part 46).

5 39 *Fed. Reg.* 30648 (Aug. 23, 1974).

6 Pub. L. No. 93-348, § 202(b), 88 Stat. 342, 348 (1974) (National Research Act).

7 40 *Fed. Reg.* 33526, 33530 (Aug. 8, 1975) (establishing Subpart B of 45 CFR Part 46).

8 Id. at 33526.

9 Id. at 33530 (codified at 45 CFR § 46.210) (1998). In May 1998, DHHS issued a proposed rule amending its regulations for the protection of human subjects. 63 *Fed. Reg.* 27794 (May 20, 1998). The proposed rule would replace the phrase "dead fetus" with "dead newborn" and would clarify that the provision concerning dead fetuses and fetal materials includes placentas after delivery. It would also add a new section stating that if information associated with such materials is "recorded for research purposes in a manner that living persons can be identified…those persons are research subjects" and the other regulations concerning protection of research subjects would apply. Id. at 27804 (proposed § 46.206).

10 45 CFR § 46.203(f) (1998).

11 40 *Fed. Reg.* at 33530 (§ 46.210), 33532 (definition of dead fetus).

12 40 *Fed. Reg.* at 33528.

13 In 1985, Congress codified the DHHS requirements governing activities with *living* fetuses as part of Title IV of the Public Health Service Act. Pub. L. No. 99-158, § 498, 99 Stat. 822, 877-878 (1985), codified at 42 USC § 289g; H.R. Rep. No. 103-28, 103d Cong., 1st Sess. 59 (1993).

14 See H.R. Rep. No. 103-28, supra note 13, at 57.

15 Id.

16 Statement by Louis W. Sullivan, Secretary of Health and Human Services, DHHS News, November 2, 1989; Letter from Louis W. Sullivan, Secretary, DHHS, to Dr. William F. Raub, Acting Director, NIH (Nov. 2, 1989).

17 Statement by Louis W. Sullivan, supra note 16.

18 Id. (stating that "in the specific area of transplantation to humans involving fetal tissue from induced abortions, it is not appropriate that federal support be provided").

19 *Finding Medical Cures: The Promise of Fetal Tissue Transplantation Research,* Hearing on S.1902 Before the Senate Comm. on Labor and Human Resources, 102d Cong., 1st Sess. 4 (1991) (statement of Assistant Secretary of Health James O. Mason).

20 Statement by Louis W. Sullivan, supra note 16.

21 National Institutes of Health, *Human Fetal Tissue Research Supported by the NIH in Fiscal Year 1990,* 1–4 (1991); Helen M. Maroney, "Bioethical Catch 22: The Moratorium on Federal Funding of Fetal Tissue Transplantation Research and the NIH Revitalization Amendments," 9 J. Contemp. Health L. and Pol'y 485, 487 n. 11 (1991) (citing NIH document).

22 Memorandum from the President to the Secretary of Health and Human Services, January 22, 1993 (*reprinted in* 58 Fed. Reg. 7457 [Feb. 5, 1993]).

23 Pub. L. No. 103-43, Title I, §§ 111, 112, 121(b)(1), 107 Stat. 122, 129, 131, 133 (1993) (codified at 42 USC §§ 289g, 289g-1, 289g-2 [Supp. 1997]).

24 42 USC § 289g-1(a) (Supp. 1997).

25 Id. §§ 289g-1(g) and 289g-2(d)(1) (emphasis added).

26 The woman providing the tissue must state in writing that (1) she is donating the fetal tissue for research on the transplantation of human fetal tissue for therapeutic purposes, (2) the donation is being made without any restriction regarding the identity of individuals who may be recipients of the transplantation of the tissue, and (3) she has not been informed of the identity of the recipient. 42 USC § 289g-1(b)(1) (Supp. 1997). The attending physician involved in obtaining the tissue also must make certain declarations in writing. Id. § 289g-1(b)(2).

27 The researcher must declare in writing that he or she:
    (1) is aware that –
        (A) the tissue is human fetal tissue;
        (B) the tissue may have been obtained pursuant to a spontaneous or induced abortion or pursuant to a stillbirth; and
        (C) the tissue was donated for research purposes;
    (2) has provided such information to other individuals with responsibilities regarding the research;
    (3) will require, prior to obtaining the consent of an individual to be a recipient of a transplantation of the tissue, written acknowledgment of receipt of such information by such recipient; and
    (4) has had no part in any decisions as to the timing, method, or procedures used to terminate the pregnancy made solely for the purposes of the research. Id. § 289g-1(c).

28 42 USC § 289g-2(b) (Supp. 1997). This section provides:
It shall be unlawful for any person to solicit or knowingly acquire, receive, or accept a donation of human fetal tissue for the purpose of transplantation of such tissue into another person if the donation affects interstate commerce, the tissue will be or is obtained pursuant to an induced abortion, and
(1) the donation will be or is made pursuant to a promise to the donating individual that the donated tissue will be transplanted into a recipient specified by such individual;
(2) the donated tissue will be transplanted into a relative of the donating individual; or
(3) the person who solicits or knowingly acquires, receives, or accepts the donation has provided valuable consideration for the costs associated with such abortion.

29 Section 113 of the Act provided that, except as authorized by the statute, "no official of the executive branch may impose a policy that the Department of Health and Human Services is prohibited from conducting or supporting any research on the transplantation of human fetal tissue for therapeutic purposes. Such research shall be carried out in accordance with section 498A of the Public Health Service Act [this section] (as added by section 111 of this Act), without regard to any such policy that may have been in effect prior to the date of the enactment of this Act" 42 USC § 289g-1 (Supp. 1997).

30 H.R. Rep. No. 103-28, supra note 13, at 77.

31 *Rabb Memorandum*, supra note 2, at 1.

32 Michael J. Shamblott, et al., "Derivation of Pluripotential Stem Cells from Cultured Primordial Germ Cells," 95 *Proc. Nat'l Acad. Sci. USA* 13726 (1998).

33 Id.

34 This memorandum does not explore whether pluripotent stem cells derived from a nonliving fetus that are grown in culture and manipulated to become tissues intended for use in transplantation would be subject to the fetal tissue transplantation restrictions within the meaning of the statute.

35 *Rabb Memorandum*, supra note 2, at 4–5.

36 The provision was enacted under "Part II—Research on Transplantation of Human Tissue." See H.R. Rep. No. 103-28, supra note 13, at 2–3.

37 H.R. Rep. No. 103-28, supra note 13, at 76.

38 *Rabb Memorandum*, supra note 2, at 1.

39 Pub. L. No. 105-277, 112 Stat. 2681 (1998).

40 Id. Div. A, § 101(f), Title V, § 511(a), 112 Stat. 2681-386. Division A, § 101(f) of the bill contains appropriations for the Departments of Labor, Health and Human Services, Education, and related agencies.

41 Id. § 511(b).

42 Pub. L. No. 104-99, Title I, § 128, 110 Stat. 26, 34 (1996); Pub. L. No. 104-208, Div. A, § 101(e), Title V, § 512, 110 Stat. 3009, 3009-270 (1996); Pub. L. No. 105-78, Title V, § 513, 111 Stat. 1467, 1517 (1997); Pub. L. No. 105-277, supra notes 39 and 40.

43 *Amoco Prod. Co. v. Village of Gambell*, 480 U.S. 531, 548 (1987) (where statutory language "is capable of precise definition," court "will give effect to that meaning absent strong evidence that Congress actually intended another meaning"); *United States v. Locke*, 471 U.S. 84, 95 (1985).

44 This memorandum focuses on "destruction" of the embryo because derivation of stem cells requires destroying the embryo. The analysis would not change, however, if derivation of stem cells required discarding the embryo or subjecting it to greater risk than is authorized by federal laws and regulations.

45 *Rabb Memorandum*, supra note 2, at 1–4.

46 See id. at 3 n. 4; see also Transcript, National Bioethics Advisory Commission, 26th Meeting, at 32 (January 19, 1999).

47 42 USC § 289g-1(g) (Supp. 1997).

48 This moratorium resulted from DHHS regulations requiring all research applications involving *in vitro* fertilization (IVF) to be reviewed by a DHHS-appointed Ethical Advisory Board (EAB) before federal funding could be allocated. 40 Fed. Reg. 33526, 33529 (Aug. 8, 1975) (adding 45 CFR § 46.204(d)). "*In vitro* fertilization" was defined as "any fertilization of human ova which occurs outside the body of a female, either through admixture of donor human sperm and ova or by any other means" id. (adding § 46.203(g)). Although the preamble to the regulation implementing the EAB requirement stated explicitly that research on human embryos created through IVF was not intended to be included within the scope of the regulation, id. at 33527 (discussing the "nonimplanted product" of IVF), DHHS nevertheless subsequently took the position that human embryo research was within the scope of IVF research and therefore subject to EAB review. 59 Fed. Reg. 45293 (Sept. 1, 1994). One EAB was convened between 1978 and 1980 and concluded that IVF was ethically acceptable providing certain safeguards were followed. However, no action was taken based on this recommendation, and no EABs were chartered after 1980. Consequently, NIH did not fund any IVF or embryo research protocols.

49 Pub. L. No. 103-43, § 121(c), 107 Stat. 122, 133 (1993) (codified at 42 USC 289g note); 59 *Fed. Reg.* 28276 (June 1, 1994) (rescinding 45 CFR 46.204(d)).

50 59 *Fed. Reg.* 28874, 28875 (June 3, 1994) (notice of meeting); National Institutes of Health, *Report of the Human Embryo Research Panel* at ix (1994) (hereinafter *Panel Report*).

51 *Panel Report*, supra note 50, at ix.

52 Id. at 76. In contrast, the panel determined that "embryonic stem cell research that uses deliberately fertilized oocytes" warranted further review. Id. at 79.

53 Id. at 79.

54 Id. at 27. The panel also recommended federal funding for the deliberate creation of embryos (not intended for transfer) for research under very limited circumstances. Id. at 77. In December 1994, shortly after an NIH Advisory Committee adopted the panel's recommendations, President Clinton directed NIH not to fund research involving the creation of embryos for research purposes. John Schwartz and Ann Devroy, "Clinton to Ban U.S. Funds for Some Embryo Studies," *Washington Post*, December 3, 1994, at A1.

55 *Departments of Labor, Health and Human Services, Education, and Related Agencies Appropriations for 1996: Hearings Before a Subcomm. of the House Comm. on Appropriations*, 104th Cong., 1st Sess. 1480 (1995).

56 See, e.g., Thomas S. Giles, "Test-Tube Wars; Federal Embryo Research; Bioethics," *Christianity Today*, Jan. 9, 1995, at 38; S. Boyd Robert, "Panel Backs Research on Human Embryos; Suggestion Sparks Fierce Ethical Debate," *Des Moines Register*, Sept. 28, 1994, at 5.

57 Giles, supra note 56.

58 *Report on Markup of H.R. 2127, Departments of Labor, Health and Human Services, and Education, and Related Agencies Appropriations Act, 1996, in the House Committee on Appropriations*, July 20, 1995, available on LEGISLATE.

59 Jay Dickey, "Lethal Experiments," *Washington Post*, September 29, 1995, at A27 (op-ed).

60 The dissent stated:

> This provision goes much further than President Clinton's current policy, which only prohibits federal funds for the creation of human embryos for research purposes. We feel that the President's is the wiser policy, and that the current provision will deprive the American people of the medical advancements that such research would provide.
>
> Early stage embryo research involves examining how cells develop and divide once sperm and egg meet and form an embryo. This research could lead to advancements in the prevention of pregnancy loss and infertility, the diagnosis and treatment of genetic disease, the prevention of birth defects, and the prevention and detection of childhood and other cancers. H.R. Rep. No. 104-209, 104th Cong., 1st Sess. 384–385 (1995).

61 Pub. L. No. 104-99, supra note 42. The provision was also debated on the floor of the Senate. Senator Bob Smith, a proponent of the ban, explained that it "prohibits the use of taxpayer funds to create human embryos, to perform destructive experiments on them, and ultimately, to destroy and discard them" 142 *Cong. Rec.* S429 (daily ed. Jan. 26, 1996). He also stated that "[t]axpayer dollars should be used to protect and uphold human life, not to destroy it" id. Senator Barbara Boxer opposed the ban, stating that it would "severely restrict high-quality scientific research that could lead to a variety of beneficial medical treatments," including the development of universal donor cells and tissue to treat victims of Alzheimer's disease and Parkinson's disease. Id. at S432.

62 142 *Cong. Rec.* H7339 (daily ed. July 11, 1996) (statement of Rep. Nita Lowey introducing the amendment).

63 Id. at H7339-H7343 (statements of Reps. Johnson, Fazio, Pelosi, and Waxman).

64 Id. at H7340 (statement of Rep. Porter).

65 Id. at H7339, H7341 (statements of Reps. Dickey and Smith).

66 Id. at H7340 (statement of Rep. Wicker).

67 Pub. L. No. 105-78, supra note 42.

68 Pub. L. No. 105-277, supra notes 39 and 40.

69 S. Rep. No. 105-300, 105th Cong., 2d Sess. 274 (1998).

70 Id. at 133.

71 See text accompanying notes 59, 65, and 66 supra.

72 See text accompanying notes 55 and 64 supra.

73 NIH's Human Embryo Research Panel referred to research involving "the *development* of embryonic stem cells," which could be interpreted as a reference to the *derivation* of the stem cells, or to their development into differentiated cells for therapeutic use, or both. See text accompanying notes 52–54 supra (emphasis added). NIH Director Varmus testified in 1995 that the NIH could not pursue embryonic stem cell research "until we're able to use early stage embryos," but there is no indication that he then contemplated the future scientific development in which stem cells derived from an embryo by one researcher could be made available to other researchers for their use. See text accompanying note 55 supra.

74 *Amoco Prod. Co. v. Village of Gambell*, 480 U.S. at 548.

75 *United States v. Locke*, 471 U.S. at 95. Some members of Congress might have, if they had considered the question, concluded that federal funds should not be used for research on stem cells that were derived from embryos during privately-funded research. However, the legislative history does not demonstrate that this question was considered by any member of Congress. To conclude that such stem cell research is prohibited would require that the prohibition be interpreted more expansively than was clearly articulated by Congress. See *United States v. Locke*, 471 U.S. at 95 ("[T]he fact that Congress might have acted with greater clarity or foresight does not give courts a *carte blanche* to redraft statutes in an effort to achieve that which Congress is perceived to have failed to do."); *Biodiversity Legal Found. v. Morrissey*, 146 F.3d 1249, 1254 (10th Cir. 1998) ("we are not inclined to impart meaning where Congress fails to do so itself.") (internal citations and quotations omitted).

76 *Continued Management Concerns at the NIH: Hearing Before the Subcomm. on Oversight and Investigations of the House Comm. on Commerce*, 105th Cong., 1st Sess. (1997).

77 Id. at 34 (reprinting letter from Kate A. Berg, Ph.D., Deputy Scientific Director, NCHGR, NIH, to Wendy Fibison, Ph.D., Georgetown University Medical Center (Oct. 10, 1996)).

78 According to the researcher, Dr. Mark Hughes, the DNA was extracted from embryos by personnel in IVF clinics and sent to him for his analysis. If he found no evidence of genetic defect, the embryos were implanted. Eliot Marshall, "Embryologists Dismayed by Sanctions Against Geneticist," 275 *Science* 472 (Jan. 24, 1997).

79 *Continued Management Concerns at the NIH*, supra note 76, at 1, 9 (statements of Rep. Barton and Dr. Varmus).

80 Failure of the DHHS General Counsel to explain the possible inconsistency could weaken the agency's position if its decision to fund stem cell research were challenged in court. Courts generally accord significant deference to an agency decision that is based on a reasonable interpretation of a statute it administers, *Smiley v. Citibank*, 517 U.S. 735, 739, 740–41, 744–45 (1996); *Chevron U.S.A., Inc. v. Natural Resources Defense Council, Inc.*, 467 U.S. 837, 843 (1984). Even where such deference is accorded, however, courts accord less deference to an interpretation that constitutes an unexplained departure from a prior consistently held view of the agency. *Smiley v. Citibank*, 517 U.S. at 742; *Good Samaritan Hosp. v. Shalala*, 508 U.S. 402, 417 (1993); *Motor Vehicle Mfrs. Ass'n v. State Farm Mutual Auto. Ins.*, 463 U.S. 29, 57 (1983); see also Singer, *Sutherland Statutory Construction* § 49.05 at 19 ("If an agency adopts a new rule, which constitutes a departure from past policy or practice, it must, at a minimum, explain its actions with reference to the objectives underlying the statutory scheme it purports to construe.").

We note that an agency counsel's legal opinion is not itself a formal agency policy or decision (see *Smiley v. Citibank*, 517 U.S. at 743), although it may subsequently be embodied in a formal agency decision. In addition, if challenged in court, the agency's decision must be supported on the basis of the reasons it offered at the time of the decision. *Motor Vehicle Mfrs. Ass'n v. State Farm Mutual Auto Ins.*, 463 U.S. at 43; *SEC v. Chenery Corp.*, 332 U.S. 194, 196 (1947).

81 *Continued Management Concerns at the NIH*, supra note 76, at 47–48; see also id. at 10–11.

DELIBERATING INCREMENTALLY ON HUMAN PLURIPOTENTIAL STEM CELL RESEARCH

Commissioned Paper
John C. Fletcher
University of Virginia

Introduction[1]

The National Bioethics Advisory Commission (NBAC) faces major choices about its deliberations on ethical and public policy issues of human pluripotential stem cell (PSC) research. PSCs are more specialized than the totipotent cells in the four-cell human embryo, each of which can develop into a total individual. PSCs have the potential to develop into several (but not all) of the various cells in the body. In late 1998, supported by private funds, two groups of scientists concurrently reported laboratory culture and growth from PSCs of several cell lines. One group derived PSCs from tissue of the gonadal ridge of aborted fetuses.[2] The other group derived human PSCs from the inner cell mass of excess embryos donated for research.[3] Due to controversy about the moral legitimacy of deriving PSCs from these sources, and for another reason,[4] President Clinton requested a "thorough" NBAC review of the issues "balancing all ethical and medical considerations." As NBAC's report to the President is due in June 1999, an added concern is timeliness.

"How thorough is thorough?" and "What is the right balance...[of considerations]?" are fitting questions. NBAC requested this paper to assist Commissioners and staff with these queries in public bioethics.[5] In the face of a virtual impasse between conservatives and liberals about the moral status of embryos and fetuses, NBAC must be humble about its ability to reconcile these views. In my view, the major goal of NBAC's response to the President is to motivate a spirit of compassionate compromise about the federal role in PSC research.

To this end, this paper has three major parts and an appendix. The first three parts of this paper concentrate mainly on the "public" or political aspects of PSC research, with modest attention to moral arguments. The appendix develops arguments in "bioethics" to support the policy recommendations in Part III.

- Part I discusses three moral problems or concerns in PSC research: 1) the moral legitimacy of access to four sources of PSCs, 2) various uses of PSCs in research in a context of three stages of activities aimed towards human trials of cell-replacement therapies, and 3) the cumulative moral effects of a ban on federal funding of embryo research. This part also discusses the scope of a full review and places federal funding of embryo research in a broader history of controversies and prohibitions on federal funding of research in embryonic and fetal life.

- Part II discusses the tasks of NBAC (see Table 3) on PSC research and proposes an alternative to a full review, i.e., an incremental or case-by-case approach to four sources of PSCs: 1) fetal tissue after elective abortion, 2) excess human embryos donated in the context of infertility treatment, 3) embryos to be created by somatic cell nuclear transfer (SCNT), and 4) human embryos created for research. An incremental approach has strengths and weaknesses.

- Part III is a policy argument for federal funding of embryo research, using donated excess embryos, to support pre-clinical and clinical PSC research. Specific conditions to effect this compromise are proposed in Table 4. The premise of these conditions is that federal funding to derive PSCs from donated excess embryos will be a "last resort" to gain knowledge required for Stage 2 (pre-clinical) and Stage 3 (clinical trials) studies of cell-replacement therapies in humans.

- The appendix is a moral argument to bolster the policy argument in Part III. The debate about federal funding of embryo research is frozen between two polarized positions on the moral status of the human embryo. Is there a third possibility? The argument proposes that the claims of two ethical principles, named for their major proponents, provide enough common moral ground from which to chart a public policy of incremental and regulated federal funding of PSC research, including derivation from donated excess embryos. The obligations created by these principles and supplemented by other principles are also resources for long-term reform of overly prohibitive federal policies on fetal and embryo research.

Earlier national commissions and expert panels on fetal[6] and embryo[7] research compiled an impressive record. NBAC can build on this record in a new scientific context of stem cell biology and SCNT technology. Part II begins with NBAC's tasks in relation to PSC research.

Part I. Human PSC Research: Clinical Promise and Moral Concerns

A. Human PSC Research

Human PSC research serves unprecedented scientific understanding of human cell development, gene function, and other biological questions.[8] Research with embryonic germ cells derived from fetal tissue or embryonic stem cells derived from embryos after *in vitro* fertilization (IVF) is only one part of an exploding field of stem cell biology. Another part is research with stem cells found in adult animals.[9] Scientists are moving beyond research in the mouse and higher animals[10] to design novel experiments with human PSCs.

The clinical promise of PSC research is cell-replacement therapy for disorders caused by early cell death or injury.[11] Many problems and inherent dangers must be overcome before clinical trials of PSC derived therapies will be ethically acceptable. If realized, this approach to treatment could be of profound benefit to patients and society. Scientists envision effective treatment for common diseases and maladies; e.g., Type I diabetes, Parkinson's disease, Alzheimer's disease, liver and heart disease, injuries to the spinal cord, and many more. Cell-replacement therapies may supplant[12] the "half-way" therapies (e.g., organ transplants, hemodialysis, enzyme replacement, etc.) now the standard of care. Thirty-three Nobel laureates' letter expressed these hopes to the President and members of Congress.[13]

Much basic research must precede any trials of therapy. Dr. Thomson[14] and other experts caution that scientific and pre-clinical foundations for trials of cell-directed therapy may require five years of work or longer. Consensus exist in the scientific community that support by the National Institutes of Health (NIH) and the National Science Foundation (NSF) of PSC research will enrich and hasten clinical trials of cell-replacement therapies. Learning whether therapeutic uses of embryonic stem cells to generate cell lines are irreversibly tumorigenic[15] is a high priority in the near future. Research with higher animals and human embryonic stem (ES) cell lines should resolve this issue one way or the other.

B. Moral Problems of PSC Research

Alongside these hopes, three difficult moral and public policy concerns confront scientists, policy makers, and the public: the moral legitimacy of access to sources of PSCs, considerations of uses of PSCs in research, and the moral effects of the current federal ban on embryo research.

1. Access to Sources of PSCs

Table 1 ranks access to sources of PSCs by degree of moral and legal acceptability and of moral controversy.[16] The discussion refers to access to sources as "Cases" 1, 2, 3, and 4.

Table 1. Sources for Deriving PSCs

Case 1.	PSCs derived from human fetal tissue following elective abortion (e.g., Gearhart research).[17]
Case 2.	PSCs derived from human embryos available in excess of clinical needs to treat infertility by IVF; with informed consent, parents donate excess embryos for research (e.g., Thomson research).[18]
Case 3.	PSCs to be derived from human (or chimeric) embryos generated asexually by SCNT (using enucleated human or animal ova).[19]
Case 4.	PSCs to be derived from human embryos created, with informed consent, from donated gametes for the sole purpose of research.[20,21]

Federal Policy on Funding Embryo Research. Federal agencies may not legally fund any research to derive PSCs from embryos, although "therapeutic" embryo research is permitted by the language of the ban.[22] In spite of these strictures, federal policy is changing to permit partial funding of PSC research. The General Counsel, Department of Health and Human Services (DHHS) advised[23] the NIH that it can legally fund "downstream" research with PSCs derived by private funds but not derivation of PSCs at the blastocyst stage. This legal opinion is controversial, and it was silent on the moral issues discussed in the next section. However, if a) Congress allows the legal opinion to stand and b) the NIH successfully oversees and guides such "downstream" funding,[24] this form of federal funding may yield answers to some of the earliest scientific questions required to plan for clinical trials. At that point in the future, Congress may be open to modify the ban on federal funds for embryo research to hasten clinical trials. However, it must be shown clearly that derivation of PSCs in Case 2 is the only way to proceed to human trials. Part III of the paper gives a policy argument for federal funding of embryo research based on the need to derive PSCs from excess embryos, as in Case 2. The argument makes two assumptions: first, such conditions will develop within two to five years; second, at the time that Congress must consider federal funding for deriving PSCs from excess embryos to hasten clinical trials, no less morally problematic alternative will exist.

Transition: Can Access Be Separated from Uses? Before a section on uses, the "separability" of the issue of access to embryos ought to be examined. The NIH has already officially separated the issue of access from uses. The agency's legal advice was altogether silent on ethical issues. Nonetheless, the Rabb memorandum loudly begs the question of the morality of derivation.[25]

The Rabb memorandum makes moral sense only on a premise that the legal permissibility of deriving embryos in some states with private funds is a moral floor for derivation. This paper adopts a premise, among others, that the morality of access to embryos is logically and morally prior to the issue of uses of PSCs in research. Further, a social practice of access to embryos (however created) for research requires a persuasive moral justification in a society and Congress deeply divided over the subject.

Ironically, the lack of a unified national policy for embryo research enables the NIH to take this direction. The history of congressional inaction on infertility and Congress's ban on federal funds for embryo research leaves such research in the private sector unregulated. The legality of PSC research in the various states is a complex topic.[26] Andrews writes:[27]

> In 24 states, there are no laws specifically addressing research on embryos and fetuses.[28]
> In those states, embryo stem cell research is not banned. However, other legal precedents covering informed consent, privacy, and commercialization, come into play.

Based on the constitutionality of the choice of elective abortion, in 1990 the Human Fetal Tissue Transplantation Research Panel (hereinafter, Fetal Tissue Panel)[29] argued that it could separate its deliberations on the morality of the uses of fetal tissue from the morality of abortion.[30] The Fetal Tissue Panel took no explicit position on the morality of abortion. In theory, NBAC could take the same approach with embryo research and argue that uses of PSCs are morally divisible from access to embryos.

The Fetal Tissue Panel's approach did reduce controversy enough to enable it to deliberate. However, there are important reasons for NBAC to be less confident in using this approach with embryo research. One reason is that the trustworthiness of law as a floor for morality is quite variable. Law does express moral beliefs and values. Rightly seen, law is a floor for morally permissible acts but not a ceiling for moral ideals. Nonetheless, one finds strong and weak floors in buildings and in law. Different elements contribute to these strengths and weaknesses. Absence of legal prohibition is a weak floor for moral acceptability of access to embryos compared with the strength of laws that bar unwarranted intrusions into a lawful choice, e.g., federal court decisions and state laws protecting the liberty to choose abortion. Collective moral experience and scholarly ethical reflection (on both sides of the issue) can be likened to strength-giving elements.

These elements are plentiful in the floor of law on abortion compared with federal or state law on embryo research. Congress first banned federal funds for embryo research in 1995 with no careful attention to consequences for patients, science, or society. In 1994, Andrews wrote: "embryo research *per se* has rarely been the subject of state legislative scrutiny."[31] Since that time, legislative activity in several states has increased. However, as reported by Andrews, state laws relevant to PSC research still vary widely.[32]

The second reason is that a lack of legal prohibition does not bar some activities and choices that are open to serious moral challenge, such as sex selection by prenatal diagnosis and selective abortion. Embryo research is open to moral challenge,[33] and the moral justification for embryo research is much more persuasive than almost any argument that can be given for sex selection by prenatal diagnosis and abortion. However, to use the absence of law as an argument for embryo research would invite comparisons with such dubiously defensible uses of prenatal diagnosis. For these reasons, NBAC cannot proceed as confidently on the basis of absence of state law on embryo research as prior panels and commissions were able to do from a basis in law on abortion. If NBAC did so proceed, an objection would surely be that NBAC avoided the moral debate on access but smuggled in a permissive position beneath a shaky legal argument. Access to live embryos for research requires a stronger moral defense than one afforded by an absence of law.

Also, advances in stem cell biology have dramatically changed the scientific context. NBAC can contribute an ethical analysis of embryo research in this new context and also account for several criticisms[34] of the moral perspective adopted by the NIH Human Embryo Research Panel (hereinafter, Human Embryo Panel or Panel) in 1994. Can the NBAC come to consensus on access to embryos for research? This question must be explored. Whatever the outcome, NBAC can then assess public policy recommendations that it desires to make.

2. Uses of PSCs in Research

The argument is that the morality of access to sources of PSCs is logically and morally prior to questions about uses. Concerns about uses of PSCs are beside the point if it is morally unacceptable to access any sources, e.g., in Cases 1–4. However, is it morally acceptable for scientists to access all *four* sources for PSCs? Part II offers ethical and public policy reasons why NBAC can be more confident about the moral acceptability and rationale for federal funding to derive PSCs in Cases 1 and 2 than it can be at present for access to sources in Cases 3 and 4.

The rest of this section on uses assumes the arguments of Part II and a premise that there are no overriding moral reasons why society or patients must forgo benefits from research in Cases 1 and 2. Part II offers scientific, ethical, and political reasons for NBAC to distinguish Cases 3 and 4 sharply from Cases 1 and 2.

Ideally, reflection on permissible uses would occur in a cultural framework of settled ethical and legal boundaries for embryo research, such as exist in the United Kingdom. However, the United States has two universes of science and the funding of science: public and private. Descriptively, when it comes to embryo research, these divided universes respectively adopt an overly permissive morality in the private sector and impose an overly protective morality in the public sector.[35] Nonetheless, as a bioethics commission, NBAC is obliged to address ethical issues as if it were possible for moral purposes to unify these two universes.[36] NBAC's work on the ethics of PSC research could eventually contribute to a unified public policy on embryo research. However, if NBAC's public policy recommendations are to be useful in the present context, it must account for the reality of these two dichotomous moral spheres within one nation.

Three Types of Uses of PSCs in Research. Dr. Harold Varmus[37] and others[38] describe three general areas of research uses of PSCs: 1) studying the efficiency and regulation of human PSC and cell line differentiation in culture, 2) studying toxicity and beneficial effects in the context of drug development, and 3) growing cells of different types for transplants to repair or replace patients' injured or dying cells.

Questions of scientific merit, utility, and linkage to larger disputes ought to be raised about all proposed uses of embryos in research. Deliberation ought to focus first on issues of scientific justification and utility and secondly on potential linkage to unresolved and controversial uses. If embryos are to be used in research, the scientific reasons need to be coherent and defensible in the processes of scientific and Institutional Review Board (IRB) review. The number of embryos needed for experimentation is related to an obligation to use the minimal number required to gain the desired knowledge. This issue of number is also related to the supply of embryos for research. Supply is shaped by practices in infertility treatment centers and the percentage of IVF embryos that will be eventually discarded.[39] Some proposed uses of PSCs are linked to large and unresolved controversies still facing society and policy makers. For example, if NBAC "thoroughly" reviewed Cases 3 and 4, it would need to revisit fully the debates about human cloning and human germline gene transfer.

Stages of PSC Research.[40] PSC research, in preparation for clinical trials of cell-replacement therapy for human beings, will likely proceed through three general stages:

Stage 1 [Investigative]: This stage involves basic science experiments not including studies in animals. Although knowledge gained in this stage can clearly be relevant to safety issues in humans, the work is mainly investigative, aimed at knowledge about guiding differentiation of specific cell lines and properties of human PSCs and cell lines derived from the most available sources of PSCs, i.e., embryonic stem cells and embryonic germ cells.[41] Dr. Brigid Hogan[42] noted differences in DNA modification between mouse embryonic germ (EG) and ES cells. The differences may be due to methylation, a process that protects recognition sites of DNA and plays a regulatory role in gene expression. Cells derived from EG cells may have less methylation than normal. The scientific and (possible) clinical import of these differences needs exploration. Dr. Hogan stressed the need to access both types of cells for this purpose.[43] Dr. Gearhart[44] outlined basic questions about PSCs derived from excess embryos (Case 2): i.e., about ways to assay blastocysts for their potential of yielding PSCs (perhaps by searching for genes that predispose for this capacity), to produce more cell lines than the five grown by Thomson's work, as well as other intrinsic or extrinsic factors that foster success. Presumably, using PSCs and cell lines in research on drug development will build upon prior research on differentiation and knowledge about cell lines grown from PSCs from various sources. Another task in this stage is to study differences between cell lines derived from ES and EG cells and those grown from stem cells recovered from adults or children.

Stage 2 [Preclinical]: This stage involves experiments in animals and in the laboratory with human cells to test the desirability of human trials.[45] This stage involves what the Food and Drug Administration (FDA) terms as "Preclinical Pharmacology and Toxicology." These studies go much beyond the proof of concept and include extensive examination of potential toxicities, determination of a "no effect" level to establish initial doses of the product in early clinical trials, examination of chronic versus acute effects, etc. At this stage it is mandatory to show that purified cell lines derived from embryonic stem cells and other sources are not tumorigenic in animals other than mice, as well as answers to other questions about safety which are given in Table 4 below. Stage 2 aims at consensus about the scientific feasibility and moral justification of human trials for one or more candidate diseases.[46]

Stage 3 [Clinical Trials]: If a consensus can be reached on the pre-clinical and ethical considerations of the first trial or trials in humans, Stage 3 is a series of clinical trials in humans of investigational cell-replacement therapies. Stage 3 aims to answer these questions: Is cell-replacement therapy safe and effective in human beings?[47] Stage 3 will be done under FDA regulation as an Investigational New Drug application.

Possible and Controversial Uses. If it is possible to generate human (chimeric) embryos by SCNT, and Case 3 is feasible, many of the same studies described in each stage above will need to be repeated in this case.

Will there be significant differences between the properties of PSCs derived from SCNT-generated human embryos and cell lines grown from EG and ES cells?

Eventually, studies will be needed of the feasibility of autologous cell replacement therapy to avoid the graft-vs.-host reaction. As discussed by Solter and Gearhart:

> ...at the outset, it was realized that the full therapeutic potential of ES cells will depend on using ES cell lines derived from the patient's own cells....[48]

Ought these chimeric embryos, if created by cloning technology, be regarded with the same degree of respect deserved by sexually created embryos? This question has already surfaced in NBAC discussion.[49] Solter and Gearhart also raise an ethical question as to whether creating embryos by SCNT to be used *only* to derive ES cells is permissible.[50] The answers to these questions needs careful reflection related in part to scientific information not now available.

Research embryos (Case 4) as a source of PSCs will be needed to create banks of multiple cell lines representing a spectrum of alleles for the major histocompatibility complex.[51] This goal requires that ova and sperm of persons with specific genotypes be selected to create embryos from which to derive particular PSCs. This use involves selection of specific embryos for research (Case 4) and is an activity similar to studying when alleles begin to express DNA in the embryo to understand origins of particular diseases.[52] Infertility centers, using private funds, now create embryos to study the viability of frozen ova or to improve the medium in which embryos grow after IVF. We must expect that some privately funded research with PSCs will occur in the context of Case 4, regulated only by the ethics of professionals. Such protection is at best porous in a marketplace spawning conflicts of interests that can easily overwhelm traditions of professional self-regulation in medicine and research.[53] Unification of oversight and regulation of public and private embryo and fetal research is a goal of this paper. This is a daunting long-term goal, only exceeded in difficulty by the higher goal of extending constitutionally guaranteed protections to all human subjects of research in the private as well as public realms.[54]

PSCs may have potent uses in research on human gene transfer in the hope of treating genetic disorders. Will PSC-assisted gene transfer resolve major technical problems in using exogenous vectors to introduce corrective DNA to target sites? Dr. Austin Smith's testimony to NBAC[55] and a NIH discussion paper on cloning point in this direction.[56] Pincus, et al.[57] discuss the use of neural stem cells that persist in the adult brain as a vector to do gene therapy in neurodegenerative diseases. Their review cites a successful neonatal experiment in a mouse model for mucopolysaccharidosis.[58] Uses of PSCs in the context of human somatic cell gene transfer raise no new ethical questions. However, any use of PSCs in tandem with an intent of inheritable genetic modifications in the DNA of gametes or preimplantation embryos does raise a host of old and new issues. Dr. Erik Parens's testimony to NBAC notes how PSC research will converge into experiments to treat the DNA of human embryos and prevent genetic diseases in children-to-be.[59] A truly "thorough" and far-ranging review of PSC research would examine the scientific and ethical issues in this vast topic.

Summary of Section on Uses. The most immediate uses of PSCs in federally funded research follow derivation in Cases 1 and 2. Future uses of PSCs derived from embryos in Case 3 are dependent on the pace of scientific advances. Much more successful research in animals must precede any attempt to answer the question: can human embryos be generated by SCNT? If so, there will be a need to compare the properties of PSCs derived as in Case 3 with PSCs from other sources. Any future guidelines for deriving PSCs from embryos from any source (Cases 2–4) will need a safeguard against transferring a research embryo for implantation.

Uses of PSCs derived from embryos in Case 4 will be confined to support by private funds until such time as the electorate and Congress are willing to unify oversight and regulation of embryo research. The NIH will inevitably face the question of using a "downstream" approach to fund uses of PSCs from Case 4, because such research is necessary to create banks of multiple cell lines. When this question comes to the fore, it could be an opportunity to reconsider the advice of the Human Embryo Panel's report[60] and limitations placed on federal funding for embryo research involving Case 4.

A final controversial prospective use of PSCs would be to assist in experiments involving deliberate modifications of DNA in gametes or embryos that are inheritable in offspring.[61] In the light of successful germ-line transfer in several animal species, the topic of human germ-line experimentation has received new attention and evaluation.[62] The FDA regulates somatic cell therapy and gene therapy.[63] This authority would doubtless extend to experiments in inheritable genetic modifications aimed at therapy in DNA and prevention of disease in offspring. Current public policy is that the NIH's Recombinant DNA Advisory Committee (NIH-RAC) "will not at present entertain proposals for germ-line alterations."[64] However, the NIH-RAC, a public body, and the FDA would collaborate jointly in prior review and approval of any such proposed experiments in the future. Members of the scientific community are obliged to submit any such proposals to this long-standing and successful public process.

Table 2 can serve as a checklist, along with guidelines of the NIH on funding PSC research,[65] for IRBs considering protocols for PSC research.

Table 2. Points to Consider for Proposals or Protocols for PSC Research

1. How will the PSCs be derived?
2. Does the investigator have a protocol and IRB approval for access to this source? A consent process that has been IRB approved?
3. What is the minimal number of fetuses/embryos required to do the study?
4. If the investigator will access fetal tissue, are the project and the consent process in compliance with Public Law 103-43, Part II?
5. Is the research design the best possible, given the state of the art?
6. Will the research plan yield answers to the questions being posed?
7. If the PSCs in this project will be derived by SCNT, the investigator must stipulate that no embryo made by cloning technology would be used for reproduction.
8. If the PSCs in this project are to be used in the context of a clinical trial of human gene transfer, the investigator must stipulate knowledge that approval of the FDA/NIH-RAC is required.

3. The Ban on Federally Funded Embryo Research

After the elections of 1993, Congress lifted a moratorium on federal funding of IVF research that required the approval of an Ethics Advisory Board (EAB).[66] After the elections of 1994, a new Congress banned federal funding for embryo research and ended a brief period of NIH hopes to fund improvements of IVF and other projects involving human embryos.[67] This section will discuss the history of public policy on federal participation in various forms of fetal and embryo research.

Turmoil over PSC research is a new chapter in a long public policy history. Since 1973, Congress[68] has adopted, in my view, overly protectionist policies regarding research activities involving the fetus and at the beginning of human embryonic development. No other ethical issues in research rise to the same level of public controversy.[69] Understanding this history will provide perspective for Commissioners and the public on the place and future of PSC research in the federal sector.[70] The history includes Congressional restrictions on federal funding for fetal research, human fetal tissue transplant research, a pattern of Congressional inaction regarding infertility research, and the broader effects of the ban on federal funding for embryo research that prevent NIH and NSF involvement in basic research to gain knowledge relevant to cancers and other genetic diseases, infertility, contraceptive development and other areas.

Federal Regulation and Law on Fetal Research.[71] Justice Blackmun and a Supreme Court majority ruled in *Roe v. Wade*[72] that a fetus is not a person in the context of constitutionally protected rights. In the wake of the ruling, members of Congress were concerned about possible research exploitation of fetuses to be aborted.[73] In 1974, the law mandating IRBs also created the National Commission (NC) for the Protection of Human Subjects of Biomedical and Behavioral Research. The first mandate of Congress to the NC was for ethical and public policy guidelines for fetal research. The NC completed its work on fetal research within four months,[74] and its work was the basis for federal regulations.

The ethical framework of the NC's report was a three-sided compromise between liberal and conservative views on fetal research, with an added feature (to facilitate the compromise) for a national EAB to review and resolve problems in future protocols on fetal research.[75] First, guided by the principle of beneficence, the NC encouraged fetal research because of its benefits.[76] Any reasonable liberal view on fetal research could support the first point. Second, the NC sharply restricted fetal research under an equality-of-protection principle, especially to protect fetuses to be aborted from exploitation. The second point was a bold specification of a conservative viewpoint that was incompatible with a utilitarian ethos previously dominating U.S. research practices which had guided investigative research with living fetuses *ex utero*.[77] The NC, even in the face of *Roe v. Wade*, specified that societal protection of human subjects of research ought to be extended to fetuses, including fetuses in the context of abortion.[78] To make this compromise work in actual practice and policy, the NC envisioned an ongoing EAB as a resource for local IRBs and for developing national policy on research ethics.[79]

Federal regulations on fetal research followed on July 29, 1975, and the moratorium on fetal research was lifted. Notably, these regulations (45 CFR Part 46, Subpart B)[80] have consistently reflected a higher commitment to seek the benefits that fetal research can accomplish than any ever expressed by Congress. The regulations distinguish[81] between research to "meet the health needs" of the fetus, and research to develop "important biomedical knowledge which cannot be obtained by other means." The standard for research risks in the former case is "only to the minimum extent necessary to meet such needs" and in the latter case the standard is "minimal."[82] Knowing that there would be difficult cases of investigative fetal research with more than minimal risks, the NC expected an EAB to review such protocols in fetal research (and for other areas) and make recommendations to the Secretary, Department of Health, Education, and Welfare (DHEW), to whom the regulations gave authority to "waive" the minimal risk standard in fetal research.[83]

Following the elections of 1984 in which President Reagan was returned to office,[84] Congress hotly debated federal funding for fetal research and enacted legislation that is far more protectionist than the second point of the federal regulations. Public Law 99-158[85] imposed the "Golden Rule" on all federally funded *in utero* fetal research, thus effectively nullifying the "minimal risk" standard for the second category of fetal research.[86] This law ended federal funding of any fetal research carrying any degree of risk[87] including research into normal fetal physiology that involved fetuses in the context of abortion. At this point in time, it is important to note again the major difference between federal law and federal regulation on fetal research; the latter is significantly less protectionist than the former.[88] The appendix describes some of the costs to scientific knowledge and opportunities for responsible fetal therapy of prohibitions of federal funding of fetal research.

Federal Inaction on Infertility Research. Infertility is a significant public health problem and assisted reproductive technologies raise a wealth of complex ethical, social, and legal concerns.[89] Physicians define infertility as the inability to conceive after 12 months of unprotected intercourse or to carry a fetus to term.[90] Epidemiologists distinguish between primary and secondary infertility. Primary infertility is determined by the number of infertile couples with no children. Secondary infertility is becoming infertile after having one or more children. Using these categories, the National Center for Health Statistics estimated in 1988 that 2.3 million married couples were infertile. This translates into 8 percent (1 in 12) of the total number of all married couples in the United States.[91] The National Survey of Family Growth estimated in 1995 a slight drop in infertility to 2.1 million couples or 7.1 percent of 29.7 million married couples with wives of childbearing age.[92] Primary infertility among women has doubled due to the trend in large numbers of women deferring marriage and childbirth.[93] Although primary prevention is the optimal approach to the public health problem, the major approach to treatment of infertility has been to combine fertility drugs with IVF or other methods and sites to fertilize ova. The relative successes of assisted reproductive technology has created a large industry and significant growth in infertility services.[94]

This section will show how a pattern of federal inaction regarding infertility research contributed to the ban on federal funding for embryo research. This story mainly concerns research on the safety and efficacy of IVF. A *de facto* moratorium was placed on federal funding of IVF in 1974 (along with fetal research) until an EAB could make recommendations to the Secretary, DHEW. In May 1978, the EAB reviewed a proposal from Vanderbilt University received by the NIH in 1977 and approved by a study section. An EAB was chartered in 1977 and convened in 1978. In May 1979, the EAB recommended approval[95] for federal funding on safety and efficacy of IVF and embryo transfer in the treatment of infertility. The approval was also for study of spare, untransferred embryos, provided researchers had IRB approval and the informed consent of women who would receive any transferred but studied embryos. The EAB set a 14-day cutoff for studying embryos *in vitro* on condition that gametes be obtained only from lawfully married couples. Richard McCormick, also then an EAB member, participated in this compromise. He departed from a Vatican position against *any* technologically assisted pregnancies, even in lawfully married couples. This position was later promulgated as moral dogma.[96] Secretary Califano published the EAB's report for public discussion but resigned at President Carter's request in late September 1979. No Secretary of DHEW or DHHS has approved the EAB's recommendations. No federal support of IVF, except with animals, has ever been permitted. Two causes contribute to this pattern of reluctance: a view of infertility as a condition non-deserving of government support in research and moral concern to avoid using embryos in research.[97]

DHEW Secretary Patricia Harris allowed the EAB to lapse on September 30, 1980, when its charter and funding expired. She did so, according to one view,[98] to avoid overlap with the President's Commission (PC) on Ethical Problems in Medicine and Biomedical and Behavioral Research, which was planned to succeed the NC. Congress had created the PC largely to study ethical problems in medicine, but by 1980 had not yet appropriated funds for operations. However, Ms. Harris disbanded the EAB fully aware that it was the only lawful body that could recommend "waiver" of minimal risk in research.[99] NIH Directors in the 1980s appealed to various Secretaries of DHHS to recharter the EAB and approve its recommendations for IVF research. No action was taken until, under pressure from Congress, Dr. Robert Windom, Assistant Secretary for Health, announced on July 14, 1988, that a new charter for an EAB was as to be drafted and discussed in public, and a new EAB was to be appointed.[100] A draft of a charter was published in the *Federal Register* and comments invited, but no approval was given before the transition to the Bush Administration, which never acted on the issue. As stated above, following the elections of 1993, Congress nullified the EAB requirement for IVF research, but no NIH funding of research involving human embryos in this context occurred.

Then Congressman and now Senator Ron Wyden (D-OR) has been the main champion of federal involvement in infertility research. While in the House, Wyden held hearings on consumer protection issues involving IVF clinics.[101] Long concerned with the issue of regulating ART programs, the only federal leverage available has been in the wake of the Clinical Laboratory Improvement Amendments (CLIA), which regulates a wide range of laboratory procedures.[102] Due to Sen. Wyden's efforts, the 1992 Fertility Clinic Success Rate and Prevention Act requires the Centers for Disease Control and Prevention (CDC) to develop model standards for state certification of embryo laboratories.[103] The standards concern issues of performance of procedures, quality control, records maintenance, and qualifications of employees. However, the law states that the standards do not regulate the practice of medicine in ART programs. Aside from the laboratories in which embryos are generated for IVF, the entire spectrum of infertility services and the research that is conducted within these centers is unregulated except by the canons of professional ethics.

Federal Regulation and Law on Fetal Tissue Transplant Research. In 1986 neurosurgeons at the NIH's Clinical Center designed a study to give patients with Parkinson's disease the choice of an adrenal autotransplant or a fetal neural cell transplant. After approval of the project by an IRB, Dr. James Wyngaarden, NIH director, decided in October 1987 to seek higher review of fetal tissue transplant research by the Assistant Secretary of Health. In March 1988 (there being no EAB), Dr. Windom withheld approval, placed a moratorium on federal support of this activity, and asked that the NIH convene an advisory panel to consider a list of ten questions. His main concern was that the benefits of fetal tissue transplant research would induce women ambivalent about abortion to have one.

The NIH assembled the 21-member Fetal Tissue Panel, which in December 1988 approved a report[104] by a vote of 18-3 recommending federal funding of fetal tissue transplant research. As noted in Part I, the majority argued that the use of fetal tissue to treat disease was separable from the morality of abortion. The panel's reasoning was that fetal tissue transplant research was a type of cadaveric transplantation.[105] Three panel members with conservative theological views dissented due to the activity's association with abortion. The panel shaped 12 recommendations to oversee and guide fetal tissue transplant research in the hope that these would become federal regulations. The Fetal Tissue Panel's report, submitted to the director of the NIH and approved unanimously by his advisory council, was rejected by letter[106] without any public hearings or prior notification published in the *Federal Register*. Acting unilaterally and in violation of the understandings that had created the EAB, Secretary Sullivan continued the moratorium "indefinitely." He gave as his major reason that the Bush Administration and Congress opposed any funding of activities by DHHS that "encourage or promote abortion." The press[107] and a letter from Rep. Theodore Weiss to Dr. Sullivan[108] cited a memorandum from Richard Riseberg, DHHS counsel, saying that the extension of the moratorium was on a "shaky legal base" and could be actionable as a violation of federal law[109] requiring that such decisions be published in the *Federal Register* and made the subject of rulemaking.[110]

In 1990 Rep. Henry Waxman (D-CA) introduced the Research Freedom Act (H.R. 2507) to overturn the moratorium on fetal tissue transplant research. The full law, as subsequently passed by Congress in 1993 after a long struggle, a) restrains the Secretary, DHHS from imposing a ban or moratorium on research for ethical reasons without the concurrence of an EAB convened to study the question and b) establishes an authority within current law for federal support of fetal tissue transplant research.[111] The other provisions of the law relevant to the consent process and other safeguards will be described in Part II under Case 1.

Congress's Ban on Federal Support for Embryo Research. From an ethical perspective of beneficence and utility, the Human Embryo Panel's Report in 1994[112] made a strong case for federal funding for embryo research. Before the ban was passed, the threat of strong opposition from Congress towards any embryo research inhibited NIH approval for funding several clinically relevant projects that passed NIH scientific review in 1994[113] concerned with cancer, genetic research, infertility, and contraception. The Human Embryo Panel specifically recommended these lines of research.[114] After the ban, the NIH received no proposals involving embryo research. The appendix discusses the costs to science and the nation of forgoing NIH involvement and scientific peer review in these areas.

This historical review has provided background on why the ban on federal funding for embryo research continues a public policy[115] that embryos, like fetuses, deserve virtually absolute societal protection from destruction or harm in research activities. The language of the embryo ban on "risk of injury or death" is based on earlier federal law restricting funding for fetal research.[116] Readers should also recall that Congress also acted to deny federal funding (with three exceptions) for elective abortions in the Medicaid program.[117] In a democracy, the moral beliefs of an elected majority may prevail when it acts to deny federal funding for activities it views as ethically unjustified. Federal and state government may also use denial of funding to ameliorate the divisiveness of intractable moral disputes like abortion. Such actions are more understandable in a nation with sharply divided (public and private) systems of health care and research.

Do such laws violate scientific freedom? Denial of funding does not legally violate scientists' freedom of inquiry, although poorly justified legislative and executive actions clearly infringe on the values that underlie scientific and academic freedom. Any proposition that scientific freedom is constitutionally protected is highly debatable. In my view, Robertson clarified the legality of denial of funding long ago:

> The [scientist's] freedom to select the means [of inquiry and research] may refer to the freedom to think, to read, to write, and to communicate, the freedom to observe events or interactions, or the freedom to experiment. Freedom of inquiry in any of these senses does not, however, include a claim that the government must fund any particular activity designed to advance knowledge.[118]

Signs of Change. As noted above, change has begun in federal science policy and the interpretation of the embryo ban. Harriet S. Rabb, General Counsel, DHHS, ruled that the NIH could legally fund uses of PSC research but not activities deriving PSCs from embryos.[119] Assuming that Congress allows the Rabb ruling to stand, NBAC will still need to evaluate the moral arguments for and against access to embryos for research and attempt to reach a consensus position. Two directly opposing views, expressed by the Human Embryo Panel's report and the ban on federal funding of embryo research, now squarely confront one another in the nation's life. NBAC can clarify the moral concerns on both sides and search for moral consensus on mid-level issues, especially about Case 2.

Part II. An Incremental Approach to NBAC's Tasks

A. The Tasks of NBAC

Table 3 shows NBAC's four tasks in regard to ethical and public policy issues in PSC research.

Table 3. NBAC's Tasks on PSC Research

1. To clarify the ethical considerations relevant to deriving and using PSCs in research. NBAC must choose whether to focus on derivation and use from each source or only on the sources which have been reported to date, i.e., Cases 1 and 2.
2. To articulate consensus ethical standards to guide policy; i.e., what standards ought to guide public policy for federal funding of PSC research.
3. To recommend safeguards to contain or prevent abuses that have occurred or that could occur when and if policy is implemented.
4. To educate the public on the nature, promise, and risks of PSC research.

A truly "thorough" review requires completing each task—on both access and uses of PSCs—for all four cases. The review would also include, per Parens's argument, how PSC research converges into the longstanding debate about inheritable genetic modifications. Doing an exhaustive review is problematic beyond Cases 1 and 2. No scientific information exists to use in evaluating Case 3. Cloning somatic cells does produce viable embryos in animals. Cloning somatic cells to produce a sheep[120] as well as mice and cattle[121] has been done. Fusing adult mouse cells with enucleated mouse oocytes, followed by implantation and reproduction, has also been done.[122] However, despite the obvious clinical benefits envisioned in Case 3, deriving PSCs from cloned embryos has not even been done in mice.[123] In 1984, The Council for Science and Society in the United Kingdom formulated a working rule for ethical debate on new technologies:

> ...refrain from moral judgment on unverifiable possibilities—as notational cases rooted neither in the reality of experience nor in a specific context.[124]

This oft-broken rule is relevant to NBAC's tasks with Case 3. Also, other groups[125] are now studying intentional and unintentional germline gene transfer, one of the subjects of Parens's challenge to NBAC. An alternative approach may better fit the NBAC's tasks and timeline.

B. An Incremental Approach: Strengths and Weaknesses

NBAC and the nation face a group of cases or situations in which PSCs can be derived and used in research. How should NBAC morally deliberate about these cases? This part discusses an incremental or case-by-case approach to NBAC's tasks. Familiarity with this approach in science, ethics, and law is a strength. When presented with several morally problematic cases which appear to be similar, one proceeds incrementally, or case-by-case. Rather than beginning from first principles and working down or across, one begins with a case, asking: "What is morally at stake here?" In response to this question, the principles and moral rules linked to the case can be discerned and would also be discussed in extant literature about the case. Beginning with the most "settled" case (or in science the most proven experiment), one then works outward, case by case, to complete certain tasks in moral deliberation.[126]

Comparing and contrasting moral similarities and differences among cases is both a descriptive and evaluative task of this approach. One searches especially for dissimilarities so sharp as to conclude that a case differs in kind and type and does not belong to this "family" or that "line" of cases. One finally reaches the least settled and most problematic cases in a line or sees such clear differences between cases as to create a new branch or line of cases.

Another task is to discern the moral judgment linked to the case, as well as the guiding principles for the judgment that can hold from this case to a similar case. Among methodologies in ethics, this approach is known as casuistical reasoning.[127] After considering its weaknesses, the remainder of this part illustrates its use with these cases.

Case-by-case moral deliberation invites criticism from those whose method of moral deliberation is based on "an adequate account of morality as a public system that applies to all rational persons."[128] A case-by-case approach is bound to be less certain about the right account of morality than about moral fallibility. The point is that modesty about the place of ethical theory or systematization invites criticism.

Those with sharply divergent views on fetal and embryo research will also disagree with this approach. John Harris[129] argues that the distinction between Case 2 and Case 4 based solely on intention (to procreate or to make embryos for research) is weak. He argues for an all-or-nothing position.

> ...If it is right to use embryos for research it is right to create them for this purpose. And if it is not right to use them for research, then they should not be so used even if they are not deliberately created for this purpose (p. 45).

An incremental approach distinguishes between the degree of moral acceptability of Case 2 and Cases 3 and 4. Harris criticizes this interpretation as timid and evasive of the most important issue, i.e.,

> taking responsibility for what we knowingly and deliberately bring about, not simply what we are hoping for.... (p. 46).

A view that any research use of human embryos and fetuses is morally complicit with unethical acts of destructive selection and abortion will not concede moral acceptability of any of the four cases. This view would hold that an incremental approach is fatally compromised because it begins from a wrong premise in Case 1, namely, that access to human fetuses following elective abortion is morally acceptable. NBAC should expect criticism from both positions if it takes an incremental approach.

C. Case-by-Case Approach to the Four Cases

Case 1. The moral controversies associated with fetal tissue transplantation research were hotly debated in the 1980s and 1990s. Sufficient areas of moral consensus emerged through democratic processes to embody them in P.L. 103-43, appropriately named "The Research Freedom Act."[130] Deriving PSCs from fetal tissue after elective abortion is clearly the most settled case of the four before NBAC.

Some basic moral principles and rules are embedded in Case 1 and in the law permitting fetal tissue transplant research:

a) Beneficence-Based Considerations. Although open to challenge, a sufficient moral consensus emerged and has persisted through several sessions of Congress that society ought not to forgo the biomedical knowledge and/or therapeutic benefits to patients of research on transplants with fetal tissue obtained after elective abortions. A consequentialist argument strengthens the obligations of beneficence in Case 1. Namely, society and science ought not to forgo the uses of fetal tissue, especially since it would otherwise be discarded. Most of the consequences of allowing viable fetal tissue to go to waste are bad. This option would respect the moral views of opponents of abortion, but all of the parties who could benefit from research will lose if the opportunity is forgone. Uses of fetal tissue in transplant research are sometimes good for patients but almost always good for science.

The moral consensus that prevails about access to aborted fetuses to obtain fetal tissue is properly framed in negative rather than positive terms. The consensus is not that society should vigorously pursue access to aborted fetuses. Rather, it is that no overriding reasons compel society to forgo benefits from fetal tissue research to patients and science. The evolution of the morality of fetal tissue transplant research in this society contributes to the assurance with which NBAC can be confident that Case 1 is the most settled case for PSC research. If the arguments that condemned the research uses of fetal tissue because of association with elective abortion[131] had prevailed and dominated the moral consensus that emerged, very different moral principles and rules would be embedded in Case 1. This outcome did not occur. The moral debate about research uses of fetal tissue led to a consensus composed of elements drawn from arguments on both sides of the issue. The consensus permitted limited research involving fetal tissue with safeguards that protected society's interests in upholding a principle of respect for the intrinsic value of life and discouraging abortion when there is a reasonable alternative. These various concerns were specified and expressed through a law that permitted federal funding and defined the current public process for regulating fetal tissue transplant research.

b) Respect for Autonomy. Although some contest it,[132] there is a sufficient moral consensus that society ought to respect the autonomous choices of women who have made legal abortion decisions to consent to donate fetal tissue for research. The objection to this choice is that the woman abrogates this exercise of autonomy by the evil of the act of abortion and becomes, in effect, a morally defective decision maker. The objection might make a further point, i.e., the abortion right is grounded in an interest in preserving bodily integrity—which is an interest not implicated after the fetus is no longer in a woman's body.[133] One answer to these objections is that liberty rights as well as interests in preserving bodily integrity are at stake in abortion choices. The liberty right that is morally and legally protected by the constitution carries over into choices to donate fetal tissue. Abortion is a troubling and serious moral choice, but those who make it should not be demeaned by preventing them from participating in research to support a goal of medicine, i.e., healing.

Does the opportunity to donate fetal tissue positively influence the decision for abortion? In the only empirical study to date, a small number of women said that they would be more likely to have an abortion if they could donate fetal tissue for transplants.[134] This important first study did not explore the mechanism of influence or prove that this result is generalizable to larger populations. The study ought to concern those who argued that the opportunity to donate would play no substantial role in the decisions of women about abortion. More social-psychological research is clearly needed.

c) Nonmaleficence-Based Considerations. Moral opposition to fetal tissue transplant research influenced a moral consensus about safeguards to prevent widening or encouraging the social practice of abortion. To this end, these moral rules are required: the consent process about abortion decisions must precede and be conducted separately from the consent process to donation of fetal tissue for transplant research; and prohibited are designated donation, monetary inducements to women undergoing abortion, and buying or selling fetal tissue.

d) Prudential Concerns. Payments are permitted to transport, process, preserve, or implant fetal tissue, or for quality control and storage of such tissue.

NBAC's review of Case 1 needs to cover the report of the Fetal Tissue Panel,[135] the history of the "indefinite" moratorium,[136] and the legislative history of the Research Freedom Act. Also important is the history of NIH funding of fetal tissue transplant research since the enabling legislation, which has proceeded entirely within the federal requirements and without significant incident.[137]

These considerations of Case 1 are not beyond moral challenge by a view condemning most elective abortions as unfair to the fetus and claiming that researchers who use fetal tissue are morally complicit with killing fetuses in abortions. To defend Case 1 adequately, NBAC's report must critically review the literature in the 1990s on the complicity issue.[138]

Case 2. Case 2 is similar to Case 1 in three morally important ways and different in one clear and distinguishing feature. At the outset, one must concede that Case 2 is more controversial than Case 1 because it involves use of living embryos in research. However, use must be further specified to the preimplantation stage, and further, that uses by researchers will not, under any circumstances, include human reproduction.

a) Beneficence-Based Considerations. First, similarly to Case 1, society and science can benefit in many ways by permitting research with excess embryos, as the Human Embryo Panel showed in 1994. Deriving PSCs from blastocysts and studying their potential can only add to these benefits.[139] Given research findings in the mouse, it appears likely that human beings will receive benefits from PSC research.[140] The Human Embryo Panel supported federal funding of derivation of PSCs from embryos in 1994. Today, science and society are in verifiable proximity to this goal. Advances in PSC research, stem cell biology, and cloning technology are the major new factors in the scientific context.

Lack of evidence that embryo research had yielded clinical benefits was among several criticisms of the Human Embryo Panel's position. Daniel Callahan[141] wrote that the Panel had not "cited a single actual benefit" from embryo research permitted in other nations or under private auspices in the United States. Speculating that either there were no benefits to report or the Panel "just forgot to ask," he skeptically continued, "In any case we are asked to bet on the future benefits. I wonder what odds the bookies in Las Vegas would give on this one." Whatever the odds may have been in 1995, recent PSC research dramatically increase the odds that using human embryos as a source of PSCs will lead to major scientific and clinical benefits. PSC research adds strength to the consequentialist arguments that promote the obligations of beneficence in Case 2.

Secondly, Case 1 and Case 2 are similar with respect to the issue of the inevitability of discard. Whereas all fetal tissue is discarded if not made available for research, only a certain percentage of embryos will be eventually discarded.[142] The options for couples in IVF about disposition of excess embryos are: cryopreservation for subsequent thawing and use to treat their infertility, donation to other infertile couples, or for research.[143] The same consequentialist reasoning about inevitable discard used in Case 1 also applies to Case 2 and heightens the obligation to be beneficent.[144] Reasons for patients and society to forgo such benefits must be strong enough to be overriding.

In this vein, the most compelling reason to forgo such benefits would be that a publicly supported practice of embryo research would threaten society, in the words of Hans Jonas:

> ...by the erosion of those moral values whose loss, caused by too ruthless a pursuit of scientific progress would make its most dazzling triumphs not worth having.[145]

NBAC can use this classic statement of the moral limits of biomedical research with human subjects as a baseline from which to evaluate the moral effects of practices in embryo research.[146] Jonas's query will be explored below in a section on considerations of nonmaleficence.

We have seen so far that beneficence-based arguments heightened by the consequences of inevitable discard and loss of opportunity to benefit, as in Case 1, are a first source of moral appeal to shape a consensus on access to donated embryos in research.

b) Autonomy-Based Considerations. Moral obligation based in respect for autonomy is a third moral similarity between Cases 1 and 2. If society ought to respect the autonomous and altruistic choices of donors in Case 1, it follows that the same imperative bears on Case 2, provided that the moral argument for access to embryos is strong enough to overcome objections. Parents who donate embryos want to contribute to knowledge about infertility, cancer, and genetic disorders. Such knowledge may yield solutions to relieve sickness and human suffering. These altruistic motives deserve respect as do the procreative intentions that caused the original creation of the embryos. IVF embryos are generated by decisions of couples who want to reproduce themselves. One must assume that they care about their embryos and enjoy the right to make decisions freely about options for disposition. These embryos exist within a web of caring relationships and are not isolated "research material." The federal and some state bans on IVF embryo research implicitly forbid embryo donation for research. These bans conflict with a right to make such donations that is respected in other states in the context of privately supported research.

Appeal to respect for the autonomous choices of donors of embryos is a second source of support for arguments favoring access.

c) Considerations Based in Nonmaleficence. Case 1 and Case 2 differ in one significant respect, i.e., the fetus as a source of PSCs is dead and cannot be harmed by research activities, but the donated embryo is a living organism that will die in the process of research rather than from being discarded altogether.

In moral terms, the major difference is that the abortion causes the death of the fetus, and the research causes the death of the embryo. How ought this difference be morally evaluated? Is Case 2 comparable in any way with cases of transplanting organs from the "dying but not yet dead" to benefit others and society, e.g., the case of harvesting organs from dying anencephalics prior to brain death?

Answers to these questions depend upon answers to deeper questions about moral perspective. What kind or type of case is Case 2? What are the strengths and weaknesses of varying perspectives on the moral worth of embryos? Can embryos even be "harmed" in research? How much protection ought society give to embryos in research? Finally, there is the Jonas query, i.e., will permitting embryo research, especially in the context of Case 2, so erode moral values as to make even the "dazzling" goal of cell-replacement therapy "not worth having?"

What Kind of Case Is Case 2? If viable PSCs were derivable from donated embryos that were "allowed to die,"[147] then Case 2 would clearly follow Case 1 in a line of cases of cadaveric sources of organs and tissues, including fetal tissue. Cadaveric transplants have strong moral backing. However, when an embryo at the blastocyst stage stops developing and dies, one must assume the deterioration of the inner cell mass along with the PSCs within it. Case 2 cannot be in the cadaveric line of cases.

However, given the procreative intent[148] of infertile couples and the clinicians who help them,[149] Case 2 is also not a case of creating embryos solely for research as are Cases 3 and 4. Cases 3 and 4 are in a new line of embryo cases posing an issue of whether there can be two morally acceptable reasons for *de novo* creation of embryos: procreation or research. As long as the number of ova stimulated and fertilized in individual treatment were not being manipulated in order to produce an excess number of embryos for research, Case 2 ought not to be viewed within this new line of embryo cases.

In my view, if the donative feature and the inevitability of discard in Case 2 can be authenticated, then Case 2 is more similar to Case 1 than Cases 3 or 4. The similarity is especially strong when one considers the largely bad consequences of discarding fetal and embryonic tissue suitable for research. Moral authentication of donation and discard requires two stages of the informed consent process. The first stage would be informed consent for treatment of infertility by the procedure of IVF. Patients need to understand IVF's known risks and benefits. This first discussion should include the issue of the number of viable embryos to be transferred.

The treatment stage ought to be separated from a second stage of informed consent regarding cryopreservation and options for disposition of excess embryos: i.e., continued treatment of the couple's infertility, donation to other infertile couples, and donation for research. The decision to donate embryos for research should be the last option explained with no undue influence on the choice of the couple or the woman.

Moral Status of Embryos. Views about the moral status of embryos also influence the choice about whether Case 2 belongs to the line of cases represented by Cases 3 and 4. Do "excess" embryos donated for research lose their moral worth because they have been selected for research?[150] If one views embryos as having no moral standing at all, then the "moral worth" question is moot. If one has serious moral concerns about Case 4 on the grounds that "it seems to cheapen the act of procreation and turn embryos into commodities,"[151] then one will focus strongly on the donative feature and the integrity of the consent process. Research with embryos donated by parents is easier to justify than creating embryos for research, because the parents have authority over the disposition of their embryos.[152]

The appendix proposes an approach to narrow the distance between polarized positions on the moral status of embryos. What are human embryos morally considered? What degree of social protection should be given to human embryos? The work of the Human Embryo Panel on the issue of moral status of embryos criticized "single criterion" approaches to personhood (e.g., genetic diploidy or self-concept).[153] The Panel desired to take a broader and more "pluralistic" approach. Key sections describing this approach are worth reproducing here:

> ...[it] emphasizes a variety of distinct, intersecting, and mutually supporting considerations...the commencement of protectability is not an all-or-nothing matter but results from a being's increasing possession of qualities that make respecting it (and hence limiting other's liberty in relation to it) more compelling.

Among the qualities considered under a pluralistic approach are those mentioned in single criterion views: genetic uniqueness, potentiality for full development, sentience, brain activity, and degree of cognitive development. Other qualities mentioned are human form, capacity for survival outside the mother's womb, and degree of relational presence (whether to the mother herself or to others, including genetic uniqueness, potential for full development, sentience, brain activity, and degree of cognitive development). Although none of these qualities is by itself sufficient to establish personhood, their developing presence in an entity increases its moral status until, at some point, full and equal protectability is required.[154]

The Human Embryo Panel cited similar reasoning about the ethics of embryo research by the EAB in 1979, the Warnock Committee in the U.K. in 1984, and a Canadian commission in 1993.

In an important article, Annas, Caplan, and Elias criticized the Human Embryo Panel's ethical perspective.[155] These authors found that:

> the pluralistic framework...is not convincing. This is so primarily because that framework requires a detailed analysis that explains why the particular properties cited confer moral worth, or to what degree each property cited is necessary and sufficient. Without such an underlying rationale, the framework looks like an attempt to rationalize a desired conclusion—namely, that some research on embryos ought to be permitted—rather than to derive a conclusion from an ethical analysis (p. 1330).

Critical of the Panel's discussion of the moral status of the fetus isolated from the relationship of parents with their embryos or an intent to procreate, Annas, et al. argued that:

> ...an embryo has moral standing not so much for what it is (at conception or later) but because it is the result of procreative activity (p. 1330).

They stressed that the moral standing of embryos not only derives from a "cluster of properties" that the embryo possesses but also from the "interests that potential parents and society bring to procreation and reproduction...."[156] This criticism revealed the need for a moral framework for embryo research that compensates for the weaknesses of the work of the Human Embryo Panel and draws on other ethical perspectives. NBAC's report on PSC research should aim for improved arguments. The appendix undertakes this task.

Can Embryos Be Harmed in Research? The article by Annas, et al. makes the excellent point that the interests of parents and society in procreation can be damaged by morally unjustified embryo research. But can an embryo be harmed in research? On the one hand, one may concur with the EAB's[157] position of "profound respect" for the preimplantation human embryo, due to its human origins. On the other hand, one can hold without contradiction that an experiment ending in an unimplanted embryo's death did not "harm" the embryo. The embryo is an organism with human origins but not yet sentient or having a set of interests. Physical harm could only be done to an embryo itself after implantation, gestation, and in the context of rudimentary sentience. From this perspective a clear and "bright line" difference emerges between the moral status of living children and embryos. To be sure, society does not permit comparable experiments with living children who are sentient and who have interests. However, society does permit Phase I trials in children with cancer, and these trials carry a risk of morbidity and mortality.[158]

It is possible, of course, to damage an embryo in research. The damage would become "harmful" in the moral sense only if the embryo was transferred to a human uterus and a future sentient person was harmed by the damage once done to the embryo.[159] This potential abuse can be prevented by regulation forbidding the transfer to a human uterus, after research activities, of any embryo or its equivalent.

Jonas's Query. Embryo research has proceeded in the private sector, regulated only by professional ethics. In this context, a wide diversity of practices could probably be found among researchers located in the nation's infertility centers.[160] Until more is known about the actual shape of these practices, it is difficult to answer Jonas's query in terms of whether these unregulated activities are "too ruthless a pursuit of scientific progress." One can cite damage done by lack of regulation and accountability to "society's moral values," but it is debatable whether embryo research as conducted today in the private sector is seriously "eroding" those moral values. In my view, if limited to Case 2 and under conditions presented in Table 4 of Part III, federal funding for embryo research could proceed incrementally.[161] The appendix provides a moral argument to bolster this step. It is obvious, however, that a too "ruthless" or commercially aggressive pursuit of embryo research could seriously threaten the values defended by Annas, et al. and others. For example, if researchers—without any public discussion—abruptly pursued a version of Case 3 by fusing human somatic cells with enucleated animal ova to create defective embryos in order to derive PSCs, the Jonas query would be clearly relevant. The same judgment would follow any attempt to circumvent public policy on inheritable genetic modifications.

Other Concerns Based in Nonmaleficence. The Human Embryo Panel carefully outlined a set of principles and guidelines[162] to prevent abuses and minimize harms to societal values and human beings. In brief, these were: 1) scientific competence of investigators, 2) valid research design and scientific/clinical benefits, 3) research cannot be otherwise accomplished (prior animal research required), 4) restricting number of embryos required for research, 5) informed consent of embryo donors for the specific research to be undertaken, 6) no purchase or sale of embryos for research, 7) IRB review, 8) equitable selection of embryos, 9) a 14-day limit on length of research.

Case 3. This case involves PSCs to be derived from human (or chimeric) embryos generated asexually by SCNT, using enucleated human or animal ova for fusion. The rule of the Council on Science and Society is relevant to Case 3.[163] Until more is known about this unverified possibility, moral judgments are inappropriate. Unlike Cases 1 and 2, virtually nothing is known scientifically about SCNT as a source of human PSCs. Case 3 is ranked above Case 4 due to the therapeutic potential of growing the patient's own cells to return to the patient in autologous cell-replacement therapy, in theory avoiding graft vs. host disease. Considering the prospective clinical benefits of SCNT-created PSCs, more moral support for Case 3 than for Case 4 seems predictable. A balancing and controversial factor is that the product of SCNT (using an enucleated human egg) would arguably be a human embryo which could become a human being if transferred to a uterus. The NBAC's recommendations for a ban (with sunset provision) on cloning a human being are relevant here.[164] Clearly, SCNT as a source of PSCs could not be pursued without a clear ban on making a baby by this method.

Case 3 is arguably different from all other cases due to the asexual origin of the source of PSCs, although a form of donation is involved. In Case 3, individuals donate a somatic cell and an ovum for asexual reproduction of the DNA in the nucleus of the somatic cell. Are embryos from this source of less moral worth than sexually generated embryos?[165] The answer is related in part to intent: creating embryos by SCNT would be done to promote clinically promising research to help human beings, which is a very different case from the original intent with which embryos in Case 2 were made, i.e., procreation. However, if one would not argue that embryos deliberately created for research (Case 4) are of less moral worth than "excess" embryos, then the embryos in Case 3 should not be so viewed. In U.S. public policy an embryo is an embryo, however made. However, the main point is that to go thoroughly down the SCNT road requires a full scale review that will be only speculative due to lack of information.

Considering the intent of the progenitors, Case 3 is more similar to Case 4 than it is to cases 1 and 2. The intent is to create embryos by SCNT only for the sake of research.

Case 4. In this case, PSCs would be derived from human "research" embryos created from donor gametes. Although the activity is the same in Case 4 as in Cases 2 and 3—research involving human embryos—Case 4 involves an important and morally relevant difference from Cases 1 and 2, i.e., the deliberate creation of embryos for research from donated gametes. Depending on the circumstances, the donors may be individuals unknown to one another or couples with particular genotypes of interest to researchers. Whether one views this activity as a major step in moral evolution that is justifiable for compelling scientific and clinical reasons (as I do) or as laden with "symbolism" (Robertson), there are reasons to argue that Case 4 is different and more complex morally than Cases 1 and 2. One reason is that creating embryos for PSC research is a precedent for inheritable genetic modifications of embryos. The embryos would be belong to couples at high risk for genetic disease. NBAC does not have the time or resources at present to conduct a full exploration of this topic.

In addition to arguments in support of federal funding for embryo research, the Human Embryo Panel justified federal funding (subject to additional review) of this activity specifically to generate PSCs for research. There was a debate among panelists about the moral and scientific justification of this recommendation. The issue concerned creating banks of cell lines from different genotypes that encoded different transplantation antigens, the better to respond to the transplant needs of different ethnic groups. This would require recruitment of embryos from ethnically different donors. However, the possibility of genetic alteration of genes controlling the major histocompatibility complex would obviate this step. This is a scientific question that still remains unanswered today.[166]

The discussion has shown important differences between Cases 1–2 and 3–4. Also, a review of the scientific background and need for research in Cases 3–4 would be a major undertaking which could not be completed in the time frame proposed by NBAC. In summary, an incremental approach to these cases seems to indicate that NBAC should concentrate on Cases 1–2 and include some attention to Cases 3–4 with emphasis on the similarities (these yield PSCs for research) and major differences as to means and ends.

Transition to Part III. Hopefully, Part II has presented persuasive reasons why an incremental approach to NBAC's tasks is preferable to an exhaustive approach. Also, the principles and rules embedded in Cases 1 and 2 can serve as sources of appeal to strengthen the case for consensus within NBAC as to why these are defensible situations for access to fetuses and excess embryos for PSC research. However, a fuller policy argument is necessary to justify federal funding to derive PSCs in the context of Case 2. Let us now turn to that argument.

Part III. A Policy Argument for Federal Funding of Research Involving Excess Embryos

A. IVF, Embryo Research, and Public Bioethics

Before IVF, the only purpose to generate embryos was procreation, i.e., to produce offspring. IVF added a second purpose: to view the preimplantation embryo and study a variety of biological and clinically relevant questions. Before IVF, gynecological surgery[167] was the sole means to view the preimplantation embryo or to obtain specimens. Rock and Menkin (1944), Edwards, et al. (1969), and Soupart and Strong (1974) pioneered IVF.[168] IVF provided a window of unparalleled opportunity but one clouded with moral controversy. From the outset, moral traditions that value only "natural" human reproduction challenged the moral justification for IVF and embryo research.

Embryo research has a variety of goals, i.e., to improve infertility treatment and contraception, to understand the preimplantation stages of the human embryo, to study origins of some types of cancers, genetic disorders, birth defects, etc. When DNA technology and IVA converged, scientists could pose basic questions: e.g., "When does gene expression begin in the embryo?"[169] Studies of such questions are permitted in the U.S. private sector but banned in the federal sector.

Embryo research is a major step in moral evolution. Indeed, there is not only a long-standing moral debate about creating embryos for research by fertilizing ova with sperm but a new phase of that debate about asexual creation of embryos by SCNT. One should expect a certain degree of confusion and a great need for education among the public about these matters.[170] The evolution of moral beliefs guiding social roles, practices, and institutions occurs very slowly.[171] Conflicts of loyalties and intense struggles, which are not always peaceful, mark the paths of such changes.[172] In open democracies, an electorate and a judiciary informed by a variety of moral traditions help to guide the scope and pace of moral evolution.

Furthermore, national and state commissions in bioethics can play a key role in providing guidance to policy makers and the public on controversial moral issues in research and medicine.[173] Insofar as PSC research is concerned, NBAC can contribute to present and future federal and state policy on the practice of human embryo research. Key questions are: Is it morally acceptable to use *any* embryo in research? Is there a morally relevant difference between embryos donated by infertile couples and embryos made by scientists but intended only for research? To what degree should society protect human embryos in research?[174] Answers to these questions draw on moral and political traditions, as well as policies governing the relation of science and society. The next section, bolstered by the moral argument in the appendix, proposes an approach to public policy on federal funding of embryo research in the context of PSC research.

B. Obligations of Distributive Justice in Appropriations for PSC Research

Beauchamp and Childress define distributive justice as follows:

> The term distributive justice refers to fair, equitable, and appropriate distribution in society determined by justified norms that structure the terms of social cooperation. Its scope includes policies that allot diverse benefits and burdens such as property, resources, taxation, privileges, and opportunities. Various public and private institutions are involved, including the government and the health care system.[175]

The claims of distributive justice bear directly on two political and ethical issues of PSC research that will require compromises of liberals and conservatives. The first issue is whether it is fair, in terms of distributive justice, for appropriations for federal support of PSC research to be indefinitely blocked by the ban on federal funding for embryo research. Is it fair for the embryo research ban to deprive many Americans of the opportunity to learn whether PSC research will lead to therapy and, if so, to benefit from that transition? Given the conditions in Table 4, appropriations to permit federal funding in Case 2 could hasten a transition from Stage 2 to Stage 3 of PSC research, as described above in Part I, Section 2. The focus of Congress ought to be especially on appropriations to support the derivation of PSCs from embryos only to complete Stage 2 and Stage 3, i.e., focused on the goal of clinical trials in humans of cell-replacement therapies for diseases that cause early death or severe debilitation.

A compromise between liberals and conservatives will be required to facilitate the appropriations process for federal funding for embryo research, but only in a context of Case 2, if the arguments given in Part II are persuasive. Given the history of the role of Congress on fetal and embryo research and the great need for public education on the issues, one can politically assume that 1) it is most persuasive with Congress and the public to focus federal funding for embryo research on pre-clinical and therapeutic aims, rather than on Stage 1 concerns,[176] and 2) that Congress should be aided in its assessments by the moral and public policy advice of an official ethics body.[177]

The policy argument is that federal funding for embryo research can be morally justified only under certain conditions and as a last resort to optimize a transition from Stage 2 to Stage 3 of PSC research. In my view, given the history of federal policy (but not federal regulations) on fetal and embryo research, Congress should not be asked to fund derivation of PSCs from embryos during Stage 1, the scientific exploration of the properties of PSCs and cell lines derived from various sources. The present stage of political considerations, i.e., whether Congress will permit the Rabb opinion to govern the NIH's funding of "downstream" PSC research, can be viewed as an early moral and political experiment in federal funding of PSC research.[178] If Congress permits the NIH to fund "downstream" PSC research, the effect will be to permit sufficient Stage 1 funding to learn if it is advisable to advance to Stage 2. It is precisely at this point that appropriations bills should have different language with regard to permitting derivation of PSCs from excess embryos for Stage 2 and 3 purposes.

In consideration of distributive justice issues, it is reasonable to ask, if the transition period to clinical trials of cell-replacement therapies for diseases that cause early death and debilitation is five to ten years rather than three to five years, how many Americans of all ages will have died whose lives could have been saved and enhanced by a speedier and successful transition to clinical trials? How much morbidity among the whole population could have been reduced by federal participation in the period of preparation for clinical trials to their conclusion? Would not federal funding focused on Stage 2 research have hastened the transition? These are legitimate questions of distributive justice that will be posed not only by ethicists, but by taxpayers who are voters in elections to come. Why should large numbers of Americans risk added debilitation and earlier death if Congress can take moderate steps to prevent this possibility? A positive result from the clinical trials will obviously be beneficial in terms of reduction of mortality and morbidity; a negative result will also have preventive benefits, i.e., it will refute arguments for any further trials with the same hypotheses or design and protect human subjects from risks of trials with erroneous scientific aims. This compromise will seem unfair to the liberal mind's overall assessment of justice issues, the moral status of embryos, and federal funding of embryo research. However, the chances of reaching a satisfying moral compromise to permit political action will be higher if federal funding is focused exclusively on a) therapeutic intent, albeit for living human beings rather than for embryos, than on b) uses to obtain biomedical knowledge otherwise not obtainable; i.e., knowledge of the properties of PSCs and cell lines derived from embryos, fetuses, adult stem cells, and possibly SCNT-generated embryos.[179] Given the history and the potential for divisiveness on the issue of embryo research, it is politically wiser to leave all Stage 1 funding for derivation of PSCs from any source to the private sector, with the exception of Case 3. Among NBAC's recommendations can be one to the effect that the private sector is obligated not to proceed with Case 3 research without public discussion, in a forum such as NBAC or the NIH-RAC, of the ethical rationale and goals for such activities. The private sector willingly assumed this obligation during the early years of DNA research.[180]

In regard to public policy recommendations for federal funding to derive PSCs from excess embryos, NBAC may desire to consider its recommendations in the context of some or all of the conditions presented in Table 4.[181] The conditions are presented as binding in terms of timing and scientific progress.

Table 4. Conditions to Warrant Federal Funding of Embryo Research for Transition to Clinical Trials

- NIH-NSF "downstream" peer review and funding of PSC research combined with support from the private sector can be shown to have led to understanding of cell differentiation, differences between embryonic germ cells and embryonic stem cells, and other scientific goals of Stage 1 of PSC research.
- Scientists can now describe desirable experiments with animals and human PSCs and cell lines to learn whether the tumorigenic dangers and other potential hazards[182] of using ES cells can be avoided.
- The NIH-FDA-NSF scientific peer review process is in agreement that strong scientific support exists to enter a pre-clinical period prior to clinical trials of cell-replacement therapies in humans for one or more diseases that are life-threatening or severely debilitating; e.g., Type I diabetes, Parkinsonism, Alzheimer's disease, etc.
- A qualified panel of scientific experts makes the case that federal funding of embryo research is required, in the context of Case 2, as a last resort to complete the pre-clinical period and conduct clinical trials in humans; i.e., there are no other less morally problematic alternatives to using live embryos for the purpose of deriving PSCs to develop cell lines for therapeutic purposes. The panel should be composed of representation from the NIH/NSF, FDA, other Public Health Service agencies, and academic, industry, and patient groups.
- The President and Congress have earlier received recommendations from NBAC that considerations of social justice and other ethical principles justify federal funding in the case of "excess" embryos. However, before taking the step of approving federal funding for embryo research in the context of Stage 1 and 2 activities, Congress will request and receive the moral and public policy advice of NBAC. If NBAC is not operational at this time, an EAB appointed by the Secretary, DHHS for this purpose may so advise the Congress.
- Appropriations for PSC research in Stage 2 may only be used to fund research in centers that assure the NIH or NSF that a) IRB approval has been obtained for a two-stage consent process that separates IVF decisions from decisions to donate embryos for research and a plan to protect the privacy of donors, b) that such research to be done with donated embryos conforms to the guidelines recommended by the NIH Human Embryo Panel, and c) that fairness in selection of subjects to donate excess embryos will be assured. No awards or contracts can be processed without satisfying these stipulations.
- Appropriations for PSC research in Stage 3 must include conditions for Stage 2 funding and legislation to assure fairness in selection of participants in the federally funded clinical trials of cell-replacement therapy.[183]

These conditions may be sufficient to move moderate conservatives with a "detached" (see appendix) view of government protection of embryos to agree to federal funding of embryo research in Case 2. It is true that liberal and some conservative members would approve federal funding in Case 2 today. However, support for an approach of "last resort and no less morally problematic alternative" conveys respect for sincerely held moral views of conservatives, as would limiting federal funding to a context of therapeutic intent. This compromise by liberals fits well with the obligations of the "Charo principle," as discussed in the appendix, Section C.

Further Specifications of Justice in Research Activities. If Congress were able to compromise and approve federal funding under these conditions, a second level of moral concern would also exist about obligations of justice in two contexts of selection of subjects: 1) selection of donors of embryos as sources of PSCs, and 2) selection of participants in clinical trials of cell-replacement therapy. The Belmont Report states that the claims of justice in research activities require the fair distribution of benefits and burdens of such activities over a whole population.[184] Federal noninvolvement in infertility research and the ban on federal funding for embryo research have already combined to infringe on obligations created by this principle in one actual and one future way: 1) the composition of the pool of donors of embryos in Case 2 is limited to private patients in infertility centers, and 2) if not prevented by a deliberate plan, the selection of subjects to participate in clinical trials of cell-replacement therapies could be biased by inequities that inhibit access to clinical trials, especially for poor and disadvantaged Americans. Steps to prevent biased selection of subjects in each of these contexts could be mandated by Congress as part of its appropriations process.

Consider the moral implications of the fact that the pool of donors of embryos for research in Case 2 is entirely composed of private patients in infertility treatment. Women or couples who donate embryos are not "human subjects" in a primary sense of being acted upon by researchers. The woman is acted upon by a specialist in reproductive medicine. Later, researchers may use her embryos in research activities. Such women and couples are, however, human subjects in a secondary sense. They have interests of voluntariness, comprehension, privacy, and justice that are protected by the ethics and regulations of research with human subjects. It follows that all Case 2 research ought to be submitted to IRBs for prior review. IRBs ought to be concerned with plans for a two-stage informed consent process as described in Part II in discussion of Case 2. Further, researchers should not be able to link embryos they receive by donation with identifiable donors. No identifiers should accompany the transmittal of embryos from the setting of therapy to the setting of research, in the interest of protecting donor privacy. Finally, both IRBs and Congress ought to be concerned about the disproportionate and unjust distribution of benefits and burdens that would be involved in accepting without question the present population of donors in Case 2 as sources of donation.

Consider the situation of economically disadvantaged persons who are infertile and of infertility patients in the private sector. These persons are also taxpayers. The costs of infertility treatment are prohibitively high. One of the chief causes is lack of federal scientific involvement and regulation. Economically disadvantaged persons in the United States receive very little treatment for infertility, in spite of a higher rate of infertility among African-Americans.[185] Although a small number of states have required some degree of health insurance coverage of infertility treatment,[186] no state Medicaid program reimburses for it.[187] Ideally, low-income infertile couples ought to be receiving treatment and positioned to share the benefits and burdens of experimental treatment and embryo donation. Yet, the combined effects of the history of federal abandonment of infertility research and the ban on federal funding work to impose all of the risks of embryo research upon private infertility patients. If they decide to donate, they will have borne the risks of IVF or other procedures required to produce embryos at all. These patients do receive the benefits of slowly improving techniques of infertility treatment. However, the risks they accept are of special concern due to lack of public oversight and regulation of embryo research. These are gross contradictions and inequalities in social practices of research. Present public policies are unfair to all taxpayers with a condition of infertility.

To prevent bias in the selection of subjects for embryo donation, Congress will need to take steps in advance of Stages 2 and 3 of PSC research to encourage states through the Medicaid program to fund infertility treatment for infertile couples eligible for Medicaid. Funding of the federal share of the Medicaid program for this particular purpose would go a long way towards assuring the states' funding of their share. Such action by Congress would be an appropriate remedy for past unfairness with which infertility treatment has been distributed, as well as an appropriate step to assure that the obligations of justice would be followed in selection of donors of embryos for federally funded PSC research.

The issue of equity in selection of subjects is addressed—in part—in federal regulations governing criteria for IRB approval of research.[188] Since the regulations do not address directly the issue of equity of access to the potential benefits of clinical trials, Congress will need to accompany its appropriations for PSC research with stipulations that recipients of such funding must assure their local IRBs that reasonable steps have been taken to make access to clinical trials of cell-replacement therapy available to "economically...disadvantaged persons." Additional funding for education of the public at large, including minorities traditionally wary of participating in research, ought to be appropriated to insure that recruitment of subjects occurs after appropriate public education has had a chance to succeed. By this step, Congress can avoid criticism that it did not take every reasonable step to assure that the public understands the rationale for federal funding of embryo research and that federal policy is to overcome past inequities of access to the potential benefits of clinical trials. A majority of conservatives and liberals should be able to agree readily to such a provision. This concludes the section on the policy argument to support federal funding of PSC research that derives stem cells from living but donated embryos.

Conclusions and Transition to the Appendix

This paper has discussed three moral problems or concerns in PSC research: the moral legitimacy of access to sources of PSCs, considerations of uses of PSCs in research, and the cumulative moral effects of the ban on federal funding of embryo research and other protectionist federal policies. The history of public policy restrictions and bans on federal funding of fetal research and research at the beginning of embryonic life finds a significant difference between a moderate legacy of the NC and subsequent federal regulations on fetal research and the strongly protectionist policies superimposed by Congress. The paper discussed the tasks of NBAC in reviewing PSC research, as well as the strengths and weaknesses of an incremental or case-by-case approach to four sources of PSCs. It concluded that moral justification for federal funding to derive PSCs from embryos is far easier to make in Cases 1 and 2 than in Cases 3 and 4. The policy argument in Part III can be seen as return to the legacy of the NC and its confidence in the potential of an EAB to contribute to conflict resolution and service to the nation and the branches of government in research ethics.

The appendix is not in the main body of the paper to conserve space (in an already lengthy paper) and to make two points. The first point is that the appeals to the principles embedded in Cases 1 and 2 are strong enough to support the compromises required for Congress to approve federal funding in PSC research, under the conditions proposed in Table 4. The second point is that these sources of appeal are not strong enough to support the case for long-range reform of public policy on fetal and embryo research. New principles are necessary to meet the challenges of this necessary goal. These are characterized as the Dworkin and Charo principles. Readers who are interested only in federal funding to derive PSCs from embryos may stop at this point. Those interested in long-term reforms may read on.

There are three relevant reasons for long-term reform: 1) policies on federal funding of research with the fetus and at the beginning of embryonic development are overly protectionist, 2) too many moral incongruities and contradictions result from permitting radically different moral approaches to govern embryo research activities in the private and public sectors, and 3) extending the moral and constitutional protection of human subjects of research to all Americans regardless of the sources of funding for research is a large and unfinished task of democracy. The arguments in the appendix aim to support these reforms.

Appendix: A Moral Argument for Federal Funding of Derivation of PSCs from Embryos

Overview

The argument is that the claims of two ethical principles, named for their major proponents, provide additional moral support for a public policy of incremental and regulated federal funding of research deriving PSCs from embryos and other areas of science with donated excess embryos, as well as for three areas of long-range reform.

The "Dworkin principle" creates obligations of respect for the intrinsic value of life, which is the source of government's obligation to limit and regulate research involving embryos and fetuses in all sectors of society. The "Charo principle" has two elements that inform political action to resolve immediate and long-range controversy about federal funding of embryo and fetal research. These elements are a) obligations of distributive justice and b) reciprocal compassion for the moral suffering required for the two sides to become committed to compromise rather than stalemate.

The justice element works to ameliorate the inequities that Congressional prohibitions on federal funding for embryo and fetal research have already imposed on infertile couples, parents at higher genetic risk, and economically disadvantaged persons. However, these remedies must be implemented by compromises that truly embody compassion for moral suffering. Those who believe either that embryos and fetuses ought to be treated equally with persons or that the tasks of responsibly learning to relieve burdens of pain, suffering, and early deaths of living human beings deserve highest loyalty will suffer morally by the requirements of the proposed compromise.

These two principles, augmented by principles of beneficence and utility, also create a moral foundation for long-range reform. Congress can remedy the combined moral effects of prohibitive bans on federal funding for embryo and fetal research which undermined an earlier compromise achieved by the NC. A unified and moderate body of ethics and public policy for research at the beginning of embryonic life is required to protect society's moral values and to narrow the massive gap between diagnosis and therapy created by the success of the Human Genome Project. NBAC's moral vision to promote this task would renew the legacy of the NC for fetal research and extend it to the ethics of research at the beginning of embryonic life.

A. Conflicting Moral Views and a Third Possibility

Technology does not cause the moral problems linked to IVF and embryo research. Clashing interpretations of the moral legitimacy of using and discarding embryos for research give rise to these problems. PSC research today is mainly embroiled in renewed controversy about the moral legitimacy of using live embryos in research.[189]

Two polar opposite positions appeal to the same ethical principle—respect for persons—but totally disagree on what ought to be done. These positions focus almost solely on the moral status of the human embryo. Each view appeals to biological data to settle the issue. Defenders of the first position argue that "...the human being must be respected—as a person—from the very first instance of his existence."[190] Moreover, society ought to protect human embryos because of their genetic uniqueness and potential to become persons. In this perspective, embryo research is a form of unjustified killing. Scientists cannot ethically learn whether embryo research can lead to significant scientific and clinical gains, since embryos would be destroyed in the process.

A second approach also appeals to the principle of respect for persons. However, this view holds that the biology of the human embryo counts against giving embryos any moral status that would prevent research to benefit patients, science, and society. For example, since twinning can occur in this period, "a determinate human being does not yet exist."[191] Further, without implantation and gestation to fetal viability and beyond, an embryo can have no interests that society ought to protect. In this view, embryo research poses no comparable moral claims of the type made by actual persons. This view of embryo research focuses largely on "brain life" as pivotal for authentic personhood.[192]

These conservative and liberal approaches collide sharply. Different worldviews and religious interpretations can lie in the background of these conflicts. In a commissioned essay for the Human Embryo Panel,[193] Steinbock discussed these views and their variants, noting origins in the abortion debate. She identified a third position as a "compromise between the conservative and liberal views," i.e., that "although embryos are not persons, they have moral value as a form of human life." In a section entitled "A Third Position: Embryos Have Symbolic Value," she discussed how several official panels that considered the ethics of embryo research adopted this position.[194]

The framework of "symbolic value" stems from Robertson's work on the ethics and law of embryo research. His framework for the status of the embryo is: "special respect but no rights for embryos." Respect is due, because the embryo is a "potent symbol of human life."[195] In Robertson's view, special respect takes the form of rigor of research review and fidelity to safeguards that ought to surround embryo research.

To date, NBAC's deliberations reflect a compromise view with significant differences from the Panel's perspective and recommendations. Commissioners appear ready to promote a spirit of compassionate compromise and empathy with important moral concerns within opposing positions than did the Panel's report, although its process cannot be faulted along these lines.[196] Commissioner Charo is a critic of the "symbolic value" framework, because it is dismissive of the moral concerns and suffering of opponents of embryo research. She was also critical of the Panel's moral reasoning as too exclusively "bioethical" by focusing almost entirely on issues of moral status, rather than on political ethics and on justice issues in particular.[197] She has made this same argument in other writings.[198] She was recused from NBAC's deliberations on PSC research.[199] However, her prior writings and talks on embryo and fetal research are very important to NBAC's considerations and clearly influence the arguments in this appendix. With some supplementary arguments, Charo's justice arguments are also a promising source of support for long-term reforms. President Clinton's request to balance "*all* (emphasis added) ethical and medical considerations" can be seen as reinforcement for Charo's main point. Can there be *any* degree of overlapping consensus between the two views of the morality of embryo research described above?

B. Dworkin's Partly Unifying Principle

Similar to the abortion debate, a single-minded focus on the moral status of the embryo has frozen debate on the ethics of embryo research into two polar opposites. Is there any hope for passage through these frozen straits? Ronald Dworkin made a significant effort[200] to locate common moral ground between liberal and conservative views on abortion.[201] Dworkin admits the stark polarization of the abortion debate and its baleful consequences for moral discourse. He is skeptical of the arguments given in several works[202] urging compromise that do not appreciate the moral depths of the divisions that exist or that argue for compromise while biased by one side of the debate.

> Self-respecting persons who give opposite answers to whether the fetus is a person can no more compromise, or agree to live together allowing others to make their own decisions, than people can compromise about slavery or apartheid or rape. For someone who believes that abortion violates a person's most basic interests and most precious rights, a call for tolerance or compromise is like a call for people to make up their own minds about rape, or like a plea for second-class citizenship, rather than either full slavery or full equality, as a fair compromise of the racial issue.
>
> So long as the argument is put in those polarized terms, the two sides cannot reason together, because they have nothing to reason or be reasonable about. One side thinks that a human fetus is already a moral subject, an unborn child, from the moment of conception. The other thinks that a just-conceived fetus is merely a collection of cells under the command not of a brain but of only a genetic code, no more a child, yet, than a just-fertilized egg is a chicken....[203]

However, Dworkin believes that conventional understanding of the polarized state of the debate is shrouded by intellectual confusion and can be clarified and dispelled. The confusion is due to a failure to distinguish between a "derivative" and a "detached" objection to abortion. The "derivative" type of objection—to abortion as murder—is derived from rights and interests that presumably all persons, including fetuses, can morally and legally claim. These rights begin with a right not to be killed. In this view, government has a "derivative" duty to protect fetuses from abortion.

A second type of objection to abortion, the "detached" type, objects to abortion as an assault on the sacred nature of human life in itself. Not dependent on particular rights or interests, it views abortion as wrong in principle due to its assault on the sanctity of human life at any stage. Dworkin concludes that:

> ...someone who accepts *this* objection, and argues that abortion should be prohibited or regulated by law for *this* reason, believes that government has a detached responsibility for protecting the intrinsic value of life.[204]

Having established this distinction, Dworkin goes to great lengths to argue that, despite the "scalding rhetoric" of the pro-life movement to the effect that the fetus is a moral person from the moment of conception, "very few people—even those who belong to the most vehemently anti-abortion group—actually believe that, whatever they say."[205] He also notes that few liberals view the fetus as mere tissue. He describes the views of most people about the issue of abortion and the duty of the government to protect the sacredness of life in the detached rather than derivative camp. If Dworkin is right, this distinction goes a long way to define the role of government in legislation about abortion and embryo research. The ban on federal funding for embryo research is clearly framed in a "derived" rather than a "detached" view of government's responsibility. A "derived" view will insist on virtually absolute protection of rights claimed for the fetus. A "detached" view will focus on the wrongmaking feature of abortion as a violation of the "intrinsic value, the sacred character, of any stage or form of human life."[206] Those who hold this view will believe that government ought either to prohibit abortion for this reason or, as government has done in *Roe v. Wade* and elsewhere in the states, to regulate it by law. The most liberal view of abortion would insist on minimal regulation of abortion and maximal protection against intrusions on women's choices. Those who take the middle way will permit abortion but regulate it carefully by law. Prohibitions of abortion are appropriate when the fetus is viable, except in situations where abortion will avert threats to the woman's life or health. With exceptions in a few states, abortion law in this society has clearly been a "detached" rather than a "derivative" type that would prohibit abortion by law.

Dworkin's central hypothesis is that understanding what we as a people really disagree about in the abortion debate will unite rather than divide. He proposes:

> The disagreement that actually divides people is a markedly less polar disagreement about how best to respect a fundamental idea we almost all share in some form: that individual human life is sacred. Almost everyone who opposes abortion really objects to it, as they realize after reflection, on the detached rather than the derivative ground. They believe that the fetus is a living, growing human creature and that it is intrinsically a bad thing, a kind of cosmic shame, when human life at any stage is deliberately extinguished.[207]

Dworkin proposes that the main difference between conservatives and liberals on abortion is not whether the fetus is or is not a moral person; it is in how these views interpret the claims that flow from the principle of respect for the sacredness of life. In a section on this proposition, he extends his central hypothesis:

> We should...consider this hypothesis: though almost everyone accepts the abstract principle that it is intrinsically bad when human life, once begun, is frustrated, people disagree about the best answer to the question of whether avoidable premature death is always or invariably the most serious possible frustration of life. Very conservative opinion, on this hypothesis, is grounded in the conviction that immediate death is inevitably a more serious frustration than any option that postpones death, even at the cost of greater frustration in other respects. Liberal opinion, on the same hypothesis, is grounded in the opposite conviction: that in some cases, at least, a choice for premature death minimizes the frustration of life and is therefore not a compromise of the principle that human life is sacred but, on the contrary, best respects that principle.[208]

Dworkin interprets disagreements between liberals and conservatives over abortion as arising mainly from varying interpretations of the scope and meaning of the principle of respect for the sacredness of life. Conservatives view the "natural" contribution to life as preeminent, while liberals view the "human" contribution as supreme. To the former, the gift of life is more important than anything a person can do. The premature ending of life is the greatest frustration. To the latter, since human investment in life gives particular lives their creative value, significant frustration of that investment can call for decisions that life should end to prevent even more frustration.

In addition to the moral and political relevance of the distinction between "derived" and "detached" objections to abortion, this passage is most relevant to the issue of federal funding of embryo research.

> We can best understand some of our serious disagreements about abortion [and embryo research]...as reflecting deep differences about the relative moral importance of the natural and human contributions to the inviolability of individual human lives. In fact, we can make a bolder version of that claim: we can best understand the full range of opinion about abortion, from the most conservative to the most liberal, by ranking each opinion about the relative gravity of the two forms of frustration along a range extending from one extreme opinion to the other—from treating any frustration of the biological investment as worse than any possible frustration of human investment, through more moderate and complex balances, to the opinion that frustrating mere biological investment in human life barely matters and that frustrating a human investment is always worse.[209]

Respect for the Intrinsic Value of Life. Although language about the "sacredness" or "sanctity" of life is appropriate, one does not have to embrace the religious premises underlying these terms to agree with the direction of Dworkin's argument. Dworkin recognizes that some will not want to use it because of its religious connotations. He often uses "inviolability" of life or human life interchangeably with "sacredness." This principle can also be understood as respect for the *intrinsic* value of life. The term "intrinsic" points to the value of something in and of itself, independent of its results for or relations to ourselves or other persons. Dworkin gives examples of great art, cultures, animal species, and each individual human life itself, as meriting profound respect apart from their instrumental value to us or other persons. Although we are part of the whole that includes these creative events, processes, and beings, we can observe that they have their own moral status apart from any particular interests we have in enjoying or using them.[210] This independent standing is worthy of the awe with which believers perceive the inherently "sacred" or "holy" quality of beings or of the profound respect with which others would view the same qualities.

How does Dworkin's work connect with the debate on federal funding for embryo research? First, it reduces the distance between the polarized sides and strengthens those who would take a middle way. Liberals and conservatives need not be permanently divided along lines of the confrontation between the Human Embryo Panel's report and the Congress's ban on federal funding for embryo research.

Using Dworkin's work, the debate about federal funding for embryo research becomes one about the meaning of respect for the intrinsic value of life. Conservatives and liberals are loyal to the same principle but interpret it in different ways. Conservatives desire not to interrupt an investment from either a biological or divine source (or both) in the embryo's unique life in spite of the frustration of human desires. Liberals value the good that can be done by relieving frustrated human investment in the lives of sick and suffering children and adults.

The main effect of Dworkin's argument on the politics of federal funding for embryo research is to encourage a middle way for liberals and conservatives willing to embrace a "detached" role for government in issues on embryo research. Those on either extreme of the issue may reject the argument, because they will not be persuaded to abandon "derivative" or rights-based positions. This means that liberals and conservatives nearest to the middle of the debate about federal funding for embryo research can join in protecting respect for the intrinsic value of human life by a decision to *regulate* embryo research in the United States in both the federal and private sectors.

Dworkin's principle so interpreted and specified is only partially unifying and takes us only close to the heart of the debate about moral status. It helps mainly to frame the debate in terms of loyalty to a common principle and a "detached" view of government's proper role. Despite its merits, the Dworkin principle will not yield a moral consensus for federal funding for embryo research, because it does not tell us why this practice can be ethically acceptable. Also the issue of moral status of embryos (or nonviable fetuses) cannot be resolved at the level of ethical theory. Ethical theory can clarify and frame the debate about moral status, but it lacks the authority finally to resolve such debates, because metaethical or theological sources of inspiration supply the final answers to the question about the *meaning* of life itself that underlies the debate about moral status. Ethical theory is not a source of the faith or belief in a guiding purpose in nature—or the lack of faith or belief in a guiding purpose, for that matter—that shape beliefs about the moral status of embryos or fetuses. In the realm of ultimate loyalties, the choices and commitments that persons make are hardly subject to convincing proofs.[211] In a democracy, the political process is a penultimate social resource to resolve such issues.[212] Both Congress and the several states can choose to embody in law either a "detached" or a "derivative" view of the use of embryos in research and of government's role in protection and/or regulation. So does democracy function to ameliorate the divisiveness of otherwise irreconcilable moral positions.

A second argument, more political in nature than the Dworkin principle, is required to inspire the moral compromise required to authorize federal funding for embryo research. Conservatives must compromise their beliefs to permit limited research with embryos, under strict conditions. Liberals must restrain their resolve to secure federal funding in Cases 2, 3, and 4 in order to maximize the scientific and clinical benefits of PSC research. Obligations of distributive justice are resources to explain why Congress can morally approve of federal funding for embryo research, provided that it legislates to regulate embryo research in loyalty to the principle of respect for the intrinsic value of life. Approval of federal support of embryo research without regulation would forsake the Dworkin principle, because scientists then would, in principle, be able to do anything they wished with embryos in research. This would be a flagrant violation of respect for the intrinsic value of life. The political strategy in Part III will not succeed without a moral framework and regulation of embryo research in the public and private sectors.

C. Putting Charo's Justice Principle to Work

Commissioner Charo has written about the resources of political ethics and traditions of social justice to understand and resolve disputes—such as the moral status of embryos—that do not yield to ethical analysis of the kind used in the Human Embryo Panel's report. The policy argument for federal funding, limited to Case 2, which were given in Part III, arise from claims of distributive justice. These arguments need not be repeated here, but arguments based on fairness in appropriating funds for biomedical research on PSC research only go part way in justifying federal funding for embryo research. Obligations of distributive justice can explain why it is fairer to the lives and future health of the greatest number of Americans to achieve federal funding for PSC research involving embryos. However, appeals to justice will not explain how these political decisions can be made in such a way as to avoid intractable conflicts that pose the Jonas query in a slightly different form. Would the acrimony and divisiveness created by a decision for federal funding for embryo research threaten to make the "dazzling triumphs" of PSC research "not worth having?" Compromise will entail moral suffering.[213] Can the decisions about federal funding for embryo research be made in a process that recognizes and respects such suffering? Commissioner Charo counsels, out of a wealth of political experience, that approaches to resolving problems in justice require respect—by those on both sides of issues—for the loyalty to moral principles that proponents invest in their causes. The "Charo principle" combines justice-seeking resolutions of divisive issues with compromises in which both conservatives and liberals show compassion for the moral suffering that each must endure to effect the compromises. This two-sided principle can help the political process of federal funding for embryo research and how it must be carried out with tolerance and compassion between those who must finally act for the greater social good.

In any decision to permit federal funding, conservatives will be torn between two goods: loyalty to their ultimate beliefs about the endowment that nature or God bestows to human life and loyalty to the cause of healing diseases that cause early death or disability. In any decision to delay federal funding, liberals are also caught between two goods: loyalty to the cause of healing and effective action at the earliest possible time to promote the cause of healing. Mutual consent to infringement of one treasured good will be required for both sides in a compromise to permit federal funding of embryo research, beginning with Stage 2 of PSC research and under conditions specified in Table 4. Without this contribution to compromise from the virtue of compassion, arguments from justice will be insufficient to carry the day.

A second hypothesis is that the Dworkin and Charo principles are insufficient on their own to make the moral case for long-range reform of federal policy on fetal and embryo research and regulation of infertility research. Claims of justice must be buttressed by obligations of beneficence and utility to guide this task and to overcome the gap between practices in research at the beginning of embryonic development in the private and public arenas.

D. Long-Range Reforms: Obligations of Beneficence and Utility

Several times the discussion has referred to the need for long-range reforms. Federal policy on fetal and embryo research is too protectionist to serve the public interest fairly in a scientific context of the completion of the Human Genome Project (HGP) and prospects for advances in stem cell biology. The section concludes with a discussion of the need for long-range and moderate reform of federal policy on research involving the fetus and at the start of embryonic development.

For those who hold a moderate or a "detached" view of government's role in protecting fetuses and embryos in research, strongly protectionist policies seriously infringe on moral obligations of beneficence and utility. Considerations of the claims of these principles need to be added to the considerations of distributive justice reviewed in the previous section. When the ban on federal funding for embryo research is combined with federal inaction on regulating infertility research and a virtual ban on federal funding of fetal research, the cumulative effects on the losses of basic knowledge and pre-clinical opportunities amount to grievous violations of obligations of beneficence and utility.[214]

Obligations of Beneficence and Utility. The "principle of beneficence refers to a moral obligation to act for the benefit of others."[215] Obligations flowing from beneficence shape medicine's goals of healing and promoting health,[216] tempered and balanced by commitments to avoid or minimize deliberate harm[217] and to considerations of utility. The principle of utility[218] "is limited to balancing benefits, risks, and costs (outcomes of actions), and does not determine the overall balancing of obligations."[219] Utility is a less weighty moral principle than beneficence, but loyalty to it creates obligations that shape the role of science in a modern democratic state. Loyalties flowing from both these same principles (beneficence, nonmaleficence, and utility) also shape the moral traditions of biomedical research. These research activities are not morally independent but are encompassed by the morality of medicine, which is accountable to structural values of democratic societies, including respect for persons, social justice, and liberty.[220]

In the relations between modern medicine and biomedical research, considerations of utility, outcomes studies, and practice guidelines have influenced the standard of care.[221] There is an obligation to prove whether existing treatments and procedures are safe and effective. The criterion of proof is the experimental method.

The same standard holds with new or innovative treatments or procedures; i.e., the obligation is to compare the safety and efficacy of the new with existing treatments or procedures which may be proven or unproven. The obligation to learn experimentally how best to treat and prevent disease is a requirement of utility aimed to maximize obligations of beneficence.[222]

In the case of promising prospective treatments or procedures, such as cell-replacement therapies developed from PSCs, loyalty to utility obliges investigators first to *learn* whether a clinical trial in humans is well-founded scientifically and pre-clinically. Stages 1 and 2 of PSC research, as described above, are aimed to carry out this obligation. This obligation is a necessary but insufficient basis upon which to conduct a clinical trial in humans. Prior to consideration of the ethical question, "Ought this trial in humans be done at all?" the investigators must have satisfied a scientific and a pre-clinical question: "Is there sufficient scientific understanding of the disease process and the action of the proposed treatment?" "Have results been achieved and replicated in suitable animal models that lead rationally to a prospect of benefits in humans?" With one exception, physician-investigators who bypass the scientific and pre-clinical stages of learning betray the canons of good science and loyalty to utility.

Exceptions to Obligations of Utility. There must be exceptions to the obligations of utility, e.g., when the means to learn to answer scientific or pre-clinical questions would be unethical to employ.[223] In such a case there could be an overriding reason to bypass the scientific stage of investigation. A good historical example is in the debate about the moral acceptability of uses of data obtained by German scientists who conducted experiments on political and concentration camp prisoners under the Nazi dictatorship. Is it ethical to cite or use such data in the process of science today?[224] In practice most scientists avoid using or citing this body of data out of respect for the victims of the moral horrors of the Holocaust, as well as "critical shortcomings in scientific content and credibility."[225] This oppositional moral position is not the only alternative. Freedman argues for a position that respects the intent of contemporary researchers who cite Nazi data in the interest of goals "to aid patients and to advance science in the interest of humankind."[226]

An example of an arguable exception to the obligations of utility occurred in planning for NIH Protocol 076, which tested the drug AZT in HIV-infected pregnant women to prevent transmission of the HIV virus from infected pregnant women to their fetuses.[227] Was there an obligation to learn, by doing fetal research in the context of elective abortion, how and when the HIV virus is transmitted *in utero* before giving a drug with then unknown side effects and potential to cause birth defects? Protocol 076 was fortunately successful,[228] but in my view it was based on unsound scientific foundations. When the trial began, how and when HIV transmission *in utero* truly occurred was knowable, but was allowed to remain unknown for political and ethical reasons. Ethical considerations and public policy at the time prevented these foundations from being laid. In retrospect, it is worth raising the question again as to whether fetal research in the context of abortion would have been morally justified to answer the question of how and when HIV is transmitted *in utero*. Virtually everyone would agree that it would have been unethical then or now to deliberately expose fetuses in the context of abortion at various stages of pregnancy to the virus in order to learn if it could be transmitted. However, would not the 076 Protocol have been a sounder study in science and ethics if knowledge about the pathophysiology of transmission of HIV *in utero* had been gained first by research in the context of elective abortion?

Long-range reforms of virtual bans on federal funding of fetal research are needed if obligations of beneficence and utility are to be carried out by federally funded scientists. Such bans defeat an obligation to build a knowledge base for experimental treatment. A good illustration is the effects of federal policy on fetal research required prior to complete Stage 2 research prior to *in utero* gene transfer experiments.[229] The result is a dearth of information about normal fetal physiology and development required for sound fetal therapy experiments. For example, ignorance about fetal immunocompetence was a prominent topic[230] in an NIH-RAC discussion of Dr. French Anderson's proposal for an *in utero* gene therapy experiment for adenodeaminase deficient severe combined immunodeficiency syndrome, a disorder that destroys an affected child's immune system. Moreover, a recent NIH-supported Gene Therapy Policy Conference[231] examined the scientific and ethical basis for experimental *in utero* gene therapy. The conference affirmed the ethical argument to prevent inevitable harm to the fetus and future child. However, it found inadequate scientific foundations to proceed with such experiments in the near future. Federal policy on fetal research creates an acute knowledge deficit even while the technical feasibility of ultrasound-guided fetal gene therapy steadily grows.[232]

Parents at Higher Genetic Risk. Consider the consequences of protectionist federal policies for parents at *known* higher risk of transmitting genetic disorders to their children. How many persons fit this situation? Although the total number is hard to ascertain, one can posit numbers in relation to other established facts. These are the parents of between one-fourth and one-third of all children admitted to pediatric units in Western nations.[233] These children need treatment for the complications of genetic diseases, congenital malformations, or mental retardation. These are the parents of the approximately 22 percent of newborn deaths in developed nations caused by congenital malformations or genetic disorders.[234] These are the parents who choose prenatal diagnosis to ascertain whether their fetus and wanted child-to-be has inherited a genetic or chromosomal anomaly. All of these parents are also federal taxpayers.

The moral situation of parents at higher genetic risk is fraught with controversy made worse by an incongruent federal policy. On the one hand, these parents are confronted with successes of the HGP. On the other hand, they are confronted by a federal policy forbidding funds for promising research to open avenues to treatment. With the help of their taxes, diagnosis of hundreds of genetic diseases is now possible, including in the fetus. This list will grow to thousands and include genes that create susceptibility to common disorders like cancer, heart disease, and diabetes.

The Human Genome Project. Consider the advances of the HGP in genetic diagnosis compared with the paucity of treatments for genetic disease. Genetic testing raises morally troubling questions for those with a strong family history of cancer, in part due to the perceived risks of genetic discrimination in health and life insurance.[235] However, the risk of discrimination pales in the face of the stark fact that there are few effective treatments for the genetic conditions that can be diagnosed. The wide gap between genetic diagnosis and treatment is the single greatest scientific and moral problem[236] facing the nation that largely created and funded the HGP. Closing this gap ought to be a major goal of federal science and health policy. PSC research is profoundly important in closing this gap. Protectionist policies maintain the gap and directly collide with the goals of medicine to heal and promote health. If the gap were closed, the opportunities of therapy for genetic disorders would create great pressures towards universalizing these benefits. Progress in genetic therapies could be a powerful force, working together with other pressures, towards more universal health care reform.[237]

Regulation of Embryo Research. A final reason to prefer long-range reform to continuing a ban on federal funding for embryo research is that the latter pathway will unduly postpone public policy to regulate this entire area. Privately funded embryo research is widely conducted but not transparent to public view and regulation. At best, traditions of self-regulation in science and medicine guide these activities. A worst case moral scenario is embryo research without accountability to any source of authority, public or private. To the extent that such practices exist, they are violations of the principle of respect for the intrinsic value of life.

Finally, as NBAC and the Congress consider the specific tasks ahead in making federal funding to derive PSCs possible, it is worth remembering additional benefits that may flow from research activities with donated excess embryos:

- improving clinical protocols used in IVF programs for the treatment of male and female infertility,

- improving techniques for preimplantation diagnosis of genetic and chromosomal abnormalities,

- providing high-quality information about the morphology, biochemical and biophysical properties, genetic expression, and similar characteristics of pregastrulation stage human embryos,

- enhancing knowledge of the process of fertilization,

- facilitating the design of new contraceptives,

- facilitating studies of teratology and the origins of certain birth defects, and

- increasing knowledge about cancer and metastasis, including the causes of certain reproductive cancers.[238]

Notes

1 The author is grateful for the opportunity to aid NBAC's response to the President's request to consider issues of such moral complexity and import for human health and healing. My colleague in bioethics, Franklin G. Miller, gave sound advice on the structure and content of prior drafts of this paper. Philip Noguchi and Brigid L.M. Hogan expertly reviewed the scientific adequacy of the paper. Members of NBAC's staff were very helpful: Eric Meslin and Kathi Hanna helped to clarify the paper's scope and aims; conversations with Andrew Siegel were extremely valuable for the main policy argument in Part III. Commissioners James Childress and Alexander Capron influenced my views, especially on the contents of Table 4 and the moral indivisibility of derivation of PSCs from their uses.

Legend of abbreviations in the paper:

AAAS	American Association for the Advancement of Science
DHHS	Department of Health and Human Services
DHEW	Department of Health, Education, and Welfare
EAB	Ethics Advisory Board
FDA	Food and Drug Administration
HGP	Human Genome Project
IND	Investigational New Drug (Application)
IRB	Institutional Review Board
IVF	*in vitro* fertilization
NBAC	National Bioethics Advisory Commission
NC	National Commission for the Protection of Human Subjects of Biomedical and Behavioral Research
NIH	National Institutes of Health
NIH-RAC	NIH-Recombinant DNA Advisory Committee
NSF	National Science Foundation
OTA	Office of Technology Assessment
PC	President's Commission for the Study of Ethical Problems in Medicine and Biomedical and Behavioral Research
PSCs	pluripotential stem cells
SCNT	somatic cell nuclear transfer (cloning technology)

2 Shamblott, M.J., Axelman, J., Wang S., et al. (1998). Derivation of pluripotent stem cells from cultured human primordial germ cells. *Proc Nat Acad Sci USA* 95, 13726–13731.

3 Thomson, J.A., Itskovitz-Edor, J., Shapiro, S.S., et al. (1998). Embryonic stem cell lines derive from human blastocysts. *Science* 282, 1145–1147.

4 Letter to Harold Shapiro, November 14, 1998. President Clinton also requested NBAC to include implications of a reported attempt to fuse a human cell with a cow egg. Wade, N. (1998). Researchers claim embryonic cell mix of human and cow. *New York Times*, Nov. 12, A-1. Chairman Shapiro responded for NBAC to this aspect of the President's request by letter (November 20, 1998). The letter stressed that scientific evidence was insufficient to conclude whether the product of such fusion would be a human embryo. He referred to NBAC's position that creating a child by somatic cell nuclear transfer was in the near future morally unacceptable due to the high risk of harm. If such fusion did result in a chimeric organism that was not a human embryo, he saw no "new ethical problems" in using such organisms in research. He concluded that using nonhuman ova may avoid the risks and complications of obtaining human ova to create human embryos for research.

5 Commissioners and readers will rightly be interested in the sources of the author's views in ethics. No one perspective or ethical theory can possibly satisfy the demands of the moral life. Several perspectives and methodologies in ethics, especially the dialogue between "principlism" and "casuistry," shape the author's views in this paper and the appendix. Very complex moral problems that face society and government, such as human PSC and embryo research, require resources from several ethical perspectives and several tools of ethics. In recent years, with Franklin G. Miller, Joseph J. Fins, Jonathan D. Moreno, and others, the author has sought to bring the resources of American pragmatism to bear upon the tasks of bioethics. The appendix is an argument in pragmatic bioethics to support the policy changes recommended in Part III. At this point, it is worth marking a difference between a vulgar view of pragmatism (i.e., pragmatism is concerned only with what works) and a view that embraces ethical principles but not does not treat them as fixed or timeless categories.

In 1922, John Dewey wrote a passage that could serve as a foreword to this paper: "...situations into which change and the unexpected enter are a challenge to intelligence to create new principles. Morals must be a growing science if it is to be a science at all, not merely because all truth has not yet been appropriated by the mind of man, but because life is a moving affair in which old moral truth ceases to apply. Principles are methods of inquiry and forecast which require verification by the event; and the time honored effort to assimilate morals to mathematics is only a way of bolstering up an old dogmatic authority, or putting a new one upon the throne of the old. But the experimental character of moral judgments does not mean complete uncertainty and fluidity. Principles exist as hypotheses with which to experiment. Human history is long. There is a long record of past experimentation in conduct, and there are cumulative verifications which give many principles a well-earned prestige. Lightly to disregard them is the height of foolishness. But social situations alter; and it is also foolish not to observe how old principles actually work under new conditions, and not to modify them so that they will be more effectual instruments in judging new cases." Dewey, J. (1988). *Human Nature and Conduct*. (Carbondale, IL: Southern Illinois University Press) pp. 164–5.

6 National Commission for the Protection of Human Subjects of Biomedical and Behavioral Research. *Report and Recommendations: Research on the Fetus*, 1975, U.S. Dept. of Health, Education, and Welfare. (DHEW Publication No. (OS) 76-127); National Institutes of Health, *Report of the Human Fetal Tissue Transplantation Research Panel*, vol. 1, December 1988.

7 Ethics Advisory Board, U.S. Dept. of Health, Education, and Welfare. (1979). *Report and Conclusions: Support of Research Involving Human In Vitro Fertilization and Embryo Transfer*. (Washington, DC: U.S. Government Printing Office); National Institutes of Health, *Report of the Human Embryo Research Panel*, vol. 1, Sept. 1994.

8 Keller, G., Snodgrass, H.R. (1999). Human embryonic stem cells: the future is now. *Nature Med* 5, 151–2.

9 For example, see reports on how cells within the ependymal lining of the adult mouse brain ventricles may be multipotent neural stem cells capable of generating new neurons and glia (Johansson, C.B., et al. [1999] *Cell* 96, 25–34) and how similar cells can regenerate blood tissues when transplanted into an irradiated mouse (Bjornson, C.R., et al. [1999] *Science* 283, 534–37). Bjorklund and Svendsen reviewed this work (*Nature* 397, 569–70, Feb. 18, 1999) and commented: "We do not know whether human neural cells also arise from the ependymal layer, or whether they have the capacity to turn into blood. However, similar embryonic human cells can be cloned (Flax, J.D., et al. [1998] *Nature Biotechnol* 16, 1033–1039), grown for extended periods (Svendsen, C.N., et al. [1998] *J. Neurosci Methods* 85, 141–52) and continue to reside in the adult brain (Eriksson, P.S., et al. [1998] *Nature Med* 4, 1313–1317), so it may not be long before we find out." Dr. Brigid Hogan commented: "The data showing that stem cells derived from neuronal cells can give rise to blood cells needs to be repeated before many scientists accept it." Personal communication, May 14, 1999.

Richard Doerflinger focuses on the Bjornson study with adult mouse stem cells to suggest that the flexibility of these cells may make embryo-derived PSC research "irrelevant." Doerflinger Testimony, Jan. 26, 1999; see note 37. This statement leaps to conclusions. If it is morally acceptable to learn about the properties of PSCs derived from embryos, then the responsible scientific approach is to compare the properties of PSCs from various sources, including embryos, fetuses, and adults. If the embryonic source proves to have a higher risk of harm to animals or humans than other sources, then the former should not be used because of potential harm to patients. Doerflinger's statement rests on a premise that it is unethical to learn about the properties of PSCs derived from embryos.

10 Thomson, J.A., et al. (1995). Isolation of a primate embryonic cell line. *Proc Natl Acad Sci USA* 92, 7844–48.

11 Good summaries of the clinical potential of human PSCs derived from germinal fetal tissue and blastocysts of human embryos are: Gearhart, J., (1998). New potential for human embryonic stem cells. *Science* 282, 1061–62; and Pedersen, R.A. (1999). Embryonic cells for medicine. *Sci Amer* April, 69–73.

12 These advances promote remarkable hopes (both of cures and profits). An example is William Haseltine, CEO of Human Genome Sciences, Inc., who predicts that today's leading killers—heart disease, cancer, Alzheimer's disease, and the "aging process itself"—will gradually become distant memories. He predicts that a century from now, "death will come mainly from accidents, murder, or war." Ignatius, D. (1999). The revolution within. *Washington Post*, March 8, A-19.

13 American Society for Cell Biology, Letter to the President and Members of Congress, March 4, 1999. Citing a large body of successful work with mouse PSCs, the letter states that PSC research has "enormous potential for the effective treatment of human disease" and argues that the President and Congress should permit federally funded researchers to work with PSCs.

14 Smaglik, P. (1998). Stem cell scientists caution: clinical applications remain years away. *The Scientist* 12, 1, 6, Nov. 23.

15 "Mouse ES cells are tumorigenic, growing into teratomas or teratorcarcinomas when injected anywhere in the adult mouse. There is no reason to believe that human ES cells will not be tumorigenic in humans. Whatever means we use to separate the undifferentiated ES cells from the desired, differentiated progeny to be injected, we will have to be absolutely sure that the separation is complete. As yet, we do not know the minimal number of ES cells necessary to form a tumor or the length of time necessary for tumor development. The answers to these questions will not come from experiments with mice because mice are too shortlived to provide an adequate test. It is entirely possible that we will have to provide some genetically designed fail-safe mechanism, a 'suicide' gene, which will enable us to destroy transplanted cells if they become tumorigenic." Solter, D., Gearhart J. (1999). Putting stem cells to work. *Science* 282, 1468.

16 Public opinion studies could address questions about the degree of controversy associated with different sources of PSCs. Although not a source of ethics, public opinion is crucial to public policy formation. Public policymaking that ignores public opinion courts disaster, not because it should always cravenly follow public opinion, but because it is prudent to legislate with reliable knowledge of what the public thinks about a particular issue.

17 See note 2 above.

18 See note 3 above.

19 The feasibility of Case 3 from a technical standpoint is an open question. This work has not yet been done in the mouse. Brigid Hogan, personal communication, March 12, 1999. A report has cast serious doubt on claims of Korean researchers to have cloned a human embryo by transferring the nucleus of a somatic cell into an enucleated egg cell, both from the same patient. Baker, M. (1999). *Science* 283, 617–18. A U.S. biotechnology company also disclosed a 3-year-old experiment (but no scientific report) fusing an encucleated cow's egg with a human cell. See note 4 above. (The remainder of this note was written by Philip Noguchi, personal communication, May 13, 1999.) Even if Case 3 is technically feasible, disturbing questions on the validity of SCNT-derived embryos as representing a normal organism comes from recent news and publications that SCNT to clone large animals such as cows is fraught with unexpected consequences. Weiss, R. (1999). *Washington Post*, May 10, A-1. Several publications have noted that cloned sheep and cattle have an unexpected high birth weight, called "large offspring syndrome." Gary, F.B., Adams, R., McCann, J.P., Odde, K.G. (1996). Postnatal characteristics of calves produced by nuclear transfer cloning. *Theriogenology*, 45:141–52. Young, L.E., Sinclair, K.D., Wilmut, I. (1998). Large offspring syndrome in cattle and sheep. *Reviews in Reproduction* 3:155–63. More disturbing is the recent article by Renard, et al., describing a cloned calf that died of severe anemia 51 days after birth following an unexplained, rapid decline in lymphocyte and hemoglobin count. Renard, J.P., et al. (1999). Lymphoid hypoplasia and somatic cloning. *The Lancet* 353:1489–91. An editorial note on *The Lancet* Web page notes: "*The Lancet* rarely publishes research done in animals. We make an exception this week because the experiment in question—somatic cloning—should arguably never be done in human beings. The death from severe anaemia 51 days after birth of the cloned calf described by Jean-Paul Renard and colleagues, suggests that somatic cloning may be associated with developmental abnormalities. This finding needs to be factored into the debate on whether human cloning is ethically acceptable."

20 Research embryos are created by infertility researchers in the private sector in the United States, and law in the United Kingdom permits the creation of research embryos under strict control. No research with PSCs has been reported with "research" embryos as the source.

21 The following is a speculative scenario that could become Case 5. Could there be conditions under which in Case 4 one might consider implantation following derivation of PSCs? One might not think so at the outset, since it is assumed that derivation of PSCs would entail destruction of the embryo, the organism. However, it has now been reasonably well established that one or more cells can be removed from preimplantation embryos without seeming alteration of the embryo's ultimate viability. A recent publication showed that preimplantation diagnosis of sickle cell disease could be used to discriminate embryos on the basis of their genetic normality. Xu, K., Shi, Z.M., Veeck, L.L., et al. 1999. First unaffected pregnancy using preimplantation genetic diagnosis for sickle cell anemia. *JAMA* 281, 1701–6. To do this, two cells were removed from two embryos and one cell removed from five embryos. Three embryos were transferred, one unaffected with 15 cells, one carrier with 10, and one unaffected with 8 cells. Of these, two carried to term and were unaffected. The point is that removal of one or more cells does not necessarily lead to embryo destruction and/or nonviability. As knowledge increases (presumably from animal studies), one can easily see a time when one or two cells could be grown in tissue culture into PSCs. Couples could procreate and donate but not engage in destruction of their embryos. This approach would also address the issue of autologous PSC without necessitating the unknown risks of a SCNT-derived embryo.

22 The language of the ban on federal funding of embryo research is modeled after the language of an earlier congressional ban on fetal research, which permits research designed to "enhance the well-being or meet the health needs of the fetus or enhance the probability of its survival to viability...." The Health Research Act of 1985, Sec. 498(a)(1). To conduct "therapeutic" research with embryos without a foundation of prior knowledge gained through investigative research into pathophysiological and genetic questions would be totally irresponsible. A solid pre-clinical basis must be laid for any new stage of therapy. Nonetheless, it is legal under the federal embryo ban to attempt such therapeutic experiments.

23 Memorandum. Harriet S. Rabb to Harold Varmus. Federal funding for research involving human pluripotential stem cells. Jan. 15, 1999.

24 Draft NIH Guidelines for Research Involving Pluripotent Stem Cell Research. For discussion at the meeting of the Working Group of the Advisory Committee to the Director, NIH, April 8, 1999. Without using the terms "ethical" or "moral," these guidelines state that NIH-funded investigators who use PSCs "should" do so only if suppliers of such cells have documented that: 1) the PSCs were derived from excess embryos donated in the context of infertility treatment, 2) were donated in the context of practices of informed consent with safeguards against undue or "even subtle" pressure to donate, and 3) that the PSCs were not derived from embryos created for research purposes. These carefully worded guidelines assume, without further argument, moral reasons for prescribing these special duties.

25 Some members of Congress responded to the moral as well as the legal question. Section 3 of this part of the paper discusses their position.

26 Based on her latest research on the subject, Andrews writes: "...statutory and court precedents dealing with embryo and fetal research, abortion, organ transplant, and payment for body tissue all have ramifications for work involving embryo stem cells. Yet no two states have identical laws covering these procedures."

"Some type of embryo stem cell research is permissible in virtually every state. [North Dakota has statutes that could be used to prohibit both the Thomson and Gearhart techniques.] Yet, because of differences in state laws, certain states would ban the collection of stem cells from embryos that were created through *in vitro* fertilization [Florida, Louisiana, Maine, Massachusetts, Michigan, Minnesota, North Dakota, Pennsylvania, and Rhode Island]. In other states, a prohibition would only apply to the isolation of stem cells from *aborted* embryos and fetuses [Arizona, Indiana, North Dakota, Oklahoma, and South Dakota]." See, in this volume, Andrews, L.B. (1999). State regulation of embryo stem cell research.

27 See note 26 above.

28 These states are Alabama, Colorado, Connecticut, Delaware, the District of Columbia, Georgia, Hawaii, Idaho, Iowa, Kansas, Maryland, Mississippi, Nevada, New Jersey, New York, North Carolina, Oregon, South Carolina, Texas, Vermont, Virginia, Washington, West Virginia, and Wisconsin.

29 See note 6 above, *Report of the Human Fetal Transplantation Research Panel* (1990) vol. 1, question 1, pp. 1–2.

30 The President's Commission (PC) in effect decided not to defend the morality of selective abortion in the context of a report on genetic counseling and screening. The commission was, however, very critical of the use of prenatal diagnosis for sex selection only. President's Commission for the Study of Ethical Problems in Medicine and Biomedical and Behavioral Research. (1983). *Screening and Counseling for Genetic Conditions*. (Washington, DC: U.S. Government Printing Office) p. 58.

31 Andrews, L. (1994). State regulation of embryo research. In National Institutes of Health, *Papers Commissioned for the Human Embryo Research Panel*, vol. 2, p. 298.

32 See note 26 above.

33 However, my moral viewpoint does not equate the seriousness and weight of objections to embryo research with those of the wrong-making features of prenatal diagnosis for sex selection, absent any sex-linked disease. See Wertz, D.C., Fletcher, J.C. (1998). Ethical and social issues in prenatal sex selection: a survey of geneticists in 37 nations. *Social Science Medicine* 46, 255–273. This statement about embryo research being "open to moral challenge" acknowledges the seriousness of moral views holding that society either ought to protect human embryos in research activities or prohibit such activities. The relevant issue for public policy is the warranted degree of protection.

34 See especially, Charo, R.A. (1995). The hunting of the snark: the moral status of embryos, right-to-lifers, and third world women. *Stanford Law and Policy Rev* 6, 11–27; and Annas, G.J., Caplan, A., Elias, S. (1996). The politics of human embryo research—avoiding ethical gridlock. *N Engl J Med* 334, 1329–32.

35 Insofar as this division of moralities is a political compromise to ameliorate conflict, the political advantages to the public sector are far less secure when the promise of curing diseases is tangible.

36 Such unification was indeed accomplished in the early debates about DNA research, which led to the creation the NIH's Recombinant DNA Advisory Committee (NIH-RAC).

37 Varmus, H. (1999). Testimony before the Senate Appropriations Subcommittee on Labor, Health and Human Services, Education and Related Agencies. Jan. 26, p. 3.

38 See note 8 above.

39 Part II gives some preliminary information on supply of donated excess embryos for research which ought to cause concern and requires more research for definitive results.

40 The author acknowledges the help of Philip Noguchi, M.D., in constructing this section.

41 "Many questions related to the possible therapeutic use of human ES cells have not been addressed in mouse ES cells simply because of the lack of interest. Fortunately, our understanding of the molecular pathways of differentiation and the molecules that mark specific cell types is extensive. This knowledge should help us to answer the following questions: Can human ES cells be forced to differentiate along a desired pathway? Can we make *all* ES cells in a culture simultaneously develop along that pathway? What exactly are the intermediary cell types, and how can they be defined? Which markers and which methods can be used to sort out the desired cell types? Human ES cell lines will provide many of the answers to these questions." Solter and Gearhart, see note 15 above, p. 1469.

42 Hogan, B.L.M. (1999). Statement to NBAC. Feb. 3, p. 3.

43 Ibid.

44 See note 11 above, p. 1062.

45 Can the Food and Drug Administration (FDA) regulate the preclinical stage of PSC research? The author put this question to Dr. Noguchi, who answered: "I note that FDA will regulate the clinical use of PSCs both public and private. The question has arisen whether FDA can regulate the preclinical research in this and other areas. The answer is complex. We have regulations concerning how…preclinical research is done; this in the vernacular is called Good Laboratory Practice (GLP). Normally this is a certification from the sponsor that all research has been done in accordance with the FDA regulations. Unlike clinical trials, we do not normally require examination in detail of GLP adherence before starting a clinical trial of a FDA regulated product in man, although we do review extensively preclinical data as one part of determining risk versus benefits. However, when strict adherence to GLP and any other FDA requirements for safety is necessary to assure the safety of the product in man, FDA may consider more stringent and active regulation of preclinical research. This is very rare; we have stated such a strategy in only two cases…the first being a statement on human cloning, and a second on the use of nonhuman primates in xenotransplantation." Personal communication, May 11, 1999).

46 FDA may also play a role in Stage 2 by helping sponsors to determine proactively what sorts of pharmacology/toxicology studies need to be done to allow first entry into humans. It is expected that in concert with formal FDA regulatory requirements that there will be additional requirements for public and societal discussion as is currently done by the NIH-RAC and FDA for gene therapy, and by the FDA/Centers for Disease Control and Prevention/NIH/Health Resources and Services Administration public forum for xenotransplantation.

47 Stage 3 can entail as many as four phases of clinical trials that the FDA requires to prove safety and efficacy in small or larger numbers and then in whole populations.

48 See note 15 above, p. 1469.

49 National Bioethics Advisory Commission, 26th Meeting, January 19, 1999, pp. 16–17.

50 They write: "Society must decide whether the therapeutic benefits justify denying full development to the constructed embryos." See note 15 above, p. 1469.

51 This approach to "large panels" of cell lines is envisioned by Solter and Gearhart as a way to circumvent the necessity of Case 3, "so that everybody will find a match or by eliminating or altering the histocompatibility antigens, thus creating 'universal' donor lines." See note 15 above, p. 1469.

52 An example of the study of gene expression in the embryo is Bondurand, N., et al. (1998). Expression of the SOX10 gene during human development. *FEBS Letters* 432, 168–72, Aug. 7. This gene is the key factor in Shah-Waardenburg syndrome. A paper (unpublished) was prepared for the Human Embryo Research Panel in support of a case for creating research embryos from couples whose children were at risk for cancers caused by genomic imprinting: Fletcher, J.C., Waldron, P., "Childhood Cancers and Human Embryo Research," April, 1994. The Panel's report cites the paper, see note 7 above, vol. 1, p. 43, with a notation that its arguments "are open to debate and not accepted by all experts." Current research on genomic imprinting assists counseling and prenatal diagnosis, e.g., Buiting, K., et al. (1998). Sporadic imprinting defects in Prader-Willi syndrome and Angelman syndrome: Implications for imprint-switch models, genetic counseling, and prenatal diagnosis. *Am J Hum Genet* 63, 170–80. Our point in 1994 was that understanding of genomic imprinting in these embryos could eventually be useful in diagnosis and treatment of retinoblastoma and other pediatric cancers.

53 See especially, Spece R.G., Shimm, D.S., Buchanan, A.E. (1996). *Conflicts of Interest in Clinical Practice and Research*. (New York: Oxford University Press).

54 The latter task was discussed in a report for NBAC: Fletcher, J.C. (1997). "Location of OPRR within the NIH: problems of status and independent authority," November 27 (NBAC document).

55 Smith, A., Testimony to NBAC, Jan. 19, 1999, p. 36.

56 National Institutes of Health. (1998). Cloning. present uses and promises. April 27. (Available from the Office of Science Policy.)

57 Pincus, D.W., Goodman, R.R., Fraser, R.A.R., et al. (1998). Neural stem and progenitor cells: a strategy for gene therapy and brain repair. *Neurosurgery* 42, 858–68.

58 Snyder, E.Y., Taylor, R.M., Wolfe J.H. (1995). Neural progenitor cell engraftment corrects lysosomal storate throughout the MPS VII mouse brain. *Nature* 374, 367–70.

59 Parens, E., Testimony to NBAC, Jan. 19, 1999, p. 98. As members of an American Association for the Advancement of Science (AAAS) taskforce, Erik Parens and Eric Juengst have outlined a promising approach to clarify and redefine the concept of "human germline gene therapy" within a broader and more scientifically accurate concept of "inheritable genetic modifications." Parens and Juengst point out that this intervention should not be called "therapy" for a number of reasons, including that it is incorrect to use the term "therapy" in connection with an unproven technique. In this vein, there is a strong effort underway in the NIH-RAC and the literature to use the term "gene transfer" rather than "gene therapy," even in discussing aims to alter somatic cells.

60 See note 7 above.

61 See note 125 below for reference to work of two national working groups on deliberate and unintentional inheritable genetic modifications.

62 Stock, G., Campbell, J. (1998). Engineering the human germline. Symposium. UCLA, Program on Science, Technology, and Society; for an excellent critical review of the state of the art, see Gordon, J.W. (1999). Genetic enhancement in humans. *Science* 283:2023–4.

63 Kessler, D.A., Siegel, J.P., Noguchi, P.D., et al. (1993). Regulation of somatic-cell therapy and gene therapy by the Food and Drug Administration. *N Engl J Med* 329, 1169–73.

64 Appendix M. of the NIH-RAC guidelines. Points to Consider in the Design and Submission of Protocols for the Transfer of Recombinant DNA Molecules into One or More Human Subjects (Points to Consider).

65 Cited at note 24 above. These guidelines have been submitted to the *Federal Register* for comment.

66 The NIH Revitalization Act nullified the requirement for an EAB approval for protocols involving IVF. Public Law No 103-43, Sec. 492A. (June 10,1993) This law opened the door for federal funding of embryo research, but NIH adopted a step-by-step process, beginning with the appointment of the NIH Human Embryo Research Panel in February 1994 to consider the ethical, legal, and social implications of human embryo research. The EAB (1978–79) had considered only the issues related to research designed to improve the technique of IVF and its outcomes. The report of the EAB (see note 7 above) stressed that another body would need to consider the larger implications of investigative embryo research. Although it would have been legal in 1993 for the NIH to fund human embryo research, especially studies designed to improve the composition of the culture medium for IVF embryos, NIH did not do so either before, during, or in the period between the Panel's final report and the ban on federal funding. NIH did receive protocols for this purpose, but limited funding only the aspects of the studies involving animal embryos, being aware of the opposition to funding activities involving human embryos by a sizeable number of conservative members of Congress. Personal communication, George Gaines, NICHD. April 29, 1999. Although President Clinton announced his acceptance of research with excess embryos, which would have been legal in this period, the NIH did not fund this type of research and has never done so. Title I, Subtitle A. of Public Law 103-43 is the content of the "Research Freedom Act," which establishes the authority to fund fetal tissue transplantation research.

67 Public Law No. 104-99, January 26, 1996, enacted a ban on federal support of any research "in which a human embryo…[is] destroyed, discarded, or knowingly subjected to risk of injury greater than that allowed for research on fetuses in utero…." The term "human embryo" in the statute is defined as "any organism…that is derived by fertilization, parthenogenesis, cloning, or any other means from one or more human gametes or human diploid cells." This ban was pertinent to FY 1996 funds. The Omnibus Consolidated Fiscal Year 1997 Appropriations Act (Public Law No. 104-208) adopted identical language. The ban is transitory in the sense that it is revisited each year when the language of the NIH appropriations bill is considered. Currently, the ban appears in Sec. 511 of the Conference Report on H.R. 4328, "Making Omnibus Consolidated and Emergency Supplemental Appropriations for Fiscal Year 1999," October 1998.

68 Congress was supported in this direction by the Reagan and Bush administrations. The Clinton administration rescinded the moratorium on fetal tissue transplant research and has been moderate on embryo research, in that it is willing to support Case 2 research activities.

69 These protectionist policies, so markedly different from the studied compromises of the National Commission (NC) (with a plan for ongoing conflict resolution) were adopted without deliberate and careful consideration of the moral and scientific costs. The role of public bioethics in American culture is to temper emotions and premature moral judgments that often mark political interests and to balance these interests with those of science, the public, and ethical and legal considerations. However, without a permanent presence in government of a body to work on the ethics of research, the task of creating new public bioethics bodies (like the EAB) can be overwhelmed by political considerations. NBAC's mandate to consider what national resources are necessary to optimize the protection of human subjects of research and advice to government on the ethics of research is directly related to such issues.

70 This history and subsequent discussion will show how NBAC's work on PSC research converges with other needs to which public bioethics must attend. These include long-range reform of federal policy on research involving fetuses and embryos, extension of moral and legal protection of human subjects of research beyond the federal sector to the private sector, an ongoing national EAB for research ethics, and elevation of the status and authority of the Office for Protection from Research Risks in government and the private sector.

71 A very informative history of events prior to 1988 is found in Lehrman, D. (1988). *Summary. Fetal Research and Fetal Tissue Research*. (Washington, DC: American Association of Medical Colleges).

72 *Roe v. Wade*, 410 U.S. 113 (1973).

73 News stories from abroad about research with live fetuses *ex utero* raised questions about NIH funding of these projects. At a demonstration led by Eunice Shriver, NIH's leaders denied such funding. The NIH then imposed a moratorium. Cohn, V. (1973). NIH vows not to fund fetus work. *Washington Post*, April 13, A-1. The moratorium was continued by the law creating the NC and was to remain until the NC made recommendations. The law prohibited "research (conducted or supported by DHEW) in the United States or abroad on a living human fetus, before or after the induced abortion of such fetus, unless such research is done for the purpose of assuring the survival of such fetus" (section 213).

74 See note 6 above for citation of the report, submitted on May 1, 1975.

75 A crucial aspect of the EAB's work, as envisioned by the NC, was to develop further national policy on fetal research and other issues of research involving human subjects. There is a connection between the NC's early work—and the impasses that it reached in its compromises on fetal research—and its vision of an EAB. This is the best point in the paper to relate this history.

The Belmont Report, adopted by the NC in 1979, is the most authoritative American source for ethical guidance for research with human subjects. It opens as follows: "Scientific research has produced substantial benefits. It has also posed some troubling questions." Since 1966 (Levine, R.J. [1988]. *Ethics and Regulation of Clinical Research*, 2nd ed.), a federal policy restricting research activities with human subjects has been in effect. The policy has two goals: 1) to protect human subjects and investigators, and 2) to question whether particular research activities ought to be done at all and to resolve disputes concerning these questions. The United States was the first nation with a federal law (Public Law 93-348) to support a local and national process to achieve both policy goals.

Locally, an IRB has authority to approve or reject a proposal or to alter it to reduce risks or increase benefits. IRBs were not designed to consider the "long-range effects" of research on society or morality (45 CFR 46.111 (2)) such as societal effects of fetal research. This task was to be done by "one or more" EABs to be established by the Secretary, DHEW (45 CFR 46.204 (a)). This two-tiered process of local (IRB) and national (EAB) oversight of research activities was envisioned as sufficient to protect both human subjects and the freedom of research, given a need to restrict some activities in the public interest.

Belmont's opening words rest on two premises. One was clearly stated (i.e., that research had benefited society). The second premise was unstated, regarding the role of scientific freedom in what ought to be done about the "troubling ethical questions" posed by research. This premise was that any restrictions on freedom of research would be justified only after careful study and debate, with limits openly arrived at in a democratic and legal process. Protection of scientific freedom was one of the key elements in early reforms of U.S. research ethics.

How would the "troubling ethical questions" faced by the NC be actually addressed? These were about research with vulnerable populations, like the fetus and pregnant women, children, prisoners, and institutionalized persons with mental disabilities. Belmont's approach was to discuss these questions guided by general ethical principles (i.e., respect for persons, beneficence, and justice). However the NC itself, according to its own report (see note 6, 1975:67), could agree on the "validity" of a principle but not on its application in a specific protocol of fetal research. This refers to the principle of equal treatment in research of fetuses to be aborted and to be delivered. The NC had difficulty applying the principle in the context of research at the time involving amniocentesis. Philosopher Stephen Toulmin (Appendix, 1976:15) wrote that the NC's (and the EAB's) task was to "keep a watchful eye" on the development of "case law" and "precedents" that would actually grow up in fetal research activities governed by decisions of local IRBs. This analogy to lower and higher courts gives insight into Toulmin's and the NC's reasoning, as well as that of Donald Chalkley, the founding Director of the Office for Protection from Research Risks, about the relationship between IRBs and the EAB. (On the reference to Chalkley, see talk by Charles McCarthy at "Belmont Revisited," Charlottesville, VA, April 17, 1999.)

The NC's vision for an EAB was as a permanent national resource. Its role would be to study, debate, and recommend approaches to resolve controversial research proposals referred by a local IRB or by the Secretary, DHEW. Also, Congress had created the NC against a background of government appreciation of the "benefit of a long partnership with science, not a long record of hostility." Dupree, A.H. (1957). *Science in the Federal Government*. Cambridge, MA: Belknap Press, 381. In the 1980's and beyond, influenced mainly by abortion politics, Congress reversed this tradition in respect to reproductive medicine and human genetics; it became hostile to scientific investigation of fetal and embryonic development and substituted what it could legitimately control, i.e., by imposing bans on federal funding.

76 E.g., in developing a vaccine against rubella, amniocentesis, treatment of Rh isoimmunization disease, and respiratory distress syndrome.

77 One example of such strictly utilitarian investigative research—designed to increase biomedical knowledge but not to benefit the fetus involved—was a 1963 study done after hysterotomy. U.S. scientists immersed 15 still-living fetuses in salt solution to learn if they could absorb oxygen through their skin. One fetus survived for 22 hours. The knowledge gained by the experiment contributed to the design of artificial life-support systems for premature infants. Goodlin, R.D. (1963). Cutaneous respiration in a fetal incubator. *Am J Ob and Gyn* 86, 571–79. The report that triggered the demonstration at the NIH was of an experiment in Finland. Researchers perfused the heads of eight fetuses after hysterotomy to learn if the fetal brain could metabolize ketone bodies. This study was the only way by which the researchers could confirm findings from animal research. Adam, P.A.J., et al. (1973). Cerebral oxidation of glucose and D-BOH Butyrate by the isolated perfused fetal head. *Ped Res* 7, 309–abstract.

78 How did the NC come to this second point of the compromise, especially in the legal context of *Roe v. Wade*? If the fetus is not a person in the constitutional sense, why do fetuses deserve equal protection in research activities? The answer lies in the collective views of the commissioners, who were ready to compromise for a number of reasons, and also in the influence of bioethicists of this period on the NC.

The NC's report drew upon the work of several ethicists who wrote commissioned papers and testified. In my view, Richard McCormick and LeRoy Walters had the greatest influence on Stephen Toulmin's draft of the NC's recommendations. A leading Catholic moral theologian, McCormick's view was that "the fetus is a fellow human being, and ought to be treated...exactly as one treats a child." McCormick, R. (1976). Experimentation on the fetus. policy proposals. In *Appendix to the Report and Recommendations: Research on the Fetus*. Washington, DC: U.S. Government Printing Office, DHEW Publication No. (OS) 76–128:5:4. McCormick argued for a very limited approach to fetal research with reasons he used in approving parental proxy consent for investigative research with children. McCormick R. (1974). Proxy consent in the experimentation situation. *Persp Biol and Med* 18, 2–20. He extended this reasoning to fetal research in a few examples of "tragic" abortions he found morally acceptable. McCormick, R. (1981). *How Brave a New World?* (Washington, DC: Georgetown University Press) p. 76. McCormick would permit fetal research in such a context, provided that "there is no discernible risk, no notable pain, no notable inconvenience, and...promise of considerable benefit." McCormick, 1976, op. cit., p. 8. McCormick's term "no discernible risk" later evolved into the category of "minimal risk" in the NC's deliberations about research with children. The meaning of minimal risk continues to be controversial and widely challenged.

Walters advised use of a principle of equality of protection in research, whether fetuses were destined for abortion or delivery. Under this "Golden Rule" idea, researchers could not impose a higher risk with a fetus to be aborted than they would with a fetus to be delivered. See Walters, L. (1976). Op. cit., 8:1–18.

Similar to the "Peel Report" (1972) in the U.K., Sisela Bok (1976:2:1–8) favored selectively higher research risks before 18 weeks in the context of abortion. She gave four reasons for society's protection of human life: 1) to protect victims, 2) to protect agents from brutalization and criminalization, 3) to protect a victim's family from grief and loss, and 4) to protect society from greater harm that would follow from permissive killing. She argued that up to a point well before viability, such reasons have no moral relevance to fetuses, because claims for the "humanity" of the early fetus fail to make sense.

Toulmin preferred McCormick's position to Bok's, because it opened a way conceptually for those giving primary rights to the fetus to accept fetal research. However, along with McCormick's position came his risk standard and the underlying premise that fetuses ought to be treated equally, as "fellow human beings."

In the face of this compromise, the NC grappled with the logical consequence that one ought to place fetuses to be delivered at the same risk in investigative fetal research as fetuses to be aborted. Could any study, e.g., of the complications of amniocentesis when it was a new and unproven procedure, actually have been designed with this equality feature within it? It would require randomizing the first cases equally to pregnancies to be terminated and delivered. Could newer approaches to prenatal diagnosis be so studied? Some commissioners said in debate on specific proposals that some important fetal research could not *ethically* be done without selectively assigning higher risks to fetuses to be aborted. While struggling with these questions, the NC invested great hope in a future EAB's role in these decisions on a case-by-case basis. (See note 6 above, at p. 67.) The NC's report was a compromise premised on strong hopes for the work of an EAB that functioned like a national IRB.

79 On this third point, the NC and those who support its legacy failed to reckon with the political obstacles in the way of the long-range task of creating a role for bioethics in government that protects a national resource like an EAB from the effects of clashes with the interests of various branches of government. See especially U.S. Congress, Office of Technology Assessment (1993). *Biomedical Ethics in U.S. Public Policy*, OTA-BP-BBS-105 (Washington DC: U.S. Government Printing Office).

80 All subsequent issuances of federal regulations have maintained this position on fetal research, including the proposed rules to amend this Subpart, c.f. *Federal Register* (1998). 63 (no. 97), 27794–804, May 20.

81 This distinction is a legacy of the NC's use of the terms "therapeutic" and "nontherapeutic" research, towards which Robert Levine is so critical. The strength of Levine's criticism is that the use of the term "therapeutic" distorts the fact that research aimed at therapy is still unproven. The weakness of his critique is that it obscures the fact that there are stages within the process of research projects that are purely investigative, preclinical, and clinical.

82 45 CFR 46.208 (a). The regulations define "minimal risk" to mean..."that the probability and magnitude of harm or discomfort anticipated in the research are not greater in and of themselves than those ordinarily encountered in daily life or during the performance of routine physical or psychological examinations or tests" 45 CFR 46.102 (i). Federal regulators have never clarified this definition in the context of fetal research. The risks of the "daily life" of the first-trimester fetus are significant when viewed in the context of the background risks of spontaneous abortion (20–30 percent). It is arguable that amniocentesis or chorionic villus sampling are "routine" physical tests. These medically indicated procedures carry risks of complications and fetal loss. See a discussion of the history of the minimal risk standard in relation to Subpart B. Fletcher, J.C. (1993). Human fetal and embryo research: Lysenkoism in reverse—how and why? In *Debates Over Medical Authority. New Challenges in Biomedical Experimentation*, Blank, R.H., Bonniksen, A.L., eds., vol. 2 (New York: Columbia University Press) pp. 208–10.

83 The Secretarial waiver was used only once by Joseph Califano, on the recommendation of the only EAB, in 1979 for a study of fetoscopy in the context of hemoglobinopathies. Although the EAB requested that he approve this type of research as a category, he only approved the project itself. Steinfels, M. (1979). At the EAB: same members, new ethical problems. *Hastings Cent Rep* 5,2. Secretary Patricia Harris allowed the EAB's charter to lapse in 1980, and there has never been another EAB, although federal regulations require that "one or more Ethical Advisory Boards shall be established by the Secretary" 45 CFR 46.204.

84 The composition of Congress following this election was Senate (46 DEM – 54 GOP), House (269 DEM – 166 GOP). Available from Clerk of the House: clerkweb.house.gov. In addition to a Republican majority in the Senate, Sen. Jeremiah Denton (D-AL) was a voice in opening a new phase of the debate on fetal research. See Lehrman at note 71 above, p. 8.

85 The Health Extension Act of 1985, November 20, 1985. For a good discussion of the legislative history of this Act, see Lehrman at note 65 above, pp. 7–9. This Act also adopted a provision introduced by Sen. Gore in 1983 and established the Biomedical Ethics Advisory Commission (BEAC). The history of BEAC proves the hypothesis that since the NC, any national bioethics body solely created by and accountable only to Congress has a vanishingly small chance of success. For a discussion of BEAC's history, see Cook-Deegan, R. (1994). *The Gene Wars*. (New York: W.W. Norton), pp. 256–62.

86 "...the Secretary shall require that the risk standard (published in Section 46.102(g) of such Part 46 or any successor to such regulations) be the same for fetuses which are intended to be aborted and fetuses which are intended to be carried to term." Health Law Extension Act of 1985, Sec. 498 (b) (3). Section 498 (a) (2) requires that the research "will pose no added risk of suffering, injury, or death to the fetus...." The effect of these two stipulations is to nullify the "minimal risk" standard of the federal regulations for development of important biomedical knowledge which cannot be obtained by other means.

87 See Fletcher J.C., Schulman J.D. (1985). Fetal research: the state of the question. *Hastings Cent Rep* 15, 6–12; Fletcher, J.C., Ryan K.J. (1987). Federal regulations for fetal research: a case for reform. *Law, Med and Health Care* 15(3), 126–38.

88 Those in charge of composing and redacting federal regulations to protect human subjects are aware of their role in protecting the legacy of the NC's commitment to fetal research because of its benefits. William F. Dommel, personal communication, April 29, 1999. This protective role has been symbolic, because no investigative fetal research of any consequence has been federally funded. One notes with sadness that a whole generation of researchers in reproductive medicine has been without federal support of fetal research and study of the beginning of embryonic development. The need to do limited and appropriate fetal research is even more important today that it was in 1974, if we are to have ethically and scientifically responsible fetal therapy. In any future reform of public policy on fetal research, the history of Subpart B of the federal regulations and the original intent of the NC ought to carry weight.

89 Amidst a very large literature, a very good recent review of these concerns is The New York State Task Force on Life and the Law. (1998). *Assisted Reproductive Technologies*. (Albany, NY: Health Education Services).

90 Wymelenberg, S., for the Institute of Medicine. (1990). *Science and Babies*. (Washington, DC: National Academy Press) p. 15.

91 Ibid.

92 Abma, J.C., et al. (1997). Fertility, family planning, and women's health: new data from the 1995 National Survey of Family Growth. *Vital and Health Stat* 23, 1. "However, even with the decreasing rate, the total number of infertile couples is the same as it was in 1982, because of the increasing number of married couples in the relevant age group." In *The New York State Task Force on Life and the Law*. (1998). See note 89 above, pp. 11–12.

93 New York State Task Force, see note 89 above, 12.

94 The Office of Technology Assessment (OTA) reported that from 1983 to 1987, the number of infertility centers offering IVF grew from 10 to 167. U.S. Congress, Office of Technology Assessment. (1988). *Infertility. Medical and Social Choices*. OTA-BA-358 (Washington, DC: U.S. Government Printing Office, May) p. 157. The role of the Serono Corporation and other commercial enterprises in this growth industry is described in Hotz, R.L. (1991). *Designs on Life*. (New York: Pocket Books) pp. 176–203.

95 See note 7 above, Report of the Ethics Advisory Board, p. 3.

96 Congregation of the Doctrine of the Faith. (1987). Instruction on Respect for Human Life in Its Origin and on the Dignity of Procreation: Replies to Certain Questions of the Day. Vatican City: The Congregation.

97 Two examples are provided. Patricia Harris, Secretary, DHEW, wrote in her own hand in response to a decision memo (Harris, 1979, Memorandum: to Kathy Schroeher, Executive Secretariat, DHEW, November 26) on the 1979 recommendations of the EAB to support IVF research: "I need greater justification for such research. Whether the research will take place with or without government support is not really relevant. Why should government support such an area as this! I have read the material. It is not persuasive." Harris mainly saw IVF as a procedure for the advantaged.

Secondly, in 1988, the OTA reported to Congress: "The effect of this moratorium on Federal funding of IVF research has been to eliminate the most direct line of authority by which the Federal Government can influence the development of embryo research and infertility treatment so as to avoid unacceptable practices or inappropriate uses. It has also dramatically affected the financial ability of American researchers to pursue improvements in IVF and the development of new infertility treatments, possibly affecting in turn the development of new contraceptives based on improved understanding of the process of fertilization." U.S. Congress, Office of Technology Assessment. (1988). *Infertility. Medical and Social Choices*. OTA-BA-358 (Washington, DC: U.S. Government Printing Office) May, p. 179.

For more discussion of the ethical and political implications of this pattern of federal reluctance, see also Blank, R.H. (1997). Assisted reproduction and reproductive rights: the case of *in vitro* fertilization. *Pol and the Life Sci* 16, 279–288.

98 Norman, C. (1988). IVF moratorium to end? *Science* 241, 405.

99 Ruth Hanft, former Assistant Secretary for Planning and Evaluation, DHHS, who served under Secretary Harris, confirmed this statement. Personal communication, May 17, 1999.

100 See Norman, C. article at note 98.

101 Wyden, R. (1989). Opening remarks and testimony at the Hearing on Consumer Protection Issues Involving *In Vitro* Fertilization Clinics, before the House Subcommittee on Regulation, Business Opportunities, and Energy. Washington, DC, Mar. 9.

102 42 U.S.C.A. § 263a (1997). CLIA is discussed in the New York State Task Force Report, see note 89 above, p. 410.

103 42 U.S.C.A. § 263a-1 et seq. (1997).

104 See note 6 above.

105 Childress, J.F. (1991). Ethics, public policy, and human fetal tissue transplantation. *Kennedy Inst of Ethics J* 1, 93–121.

106 Sullivan, L.B. to William F. Raub. November 2, 1989.

107 Hilts, P.J. (1990). U.S. aides shaky legal basis for ban on fetal tissue research. *New York Times*, Jan. 30, A-1.

108 Weiss, T. to Sullivan, L., January 26, 1990.

109 Administrative Procedures Act (5 U.S.C. 552, 553).

110 The term "indefinite" attached to "moratorium" was chosen to circumvent federal law and deflect legal action against DHHS. A document leaked from the Public Health Service to the press was cited by Hilts (see note 107 above, at A-1): "We have chosen to make the moratorium indefinite rather than permanent, because a permanent prohibition of this research would require formal rulemaking and this would require extensive public comment and would be rather easily susceptible to litigation which could reverse this action."

111 See note 66 above.

112 See note 7 above.

113 Although the NIH could have funded research with donated excess embryos in 1994 in the interim between President Clinton's decision not to support NIH funding of creation of research embryos and the imposition of the ban, "As of spring 1995, NIH has yet to fund any human embryo research, despite 70 pending proposals and eight proposals that have cleared scientific review." Charo, R.A. (1995). The hunting of the snark: the moral status of embryos, right-to-lifers, and third world women. *Stanford Law and Policy Rev* 6, 11–27.

114 See note 7 above, pp. 7–8.

115 The moral aspects of this policy will be discussed more fully in Part II and in the Appendix.

116 Public Law 105-277. Conference Report to Accompany H.R. 4328. Sec. 511. "None of the funds made available in this Act may be used for...(2) research in which a human embryo or embryos are destroyed, discarded, or knowingly subjected to risk of injury greater than that allowed for research on fetuses *in utero* under 45 CFR 46.208(a)(2) and section 498(b) of the Public Health Service Act (42 U.S.C. 289g(b))." Even more revealing of the moral reasoning of opponents of embryo research in Congress is language in their letter to Secretary Shalala corresponding exactly to language in federal regulations on research with living fetuses in the context of elective abortion.

117 First introduced in 1976, the Hyde Amendment, named for its sponsor, Henry Hyde (R. Il), restricts all funding of abortion for the federal share of Medicaid except for cases where two physicians attest that continuation of the pregnancy will result in severe and lasting damage to the woman's physical health and in cases of reported rape and incest. The law took effect after a Supreme Court ruling: *Harris v. McRae* 448 U.S. 297 (1980).

118 Robertson, J.A. (1978). The scientist's right to research: a constitutional analysis. *South Cal Law Rev* 51, 1203–79.

119 See note 22 above. Rabb based her opinion on a scientific definition of PSCs as neither a human "organism" as defined by the statute nor capable of developing into a human being. If PSCs are not embryos, she argued, then the statute does not prevent NIH funding PSC research "downstream" from derivation of PSCs that was privately funded. Since the ban on embryo research only follows the public dollar, there are no legal restrictions on private companies or universities funding such work, if the equipment and laboratory facilities are not purchased or operated with federal funds.

Subsequently, Secretary Shalala received two letters signed by seventy House members and five Senators. The signers implored her to correct the legal opinion and reverse Dr. Varmus' decision to fund PSC research. The House letter (From Jay Dickey, et al. to Donna Shalala, Feb. 11, 1999) argued that the Rabb opinion evaded the linkage to and complicity in prior destruction of embryos. It also advanced a key legal interpretation, i.e., that Congress intended the scope of its ban to bar any tax dollars being spent on research which "follows or depends on the destruction of or injury to a human embryo." The key sentence was: "in the embryonic stem cell research which NIH proposes to fund, the timing, method, and procedures for destroying the embryonic child would be determined solely by the federally funded researcher's need for useable stem cells." This language repeats identical language in federal regulations on fetal research (45 Code of Federal Regulations § 46. 206 (3)) and law on fetal tissue transplant research (Public Law 103-43, § 112 (c) (4)). The effort is to frame access to embryos for research in the same legal and moral context as access to the living fetus in the context of abortion. A choice of words often reflects a moral choice. "Embryonic child" shows how the dispute is joined. One does not have to agree with their premises to agree with the point that the morality of access to embryos cannot be separated from the morality of uses of PSCs.

Secretary Shalala answered that the legislative history showed that the ban does not prevent federal funding of research "preceding or following" banned research in which embryos would be discarded or harmed. Her position was: "Proceeding cautiously with research on existing pluripotential stem cells is both legal and appropriate." Letter from Donna Shalala to Jay Dickey, et al., Feb. 23, 1999.

120 Wilmut, I., Schnieke, A.E., Kind, A.J., Campbell, K.H.S. (1997). Viable offspring derived from fetal and adult mammalian cells. *Nature*, 385, 810–13.

121 Wilmut, I. "Cloning for Medicine." *Scientific American*, Dec. 1998.

122 Wakayama, T., Perry, A.C.F., Zuccotti, M. (1998). Full-term development of mice from enucleated oocytes injected with cumulus cell nuclei. *Nature*, 394, 369–73. Obtaining PSCs from embryos that resulted from fusion of adult cells with enucleated human oocytes could become a Case 5 or a Case 6, depending on the feasibility of the approach described in note 21 above.

123 See note 19 above.

124 Council for Science and Society. (1984). *Human Procreation*. (Oxford: Oxford University Press) p. 7.

125 A Task Force of the AAAS is studying the ethical, legal, and social issues in intentional germline gene transfer; the NIH-RAC is presently examining unintentional germline effects of somatic cell gene therapy.

126 A discussion of the key elements in such an approach that focuses on clinical cases is in Fletcher, J.C., Lombardo, P.A., Marshall, M.F., Miller, F.G. (1997). *Introduction to Clinical Ethics*. 2d ed. (Frederick, MD: University Publishing Group) pp. 21–38. The approach of "clinical pragmatism" discussed here is a hybrid that combines elements within casuistry, the dialectical method of moral reasoning used by Beauchamp, T.L., Childress J.F. (1995). *Principles of Biomedical Ethics*, 4th ed. (New York: Oxford University Press), and virtue ethics. The latter element is particularly important as a component of the "Charo principle" discussed below, i.e., the virtue of compassionate response to moral suffering, even if one disagrees with the premises and worldview of those who suffer from losses to their moral values when compromises are necessary. A strong feature of clinical pragmatism is that it will be concerned as much with the issues of "who decides?" and "how ought the decision to be carried out?" as with "what ought the decision to be?" These questions are as relevant to public policy choices as they are to decisionmaking in the clinical setting.

127 The renewal of casuistry in a historical perspective is best discussed by Jonsen, A.R., Toulmin, S. (1988). *The Abuse of Casuistry: A History of Moral Reasoning*. (Berkeley: University of California Press). For an evaluation of the contribution of casuistry to biomedical ethics, see Beauchamp and Childress, at note 31 above, pp. 95–100. A valuable text in "pluralistic casuistry" is Brody, B. (1988). *Life and Death Decision Making*. (New York: Oxford University Press). Brody uses a model of "conflicting appeals" with complex clinical cases to gain insight about how the case should be resolved. Also, for an expert philosophical evaluation of the case-by-case approach, see Arras, J.D., (1991). Getting down to cases. *J Phil and Med* 16, 29–51.

128 Clouser, K.D., Gert, B. (1990). A critique of principlism. *J Med and Phil* 15, 234.

129 Harris, J. (1992). *Wonderwoman and Superman*. (New York: Oxford University Press) pp. 45–46.

130 Subsequently embodied in the NIH Revitalization Act of 1993.

131 Burtchaell, J.T. (1989). Use of aborted fetal tissue in research and therapy. In Burtchaell, J.T., ed. *The Giving and Taking of Life* (South Bend, IN: Notre Dame University Press) pp. 155–87.

132 Burtchaell, J.T. (1989). Use of aborted fetal tissue: a rebuttal. *IRB* (Mar-April) 11, 9–12.

133 Andrew Siegel's comments were important to developing these points.

134 Of 266 respondents, 32 (12%) reported that they would be more likely to have an abortion if they could donate tissue for fetal tissue transplantation. 178 (66.9%) stated that they would not be more likely to do so, and 56 (21.1%) were uncertain. Martin D.K., Maclean, H., Lowy, F.H., et al. (1995). Fetal tissue transplantation and abortion decisions: a survey of urban women. *Canad Med Assoc* 153, 545–52.

135 See note 6 above.

136 Fletcher J.C. (1990). Fetal tissue transplantation research and federal policy: a growing wall of separation. *Fetal Diagnosis and Therapy* 5, 211–225.

137 U.S. General Accounting Office, (1997). NIH-Funded Research: Therapeutic Human Fetal Tissue Transplantation Projects Meet Federal Requirements. Report to the Chairmen and Ranking Minority Members, Committee on Labor and Human Resources, U.S. Senate, and Committee on Commerce, House of Representatives. US-GAO, Washington, DC, March.

138 See Siegel, A. (1999). *Complicity and Consent* (draft). NBAC Document, April 8, 1999.

139 In the unlikely event that research proves that PSC research will not lead to cell-replacement therapy, science and society will be better off. A negative finding benefits science and prevents harmful experimentation.

140 Rathjen, P.D., Lake, J., Whyatt, L.M., et al. (1998). Properties and uses of embryonic stem cells: prospects for applications to human biology and gene therapy. *Reprod, Fertil, and Devel* 10, 31–47.

141 Callahan, D. (1995). The puzzle of profound respect. *Hastings Cent Rep* 25, 39–43.

142 Research to date by the NBAC staff on the question of "discard" shows: 1) a wide variation of practices regarding consent for cryopreservation of excess embryos and choices about disposition of embryos, 2) only 10–25 percent of frozen embryos are truly considered excess, 3) patients are more likely to discard embryos than to donate to other couples, 4) at clinics where the option to donate embryos to research is given, couples are equally as likely to donate as to discard, and most significantly, 5) new technology allows longer culture of embryos (up to five days) and permits more quality assurance; embryos that do not appear normal and implantable are discarded, and the remaining desirable embryos are frozen. The preliminary picture, which calls for more research, is that there are several pressures that will reduce the supply of excess embryos for research.

143 The options to shape an optimal process for informed consent must be examined to heighten assurance that the embryos donated for research in Case 2 are ones that will be discarded and die.

144 Parens concentrates on the problems of ascertaining the "intentions of embryo makers" at the time of creation of embryos (i.e., to reproduce or to use for research) and is skeptical about the validity of a morally relevant difference between Cases 2 and 4. However, Parens does not take account of the similarity of Cases 1 and 2 in terms of the consequences of discarding fetal or embryonic tissue. Parens, E. (1999). *What Has the President Asked of NBAC? On the Ethics and Politics of Embryonic Stem Cell Research,* available in this volume.

145 Jonas, H. (1969). Philosophical reflections on experimenting with human subjects. *Daedalus* 98, 245.

146 This is not an argument that an embryo is a human subject. It is a thought experiment using Jonas' moral wisdom as a mirror for reflection on the ethics of embryo research.

147 The author asked two experts, Ted Thomas (University of Virginia) and Mark Hughes (Wayne State University), for an opinion on the question of whether usable PSCs would survive embryo death. Each viewed it as highly improbable but knew of no research on the specific question. Dr. Hughes referred to the nonintegrity of DNA samples taken from 4–5 day old embryos in the process of dying. Personal communication, Feb. 25, 1999.

148 The moral relevance of parental intent to procreate as well as their active concerns for their embryos is discussed in Annas, G.J., Caplan, A., Elias, S. (1996). The politics of human embryo research—avoiding ethical gridlock. *N Engl J Med* 334, 1329–32.

149 Parens's points about the difficulties of oversight bodies in discerning intentions of "embryo makers" are well taken. See note 98 above, p. 11. However, oversight bodies ought to be more concerned with the authenticity of the consent process for parents to donate embryos for research than with discerning their intentions.

150 This is a complex question that is related to the issue of moral worth of fetuses in the context of abortion. U.S. public policy is that there should be no difference in the degree of research protection owed to fetuses in the abortion context than in a context of continued gestation to delivery of the infant. This "Golden Rule" approach to fetal research is repeated in the embryo ban. The point is that the policy history within which NBAC is working assumes that there ought to be no differences between the moral worth of embryos, regardless of their source. This policy framework is open to challenge, but it is the prevailing framework.

151 Annas, et al., at note 148 above, p. 1331.

152 In my view, the decisive factors in Cases 3 and 4 combine the degree of weight given to the moral status of embryos with proximity to scientific and clinical benefits.

153 The letter of transmittal from the Chairman of the Human Embryo Research Panel to the Advisory Committee to the Director, NIH, reveals a degree of equivocation about the use of the term "pluralistic" to describe the Panel's ethical perspective on embryo research. Dr. Muller states: "The panel began from the position that it was not called upon to decide which among the wide range of views held by American citizens on the moral status of preimplantation embryos is correct, but rather that its task was to make recommendations that would assist the NIH in developing guidelines for preimplantation human embryo research that took full account of generally-held public views regarding the beginning and development of human life." Stephen Muller to Ruth Kirschstein, see note 7 above, p. v. The report describes two approaches to debates on the issue of moral status: one proposes some single criterion, a second approach is "pluralistic." "It sees moral respect and personhood as deriving not from one or even two criteria from a variety of different and interacting criteria." See note 7 above, at pp. 35–36. A fair reading of the report would lead one to conclude that the Panel took a clearly liberal position on moral status, did not defend it as such, and then used the term "pluralistic" as a surrogate for what the liberal majority (including the liberal majority on the Human Embryo Research Panel) might approve.

154 See note 7 above, pp. 38–39.

155 See note 148 above.

156 See note 148 above, p. 1131.

157 Ethics Advisory Board, see note 7 above, p. 35–36.

158 Furman, W.L., Pratt, C.B., Rivera, G.K. (1989). Mortality in pediatric phase I clinical trials. *J Nat Cancer Inst* 81, 1193–94.

159 This point is made by Helga Kuhse and Peter Singer in "Individuals, Humans, and Persons," in Singer, P., Kuhse, H., Buckle, S., et al., eds. (1990). *Embryo Experimentation. Ethical, Legal, and Social Issues*. (Cambridge: Cambridge University Press) p. 73.

160 What is known about these practices? Has any research been done about whether the guidelines for embryo research recommended by the American Society for Fertility and Sterility are followed? Is it known whether researchers supported by private funds submit their protocols to local IRBs?

161 This step could be taken after the NIH's "downstream" approach to funding PSC research has had a chance to be tested. There is a precedent in the U.K., as the Human Embryo Authority allows derivation of PSCs from excess embryos.

162 See note 7 above, vol 1, pp. x–xi.

163 See note 124 above.

164 National Bioethics Advisory Commission. (1997). *Cloning Human Beings*. vol. 1. (Rockville, MD: U.S. Government Printing Office) p. iv.

165 Julian Savulescu sees no morally relevant differences between "a mature skin cell, the totipotent stem cell derived from it, and a fertilized egg. They are all cells which could give rise to a person if certain conditions obtained." Savulescu, J. (1999). Should we clone human beings? cloning as a source of tissue for transplantation. *J Med Ethics* 25, 87–95. I am inclined to agree with this reasoning, although the verifiable possibilities of creating implantable human embryos by cloning must still be established.

166 See note 11 above, p. 1061.

167 Dr. John Biggers informed the EAB in 1979 that, prior to IVF, the total body of information about human ova and embryos was comprised of 15 specimens in the world's literature. Ethics Advisory Board, *Appendix*, No. 8, pp. 7–18. Discussed in Grobstein, C. *From Chance to Purpose* (Reading, MA: Addison-Wesley, 1981) p. 36.

168 Van Blerkom, J. (1994). The history, current status, and future direction of research involving human embryos. In National Institutes of Health. *Papers Commissioned for the Human Embryo Research Panel*. vol. 2, p. 9.

169 Braude, P., Bolton, V., and Moore, S. (1988). Human gene expression first occurs between the four-and eight-cell stages of preimplantation development. *Nature* 332, 459.

170 Although she agreed in principle with the Human Embryo Panel's recommendation to approve federal funding of embryo research (as in Case 4), Patricia King's partial dissent to the report stated: "Allowing fertilization of oocytes expressly for research purposes offers potential for benefit to humankind, but it also raises fundamental ethical concerns. The prospect that humanity might assume control of life creation is unsettling and provokes great anxiety. The fertilization of human oocytes for research purposes is unnerving because human life is being created solely for human use. I do not believe that this society has developed the conceptual frameworks necessary to guide us down this slope. My concerns are heightened in the context of research activities where practices cannot be monitored easily by the public and where it is difficult to ascertain whether the research is being conducted responsibly." See note 7 above, vol. 1, p. A-3. Given the strong reactions to the Panel's report, Prof. King's dissent was visionary and perceptive. Reflecting on her dissent and participating in the NBAC process has also moderated my earlier uncritical support for the Panel's report: Fletcher, J.C. (1995). U.S. public policy on embryo research: two steps forward, one large step back. *Human Reproduction* 10, 1875–8. The Panel could have taken a more moderate and incremental approach. The argument in the appendix was developed in part to respond to Prof. King's dissent. Her words are just as fitting today as in 1994 and should serve as a caution to NBAC and the scientific community.

171 The Warnock Commission (1984) in the U.K. was the first public bioethics body to address the questions of embryo research. Their report eventually led to the Embryo Research Act (1990) which permitted embryo research under the careful scrutiny of a public authority that grants licenses for this activity. See Department of Health and Social Security. (1984). Report of the committee of inquiry into human fertilization (London: Her Majesty's Stationery Office) 1987.

172 Rachaels, J. (1990). *Created from Animals* (New York, Oxford University Press) is an excellent discussion of the slow pace of cultural change in the context of the moral implications of Darwin's discovery of evolution by natural selection.

173 Of relevance here is the OTA study on the history and various types of commissions and panels in biomedical ethics in the federal sector. See note 79 above. Also see Fletcher J.C., Miller F.G. (1996). The promise and perils of public bioethics. In *The Ethics of Research Involving Human Subjects: Facing the 21st Century*, H.Y. Vanderpool, ed. (University Publishing Group, Frederick, MD) pp. 155–184.

174 President Clinton's own response to the Human Embryo Panel's recommendations in 1994 illustrates this point. He could accept research with excess embryos but not with embryos created only for research. Marshall, E. (1994). Human embryo research. Clinton rules out some studies. *Science*, 266, 1634–35. An editorial, "Embryo research: drawing the line," *Washington Post*, Oct. 2, 1994, A-21, had earlier expressed the same view.

175 See note 126 above, p. 327.

176 The distinction between the "investigative" first stage of research and the other two stages will not be as persuasive for scientists. Many, if not most, scientists would argue that all data developed at any stage of PSC research are relevant to the use of PSCs in humans. Nonetheless, there is a practical difference between basic research to answer foundational questions in a new field and research with animals and humans to answer preclinical and clinical questions. Also, the public and Congress would recognize the distinction between stages and what is being asked of them.

177 My recommendation is that if NBAC is not still operational at such a point in time, an EAB should be appointed by the Secretary, DHHS, for the specific purpose of advising the Secretary and Congress on the question of federal funding of deriving PSCs from embryos. Existing federal regulations enable such a step to be taken. 45 CFR 46.204.

178 NBAC and members of the administration and Congress can help to create a vision of the moral and political implications of where permitting (with eyes wide open to the derivation of PSCs from embryos using private funds) NIH "downstream" funding can lead, i.e., to creative and moderate reforms of the polarized and incoherent extremes of American research practices in the private and public arenas.

179 Note that the compromise fits with the two categories of research protected by federal regulations on fetal research (described in Part I, Section 3) and extended to federal funding of embryo research. The compromise also assumes a "detached" view of the government's role in protecting embryos in research, i.e., expressed by a carefully regulated practice with safeguards to prohibit or minimize abuses. Most importantly, any compromise that permits federal funding of embryo research (for the first time, even in the context of therapeutic intent) must assume the construction of federal regulations of embryo research that unify practices in both the public and private sectors. In this respect, federal funding of embryo research, limited to Case 2, could be the opening chapter in a future history of moderating all of the overly protectionist public policies on fetal and embryo research that are the legacies of the political cultures of the 1980s and 1990s. What better way (in research activities) to show respect for the intrinsic value of life than to reform by *the politics of compromise* the overly permissive morality of the private sector and the overly protective morality of the public sector into a new and unified public policy for embryo and fetal research? Such an effort to reform research ethics at the beginning of human life could also be the beginning of efforts to expand the circle of moral and constitutional protections to all human subjects of research, regardless of the source of financial support of research.

180 President's Commission for the Study of Ethical Problems in Medicine and Biomedical and Behavioral Research. (1982). *Splicing Life*. (Washington, DC: U.S. Government Printing Office) Chapters 1 and 2.

181 These conditions were expressed, in principle, in remarks made by Commissioner Childress, during the NBAC meeting in Charlottesville, VA, on April 16, 1999.

182 Some of these potential hazards include: genetic abnormality, aberrant developmental processes, functional aberrations (e.g., inappropriate or uncontrolled insulin secretion). Hazards related to specific diseases need to be investigated. For example, if neuronal cells become adrenergic instead of cholinergic, it can greatly impact the safety of a clinical study.

183 The FDA and the NIH have increasingly mandated equity in access to clinical trials, notably to require gender equality and (in the case of the FDA) to redefine pediatric labeling and study requirements. FDA/NIH could be meaningful co-partners in applying access requirements equally to both publicly and privately funded studies.

184 National Commission for the Protection of Human Subjects of Biomedical and Behavioral Research. (1979). *The Belmont Report*. FR Doc. 79-12065, p. 5. This classic statement of the principles that govern biomedical research with human subjects discusses justice primarily in terms of the fair distribution of risks and vulnerability of certain groups for recruitment. It does stress that "whenever research supported by public funds leads to the development of therapeutic devices and procedures, justice demands that these not provide advantages only to those who can afford them…."

185 "The incidence of infertility is 10.5 among married couples with non-Hispanic black women, roughly 1.5 times greater than among Hispanic or non-Hispanic white women." Cited in New York Task Force, see note 89 above, p. 11.

186 New York State Task Force, see note 89 above, p. 432.

187 Personal communication, American College of Obstetrics and Gynecology, April 15, 1999.

188 IRBs are required, among other things, to determine if "selection of subjects" for the proposed project "is equitable" 45 CFR 46.111(3). The IRB is to take into account the purposes of the research, the setting in which it is to be conducted, and be especially "cognizant of the special problems of research involving vulnerable populations, such as children, prisoners, pregnant women, mentally disabled persons, or economically or educationally disadvantaged persons." The history of these provisions is marked by

protecting such vulnerable persons from exploitation in research or biases in selection that would expose them to higher risks. IRBs should be equally concerned with the issue of fairness in selection of subjects to improve *access* to the potential benefits of clinical trials, especially for "economically...disadvantaged persons," since participation in clinical trials is heavily biased already in favor of more advantaged groups.

The role of the FDA and the NIH in requiring standards of equity in access to trials is described above in note 183.

189 This appendix would be richer in content if it discussed the full spectrum of theological and secular views on embryo research and its meaning for human life. An important recent contribution was made in a statement about the ethics of PSC research by the EAB of the Geron Corporation, which is composed of members of diverse religious bodies. A symposium on the statement was organized by the Hastings Center; see Symposium: human primordial stem cells. (1999). *Hastings Cent Rep* (Mar-April), 30–48.

190 Vatican, Congregation for the Doctrine of the Faith. (1992). Instruction on respect for human life in its origin and on the dignity of procreation. In Alpern, D., ed. (1992). *The Ethics of Reproductive Technology* (New York: Oxford University Press) p. 85; see also Doerflinger, R.M. (1999). Testimony before the Senate Appropriations Subcommittee on Labor, Health and Education, Jan. 26.

191 Lockwood, M. (1995). Human identity and the primitive streak. *Hastings Cent Rep* 25 (Jan-Feb), 45.

192 Lockwood, M. (1985). When does a life begin? In *Moral Dilemmas in Modern Medicine*, Lockwood, M., ed. (Oxford: Oxford University Press) pp. 9–31.

193 Steinbock, B. (1994). Ethical issues in human embryo research. In National Institutes of Health, *Papers Commissioned for the Human Embryo Research Panel*, 30–32.

194 See note 3. The EAB (1979) found that "the human embryo is entitled to profound respect; but this respect does not necessarily encompass the full legal and moral rights attributed to persons" (pp. 35–6). The Warnock Committee's position was that "the embryo of the human species ought to have a special status," and "should be afforded some protection in law" (pp. 63–4). The NIH Human Embryo Research Panel stated that "although the preimplantation human embryo warrants serious moral consideration as a developing form of human life, it does not have the same moral status as an infant or child" (p. x).

195 "Special respect but no rights for embryos makes sense if one views the underlying ethical and policy question as one of demonstrating respect for human life. If the embryo is too rudimentary in development to have interests, it may nevertheless be a potent symbol of human life." Robertson, J.A. (1995). Symbolic issues in embryo research. *Hastings Cent Rep* 37 (Jan-Feb), 37. Also, see his testimony before NBAC. Jan. 19, 1999, pp. 81–87.

196 "The Panel held six extensive meetings, heard 46 oral presentations, and received more than 30,000 letters, cards, and signatures on petitions as a panel, plus uncounted hundreds of items of correspondence addressed individually to panel members. From the first to the last day of the panel's work, there was constant and profound awareness of the high level of public concern about the sensitive and complex issues involved." Stephen Muller to Ruth Kirschstein, October 12, 1994. See note 7 above, at p. v.

197 Charo, R.A. (1995). The hunting of the snark: the moral status of embryos, right-to-lifers, and third world women. *Stanford Law and Policy Rev* 6, 11–27.

198 See note 20 above; also Charo, R.A. (1995). "La penible valse hesitation": fetal tissue research review and the use of bioethics commissions in France and the United States. In Bulger, R.E., Bobby, E.M., Fineberg, H.F., eds., *Society's Choices: Social and Ethical Decision Making in Biomedicine*. (Washington, DC: National Academy Press) pp. 477–500; Charo, R.A. (1996). Principles and pragmatism. *Kennedy Institute of Ethics J* 6, 319–22.

199 Commissioner Charo was recused because she is employed by the University of Wisconsin, which has a financial interest in PSC research (Thomson study) and issues that NBAC will address in its recommendations. NBAC Proceedings, Feb. 2, 1999, p. 2.

200 Dworkin, R. (1993). *Life's Dominion. An Argument About Abortion, Euthanasia, and Individual Freedom*. (New York: Knopf).

201 This section cites several key passages of this important work. Dworkin's insights are more appreciable in his own well-chosen words.

202 Tribe, L.H. (1990). *Abortion, The Clash of Absolutes*. (New York: W.W. Norton); Rosenblatt, R. (1992). *Life Itself*. (New York: Random House).

203 See note 200 above, p. 10.

204 See note 200 above, p. 11.

205 See note 200 above, p. 13.

206 See note 200 above, p. 11.

207 Ibid.

208 See note 200 above, p. 90.

209 See note 200 above, p. 91.

210 Without adopting the theological premises that gave rise to his work, one can appreciate that Martin Buber's reflections on the distinction between "I-Thou" relations and "I-It" relations point to the same perception of the independent standing of entities (especially persons) entitled to such respect.

211 The issue of "wrongful life" in court cases and the issue of "personhood" in the context of embryo and fetal research can illustrate this point. There is a point in the arguments involved past which judicial authorities acknowledge they cannot go with the confidence that they are still in the realm of human and public affairs.

212 In a democracy that prizes separation of church and state, there is no final arbiter of such issues as the moral status of embryos or fetuses, which raise questions about the ultimate meaning of life itself. This is properly viewed as a religious or philosophical pursuit, best addressed in the context of worship or other forums that have evolved to permit the free pursuit and expression of answers to the question.

213 I defined moral suffering, in a 1972 report about the moral experience of parents in genetic counseling and amniocentesis, as follows: "...the state of being threatened by normlessness, even as one is caught between two forces or principles, both of which are right. Moral suffering is not the direct effect of *anomie*, relative cultural normlessness, on the individual concerned, but rather the opposite, as when a person is caught in a dilemma between two goods. Moral suffering occurs when highly motivated parents who desire children intensely, even desperately, are caught between the rightness of protecting their families from the great strains which genetic disease may place upon them, and the rightness of unconditional caring for the life of their conceived child." Fletcher, J.C. (1972). The brink: the parent-child bond in the genetic revolution. *Theological Studies* 33, 457–85.

214 There were no hearings or congressional testimony regarding the scientific or clinical consequences of the ban on embryo research prior to its passage.

215 Beauchamp, T.L., Childress, J.F. (1994). *The Principles of Biomedical Ethics*. (New York: Oxford University Press) p. 260.

216 Medicine is a goal oriented profession. Leon Kass argues that medicine has one absolute end: healing. (1985). *Towards A More Natural Science*. (New York: Free Press). His claim is overstated, because it is clearly problematic to fit other valid activities that serve the goals of medicine (e.g., prevention and research) under healing. Actual experience recommends viewing medicine as having multiple, complex, and sometimes competing goals: e.g., healing, promoting health, and helping patients achieve a peaceful and dignified death. This more complex view is reflected in Miller F.G., Brody, H. (1995). Professional integrity and physician-assisted death. *Hastings Cent Rep* 25, 8–17.

217 A traditional ethical norm of medicine is "Above all (or first) do no harm" which is transmitted in the ethical principle of nonmaleficence. See Beauchamp, T.L., Childress, J.F. (1995). *Principles of Biomedical Ethics*. 4th ed. (New York, Oxford University Press) p. 189.

218 This principle is also called proportionality.

219 See note 217 above, p. 261.

220 It is this understanding of the role of medicine and biomedical research within the structural values of society, i.e., as accountable to moral principles expressing those values, that empowers public bioethics to address moral problems without qualification as to whether these occur in the public or private spheres.

221 Brennan, T.A., Berwick D.M. (1996). *New Rules. Regulation, Markets, and the Quality of American Health Care*. (San Francisco: Jossey-Bass) p. 197.

222 Alta Charo discussed a "civic duty" to volunteer for research, which must be balanced with principles of good government and distributive justice (Belmont Revisited Conference, April 17, 1999). In biomedical research, there is a corresponding "scientific duty" to learn how best to treat disease by the scientific method that must also be balanced by principles of nonmaleficence and distributive justice.

223 Virtually everyone would now agree, e.g., that keeping (after hysterotomy) fetuses *ex utero* alive as a means to answer vital questions about absorption of oxygen through skin or brain metabolism is unethical. See note 77 above for an account of these prior experiments promoted by an unmodified utilitarian ethos. Today, these experiments would be viewed as grievous assaults on respect for the intrinsic value of life.

224 Moe, K. (1984). Should the Nazi data be cited? *Hastings Cent Rep* 14, 5–7.

225 Berger, R.L. (1990). Nazi science—the Dachau hypothermia experiments. *N Engl J Med* 322, 1435–40.

226 Freedman, B. (1992). Moral analysis and the use of Nazi experimental results. In *When Medicine Went Mad*, Caplan, A., ed., p. 151.

227 At the time the 076 Protocol was first discussed at the NIH in 1986, the author was Chief of the Bioethics Program at the Magnuson Clinical Center. He raised questions about the scientific and moral adequacy of a trial of a toxic drug *in utero* in advance of gaining understanding of how and when the HIV virus was actually transmitted *in utero*. The only evidence available about HIV transmission was from the tissues of abortuses suggesting that transmission was later in gestation rather than earlier. However, this evidence was only observational and did not prove how and when transmission occurred. A way to answer the question would have been to conduct a serial study of fetal blood drawn before and after elective abortions in the first and second trimesters of pregnancy. During the next year, the author joined the faculty at the University of Virginia but remained in dialogue with NIH scientists about the 076 Protocol and the issue of scientific and preclinical obligations. The regulations on fetal research and the "Golden Rule" law as passed by Congress clearly prohibited such a study without a "waiver" from the Secretary, DHHS. There being no EAB, Paul Rogers, former member of Congress, advised the author to approach members of Congress with a request to the Secretary to waive the minimal risk requirement. Support was sought for this idea from several officials at the NIH, who were reluctant to support or accompany the author on this mission. In point of fact, the 076 Protocol succeeded in reducing transmission, and the investigators were lucky in their guess that HIV was transmitted later rather than earlier in pregnancy.

228 Connor, E.M., Sperling, R.S., Gelber, R., et. al. (1994). Reduction of maternal-infant transmission of human immunodeficiency virus type I with zidovudine treatment. *N Engl J Med* 331, 1173–80.

229 The language of the embryo ban reflects prior federal policy on fetal research extended onto embryo research.

230 Remarks of Dr. Roberta Buckley. NIH Recombinant DNA Advisory Committee (RAC) Meeting, September 24–25, 1998, p. 4.

231 National Institutes of Health. Prenatal Gene Transfer: Scientific, Medical, and Ethical Issues. Third Gene Therapy Conference, January 7–8, 1999.

232 Schneider, H., Coutelle, C. (1999). *In utero* gene therapy: the case for. *Nature Med* 5, 256–57.

233 Brent, R.L. (1985). The magnitude of congenital malformations. In *Prevention of Physical and Mental Congenital Defects. Part A. The Scope of the Problem.* (New York: Alan R. Liss) p. 55.

234 Galjaard, H. (1984). Early diagnosis and prevention of genetic disease. In Galjaard, H., ed. *Aspects of Genetic Disease.* (Basel, Switzerland: Karger) p. 1.

235 Collins, F.S. (1996). BRCA1—lots of mutations, lots of dilemmas. *N Engl J Med* 334, 186–88; Parens, E. (1996). Glad and terrified: on the ethics of BRCA1 and 2 testing. *Cancer Invest* 14, 405–11.

236 The gap is a moral problem because treatment for genetic disease is more fitting with the goals of medicine than selective abortion to prevent or avoid it. If the gap ought morally to be closed, and it can be closed more quickly with congressional action, then it follows that there is a significant moral problem affecting the Congress and the whole nation.

237 Fletcher, J.C. (1998). The long view: how genetic discoveries will aid healthcare reform. *J Women's Health* 7, 817–23.

238 As recommended for federal funding by the NIH Human Embryo Panel, see note 7 above, at p. 2.

Bioethical Regulation of Human Fetal Tissue and Embryonic Germ Cellular Material: Legal Survey and Analysis

Commissioned Paper
J. Kyle Kinner
Presidential Management Intern
National Bioethics Advisory Commission

A. Historical Overview

Since at least the 1930s, American biomedical research has involved *ex utero* fetal tissue as both a medium, and increasingly, an object for experimentation.[1] The 1954 Nobel Prize for Medicine, for example, was awarded to American immunologists using cell lines obtained from human fetal kidney cells to grow polio virus in cell cultures other than nerve tissue.[2] It was not until 1972, in a period that coincided with a larger societal debate over elective human abortion, that the use of *ex utero* fetal tissue for research (along with research involving fetuses generally) became controversial.[3] In 1974, following the imposition a year earlier of a moratorium on federally-funded research on live fetuses, Congress established the National Commission for the Protection of Human Subjects of Biomedical and Behavioral Research.[4] The Commission recommended guidelines applicable to research conducted or funded by the Department of Health and Human Services (DHHS) (then the Department of Health, Education, and Welfare), and in 1975, the Department adopted regulations governing fetal research; Congress also passed similarly directed legislation.[5]

Controversy erupted again in October 1987, when National Institutes of Health (NIH) scientists presented Director James B. Wyngaarden with a request to fund research on Parkinson's disease involving fetal brain tissue transplantation, already approved by an internal NIH review body.[6] Director Wyngaarden sought an opinion from DHHS Assistant Secretary Robert Windom, who responded by declaring a temporary moratorium on federally funded transplantation research involving fetal tissue from induced abortions.[7] In March 1988, the Assistant Secretary asked NIH to establish an advisory committee to consider whether such research should be conducted and under what conditions.[8] The 21-member Human Fetal Tissue Transplantation Research Panel, composed of a cross-section of research investigators, lawyers, ethicists, clergy, and politicians, deliberated until the fall of 1988. The panel voted 19-2 to recommend continued funding for fetal tissue transplantation research, including guidelines to assure the ethical integrity of any experimental procedures.[9] In November 1989, DHHS Secretary Louis Sullivan extended the moratorium indefinitely, adopting the position of minority panel members who believed that such research would increase the incidence of elective abortion.[10] Attempts by Congress to override the Secretary's decision were vetoed by President Bush or were not enacted into law.[11]

In October 1992, a consortium of disease advocacy organizations filed suit against DHHS Secretary Sullivan, alleging that the Hyde Amendment[12] (banning federal funding of abortion) did not apply to research or transplantation involving fetal tissue. The plaintiffs argued, moreover, that applying a fetal tissue transplantation research ban was beyond the Department's statutory authority under the law.[13] The suit was preempted on January 22, 1993, when the new administration shifted national biomedical policy and directed DHHS Secretary Donna Shalala to remove the ban on federal funding for human fetal tissue transplantation research.[14] On February 5, 1993, Secretary Shalala officially rescinded the moratorium, and in March 1993, NIH published interim guidelines for research involving human fetal tissue transplantation.[15] Governing legislation was quickly proposed in Congress, and President Clinton signed the NIH Revitalization Act of 1993 into law on June 10, 1993.[16]

It is important to note that, throughout the period of controversy over transplant research involving tissue from induced abortions, other areas of fetal research continued to receive governmental funding and attention. One journalist has observed that "during the period of the moratorium, NIH—except for studies involving fetal material obtained from elective abortions [used in transplantation research]—continued to support human fetal tissue research. In 1992, this support totaled some $12.4 million, more than 90 percent of which went toward extramural projects."[17]

B. Federal Statutes

I. NIH Revitalization Act of 1993

Codified at 42 USC §§ 289g-1 and g-2, the NIH Revitalization Act of 1993 includes most prior statutory and regulatory provisions on research involving fetal tissue transplantation. In general, the Act states that any tissue from any type or category of abortion may be used for research on transplantation, but only for "therapeutic purposes."[18] The investigator's research scope is not unlimited. Such activity must be conducted in accordance with applicable State and local law.[19] The investigator must further obtain a written statement from the donor verifying that (a) she is donating fetal tissue for therapeutic purposes; (b) no restrictions have been placed on the identity of the recipient; and (c) she has not been informed of the identity of the recipient.[20] Finally, the attending physician must sign a written statement affirming five additional requirements about the abortion, effectively erecting a "firewall" between the donor's decision to abort and her decision to provide tissue for fetal research.[21] The person principally responsible for the experiment must also affirm his or her own knowledge of the source of the tissue, that others involved in the research are aware of the tissue's original status, and that he or she had no part in the decision or timing of the abortion.[22] The drafters included no specific penalties in 42 USC § 289g-1.

By contrast, 42 USC § 289g-2 provides significant criminal penalties for violation of four prohibited acts (relating to interstate commerce, for purposes of jurisdiction): (1) purchase or sale of fetal tissue "for valuable consideration" beyond "reasonable payments [for] transportation, implantation, processing, preservation, quality control, or storage..."; (2) soliciting or acquiring fetal tissue through the promise that a donor can designate a donee; (3) soliciting or acquiring fetal tissue through the promise that the material will be implanted into a relative of the donor; or (4) soliciting or acquiring fetal tissue after providing "valuable consideration" for the costs associated with the abortion itself.[23]

II. Human Research Extension Act of 1985

Codified at 42 USC § 289g, the Human Research Extension Act provides guidance on fetal research generally, directing that no federal research or support may be conducted on a nonviable living human fetus *ex utero* or a living human fetus *ex utero* for whom viability has not been determined, unless (a) the research or experimentation may enhance the health, well-being, or probability of survival of the fetus itself; or (b) the research or experimentation will pose no added risk of suffering, injury, or death to the fetus where the activity is intended for "the development of important biomedical knowledge which cannot be obtained by other means."[24] In either instance, the degree of risk must be the same for fetuses carried to term as for those intended to be aborted.[25]

III. National Organ Transplant Act

The National Organ Transplant Act (NOTA), 42 USC §§ 273-274e, prohibits the sale of any human organ for "valuable consideration for use in human transplantation" if the sale involves interstate commerce.[26] In 1988, the Congress amended NOTA to include fetal organs within the definition of "human organ," prohibiting—to the extent possible—the sale of fetal tissue within interstate commerce.[27]

C. Federal Regulations[28]

I. 45 CFR §§ 46.201-211, Subpart B

Located within the general protections for biomedical research subjects provided by federal regulation, 45 CFR §§ 46.201-211, Subpart B speaks directly to research involving the human fetus.[29] First promulgated in 1975, this regulatory subpart covers research on "(1) the fetus, (2) pregnant women, and (3) human *in vitro* fertilization" and applies to all DHHS "grants and contracts supporting research, development, and related activities directed towards those subjects.[30] The regulation states explicitly that "the purpose of this subpart [is] to...assure that [applicable research] conform[s] to appropriate ethical standards and relate[s] to important societal needs."[31] Like its statutory counterpart at 42 USC §§ 289g, 289g-1 and g-2, 45 CFR § 46.201-211 attempts to address the particular concerns inherent in fetal research and to reduce attendant risks. These protections include (1) provision for stringent IRB review;[32] (2) pre-studies on animals and nonpregnant individuals;[33] (3) an assessment of minimal risk to the fetus (except where the research purpose is intended "to meet the health needs" of the mother or the fetus);[34] (4) exclusion of investigators from any involvement in the decision to terminate a pregnancy or any assessment of fetal viability;[35] and (5) prohibition on inducements to terminate a pregnancy for research purposes.[36] Specific restrictions are imposed on the inclusion of pregnant women or fetuses *in utero* in research activities.[37]

Of special relevance to fetal tissue research, 45 CFR §§ 46.209 and 210 address requirements for federal funding of activities directed towards the fetus *ex utero*, including nonviable (but still living) fetuses.[38] Section 46.209 focuses on viable and nonviable (but still living) fetuses.[39] Until a determination has been made of fetal viability, no research may occur unless (1) there is no additional risk to the fetus and the purpose is the development of important biomedical knowledge that cannot be obtained by other means; or (2) the purpose is to enhance the viability of the particular fetus to the point of survival.[40] Once viability is determined, the regulation specifies that research on a nonviable fetus may only occur where (1) vital functions of the fetus are not artificially maintained; (2) experimental activities that would themselves terminate heartbeat or respiration are not employed; and (3) the underlying purpose of the research is the development of important biomedical knowledge that cannot be obtained by other means.[41] Where a fetus *ex utero* is determined to be viable, its status is protected under 45 CFR § 46.101 et seq. as a human subject.[42] Note finally that research on fetuses for which viability has not been determined, or fetuses that have been deemed nonviable, may occur only where the mother and father are legally competent and have given their informed consent, or where only the mother consents if the father's identity or whereabouts cannot be ascertained; he is not reasonably available; or the pregnancy resulted from rape.[43]

45 CFR § 46.210 provides fewer limitations and deals exclusively with research involving the dead fetus, fetal material derived from dead fetuses, or the placenta.[44] The regulation states that "activities involving the dead fetus, macerated fetal material, or cells, tissue, or organs excised from a dead fetus shall be conducted only in accordance with any applicable state or local laws regarding such activities."[45] As commentary infra suggests, at least some analysts, perhaps including the DHHS General Counsel, conclude that 45 CFR § 46.210 is the *only* regulatory component of Subpart B that is applicable in the context of fetal tissue transplant research or research in which fetal cellular material or tissue is separated from the fetus as a whole for experimentation. (This position is supported by commentators like Judith Areen arguing from textual analysis of Subpart B, while others conclude on a practical basis that it is simply not possible to extract tissue for research purposes from living, nonviable or viability undetermined fetuses in a way that involves "minimal risk.") Interested members of the NIH community, in commenting on this paper in its draft form, argue that living, nonviable, or viability-undetermined fetuses should not be considered a source of fetal material for research of the type contemplated here (for the practical reasons described supra), but still support the application of other protective provisions within Subpart B for fetal tissue research in addition to Section 46.210.[46] Uncertainty over whether and to what degree individual sections of Subpart B are controlling for fetal-derived stem cell research may cloud efforts to pursue study in this area.

II. 45 CFR §§ 201-210, Subpart B – Proposed Rule

On May 20, 1998, DHHS released for public comment its proposal to revise and rewrite 45 CFR §§ 46.201-210, Subpart B, "Additional DHHS Protections for Pregnant Women, Human Fetuses, and Newborns Involved as Subjects in Research, and Pertaining to Human *In Vitro* Fertilization."[47] The changes contained in the proposed rule are the product of an intensive 14-month review of existing Subpart B regulations by the NIH Office for Protection from Research Risks (OPRR) Public Health Service Human Subject Regulation Drafting Committee.[48] The newly revised regulations have not been finalized by the Department and do not presently supercede the original 1975 Subpart B (discussed supra). While the regulations have been substantially reorganized under the proposed rule, in general there are only a few changes that are material to fetal tissue research as it is discussed here.

Four proposed revisions are worth noting. First, new 45 CFR § 46.201(b) applies the categories of noncontroversial research deemed exempt from regulation in Subpart A at 45 CFR § 46.101(b)(1)-(6) to Subpart B.[49] Second, the regulations governing fetal research have been rewritten, but not substantively changed, to reflect (i) "fetuses may be involved in research where the risk is not greater than minimal;"[50] or (ii) where "any risk to the fetus which is greater than minimal is caused solely by activities designed to meet the health needs of the mother or her fetus;"[51] (iii) "any risk is the least possible for achieving the objectives of the research;"[52] (iv) Institutional Review Boards (IRBs) are no longer required to determine the beneficial purpose of the research (e.g. whether the protocol involves "the development of important biomedical knowledge which cannot be obtained by other means");[53] (v) "consent of the father is not required;" rather, "consent of the mother *or* her legally authorized representative is required" [after she is]..."informed of the reasonably foreseeable impact of the research on the fetus."[54] Third, parental consent requirements for *ex utero* living nonviable and viability-undetermined fetuses were amended somewhat.[55] Finally, a revised 45 CFR § 46.206 relating to dead fetal and placental material, like its predecessor subsection 46.210, retains the controlling status of State regulation codified in the earlier 1975 rule, but also includes a new and unrelated paragraph (b) directing that any living individual who becomes personally identified as a result of research on dead fetal or placental material must be treated as a research subject and accorded the protections of 45 CFR §§ 46.101 et seq.[56]

III. Federal Common Rule – 45 CFR §§ 46.101, Subpart A

45 CFR § 46.101, Subpart A, also termed the "federal Common Rule," governs federally supported human subjects research generally.[57] Important bioethical protections like IRB review, informed consent, and subject privacy are regulated in detail.[58] 45 CFR § 46.101(a) provides that "this [Common Rule] applies to all research involving human subjects conducted, supported or otherwise subject to regulation by any federal Department or Agency which takes appropriate administrative action to make the [Common Rule] applicable to such research." At 45 CFR § 46.102(f), "human subject" is defined as "a living individual about whom an investigator (whether professional or student) conducting research obtains (1) data through intervention or interaction with the individual, or (2) identifiable private information."[59] This definition leaves considerable room for debate over whether fetal tissue research, the derivation of fetal stem cells, and research involving successor stem cells from fetal sources are governed by the federal Common Rule.[60]

In considering this question, OPRR has devised a series of decisional charts intended to help researchers determine when to apply human subjects protections.[61] According to OPRR, the investigator must first ask: "Is there an intervention or an interaction with a living person that would not be occurring or would be occurring in some other fashion, but for this research?"[62] If the answer is yes, "human subjects [are] involved. Follow 45 CFR 46 or meet criteria for exemptions."[63] If the answer is no, the chart directs the investigator to ask: "Will identifiable private data/information be obtained for this research in a form associable with the individual?"[64] The term "associable" is defined as meaning that "the identity of the subject is or may be readily ascertained or associated with information."[65]

The outcome of OPRR's first decision point is somewhat unclear. Certainly the decision to abort and the abortion itself are separate from the research activity (not an "interaction"). Removal of cellular material from the dead abortus, while qualifying as an interaction, would not involve any "living individual."[66] The best argument seems to be that the obtaining of consent to use donated fetal tissue for research purposes constitutes an interaction with the mother/donor (a "living individual") sufficient to invoke the Common Rule. However, the obtaining of consent is not itself a part of the research, and further, researchers are under no obligation to obtain consent pursuant to 42 USC § 289g-1(b) where the underlying activity does not involve "research on the transplantation of human fetal tissue for therapeutic purposes."[67]

On the latter decision point, OPRR's question regarding identifiable private data or information raises similar concerns. Does any identifying information accompany the delivery to investigators of fetal material used in the derivation of stem cells (or the resulting stem cells themselves if derivation took place off-site)? Certainly the tissue mass itself, divorced from any medical record, would not qualify as an identified sample. The abortion provider would probably retain identifying records about its clients and their decision to donate (useful unless tissue samples from multiple abortuses were aggregated), but perhaps a third party who dissected the fetus, extracted the necessary germ cells, or derived usable stem cells would not. The abortion provider would not be obliged to maintain identifying records to pass on to researchers pursuant to 42 USC § 289g-1(d)'s audit requirement unless that experimental activity qualified under the enabling scope of therapeutic transplantation research at 42 USC § 289g-1(a). Even so, it is not clear from 42 USC § 289g-1(c)'s enumeration of researcher duties that information regarding the "identity of the subject" would ever be passed on to the project investigator (or, for that matter, what the term "subject" actually means in a fetal experimentation context [the mother/donor or the fetus]).

Finally, recall that OPRR references statutory exemptions to the Common Rule at 45 CFR § 46.101(b)(4). In order to be considered exempt, "this research [must] use solely existing data or specimens," those materials must be "publicly available," and the investigator cannot record information "in such a way that it can be linked to the subject."[68] The term "existing" is defined as meaning "collected (i.e. on the shelf) prior to the research for a purpose other than the proposed research. It includes data or specimens collected in research and nonresearch activities."[69] The definition's emphasis on the preexisting status of specimen material may make it unlikely that stem cell research will be exempted. However, to the extent that fetal material is harvested by a clinic or other third-party processor for general "research use" and only then specifically directed to a project or investigator doing fetal stem cell derivation or stem cell research, it may not be difficult to establish that fetal material was "on the shelf" for some short duration necessary to qualify for exemption from the Common Rule. Compounding the uncertainty, 45 CFR § 46.101(i), footnote 1 provides inter alia that "the exemptions at 45 CFR 46.101(b) do not apply to research involving prisoners, fetuses, pregnant women, or human *in vitro* fertilization, Subparts B and C." This footnote is presumably explanatory and is not a part of the regulation itself. Its language seems to take "fetuses" and "human *in vitro* fertilization" out of the exempted category, but the overall effect is not entirely clear given the text of new 45 CFR § 46.201(b)-(c).[70]

Relevant to the question, supra, of which individual subsections of 45 CFR 46, Subpart B apply to fetal stem cell research, 45 CFR § 46.201 ("Applicability") provides that "the requirements of this subpart [B] are in addition to those imposed under the other subparts [A, C, and D] of this part [46—Human Subjects Protections]." This invoking reference to the Common Rule through Subpart B may mean that by clearly falling under 45 CFR § 46.210 ("Activities involving the dead fetus, fetal material, or the placenta"), research involving fetal material will qualify for human subjects protections under Subpart A *even* if such material presents conceptual difficulties that would make it unlikely to qualify independently for inclusion under the Common Rule. In the alternative, it may instead be argued that failure to qualify for inclusion under the Common Rule at 45 CFR §§ 46.101 and 46.102(f) effectively precludes *any* Common Rule application whatsoever for fetal stem cell research, avoiding Subpart B no matter what 45 CFR § 46.210 may provide (and rendering section 46.210 meaningless). Finally, it may simply be the case that fetal-derived stem cell research can fail to qualify under Subpart A but nevertheless satisfy Subpart B (accepting section 46.201 but concluding that no additional requirements are imposed under other subparts). There appears to be no textual basis for preferring one interpretation over another; OPRR may wish to clarify this question.

D. Uniform Acts

Uniform Anatomical Gift Act[71]

Originally promulgated to encourage organ donation for transplantation, the Uniform Anatomical Gift Act (UAGA) has been widely enacted into law by the states.[72] First proposed in 1968 in a version enacted by all fifty states and the District of Columbia, a 1987 revision has been enacted by twenty-two states.[73] The UAGA is relevant not only because federal fetal tissue statutes and regulations explicitly condition funding and authority on compliance with state and local laws, but also because private researchers are bound by state statute even absent federal authority. In its 1987 definition, UAGA defines a "decedent" as a "deceased individual [that] includes a stillborn infant or fetus."[74] The law permits the use of human tissue for the purposes of education, research, or the advancement of science.[75] It requires that an attending physician determine the time of death, and like 42 USC § 289g-1(b), the Act provides that informed consent must be obtained prior to the donation of any tissue.[76] Like 45 CFR § 46.209(d), parents of the fetus retain ultimate authority to decide whether to make a donation.[77]

Several sections of UAGA may be materially different from existing federal law and regulation. For example, an entire body or parts of a body may be donated as an "anatomical gift" to a recipient, including individual donees.[78] This section is consistent with 42 USC §§ 289g-1(b)(1)(B)-(C) and 289g-2(b)(1) only where designation by the donor of a recipient for aborted fetal tissue means a designated researcher or research facility (since designation or even knowledge of an individual transplant recipient is prohibited), or where fetal tissue is donated for research not involving transplantation.[79] In addition, the Act provides that "neither the physician or surgeon who attends the donor at death nor the physician or surgeon who determines the time of death may participate in the procedures for...transplanting a part."[80] This section, although waivable, appears slightly more stringent than statutory and regulatory restrictions at 42 USC § 289g-1(c)(4) and 45 CFR § 46.206(3) on researcher involvement in the decision or act of abortion, prohibiting researchers' physical presence or assistance at the clinical procedure from which fetal tissue for research is derived.[81] Finally, commentators have noted that, unlike 45 CFR § 46.209(b), which addresses the issue of living but nonviable fetuses, UAGA apparently does not. "UAGA does not apply to tissue donations from live persons, such as blood donations, skin donations, bone marrow, or kidney donations, so there may be no applicable law for fetal donation in such cases."[82] The authors suggest that "UAGA is probably best applied by analogy until an amendment can resolve this point."[83]

In other areas, the UAGA closely tracks federal statutory provisions and, as a result, may share similar difficulties. Sections 10(a)-(b) of the uniform Act, included in the 1987 revisions, prohibit the actual sale or purchase of any human body parts for any consideration beyond the amount necessary to pay for expenses incurred in removal, processing, and transportation of the tissue.[84] This is essentially the same proscription included at 42 USC § 289g-2(a) barring the acquisition or transferal of fetal tissue for "valuable consideration," with the same exceptions.[85] One commentator has argued that the federal provision (and by extension UAGA) is unenforceably vague in its definition of reasonable processing fees, "leav[ing]...room for unscrupulous tissue processors to abuse the law...."[86] Drafters on the federal level and in the states that have enacted UAGA's 1987 no-sale provision have attempted to address this concern by making violation of the section a felony with substantial penalties.[87] Some states have added a further clarification in their enactment to indicate that the donation of human tissue for transplantation is to be understood as a service and not a sale.[88]

E. Case Law

Any consideration of court-derived law in the area of fetal tissue research and transplantation will conclude that litigation directly related to this subject is relatively uncommon. A wide range of cases deserve mention however, including those affecting privacy and reproductive freedom generally;[89] donative autonomy;[90] informed consent;[91] self-determination and surrogate decision-making;[92] determination of brain death and/or viability;[93] allocation of parental authority among competing parties;[94] fetal tort rights;[95] biological property interests;[96] and the ability of Congress to regulate fetal tissue use or transfer in interstate commerce, among others.[97]

Only a few cases touch directly on the subject of fetal tissue research itself.[98] While numerous states have enacted laws affecting or regulating fetal experimentation, an important statute to face scrutiny on that issue was enacted by Louisiana in 1978. La. Rev. Stat. Ann. § 40:1299.31 et seq. forbade virtually all research or study involving the fetus or fetal tissue ("a live child or unborn child") not "therapeutic" to that child.[99] The statute protected the fetus *in utero*, but did not address *ex utero* fetal tissue except by implication. In 1981, the Louisiana legislature expanded the Act's scope to include aborted fetal tissue in its prohibition.[100] In its initial review, the court considered only the pre-1981 act (without the fetal tissue amendment), holding that plaintiffs' failure to demonstrate the statute's negative impact on the right of privacy left the court to conclude that "no obstacle has been placed in the path of the woman seeking abortion."[101] Under the resulting rational basis analysis, the court found that the statute was rational because it tended to protect the public from the "dangers of abuse inherent in any rapidly developing field."[102] The statute was challenged again after the 1981 amendment and included a showing by plaintiffs that a prohibition on research did burden the right of privacy.[103] Reversing its earlier decision, the court found that the revised statute infringed on the fundamental right of privacy and applied strict scrutiny analysis. The court concluded that a research ban did not further the state's compelling interest in protecting the health of the woman and that its interest in the potential life of the unborn did not continue past the death of the fetus.[104] Finally, the district court addressed the statute's vagueness, noting that it was not possible, *ex utero*, to distinguish between fetal and maternal tissue or the products of spontaneous and induced abortions.[105] On appeal to the Fifth Circuit, the court ignored the district court's analysis entirely, finding instead that the term "experiment" as used in the statute's prohibition against fetal experimentation was unconstitutionally vague.[106] "The whole distinction between experimentation and testing, or between research and practice, is…almost meaningless," such that "experiment" is not adequately distinguishable from "test."[107] As a criminal prohibition without effective standards, the statute was deemed void.[108]

A less stringent Utah statute was examined by the Tenth Circuit in 1995 in *Jane L. v. Bangerter* and rejected on similar grounds.[109] Unlike the Louisiana law, Utah Code Ann. § 76-7-310 permitted discretionary experimentation aimed at acquiring genetic information about the embryo or fetus. A lower court upheld the statute by narrowly interpreting the term "experimentation" to mean "tests or medical techniques which are designed solely to increase a researcher's knowledge and are not intended to provide any therapeutic benefit to the mother or child."[110] The Tenth Circuit disagreed, arguing that the district court "blatantly rewrote the statute, choosing among a host of competing definitions for 'experimentation.'"[111] The court further concluded that the word "benefit" was itself ambiguous: "If the mother gains knowledge from a procedure that would facilitate future pregnancies but inevitably terminate the current pregnancy, would the procedure be deemed beneficial to the mother? Does the procedure have to be beneficial to the particular mother and fetus that are its subject?"[112] Without clear boundaries between permissible action and criminal conduct, the statute was deemed unconstitutionally vague and invalid.[113]

Finally, an Illinois district court in *Lifchez v. Hartigan*[114] considered a claim by a class of reproduction and infertility specialists seeking to invalidate a criminal misdemeanor statute that prohibited "experiment[ation] upon a fetus produced by the fertilization of a human ovum by a human sperm unless such experimentation is therapeutic to the fetus thereby produced."[115] Plaintiffs claimed that the words "experimentation" and "therapeutic" rendered the statute vague and unconstitutional, and the district court agreed. "[P]ersons of common intelligence will be forced to guess at whether or not their conduct is unlawful....[T]here is no single accepted definition of 'experimentation' in the scientific and medical communities."[116] The court observed that experimental procedures evolve quickly into routine diagnostic and therapeutic interventions, and tests to obtain information about a fetus' development often are not therapeutic to the fetus in the sense meant by that term.[117] The court was also troubled by the possibility that a term like "therapeutic" might prohibit assisted-reproduction technologies generally, or impede the detection or novel treatment of disorders that are considered life-threatening to the mother.[118] Under this analysis, the court decided that a scienter (knowledge or intent) requirement included in the Illinois statute did not mitigate vagueness where the law "has no core meaning to begin with."[119] Rather, the court expanded its vagueness argument to conclude that potential restrictions on a woman's reproductive decision arising from the law's broad effect and definitional uncertainty were an encroachment on the essential right of privacy as outlined by the Supreme Court in *Griswold* and *Roe*.[120]

F. State Law

The National Bioethics Advisory Commission recently commissioned Lori B. Andrews, J.D., a professor at the Chicago-Kent College of Law, to examine legislative activity on the state level, surveying and reporting applicable state law and regulation that affects research on stem cells (including fetal-derived embryonic germ [EG] cells.)[121] Andrews reports that 24 states do not have laws specifically addressing research on embryos or fetuses. Of the remaining 26 states and the District of Columbia, variation among statutes is the norm. Twelve states within the enacting group apply regulations only where the research involves fetal material derived from elective abortion (by extension, permitting research involving fetal material from stillborn or miscarried abortuses). Six states ban research altogether if it involves dead aborted fetuses. By contrast, six states permit fetal research where the fetus is dead, but mandate that the donor must provide consent. At least one state bans research involving "live" fetuses. Some states also extend their statutory reach beyond the abortus itself to regulate the resulting organs or tissues.

In the area of informed consent, Andrews notes that while six states require donor consent,[122] none of the state statutes affecting embryo and fetus research (other than New Mexico's "live fetus" statute) specifically address the kind of information that must be provided to the progenitors before they are asked for consent. New Mexico requires that the donor who accedes to research use of her fetus be fully informed regarding possible impact on the fetus, but applies that requirement only in the context of "live" fetuses. Note that Pennsylvania reiterates 42 USC § 289g-1(c) by requiring that research investigators (rather than donors) be informed of how the tissue on which they work was procured. Finally, while Andrews does not include UAGA-derived consent in her survey, every state and the District of Columbia has enacted a version of proxy-donor consent that would presumably apply in those cases where an investigator seeks fetal tissue from an individual donor rather than approaching a clinic (legal insofar as the donation does not carry a recipient designation).[123]

Addressing the UAGA, Andrews highlights the conflict in many states' donor designation statutes with existing federal law. She also identifies state laws whose restrictions regarding choosing tissue recipients are broader than federal law, and may have implications for stem cell research." Andrews cites Pennsylvania, where "no person who consents to the procurement or use of any fetal tissue or organ may designate the recipient of that tissue or organ, nor shall any other person or organization act to fulfill that designation."[124] On the issue of commercialization, Andrews' examination of UAGA-related statutes surfaces the possibility of fetal tissue sale for research purposes (currently prohibited under the post-1987 UAGA revision). Eight states ban payment for fetuses for any purpose. Only Minnesota prohibits sale of living fetuses or nonrenewable organs while permitting "the buying and selling of a cell culture line or lines taken from a [dead fetus]...."[125] Twenty states have enacted UAGA statutes that include the fetus in the definition of "decedent" and also prohibit the sale of decedent organs and/or organ parts.[126] Note that the range of statutory language and problems inherent in the UAGA itself make it difficult to characterize even these states' statutes as absolutely prohibiting the sale of fetal tissue for research purposes (a definitive conclusion is not possible without litigation to test the statutes). Only Virginia unambiguously prohibits the sale of "any natural body part for any reason including but not limited to medical and scientific uses such as transplantation, implantation, infusion, or injection."[127] Finally, six states' statutes prohibit the sale of human organs but fail to include a definition of "valuable consideration" that stipulates an exemption for miscellaneous overhead expenses. Sixteen states provide such an exemption.[128]

G. Commentary

I. Directed Donation

The issue of recipient-specified (or "directed") donation of fetal tissue for research or transplantation is an area of ambiguity under the law that deserves further consideration. Under statute, 42 USC § 289g-2(c)(1) provides a fine and/or incarceration for any person who "solicit[s] or knowingly acquire[s], receive[s], or accept[s] a donation of human fetal tissue for the purposes of transplantation…into another person [where] the donation will be or is made pursuant to a promise…that the donated tissue will be transplanted into a recipient specified by such individual" (42 USC § 289g-2(b)(1)). The law also proscribes any inducement to donate on the basis of a promise that "the donated tissue will be transplanted into a relative of the donating individual" (42 USC § 289g-2(b)(2)). Three qualifying criteria are necessary for this statute to apply: the solicitation or acquisition must be "for the purpose of transplantation…into another person" (42 USC §§ 289g-2(b)); the donation "affects interstate commerce" (42 USC § 289g-2(b)); and the tissue "will be or is obtained pursuant to an induced abortion" (42 USC § 289g-2(b)).

In its enactment, Congress has attempted to isolate the decision to abort from any consideration that the aborted tissue may benefit someone known to the donor, fearing that such knowledge would influence the initial termination decision.[129] This is demonstrated more generally at 42 USC § 289g-1(b)(1), requiring that the donor, prior to donation, affirm in writing that such decision "is made without any restrictions regarding the identity of individuals who may be the recipients of transplantations of the tissue; and [that] the [donor] has not been informed of the identity of any such individuals." A written statement must be supplied from the attending physician affirming, inter alia, that the donor's decision to abort was separated from (and prior to) the decision to donate tissue and that the statements contained in the donor's written affirmation were in fact true (42 USC § 289g-1(b)(2)). Researchers involved in fetal tissue transplantation are permitted to use donated material from any source on the condition that they also affirm, inter alia, that the researcher "has had no part in any decisions as to timing method, or procedures used to terminate the pregnancy made solely for the purposes of the research" (42 USC § 289g-1(c)).[130] Researchers are also subject to the aforementioned criminal penalties for procuring fetal tissue for directed donations at 42 USC § 289g-2(b).

The real difficulty with directed donation resides not so much in the text of 42 USC § 289g-2 or in federal law generally, but in the apparent conflict both in language and in spirit with Section 6(a)(3) of the UAGA.[131] The UAGA specifies that permissible donees for organ transplant or research are "(a)(1) a hospital, physician, surgeon, or procurement organization for transplantation, therapy, medical or dental education, research, or advancement of medical or dental science; (a)(2) an accredited medical or dental school, college, or university for education, research, advancement of medical or dental science; or (a)(3) a designated individual for transplantation or therapy needed by that individual." No restrictions comparable to 42 USC §§ 289g-1(b) or 289g-2(b) are present anywhere in the UAGA, a widely adopted model act whose core tenet "require[s] that the intentions of a donor be followed."[132] The conflict between federal statutes and regulations and state law incorporating the UAGA is not an easy one. Ordinarily, one or more theories of legislative preemption provide that federal statutes supercede those of the subordinated states. However, both 42 USC § 289g-1(e) (applicable to DHHS and its grantees) and 45 CFR § 46.210 (applicable to DHHS and its grantees whose research involves dead fetuses or fetal material) explicitly condition research access to fetal tissue on compliance with "applicable State or local laws regarding such activities."

A review of presently enacted UAGA statutes reveals only one state, Vermont, that has omitted the conflicting subsection (a)(3) from its enactment.[133] The remaining jurisdictions, including the District of Columbia, permit designated donation of organs or tissue for transplant, research, or other purposes. It is not necessarily the case, however, that the apparent conflict in law is irremediable. Rather, one could reasonably argue that what state statutes uniformly permit (recipient designation), is nevertheless prohibited in the limited context of fetal tissue donations for transplantation research or therapy conducted by agency scientists or extramural grantees. The explicit statutory subordination to state law at 42 USC § 289g-1(e) and 45 CFR § 46.210 may create the perception of some analytic circularity in this approach, but it seems extremely unlikely that permission from countervailing state statutes could enable agency conduct that federal statutory language expressly forbids. Commentators who have considered the question have rightly noted that this statutory conflict would appear to be ripe for resolution.[134] Given the wide prevalence of enacted UAGA statutes throughout the states, it would be significantly more feasible for Congress to amend 42 USC § 289g-1(e) in a way that clarifies the preemptive status of federal jurisdiction on the narrow question of directed donation.[135]

II. Statutory and Regulatory Research Limitations

The scope of federal authority to pursue various types of research involving human fetal tissue is not altogether straightforward under the statutes or accompanying regulations. For analytical purposes, it is useful to divide fetal-involved research activities into three broad categories: (a) therapeutic fetal tissue transplant research; (b) fetal tissue research not involving transplantation; and (c) nontherapeutic fetal tissue transplant research. We will consider each category in turn.

(a) Therapeutic fetal tissue transplant research

Federal statutory and regulatory guidance is somewhat less ambiguous in this research category. For pragmatic purposes, the NIH Revitalization Act of 1993 divides research science relating to transplantation into public and private sectors. 42 USC § 289g-1 is directed exclusively at those activities which the DHHS may in its discretion conduct through its own scientists or through extramural grantees. The statute permits "the [DHHS] Secretary…[to] conduct or support research on the transplantation of human fetal tissue for therapeutic purposes…regardless of whether the tissue is obtained pursuant to a spontaneous or induced abortion or…stillbirth" (42 USC § 289g-1(a)). Accordingly, the mechanisms included in 42 USC § 289g-1 to ensure that a firewall exists between donor and researcher are applied only in those instances where DHHS or its subordinates perform research of a type described in the Act. Regrettably, the statute fails to define the term "therapeutic" or to articulate the scope of what it means by "transplantation."[136]

By contrast, 42 USC § 289g-2 applies more broadly to include both public and private-sector research, covering "any person," (42 USC § 289g-2(a)-(b)) irrespective of funding status or governmental affiliation, who acquires fetal tissue. The prohibition on sale of tissue at Section 2(a) covers *all* transactions involving human fetal tissue, irrespective of purpose or category, and is applicable where a transfer occurs "[1] for valuable consideration [2] if the transfer affects interstate commerce." Section 2(b) imposes additional donative limits and covers "any person" who acquires human fetal tissue "[1] for the purpose of transplantation of such tissue into another person [2] if the donation affects interstate commerce, [and] [3] the tissue will be or is obtained pursuant to an induced abortion…" (42 USC § 289g-2(b)). Inasmuch as most or all research transactions will involve interstate commerce, the determinative criterion for 42 USC § 289g-2(a) is (1) valuable consideration; and the criteria for 42 USC § 289g-2(b) are (1) transplantation into a human recipient; and (2) induced abortion as the tissue source. No mention was included by the drafters in §§ 289g-2(a) *or* (b) of any restrictions on the "therapeutic" purpose or intent of research conducted under 42 USC § 289g-2.

The question of regulatory limitations is somewhat clouded. In her 1988 study for the Human Fetal Tissue Transplantation Research Panel, Judith C. Areen documents "four quite different interpretations of the current regulations…."[137] The problem Areen identified, and one which remains unresolved, is how to determine whether individual subsections of 45 CFR Section 46, Subpart B are applicable to human fetal tissue transplantation research. Most obviously, there is 45 CFR § 46.210 which covers activities involving "the dead fetus, macerated fetal material, or cells, tissue, or organs excised from a dead fetus," and provides that such activities are controlled by state or local laws where applicable.[138] This "single section interpretation" is contrasted with "double," "triple," and "four section" interpretations that purport to include, respectively, sections 46.206 (general limitations); 46.205 (additional duties for the IRB); and 46.209 (regulations applicable to viable and viability-undetermined fetuses *ex utero*).[139] On the basis of her own textual analysis, Areen concludes that the "one section" theory is correct and that only 45 CFR § 46.210 applies to fetal tissue transplantation research.[140] This position was recently supported, at least by implication, by the DHHS General Counsel's office.[141] Areen notes that the Association of American Medical Colleges had conversely adopted the "double section" theory.[142] It appears from available commentary that other writers have not considered the issue carefully, with some adopting a modified "double section" theory that applies as relevant only sections 46.209 and 46.210, while others appear to agree with Areen.[143] The author aptly concludes that "it is imperative that the Department clarify which of the [interpretations] is correct."[144]

(b) Fetal tissue research not involving transplantation

This is an area of research that has not been particularly controversial. DHHS and NIH have funded and continue to fund research of this type with congressional oversight and seeming approval. Research involving human fetal tissue that excludes transplantation is not addressed by 42 USC § 289g-1 and 289g-2(b) and is otherwise more generally regulated at 42 USC §§ 289g, 289g-2(a), 274e and 45 CFR Section 46, Subpart B. While this conclusion is not stipulated in statutory or regulatory text (other than DHHS' broad enabling statutes), it was accepted practice even during the fetal tissue transplant moratorium that other types of non-transplant research involving fetal tissue would still receive funding on an uninterrupted basis.[145] Inasmuch as it is widely believed that Congress did not act to constrict the scope of permissible fetal tissue research in this area when it enacted the NIH Revitalization Act of 1993, it can be reasonably argued that those activities which NIH funded or conducted prior to the moratorium's having been lifted could legitimately be funded or conducted thereafter.

From this analysis, one might conclude that research of the type conducted by Drs. Shamblott and Gearhart et al.—in which primordial germ cells were cultured from "gonadal ridges and mesenteries of 5- to 9-week postfertilization human [fetal] embryos (obtained as a result of therapeutic termination of pregnancy….)"[146] but not intended for transplant—would not be regulated by 42 USC § 289g-1's transplant "firewall" or the donative limitations at 42 USC § 289g-2(b) (excepting 42 USC § 289g-2(a)'s general prohibition of tissue sale, which is applicable). Rather, researchers pursuing this type of basic science in the future could proceed *outside* 42 USC § 289g-1 and 289g-2(b), subject to regulations that limit their ability to use cells extracted from a living, non-viable fetus (42 USC § 289g; 45 CFR § 46.209); by state law where applicable (42 USC § 46.210); or where their research matures to a point that it can be described as intended for transplantation. Seeming to support this analysis, NIH Director Dr. Harold Varmus has stated that the development of transplantation science is only one reason for conducting fetal-derived stem cell research; other areas include "how stem cells differentiate into specific types of cells...which, in turn, could lead to the discovery of new ways to prevent and treat birth defects and even cancer"; and "pharmaceutical development...study[ing] the beneficial and toxic effects of candidate drugs in many different cell types and potentially reduc[ing] the numbers of animal studies and human clinical trials required for drug development."[147]

In formulating Commission policy on this question, it is worth considering that the Gearhart exclusion argument is not uniformly accepted. DHHS General Counsel Harriet Rabb, in a January 1999 memorandum to NIH Director Varmus, concludes somewhat inexplicably that "[t]o the extent human pluripotent stem cells are considered human fetal tissue by law, they are subject to...the restrictions on fetal tissue transplantation research that is conducted or funded by DHHS, as well as to the federal criminal prohibition on the directed donation of fetal tissue."[148] General Counsel Rabb examines the definition of "fetal tissue" at 42 USC § 289g-1(g) which "means tissue or cells obtained from a dead human embryo or fetus after a spontaneous or induced abortion, or after a stillbirth" and observes that "some stem cells, for example those derived from the primordial germ cells of nonliving fetuses, would be considered human fetal tissue for purposes of [federal law]."[149] Having concluded (we think correctly) that primordial germ cells extracted from nonliving fetuses are a type of fetal tissue, the General Counsel goes on to apply, without explanation, the prohibition on sale of fetal tissue at 42 USC § 289g-2(a); firewall restrictions at 42 USC § 289g-1; the remaining donative limitations at 42 USC § 289g-2(b); and the single section 46.210 of 45 CFR.[150] While it is clear from the text of the statute that 42 USC § 289g-2(a)'s prohibition on fetal tissue sale was intended to apply broadly, both 42 USC § 289g-1 and 289g-2(b), like NOTA's prohibition at 42 USC § 274e(a), are explicitly limited to the narrow research context of transplantation. It would appear that the General Counsel believes that research of the type conducted by Drs. Shamblott and Gearhart et al. represents transplantation research of sufficient maturity to qualify for regulation under the statutes, but no argument or analysis is presented to support this conclusion. Nor, finally, is it clear why the General Counsel chose to apply only 45 CFR § 46.210 (she adopts the "one section" theory without explanation).

(c) Nontherapeutic fetal tissue transplant research

Without question, this category represents an area of potentially important research science that deserves greater clarity. The statutes and regulations themselves are essentially silent on nontherapeutic fetal tissue transplantation research and the scope of research activity that can be considered permissible. The same broad regulations that affect the prior category (42 USC § 289g; 289g-2(a); one or more subsections of 45 CFR Section 46, Subpart B; and state and local laws where appropriate, see 45 CFR § 46.210) are almost certainly applicable here. Further, given the language of 42 USC § 289g-1 and 289g-2, it seems clear that scientists in the private sector are able to pursue nontherapeutic transplant research, including transplantation into a human recipient, subject only to the tissue sale prohibition at 42 USC § 289g-2(a) and the donative restrictions at 42 USC § 289g-2(b). However, the structure and language of § 289g-1 and 289g-2 leave unanswered the question of whether federal intramural scientists or extramural grantees could conduct research (or even ignore firewall restrictions) where the transplantation research is *not* "for therapeutic purposes." This uncertainty is particularly acute given the absence of criminal penalties at 42 USC § 289g-1 concomitant with those provided at 42 USC § 289g-2(c). It might be argued that 42 USC § 289g-1 permits one type of transplant research ("therapeutic"), is silent as to the other ("nontherapeutic"), and lacks an enforcement mechanism to police the difference even if it is assumed that the latter research activity was intended to be forbidden.

This is almost certainly an anomalous interpretation given the statutes' origin as a congressional action intended to revoke the predecessor moratorium and to provide comprehensive guidelines for transplantation research involving human fetal tissue.[151] Normal rules of construction would generally dictate that use of a specific term like "therapeutic" to permit what had otherwise been forbidden limits the scope of permissible action to the general meaning of that term. Congress appears to have intended to apply more stringent regulations to governmental or grant-supported transplantation investigators than to researchers working in the private sector. By including the term "therapeutic," Congress expressed its desire to prohibit federal or federally-funded *nontherapeutic* transplantation research, irrespective of tissue source (induced, spontaneous, or stillbirth) while failing entirely to explain what this means. The provision seemingly limits the scope of permissible transplant research on the federal level to only those procedures whose primary or intended purpose is to provide therapeutic benefit to the transplant recipient herself. Congress' failure to include penalties in 42 USC § 289g-1 like those applied in § 289g-2 should rather be attributed to its assumption that sanctions were unnecessary to ensure compliance by the executive agency and its extramural research grantees.

It appears that NIH and the Department have attempted to interpret for themselves this puzzling statute, and may arguably have received tacit support both from Congress and from the General Accounting Office (GAO) in their interpretation. In its reports to Congress on fetal tissue transplant research (FY 1996–97 and FY 1993–95), NIH describes three categories of research involving human fetal tissue:

> [1] Basic research involving human fetal tissue focused on the advancement of knowledge of basic biological processes....[2] Pre-clinical investigations aim[ed] [at]...further therapeutic research through transplantation studies in animals or the development of improved methodologies for processing and preserving tissue [and]....[3] Projects involving the transplantation of human fetal tissue into humans [and] classified as therapeutic or clinical research if they are conducted on human subjects and are aimed at the development of therapeutic approaches for the cure or amelioration of diseases and disorders.[152]

The reports are mandated by 42 USC § 289g-1(f) and require the Department to tell Congress what research it supports pursuant to 42 USC § 289g-1(a) (the subsection that includes the Department's enabling language and the terms "therapeutic" and "transplantation"). All of the research grants that NIH reported to Congress under section 289g-1(a) were characterized as falling into the third category, either actual transplantation of human fetal tissue into other humans or follow-up studies on patients whose treatment had included fetal tissue transplants.[153] In its FY 1993–95 combined report, NIH described a quality-of-life study on Parkinson's patients that it had funded at three universities, stating that "since this research does not involve *the actual transplantation of tissue*, the funds for the study are not included in the total [fetal tissue transplant] funding figure for the year."[154] By its own reporting, its earlier definitional categories, and this admission, NIH seems to imply that (a) it presently supports what may be variously described as therapeutic or nontherapeutic transplant research; and (b) that such research is only deemed to be controlled by the enabling language and term "therapeutic" as a limitor where the "research...involves the actual transplantation of tissue."[155] After reporting twice to the House Committee on Commerce and the Senate Committee on Health, Education, Labor, and Pensions, it may be assumed that the committees have at least implicitly concurred in the NIH's definition of its authority. Finally, the GAO, in its March 1997 report to the Chairmen and Ranking Minority Members, considered NIH's compliance with 42 USC § 289g-1 and 289g-2 and concluded that "the requirements of the act are being complied with." It must be admitted that the GAO did not consider the definitional issue in the form described here, but it read the law closely, and its assessment of NIH's performance was favorable.

The net effect of all of this seems to be that NIH as a research and granting agency can and will fund (a) any *therapeutic* fetal tissue transplant research, subject to statutory and regulatory restrictions and reporting obligations (where the "research...involves the actual transplantation of tissue...conducted on human subjects and...aimed at the development of therapeutic approaches for the cure or amelioration of diseases and disorders"); (b) any fetal tissue research (therapeutic or nontherapeutic) *not involving transplantation*, subject to substantially fewer statutory and regulatory restrictions; and theoretically exclude (c) *nontherapeutic* fetal tissue transplant research. In simpler terms, almost any project that involves "the actual transplantation of tissue" will generally be styled as "therapeutic transplantation" and treated under the aforementioned section (a). (NIH's narrow definition of "transplantation" would exclude every procedure not involving the physical implantation of fetal tissue into a human.) The NIH definition of "therapeutic" is quite broad, and practically speaking, any injection of human fetal cells into a human recipient is not likely to be approved if it is not also at least "therapeutic" in the inclusive sense that NIH understands the term.[156] As a result, a narrow definition of "transplantation," a broad definition of "therapeutic," and the reality of human subjects' protections will effectively avoid the application of an inferred prohibition against nontherapeutic fetal tissue transplantation for almost any conceivable research proposal involving human fetal tissue. Following this logic, any project that does not involve "the actual transplantation of tissue" should be categorized as *not transplantation* under the NIH definition (since no cell matter is actually injected into a human recipient) and may be legitimately funded outside the restrictions established in 42 USC § 289g-1 and 289g-2(b). Note finally that this interpretation makes the General Counsel's sweeping analysis even more difficult to understand or to reconcile; the research conducted by Drs. Shamblott and Gearhart et al. did not involve "the actual transplantation of tissue" (a fact that should remove it from the transplant category entirely, according to NIH).

III. Prohibition on Sale of Fetal Tissue

A number of issues related to the federal statutory, regulatory, and UAGA prohibitions on sale of human fetal tissue present areas for further consideration. 42 USC § 289g-2(a) broadly states that "it shall be unlawful for any person to knowingly acquire, receive, or otherwise transfer any human fetal tissue for valuable consideration if the transfer affects interstate commerce." Fetal tissue is defined to mean "tissue or cells obtained from a dead human embryo or fetus after a spontaneous or induced abortion, or after a stillbirth" (42 USC § 289g-2(d)(1) [by reference to 42 USC § 289g-1(g)]. "Valuable consideration" is defined by its negation at 42 USC § 289g-2(d)(3) as "not includ[ing] reasonable payments associated with the transportation, implantation, processing, preservation, quality control, or storage of human fetal tissue." A complementary prohibition at 42 USC § 289g-2(b)(3), more narrowly limited to "donation…for the purpose of transplantation," makes it unlawful to solicit or acquire fetal tissue by "provid[ing] valuable consideration for the costs associated with…abortion." DHHS departmental regulations at 45 CFR § 46.206(b), applicable to DHHS scientists and extramural grantees, provide that "no inducements, monetary or otherwise, may be offered to terminate pregnancy for purposes of the [research] activity." NOTA 42 USC § 274e prohibits acquisition or transfer of "any human organ for valuable consideration for use in human transplantation if the transfer affects interstate commerce." NOTA states that "the term 'human organ' means the human (*including fetal*) kidney, liver, heart, lung, pancreas, bone marrow, cornea, eye, bone, and skin or any subpart thereof and any other human organ (or any subpart thereof, *including that derived from a fetus*) specified by the [DHHS] Secretary by regulation" (42 USC § 274e(c)(1)). The statute's definition of "valuable consideration" at 42 USC § 274e(c)(2) is effectively the same as 42 USC § 289g-2(d)(3).

The 1987 revision of the UAGA at section 10 directs that "(a) a person may not knowingly, for valuable consideration, purchase or sell a part for transplantation or therapy, if removal of the part is intended to occur after the death of the decedent." Further, "(b) valuable consideration does not include reasonable payment for the removal, processing, disposal, preservation, quality control, storage, transportation, or implantation of a part." The UAGA § 10(c), like its federal counterparts at 42 USC § 289g-2(c) and § 274e(b), provides substantial criminal penalties for violation. A review of the enacted UAGA legislation discloses two substantive changes by state legislatures: New Mexico and Nevada omit the phrase "if removal of the part is intended to occur after the death of the decedent" from § 10(a), and both Oregon and Connecticut omit entirely the section prohibiting sale of human organs. The effect of the former has little bearing on fetal tissue research except to the extent that it may affect work involving living, nonviable fetuses in those states adopting that redaction, and the latter reflects the apparent consensus among states (only 20 having enacted the necessary post-1987 UAGA revisions) that organ sale not be made illegal.

The unwillingness of states such as Oregon and Connecticut to include prohibition of organ sale in their statutes reflects a wider and unexplained state practice generally on this subject. Whether the legislatures of the remaining 30 states have consciously chosen to forgo state-enacted bans on organ sale is not known; it is quite possible that, given an existing version of the UAGA that predates the 1987 revision, inertia has precluded enactment of the newer prohibition, or simply that the state legislators believe federal prohibitions are sufficient. On the latter justification, there is some question as to whether this rationale actually works. Without getting deeply enmeshed in the current legal debate over Congress' ability to regulate state activity through the commerce clause, it is certainly true that the Supreme Court has begun to place limits on this regulatory power that did not exist when NOTA or the NIH Revitalization Act of 1993 were enacted.[157] Available commentary on this subject does not adequately address whether fully *intra*state research activity (and the associated purchase or sale of fetal tissue) would be prohibited under federal law, nor has any court considered the question.[158] Combining this potential for federal jurisdictional incapacity with the majority of state legislatures' not having enacted prohibitions on organ sale, it is at least possible that the sale of human fetal tissue for research or transplant may legally occur now or in the future in at least some states.[159]

Finally, it is probably worth noting what numerous commentators have already observed: the statutes are not clearly written.[160] For example, the reference in 42 USC § 289g-2 and other statutes to "valuable consideration" lacks meaningful content. While its purpose is generally understandable, the statutes' use of this ambiguous term leaves real questions unanswered about the scope of permissible activity under the law. And as cases like *Margaret S., Jane L.,* and *Lifchez* demonstrate, definitional ambiguity in this area can be fatal.[161] One tactical difficulty inherent in the definition is the degree to which various entities, including a range of intermediaries, are permitted to profit from their participation in the harvest of tissue or its subsequent use in research or transplantation. On its face, the statutes appear to exclude payments to the donor/mother for any commercial value the tissue may hold, nor may researchers under 42 USC § 289g-2(b)(3) provide reimbursement for the cost of the induced abortion. Beyond the donor/mother, however, the issue becomes much less clear. Are the intermediaries who collect and process the tissue, heretofore not-for-profit foundations, expected to recoup costs only, or are future commercial entities permitted to charge more? How much more? May research investigators or their institutional partners who immortalize valuable cell lines from fetal tissue legally transfer them for profit? May cell banks?

In an ambiguity specific to NOTA at 42 USC § 274e, one commentator aptly observed that "the statute…fails to state clearly what exactly is included in fetal organs and 'any subparts thereof.' At the time of transplantation, many organs, such as the pancreas and liver, are not yet developed. Thus, what is being transplanted, from a medical standpoint, are not organs or organ subparts, but precursor cells and tissues."[162] Moreover, "because NOTA does not specifically list, for instance, whether the brain is a controlled organ, it is unknown whether it is exempt."[163] It could easily be argued that the gonadal ridges and mesenteries of 5- to 9-week postfertilization embryos used by Drs. Shamblott and Gearhart et al., in their research on the primordial germ cell would be difficult to categorize developmentally as organs,[164] nor are they listed as such by 42 USC § 274e or by any departmental guidelines promulgated pursuant to 42 USC § 274e(c)(1). It may, however, one day be the case that differentiated successors to pluripotent stem cell cultures will be made to specialize for transplantation in a way that qualifies them for organ tissue status. Under the circumstances, congressional action to clarify the definitional issues brought about by medical and technological advances would appear to be appropriate.[165]

Notes

1 See e.g., Gregory Gelfand and Toby R. Levin, "Fetal Tissue Research: Legal Regulation of Human Fetal Tissue Transplantation," *Washington and Lee Law Review* 50 (1993):668; Cory Zion, "Comment, The Legal and Ethical Issues of Fetal Tissue Research and Transplantation," *Oregon Law Review* 75 (1996):1282.

2 See Gelfand and Levin, "Fetal Tissue Research," 668; Edgar Driscoll, Jr., "Nobel Recipient John Enders, 88; Virus Work Led to Polio Vaccine," *Boston Globe*, 10 September 1985: sec. Obituary; and Duke, Statement of the Population Crises Committee in Consultants to the Advisory Committee to the Director. National Institutes of Health, *Report of the Human Fetal Tissue Transplantation Research Panel* D112, D114 (vol II 1988) (hereinafter *HFTTRP II*) ("For many years, the production and testing of vaccines, the study of viral reagents, the propagation of human viruses, and the testing of biological products have been dependent on the unique growth properties of fetal tissue.").

3 Proposed guidelines for fetal tissue research were released by NIH, DHHS (then DHEW) in 1973, 38 *Fed. Reg.* 31,738 (1973). See Gelfand and Levin, "Fetal Tissue Research," 668. NIH, DHHS also imposed a temporary moratorium on federally funded research on live fetuses, id.

4 See *National Research Act*, Public Law 93-348, Section 201(a), 88 Stat. 348 (1974).

5 National Commission for the Protection of Human Subjects of Biomedical and Behavioral Research. *Research on the Fetus: Report and Recommendations* (Washington, D.C.: 1975), reprinted in 40 *Fed. Reg.* 33,526 (1975); 45 CFR §§ 46.201-.211 (1997); 42 USC 274e (1997); 42 USC § 289g(a) (1997).

6 See James E. Goddard, "Comment, The NIH Revitalization Act of 1993 Washed Away Many Legal Problems with Fetal Tissue Transplantation Research but a Stain Remains," *SMU Law Review* 49 (1996): 383–84.

7 See *Memorandum from Robert E. Windom, M.D., Assistant Secretary for Health, DHHS, to James B. Wyngaarden, M.D., Director of NIH, DHHS* (22 March 1988) in A3 *HFTTRP II*. See also Kenneth J. Ryan, "Symposium on Biomedical Technology and Health Care: Social and Conceptual Transformations. Technical Article: Tissue Transplantation from Aborted Fetuses, Organ Transplantation from Anencephalic Infants and Keeping Brain-Dead Pregnant Women Alive Until Fetal Viability," *Southern California Law Review* 65 (1991):687 ("Although such approval [from the Assistant Secretary] was not required, the Assistant Secretary was consulted because of the scientific and ethical implications of the study.").

8 Id. at 687 ("In the meantime the protocol was shelved and a moratorium placed on any use of NIH funding for such activities.").

9 See *Letter from Arlin M. Adams, Panel Chair, to James B. Wyngaarden, M.D., Director of NIH, DHHS* (12 December 1988) in Consultants to the Advisory Committee to the Director. National Institutes of Health, *Report of the Human Fetal Tissue Transplantation Research Panel* A3 (vol I 1988) (hereinafter *HFTTRP I*). See also Pedro F. Silva-Ruiz, "Section II: The Protection of Persons in Medical Research and Cloning of Human Beings," *American Journal of Comparative Law* 46 (1998):278–279; Kimberly Fox Duguay, "Fetal Tissue Transplantation: Ethical and Legal Considerations," *Buffalo Women's Journal of Law and Social Policy* 1 (1992):36.

10 See *Letter from Louis Sullivan, Secretary, DHHS, to William Raub, M.D., Acting Director of NIH, DHHS* 3 (2 November 1989); Goddard, "NIH Revitalization Act," 384; a useful analysis of this debate is found at John A. Robertson, "International Symposium on Law and Science at the Crossroads—Biomedical Technology, Ethics, Public Policy, and the Law: Abortion to Obtain Tissue for Transplant," *Suffolk University Law Review* 27 (1993):1362–69.

11 See H.R. 2507, 102d Cong., 1st Sess. (1991) (amending Part G of Title IV of the Public Health Service Act); see also H.R. 5495, 102d Cong., 2nd Sess. (1992) (amending Part G of Title IV of the Public Health Service Act and incorporating the establishment of a federally operated national tissue bank as provided by Exec. Order No. 12,806 (1992)). During this period of controversy, in an apparent attempt to identify alternatives to fetal tissue derived from elective abortion, the administration established (without success) a tissue bank to collect cellular material for research from ectopic pregnancies and miscarriages. Exec. Order No. 12,806, 57 *Fed. Reg.* 21,589 (1992). Because spontaneously aborted tissue may contain viral infections or pathological defects, the use of ectopic and miscarried abortuses is disfavored for transplantation and most other research.

12 *Departments of Labor, Health, Education, and Welfare Appropriations Act of 1977*, Public Law 94-439, 209, 90 Stat. 1418, 1434 codified at USC, vol. 42, secs. xx (1997).

13 Nikki Constantine Bell, "Regulating Transfer and Use of Fetal Tissue in Transplantation Procedures: The Ethical Dimensions," *American Journal of Law and Medicine* 20 (1994):280.

14 See *Memorandum on Fetal Tissue Transplantation Research*, 29 Weekly Comp. Pres. Doc. 87 (22 January 1993), reprinted in 58 *Fed. Reg.* 7457 (22 January 1993).

15 DHHS. OPRR Reports, *Human Subjects Protections: Fetal Tissue Transplantation—Ban on Research Replaced by New Statutory Requirements*, by Gary B. Ellis, Director, OPRR (Bethesda, MD: 1994), 1–2. The NIH guidelines were withdrawn upon passage of Public Law 103-43 ("NIH Revitalization Act of 1993"), codified at *NIH Revitalization Act of 1993*, USC, vol. 42, secs. 289g-1 and 289g-2 (1997) as Section 498A of the Public Health Service Act.

16 Id. The administration's policies on fetal tissue transplantation did not entirely quell public controversy or congressional interest. See e.g., U.S. General Accounting Office. *Report to the Chairmen and Ranking Minority Members, Committee on Labor and Human Resources, U.S. Senate, and Committee on Commerce, House of Representatives: NIH-Funded Research: Therapeutic Human Fetal Tissue Transplantation Projects Meet Federal Requirements* (Washington, D.C.: GPO, March 1997), 1–8.

17 Myrna E. Watanabe, "Research: With Five-Year Ban on Fetal Tissue Studies Lifted, Scientists Are Striving to Make Up for Lost Time," *The Scientist* 7 (4 October 1993):1.

18 42 USC § 289g-1(a)(1) (1997). See Goddard, "NIH Revitalization Act," 390, 394 ("a huge problem...Congress neglected to define the term 'therapeutic purposes'....").

19 42 USC § 289g-1(e) (1997).

20 42 USC § 289g-1(b)(1) (1997).

21 42 USC § 289g-1(b)(2) (1997).

22 42 USC § 289g-1(c) (1997).

23 42 USC § 289g-2(a)-(c) (1997). But see Goddard, "NIH Revitalization Act," 393–94 (criticizing vague terms; questioning constitutionality of ban on recipient designation).

24 *Human Research Extension Act of 1985*, USC, vol. 42, sec. 289g (1997).

25 42 USC § 289g(b) (1997).

26 *National Organ Transplant Act*, USC, vol. 42, sec. 274e(a) (1997). "Valuable consideration" is defined at 42 USC § 274e(c)(2) (1997) by its negation: "'valuable consideration' does not include the reasonable payments associated with the removal, transportation, implantation, processing, preservation, quality control, and storage of a human organ or the expenses of travel, housing, and lost wages incurred by the donor of a human organ in connection with the donation of the organ." A similar definition (excluding donor costs) is provided at 42 USC § 289g-2(d)(3) (1997).

27 *Organ Transplants Amendment Act of 1988*, USC, vol 42, sec. 274(e)(c)(1) (1997); Senator Gordon Humphrey's amendment was specifically intended to prevent the "sale or exchange for any valuable consideration" of fetal organs and tissue. 134 *Cong. Rec.* S10, 131 (daily ed. July 27, 1988).

28 Federal regulations are applicable only to federal agencies and the expenditure of federal research funds. Cf. 45 CFR § 46.123(b) (regulation permitting DHHS Secretary to terminate all federal funding to any institution if the Secretary determines that researchers have "materially failed [their] responsibility for the protection of the rights and welfare of human subjects").

29 *Subpart B—Additional Protections Pertaining to Research, Development, and Related Activities Involving Fetuses, Pregnant Women, and Human In Vitro Fertilization*, CFR, Title 45, Vol. 1, Part 46.201-211 (1997).

30 45 CFR § 46.201(a) (1997).

31 45 CFR § 46.202 (a) (1997).

32 45 CFR § 46.205(a) (1997).

33 45 CFR § 46.206(a)(1) (1997).

34 45 CFR § 46.206(a)(2) (1997).

35 45 CFR § 46.206(a)(3) (1997).

36 45 CFR § 46.206(a)(4) (1997).

37 45 CFR § 46.207 (1997); 46.208 (1997).

38 45 CFR § 46.209-210 (1997).

39 45 CFR § 46.209 (1997). According to 45 CFR § 46.203(d) (1997), "viable as it pertains to the fetus means being able, after either spontaneous or induced delivery, to survive (given the benefit of available medical therapy) to the point of independently maintaining heart beat and respiration….If a fetus is viable after delivery, it is a premature infant." At 45 CFR § 46.203(e) (1997) "nonviable fetus means a fetus *ex utero* which, although living, is not viable," and at 45 CFR § 46.203(f) (1997) "dead fetus means a fetus *ex utero* which exhibits neither heart beat, spontaneous respiratory activity, spontaneous movement of voluntary muscles, nor pulsation of the umbilical cord (if still attached)."

40 45 CFR § 46.209(a)(1)-(2) (1997); this language is essentially equivalent to 42 USC § 289g(a)(1)-(2) (1997).

41 45 CFR § 46.209(b)(1)-(3) (1997).

42 45 CFR § 46.209(c) (1997).

43 45 CFR § 46.209(d) (1997).

44 45 CFR § 46.210 (1997). See note 38 supra for definitions.

45 Id.

46 They point to a series of reports prepared by the NIH Office of Science Policy in 1987–92 that purport to show that NIH has applied 45 CFR § 46, Subpart B restrictions broadly against a range of categories of fetal research. This position may be supported in a 1988 memorandum from Director Wyngaarden to Assistant Secretary for Health Robert E. Windom ("as you know, the NIH conducts all human fetal tissue research in accordance with Federal Guidelines (45 CFR 46)") and the accompanying 1987 summary of fetal tissue research at NIH ("NIH-supported human fetal tissue research is conducted in compliance with all federal…regulations regarding the use of human fetal tissue. These [Subpart B] regulations include restrictions on tissue procurement that are intended to prevent possible ethical abuses."). The paragraph further cites as applicable 45 CFR § 46.206(a)(3) and 46.206(b). See *Memorandum from NIH Director James B. Wyngaarden, M.D. (signed by William F. Raub, Ph.D.) to DHHS Assistant Secretary for Health Robert E. Windom, M.D.* (2 February 1988):2; *National Institutes of Health, Summary Highlights of FY 1987 Human Fetal Tissue Research Supported by the NIH* (1987):1.

47 See Fed. Reg. 27794–27804 (20 May 1998).

48 Id. at 27794.

49 Id. at 27803, proposed rule 45 CFR § 46.201(b). "(1) Research conducted in established or commonly accepted educational settings, involving normal educational practices, such as (i) research on regular and special education instructional strategies, or (ii) research on the effectiveness of or the comparison among instructional techniques, curricula, or classroom management methods. (2) Research involving the use of educational tests (cognitive, diagnostic, aptitude, achievement), survey procedures, interview procedures or observation of public behavior, unless: (i) Information obtained is recorded in such a manner that human subjects can be identified, directly or through identifiers linked to the subjects; and (ii) any disclosure of the human subjects' responses outside the research could reasonably place the subjects at risk of criminal or civil liability or be damaging to the subjects' financial standing, employability, or reputation. (3) Research involving the use of educational tests (cognitive, diagnostic, aptitude, achievement), survey procedures, interview procedures, or observation of public behavior that is not exempt under paragraph (b)(2) of this section, if: (i) The human subjects are elected or appointed public officials or candidates for public office; or (ii) federal statute(s) require(s) without exception that the confidentiality of the personally identifiable information will be maintained throughout the research and thereafter. (4) Research, involving the collection or study of existing data, documents, records, pathological specimens, or diagnostic specimens, if these sources are publicly available or if the information is recorded by the investigator in such a manner that subjects cannot be identified, directly or through identifiers linked to the subjects. (5) Research and demonstration projects which are conducted by or subject to the approval of department or agency heads, and which are designed to study, evaluate, or otherwise examine: (i) Public benefit or service programs; (ii) procedures for obtaining benefits or services under those programs; (iii) possible changes in or alternatives to those programs or procedures; or (iv) possible changes in methods or levels of payment for benefits or services under those programs. (6) Taste and food quality evaluation and consumer acceptance studies, (i) if wholesome foods without additives are consumed or (ii) if a food is consumed that contains a food ingredient at or below the level and for a use found to be safe, or agricultural chemical or environmental contaminant at or below the level found to be safe, by the Food and Drug Administration or approved by the Environmental Protection Agency or the Food Safety and Inspection Service of the U.S. Department of Agriculture" 45 CFR § 46.101(b)(1)-(6). Note proposed 45 CFR § 46.101(b)(4): read literally, this subsection may have the unintended effect of exempting fetal stem cell derivation and/or successor stem cells thereby derived from Subpart B (and the Common Rule generally). It is not clear how this subsection will be interpreted by the Department.

50 See Fed. Reg. 27803–04, proposed rule 45 CFR § 46.204 and 204(b).

51 Id.

52 Id. at 27804, proposed rule 45 CFR § 46.204(c).

53 Language in 45 CFR § 46.209 referencing this standard has been omitted in the proposed rule 45 CFR § 46.204-206.

54 See Fed. Reg. 27804, proposed rule 45 CFR § 46.204(e) *and* Table 1, "Current and Proposed 45 CFR 46, Subpart B," at 27798, explanatory text.

55 See Fed. Reg. 27804, proposed rule 45 CFR § 46.205(a)(2), 205(b)(4) and Table 1, "Current and Proposed 45 CFR 46, Subpart B," at 27798, explanatory text ("Research involving newborns of uncertain viability": "Consent of the mother *or* the father is required, *or* that of a legally authorized representative of the mother *or* father if both parents are unavailable, temporarily incapacitated, or incompetent"; "Research involving nonviable newborns"; "Consent of the mother and father are required, *unless* one is unavailable, incompetent, or temporarily incapacitated. Consent of a legally authorized representative is prohibited.").

56 See Fed. Reg. 27804, proposed rule 45 CFR § 46.206.

57 See 45 CFR § 46.101 et seq. Note that 45 CFR Part 46 ("Protection of Human Subjects"), Subparts A–D comprise the full complement of human subjects regulation as it affects DHHS, but the term "Common Rule" is commonly used to refer to those regulations contained in Subpart A ("Federal Policy for the Protection of Human Subjects (Basic DHHS Policy for Protection of Human Research Subjects)").

58 See e.g., 45 CFR §§ 46.107-109; 46.111-115; 46.116-117; 46.111(7).

59 45 CFR § 46.102(f) ("*Intervention* includes both physical procedures by which data are gathered (for example, venipuncture) and manipulations of the subject or the subject's environment that are performed for research purposes. Interaction includes communication or interpersonal contact between investigator and subject. *Private information* includes information about behavior that occurs in a context in which an individual can reasonably expect that no observation or recording is taking place, and information which has been provided for specific purposes by an individual and which the individual can reasonably expect will not be made public (for example, a medical record). Private information must be individually identifiable (i.e., the identity of the subject is or may readily be ascertained by the investigator or associated with the information) in order for obtaining the information to constitute research involving human subjects.").

60 Fetal tissue for research is taken from dead abortuses; but note 45 CFR §§ 46.101(c) (1997) ("Department or Agency heads retain final judgment as to whether a particular activity is covered by this [Common Rule]"); 45 CFR §§ 46.101(d) (1997) ("Department or Agency heads may require that specific research activities or classes of research activities conducted, supported, or otherwise subject to regulation by the Department or Agency but not otherwise covered by this [Common Rule], comply with some or all of the requirements of this policy."). DHHS has not acted to vary the terms of the Common Rule to apply its terms to fetal tissue transplant or stem cell derivation or research.

61 See OPRR, *Human Subject Regulations Decision Charts* (February 1998):1 ("Chart 1: Definition of Human Subject at Section 46.102(f)").

62 Id.

63 Id.

64 Id.

65 *Decision Charts* at 1, n. 1.

66 There may be some involvement of living third parties, e.g., the abortion provider, any physician or pathologist responsible for dissecting the fetus to remove EG cells, and any preresearch entity responsible for deriving usable stem cells from EG cells (none of whom fall within the intended meaning of "living individual" for purposes of human subjects protections).

67 See 42 USC § 289g-1(a) (enabling language for fetal tissue research, discussed supra).

68 *Decision Charts* at 2.

69 *Decision Charts* at 2, n. 1.

70 See Fed. Reg. 27803, proposed rule 45 CFR § 46.201(b) ("The exemptions at § 46.101(b)(1) through (6) are applicable to this subpart"); and proposed rule 45 CFR § 46.201(c) ("The additions, exceptions, and provisions for waiver as they appear in § 46.101(c) through (i) are applicable to this subpart...."). The footnote in question is attached to 45 CFR § 46.101(i)—though the provision at § 46.101(i) clearly applies, it is not certain from the language whether the footnote also carries over to the new Subpart B, or if the footnote has the effect of regulation if it is carried over (or what effect, since the application of the exemptions seems to directly contravene the qualification set out in the footnote).

71 *Uniform Anatomical Gift Act (UAGA)* secs. 1–17, 8A U.L.A. 29 (1994).

72 See National Conference of Commissioners on Uniform State Laws, "A Few Facts About the Revised Uniform Anatomical Gift Act (1987)," *Fact Sheet* 1 (1998); Zion, "Legal and Ethical Issues," 1292.

73 According to NCCUSL *Fact Sheet*, 1, the following states have enacted the 1987 revision: Arizona, Arkansas, California, Connecticut, Hawaii, Idaho, Indiana, Iowa, Minnesota, Montana, Nevada, New Hampshire, New Mexico, North Dakota, Oregon, Pennsylvania, Rhode Island, Utah, Vermont, Virginia, Washington, and Wisconsin. The two important new sections in the 1987 revision are Section 10 (criminal prohibition on the purchase or sale of body parts, discussed infra) and Section 4 (presumption of willingness to donate tissue or organs in the absence of known objection after reasonable efforts to discern patient/next-of-kin intent, not applicable in the context of fetal tissue derived from elective abortion). Gelfand and Levin, "Fetal Tissue Research," 671–75.

74 UAGA sec. 1(3). But note Zion, "Legal and Ethical Issues," 1293 ("UAGA...does not differentiate between a fetus donated from a miscarriage or one given through an elective abortion. Presumably, either type of donation is included, but a certain determination is difficult.").

75 UAGA sec. 6(a)(1)-(2).

76 UAGA secs. 8(b); 5.

77 UAGA sec. 3 (preceded in order of proxy donation by spouse or adult child of the decedent); See Gelfand and Levin, "Fetal Tissue Research," 679 ("UAGA makes the mother's consent determinative unless the father objects, and...does not provide for notice to the father. The federal regulations [at 45 CFR § 46.209(d)] require the father's consent, unless he is 'unavailable' to consent."). Note that members of the NIH community who reviewed this paper in draft form have argued that no use of fetal tissue for research purposes from living, nonviable, or viability-undetermined fetuses is possible under the regulations (no tissue extraction could be deemed "minimal risk"). As a result, parental consent at 45 CFR § 46.209(d) is not operative.

78 UAGA secs. 1(a); 6.

79 See Zion, "Legal and Ethical Issues," 1293 ("The act may need an amendment that prohibits specification of a donee...."); accord Gelfand and Levin, "Fetal Tissue Research," 673.

80 UAGA sec. 8(b).

81 See e.g., 45 CFR § 46.206(3) ("Individuals engaged in the activity [of research] will have no part in: (i) Any decisions as to the timing, method, and procedures used to terminate the pregnancy, and (ii) determining the viability of the fetus at the termination of the pregnancy"); see also Zion, "Legal and Ethical Issues," 1294 ("These provisions create a 'Chinese Wall' between the individuals effecting the abortion and those conducting fetal tissue research and transplantation....While this language standing alone would likely preclude most undue influence, the UAGA also provides for the waiver of the 'Chinese Wall'....[R]evision may be necessary.").

82 See Gelfand and Levin, "Fetal Tissue Research," 671. At least one commentator has suggested that UAGA may not govern *any* fetal tissue donation, Jonathan Hersey, "Comment, Enigma of the Unborn Mother: Legal and Ethical Considerations of Aborted Fetal Ovarian Tissue and Ova Transplantations," *UCLA Law Review* 43 (1995):174 ("[I]n the vast majority of abortion procedures, the woman is alive. Therefore, if one believes that a fetus maintains few or no rights independent of the woman, the UAGA statutes are inapplicable to fetal tissue donations.").

83 Id.

84 UAGA secs. 10(a)-(b).

85 42 USC § 289g-2(a) (1997); NOTA prohibits sale of organs for "transplantation," while UAGA's somewhat broader proscription includes "transplantation *or therapy*, if the removal of the part is intended to occur after the death of the decedent" [italics added].

86 See Goddard, "NIH Revitalization Act," 394.

87 42 USC § 289g-2(c) (1997); discussion of state laws, infra. But see Hersey, "Enigma," 113 ("Only the 1987 version of the UAGA explicitly prohibits sales of procured organs. Thus, unless the states still enforcing the 1968 version have supplementary statutes banning the purchase and/or sale of fetal tissue and organs, the specter of a cottage industry of fetal reproductive organs looms....").

88 Defining the transaction as a service rather than a sale may assist regulators and the courts in better distinguishing between reasonable overhead (permitted under 42 USC § 289g-2(d)(3) and UAGA 10) and profit (not permitted). It would certainly still be the case under UAGA that the mother/donor could not be compensated beyond reasonable expenses for donation of fetal tissue, although such payments may be permissible under federal law for research separated from interstate commerce (42 USC §§ 289g-2(a); 274(e)(a)).

89 See e.g., *Griswold v. Connecticut*, 381 U.S. 479 (1965); *Roe v. Wade*, 410 U.S. 113 (1973); *Planned Parenthood of Southeastern Pennsylvania v. Casey*, 505 U.S. 833 (1992).

90 See e.g., *Schloendorff v. Society of New York Hospital*, 211 N.Y. 125, 105 N.E. 92 (1914); *McFall v. Shrimp*, 10 Pa. D. and C.3d 90-92 (1978); *Matter of Storar* and *Matter of Eichner*, 420 N.E.2d 64 (N.Y. 1981).

91 See e.g., *Salgo v. Leland Stanford Jr. University Board of Trustees*, 154 Cal. App. 2d 560, 317 P.2d 170 (1957); *Canterbury v. Spence*, 464 F.2d 772 (D.C. Cir. 1972), cert. denied, 409 U.S. 1064 (1972).

92 See e.g., *Matter of Conroy*, 98 N.J. 321, 486 A.2d 1209 (1985); In re Quinlan, 70 N.J. 10, 355 A.2d 647 (1976), cert. denied, 429 U.S. 922 (1976); *Brophy v. New England Sinai Hospital, Inc.*, 497 N.E.2d 626 (Mass. 1986); Matter of Jobes, 108 N.J. 294, 429 A.2d 434 (1987); *Cruzan v. Director, Missouri Department of Health*, 497 U.S. 261 (1990); In re A.C., 573 A.2d 1235 (D.C. App. 1990); *Matter of Baby K*, 16 F.3d 590 (4th Cir. 1994), cert. denied, 115 S.Ct. 91 (1994); *Carey v. Population Services International*, 431 U.S. 678 (1977); *Brotherton v. Cleveland*, 923 F.2d 477 (6th Cir. 1991); *State v. Powell*, 497 So.2d 1188 (Fla. 1986), cert. denied, 481 U.S. 1059 (1987).

93 See e.g., *Youman v. McConnell and McConnell, Inc.*, 7 La. App. 315 (1927); *Porter v. Lassiter*, 91 Ga. App. 712, 716, 87 S.E.2d 100, 103 (1955); *Hogan v. McDaniel*, 204 Tenn. 235, 319 S.W.2d 221 (1958); *West v. McCoy*, 233 S.C. 369, 105 S.E.2d 88 (1958); *Endresz v. Friedberg*, 24 N.Y.2d 478, 248 N.E.2d 901, 301 N.Y.S.2d 65 (1969); *Presley v. Newport Hosp.*, 117 R.I. 177, 365 A.2d 748 (1976) (Kelleher, J., dissenting); *Wascom v. American Indem. Corp.*, 348 So. 2d 128 (La. App. 1977); *Wallace v. Wallace*, 120 N.H. 675, 421 A.2d 134 (1980); *Danos v. St. Pierre*, 402 So. 2d 633 (La. 1981); *Witty v. American Gen. Capital Distribs., Inc.*, 727 S.W.2d 503 (Tex. 1987); *Webster v. Reproductive Health Services*, 109 S.Ct. 3040 (1989); In re T.A.C.P., 609 So.2d 588 (Fla. 1992); *Planned Parenthood of Southeastern Pennsylvania v. Casey*, 505 U.S. 833 (1992).

94 See e.g., *In the Matter of Baby M*, 109 N.J. 396, 537 A.2d 1227 (1988); *Johnson v. Calvert*, 5 Cal. 4th 84, 851 P.2d 776, 19 Cal. Rptr. 2d 494 (1993); *Davis v. Davis*, 842 S.W.2d 588 (Tenn. 1992), cert. denied, 113 S.Ct. 1259 (1993); *Hecht v. Superior Court (Kane)*, 16 Cal. App. 4th 836, 20 Cal. Rptr. 2d 275 (1993).

95 See e.g. *Dietrich v. Northampton*, 138 Mass. 14 (1884); *Allaire v. St. Luke's Hosp.*, 184 Ill. 359, 56 N.E. 638 (1900) (Boggs, J., dissenting); *Amann v. Faidy*, 415 Ill. 422, 114 N.E.2d 412 (1953) (overruling *Allaire*); *Bonbrest v. Kotz*, 65 F. Supp. 138 (D.D.C. 1946); *Verkennes v. Corniea*, 229 Minn. 365, 38 N.W.2d 838 (1949); *Wallace v. Wallace*, 120 N.H. 675, 421 A.2d 134 (1980); *Presley v. Newport Hosp.*, 117 R.I. 177, 365 A.2d 748 (1976).

96 See e.g., *York v. Jones*, 717 F. Supp. 421 (E.D. Va. 1989); *Davis v. Davis*, 842 S.W.2d 588 (Tenn. 1992), cert. denied, 113 S.Ct. 1259 (1993); *Hecht v. Superior Court (Kane)*, 16 Cal. App. 4th 836, 20 Cal. Rptr. 2d 275 (1993); *Moore v. Regents of the University of California*, 51 Cal.3d 120, 793 P.2d 479, 271 Cal. Rptr. 146 (Cal. 1990), cert. denied, 499 U.S. 936 (1991); *Brotherton v. Cleveland*, 923 F.2d 477 (6th Cir. 1991); *Georgia Lions Eye Bank v. Lavant*, 255 Ga. 60, 335 S.E.2d 127 (1985); *State v. Powell*, 497 So.2d 1188 (Fla. 1986), cert. denied, 481 U.S. 1059 (1987); *Tillman v. Detroit Receiving Hospital*, 138 Mich. App. 683, 360 N.W.2d 275 (1984).

97 See 42 USC §§ 289g-2(a); 273-274; S.H.D., "Note, Regulating the Sale of Human Organs," *Virginia Law Review* 71 (1985):1015, 1025 (suggesting that "the courts...will probably continue their broad construction of Congress' Commerce Clause power and will find that intrastate organ sales do 'affect interstate commerce'") citing generally, *Heart of Atlanta Motel v. United States*, 379 U.S. 241 (1964); *Katzenbach v. McClung*, 379 U.S. 294 1964); *Wickard v. Filburn*, 317 U.S. 111 (1942); but see *U.S. v. Lopez*, 514 U.S. 549 (1997) (restricting Congress' ability to regulate pursuant to the Commerce Clause).

98 See e.g., *Doe v. Rampton*, 366 F. Supp. 189, 194 (D.Utah 1973) (suggesting in dicta that statute provision prohibiting research on live fetus may not be otherwise unconstitutional), vacated and remanded, 410 U.S. 950 (1973) (directing further consideration in light of *Roe*); *Wolfe v. Schroering*, 388 F. Supp. 631, 638 (W.D. Ky. 1974), aff'd. in part, rev'd. in part on other grounds, 541 F.2d 523 (6th Cir. 1976) (upholding prohibition on experimentation on a viable fetus due to state's interest in fetus after viability); *Planned Parenthood Association v. Fitzpatrick*, 401 F. Supp. 554 (E.D. Penn. 1975), aff'd. without opin sub nom., *Franklin v. Fitzpatrick*, 428 U.S. 901 (1976) (affirming legitimate state interest in disposal of fetal remains); *Wynn v. Scott*, 449 F. Supp. 1302, 1322 (N.D. Ill. 1978) (holding that medical researchers have no fundamental rights under the Constitution to perform fetal experiments), aff'd. on other grounds sub nom. *Wynn v. Carey*, 599 F.2d 193 (7th Cir. 1979) (upholding state's rational interest in regulating medicine as to viable fetus); *Leigh v. Olson*, 497 F. Supp. 1340 (D.N.D. 1980) (striking fetal disposal statute as vague where it left "humane disposal" undefined and required mother to determine method of disposal); *Akron v. Akron Center for Reproductive Health, Inc.*, 462 U.S. 416 (1983) (struck down local ordinance that, inter alia, mandated humane and sanitary disposal of fetal remains, finding the provision impermissibly vague because it was unclear whether it mandated a decent burial of the embryo at the earliest stages of formation); *Planned Parenthood Association v. City of Cincinnati*, 822 F.2d 1390, 1391 (6th Cir. 1987) (struck down on other grounds, the court noted in dicta that the wording used by the municipal code regulating disposal of aborted fetal tissue might be precise enough to survive scrutiny); *Planned Parenthood of Minnesota v. Minnesota*, 910 F.2d 479 (8th Cir. 1990) (upholding Minnesota's fetal disposal statute against challenge of vagueness and infringement of privacy).

99 La. Rev. Stat. Ann. § 40:1299.35.13 (1998).

100 La. Rev. Stat. Ann. § 40:1299.35.13 (1998). See Marilyn Clapp, "Note, State Prohibition of Fetal Experimentation and the Fundamental Right of Privacy," *Columbia Law Review* 88 (1988):1076–77 ("The Louisiana statute effectively prohibits any research, experimentation, or even observational study on any embryo, fetus, or aborted fetal tissue. The ban encompasses a range of activities, including studies of the safety of ultrasound and pathological study of fetal tissues removed from a woman for the purpose of monitoring her health. Research on *in vitro* fertilization is likewise barred. Since the aborted pre-viable fetus is not living or cannot survive for long, no procedure performed upon it could be considered 'therapeutic,' and therefore use of this tissue is likewise prohibited. If performed on tissues from a miscarriage, such experimentation would be acceptable under the statutory scheme." [footnotes omitted]).

101 *Margaret S. v. Edwards*, 488 F. Supp. 181, 220 n.124 (E.D. La. 1980) [hereinafter *Margaret S. I*]. This suit was a class action brought on behalf of pregnant women who sought abortions, three physicians who performed abortions, and five clinics that provided abortion facilities.

102 Id. at 221.

103 *Margaret S. v. Treen*, 597 F. Supp. 636 (E.D. La. 1984), aff'd. sub nom. *Margaret S. v. Edwards*, 794 F.2d 994 (5th Cir. 1986) [hereinafter *Margaret S. II*]. See Clapp, "State Prohibition," 1078–79 (the court "specifically not[ed] that reproductive choice was 'not limited to abortion decisions…but extends to both childbirth and contraception.' Prohibiting experimentation on fetal tissues could deny a woman knowledge that would influence her own future pregnancies, as well as prohibit procedures of immediate medical benefit such as pathological examination of tissues. The court also found that the prohibition curtailed the development and use of prediagnostic techniques, including amniocentesis. This result constituted a 'denial of health care' and a 'significant burden' on choice made during the first trimester" [footnotes omitted]).

104 *Margaret S. II*, at 674–75. See Clapp, "State Prohibition," 1079, n. 48 ("The court further suggested the statute would fail even a rational relation test because it failed to serve its own stated purpose of treating the fetus like a human being, since it treated fetal tissue differently from other human tissue" id. at 674–75).

105 *Margaret S. II*, at 675–76.

106 *Margaret S. v. Edwards*, 794 F.2d 994, 999 (5th Cir. 1986) [hereinafter *Margaret S. III*]. Note the court's concurring opinion that "criticized the majority for avoiding the real constitutional issue raised—that any statutory ban on experimentation would inevitably limit the kinds of tests available to women and their physicians and thus could not help but infringe on fundamental rights." Id. at 999–1002 (Williams, J., concurring). Clapp, "State Prohibition," 1080, n. 50.

107 *Margaret S. III* at 999 ("every medical test that is now 'standard' began as an 'experiment'"). But see Clapp, "State Prohibition," 1080, n.54 ("the court hypothesized that the statute was intended 'to remove some of the incentives for research-minded physicians …to promote abortion' and was therefore 'rationally related to an important state interest.' This language suggests that if the statute had not been vague, the court would have applied less than strict scrutiny to a ban on fetal research. The court also implied, in dicta, that the rationale was based on the 'peculiar nature of abortion and the state's legitimate interest in discouraging' it, relying on *H.L. v. Matheson*, 450 U.S. 398, 411–13 (1981)").

108 *Margaret S. III* at 999.

109 61 F.3d 1493 (10th Cir. 1995).

110 *Jane L. v. Bangerter*, 794 F. Supp. 1528 (D. Utah 1992).

111 Id. at 1501.

112 Id. at 1502.

113 Id.

114 735 F. Supp. 1361 (N.D. Ill. 1990).

115 Ill. Rev. Stat., Ch. 38 para 81-26, § 6(7) (1989).

116 *Lifchez* at 1364.

117 Id. at 1366–67.

118 Id. at 1367–70.

119 *Lifchez* at 1372.

120 Id. at 1376–77. See also note 89, supra.

121 See generally Lori B. Andrews' paper, "State Regulation of Embryo Stem Cell Research," commissioned by NBAC and available in this volume.

122 See note 6, supra.

123 See UAGA discussion, supra. Recipient designation for transplant is prohibited at 42 USC § 289g-1(b) and 289g-2(b). The researcher may also be limited to some degree by UAGA § 8(b), a section that prohibits "the physician or surgeon who attends the donor at death or the physician or surgeon who determines the time of death [from] participat[ing] in the procedures for...transplanting a part." Federal or federally funded researchers are prohibited from contributing to "(i) any decisions as to the timing, method, and procedures used to terminate the pregnancy, [or] (ii) determining the viability of the fetus at the termination of the pregnancy" 45 CFR § 46.206(a)(3).

124 Andrews, "State Regulation of Embryo Stem Cell Research," at 8.

125 Id. at 9.

126 Note that Andrews counts 21 states as "us[ing] language which is broad enough to forbid payment for fetal tissue which will be used in transplant or therapy." She includes in her total Minnesota statute § 525.921 prohibiting sale of decedent organ parts and defining "decedent" as "a deceased individual [that] includes a stillborn infant or an embryo or fetus that has died of natural causes *in utero*." Because it is not clear that the Minnesota definition of "decedent" necessarily includes fetal tissue derived from elective abortion (the most common source for research purposes), Minnesota should probably be excluded from the total number of states in which sale of fetal tissue is more likely than not prohibited.

127 Andrews, "State Regulation of Embryo Stem Cell Research," at 9.

128 Id.

129 See Robertson, "Symposium," at 1369 ("The federal law...reflects the prevailing consensus that the NIH advisory panel and other ethical review bodies have reached: fetal tissue transplantation research or therapy is acceptable as long as the donated fetal tissue is not the product of an abortion induced for donation purposes, but is the by-product of abortions that would be occurring anyway.").

130 This statutory subsection is also the subject of an expression of congressional concern over DHHS General Counsel Harriet Rabb's memorandum to NIH Director Harold Varmus analyzing various legal issues related to stem cell research (see discussion, infra). ("The Rabb memo also ignores the policy reflected in current law on fetal tissue transplantation research using tissue from intentionally aborted children. While that law is itself open to criticism, it at least bans the use of fetal tissue in federally funded research if abortion was induced for the purpose of providing the tissue. Under current law, federal funds may not be used for fetal tissue transplantation experiments following an abortion if the timing and method of the abortion were altered solely for the purpose of providing usable tissue for research. Yet, in the embryonic stem cell research which NIH proposes to fund, the timing, method, and procedures for destroying the embryonic child would be determined solely by the federally funded researchers' need for usable stem cells.") *Letter from Seventy Members of the U.S. House of Representatives to Donna E. Shalala, Secretary of Health and Human Services* (11 February 1999) at 2.

131 Christie A. Seifert, "Comment, Fetal Tissue Research: State Regulation of the Donation of Aborted Fetuses Without the Consent of the 'Mother,'" *John Marshall Law Review* 31 (1997):290 ("All fifty states have adopted a version of the UAGA. Approximately half of the states, however, have chosen to supplement the UAGA with laws that specifically address issues revolving around the use of fetal tissue for research. Of these states which have chosen to further regulate the use of fetal tissue research, roughly half have laws requiring the consent of the aborting woman prior to the donation and subsequent use of the fetus. Although these supplementary laws exist, the controlling law with respect to the donation of fetal tissue for research purposes is still the state-adopted UAGA." [footnotes omitted]).

132 UAGA prefatory note, 1994 main volume.

133 Id. at sec. 6, "Historical Notes—Action in Adopting Jurisdictions." See also, UAGA (1968), sec. 3(4) [predecessor to sec. 6(a)(3) (1987)] "Historical Notes—Action in Adopting Jurisdictions."

134 See generally Bell, "Regulating Transplantation," at 282 ("Many proponents of fetal tissue transplantation suggest that regulations should not allow the woman who donates fetal remains to designate the recipient of the fetal tissue. Additionally, it would be desirable to maintain the anonymity of both the pregnant woman and the tissue recipient. This precaution would eliminate the possibility that a woman would conceive a fetus in order to terminate the pregnancy and donate the fetal tissue. This measure would also ensure against the recipient seeking to reward the woman with gifts or fervent displays of gratitude after transplantation. Likewise, it would discourage tissue donation by a woman who might otherwise donate tissue hoping for a grateful response from the recipient." [footnotes omitted]); accord Duguay, "Fetal Tissue," at 41, citing Childress, "Disassociation from Evil: The Case of Human Fetal Tissue Transplantation Research," *Social Responsibility* (1990):32, 37 (designated donation creates risk of exploitation

and pressure); *HFTTRP I*, at 2, 3, 8; Joanna H. Kinney, "Restricting Donative Choice: Fetal Tissue Transplantation and Respect for Human Life," *Journal of Law and Health* 10 (1995):261; S.C. Hicks, "The Regulation of Fetal Tissue Transplantation: Different Legislative Models for Different Purposes," *Suffolk University Law Review* 27 (1993):1623–1629; T.M. Hess-Mahan, "Human Fetal Tissue Transplantation Research: Entering a Brave New World," *Suffolk University Law Review* 23 (1989):818.

135 But see, supporting a state-initiated UAGA amendment, Frankowska, "A Proposal to Amend UAGA," at 1116 ("The states should amend the UAGA to prohibit either parent from designating the recipient of tissue from an electively aborted fetus"); accord Mark W. Danis, "Fetal Tissue Transplants: Restricting Recipient Designation," *Hastings Law Journal* 39 (1988):1079; Seifert, "State Regulation," at 296; Zion, "Legal and Ethical Issues," at 1293–94; Beverly Burlingame, "Note, Commercialization in Fetal-Tissue Transplantation: Steering Medical Progress to Ethical Cures," *Texas Law Review* 68 (1989):236–37; Hersey, "Enigma," at 206; Gelfand and Levin, "Fetal Tissue Research," at 673; and see opposing designated donation restrictions generally Shirley K. Senoff, "Focus: Issues in Bioethics: Canada's Fetal-Egg Use Policy, The Royal Commission's Report on New Reproductive Technologies, and Bill C-47," *Manitoba Law Journal* 25 (1997):29 ("Women who have undergone abortions may, in some cases, be allowed to designate recipients for fetal eggs....While mandatory counseling for both the donating woman and the recipient would safeguard against the possibility that a woman may be coerced into aborting, the decision to designate a recipient would ultimately be that of the woman"); Robertson, "Symposium," at 1361 ("the ban on designation of fetal tissue recipients unconstitutionally infringes upon the fundamental right to abortion...[but] a set of procedural safeguards should be required to ensure that women freely consent to such donations").

136 In its FY 1996–97 and FY 1993–95 reports to Congress on fetal tissue transplantation (filed pursuant to 42 USC § 289g-1(f)), NIH adopts the interpretation that "projects involving the transplantation of human fetal tissue into humans are classified as therapeutic or clinical research if they are conducted on human subjects and are aimed at the development of therapeutic approaches for the cure or amelioration of diseases and disorders." "Therapeutic research" is the third in a spectrum of activities that NIH groups with "basic research," ("advancement of knowledge of basic biological processes") and "pre-clinical investigation," (that "further[s] therapeutic research through transplantation studies in animals or the development of improved methodologies for processing and preserving tissue"). See Department of Health and Human Services. National Institutes of Health, *Therapeutic Human Fetal Tissue Transplantation Research Activities Funded by the National Institutes of Health in FY 1996–97: Report to Congress* (Bethesda, MD: 1998), 1 [hereinafter *Department II*]; Department of Health and Human Services. National Institutes of Health, *Therapeutic Human Fetal Tissue Transplantation Research Activities Funded by the National Institutes of Health in FY 1993–95: Report to Congress* (Bethesda, MD: 1997), 1 [hereinafter *Department I*].

137 Judith C. Areen, "Statement on Legal Regulation of Fetal Tissue Transplantation for the Human Fetal Tissue Transplantation Research Panel," D21 in *HFTTRP II*.

138 Id. at D21.

139 Id. at D22–23.

140 Id. at D23–24.

141 See *Memorandum from Harriet S. Rabb, HHS General Counsel, to Dr. Harold Varmus, M.D., NIH Director* (15 January 1999): 5 ("45 CFR [§] 46.210, Subpart B...would apply to certain human pluripotent stem cells, including those derived from the primordial germ cells of nonliving fetuses").

142 Areen, at D22 in *HFTTRP II*. It appears from the *Summary Highlights of FY 1987 Human Fetal Tissue Research Supported by NIH*, and accompanying memorandum, see note 46 supra, that NIH may have supported this "double section" interpretation in 1987–88. (Note that NIH's present position is undetermined.)

143 See e.g., Goddard, "NIH Revitalization Act," at 387 (modified "double section"); Danis, "Restricting Recipient Designation," at 1086 (inexplicit agreement with Areen); Ania M. Frankowska, "Note, Fetal Tissue Transplants: A Proposal to Amend the Uniform Anatomical Gift Act," *University of Illinois Law Review* 1989 (1989):1105 (also seems to agree with Areen); but see Gelfand and Levin, "Fetal Tissue Research," 670 (highlighting 45 CFR §§ 46.203 [definitions]; 46.209; 46.210; and 46.206(a)(3) [donor-researcher separation provisions]).

144 Areen, at D24 in *HFTTRP II*.

145 See note 17, supra.

146 Michael J. Shamblott, John Gearhart, et al., "Derivation of Pluripotent Stem Cells from Cultured Human Primordial Germ Cells," *Proceedings of the National Academy of Sciences* 95 (November 1998):13726.

147 *Statement of Harold Varmus, M.D., NIH Director, Before the Senate Appropriations Subcommittee on Labor, Health, and Human Services, Education, and Related Agencies* (26 January 1999) at 1. It is unclear from his public statements whether Dr. Varmus regards 42 USC § 289g-1 and 289g-2(b) as applicable to stem cell research involving fetal tissue. See e.g., id. at 1 ("Dr. Gearhart's work could have been supported with federal funds....Federal laws and regulations already exist that govern research on fetal tissue"); *Testimony of Harold Varmus, M.D., NIH Director, Before the National Bioethics Advisory Commission* (19 January 1999) at 15, 16, 18 ("[quoting the Rabb memorandum] '[fetal-derived] stem cells...are subject to the existing statutes'; that is, there will be no reason to exclude Federal support or work with those cells as long as statutes and laws are obeyed...there are restrictions on fetal tissue transplantation research, and those restrictions are good....At present the Federal government can support derivation of pluripotent stem cells from fetal germ cells....I would point out the Administration does not at this time seek any changes in the law").

148 *See Memorandum from Harriet S. Rabb, HHS General Counsel, to Dr. Harold Varmus, M.D., NIH Director* (15 January 1999): 1.

149 Id. at 4.

150 Id. at 4–5. For 45 CFR § 46.210 (1997), see discussion, supra.

151 See historical discussion, supra.

152 See *Department I* at 1, and *Department II* at 1.

153 Id.

154 See *Department I* at 4 (italics added).

155 Id.

156 Recall that NIH describes "therapeutic" somewhat elliptically in its qualifying definition as including projects that "are conducted on human subjects and are aimed at the development of therapeutic approaches for the cure or amelioration of diseases and disorders." *Department I*, at 1; *Department II*, at 1. For human subjects protections, see generally 45 CFR Section 46, *Protection of Human Subjects*, and 45 CFR § 46.111(a), *Criteria for IRB Approval of Research* ("risks to subjects are minimized...do not unnecessarily expose subjects to risk...whenever appropriate by using procedures already being performed on the subjects for diagnostic or treatment purposes;...risks to subjects are reasonable in relation to anticipated benefits, if any, to subjects, and the importance of the knowledge that may reasonably be expected to result....").

157 See note 97, supra.

158 See e.g., note 97, supra, S.H.D. (suggesting that "the courts...will probably continue their broad construction of Congress' Commerce Clause power and will find that intrastate organ sales do 'affect interstate commerce.'" Note, however, that the commentator's observation was written well before the Court began to revise its understanding of the commerce clause).

159 See Hersey, "Enigma," at 113 ("Only the 1987 version of the UAGA explicitly prohibits sales of procured organs. Thus, unless the states still enforcing the 1968 version have supplementary statutes banning the purchase and/or sale of fetal tissue and organs, the specter of a cottage industry of fetal reproductive organs looms....").

160 See e.g., Hersey, "Enigma," at 117; Zion, "Legal and Ethical Issues," at 1291; Burlingame, "Commercialization in Fetal-Tissue Transplantation," at 226.

161 See generally, discussion supra.

162 See Zion, "Legal and Ethical Issues," at 1292 [footnotes omitted]. See also Marjory Spraycar, ed., *Stedman's Medical Dictionary, 26th Edition* (Baltimore: Williams and Wilkins, 1995) at 1257 ("Organ...Any part of the body exercising a specific function, as of respiration, secretion, digestion").

163 See Zion, "Legal and Ethical Issues," at 1292 [footnotes omitted].

164 See Shamblott and Gearhart, et al., "Derivation of Pluripotent Stem Cells," at 13726.

165 See Zion, "Legal and Ethical Issues," at 1292 ("Unanswered questions such as these suggest that NOTA requires further amendments.").

Regulating Embryonic Stem Cell Research: Biomedical Investigation of Human Embryos

Commissioned Paper
J. Kyle Kinner
Presidential Management Intern
National Bioethics Advisory Commission

A. Introduction

I. Historical Overview

In sharp contrast to the regulatory complexity of existing federal and state laws governing human fetal tissue research,[1] the schema for human embryo research is relatively straightforward. With the exception of a few unenforced state statutes, no viable regulatory system currently exists to guide or control the practice of human embryo research in the United States.[2] Among federally-supported scientists, such experimentation is prohibited by law, while research conducted in the private sector takes place without any medical or bioethical oversight specific to the human embryo.[3] Given the pervasive degree of control applied by Congress and the states to fetal tissue or to research involving human subjects, the regulation applied to human embryos is unusually brief.[4] Yet, however minimal, the regulatory model in its present state will substantially inform stakeholder and bioethical analysis for embryology and the rapidly developing field of embryonic stem cell research.

In 1973, the Department of Health and Human Services (DHHS) (then DHEW) imposed a temporary moratorium on federally funded research on live fetuses.[5] The following year, Congress applied its own moratorium to all human fetal and human *in vitro* fertilization (IVF) research until DHHS could adopt comprehensive regulations to address those issues.[6] Congress concurrently established the National Commission for the Protection of Human Subjects of Biomedical and Behavioral Research to examine bioethical issues and to make recommendations.[7] In August 1975, DHHS wrote regulations necessary to lift the moratoria, incorporating in most cases the provisions recommended by the commission.[8] Subpart B of the current DHHS version of the federal Common Rule (45 CFR 46)—Additional Protections Pertaining to Research, Development, and Related Activities Involving Fetuses, Pregnant Women, and Human *In Vitro* Fertilization—reflects almost unchanged the work of this period.[9] Congress also enacted similarly protective legislation, directed primarily towards fetal research.[10]

Of particular interest to the question of human embryo research was a provision in the Department's 1975 regulations allowing IVF experimentation upon the approval of an Ethics Advisory Board (EAB).[11] Although legally permitted to act, the Secretary of DHHS did not direct the EAB to perform a funding review of proposed research until September 1978.[12] The National Institutes of Health (NIH) submitted a funding request pursuant to regulation, prompting the EAB to issue a report in March 1979 supportive of the grant proposal, under certain conditions.[13] No action was ever taken by the Secretary with respect to the board's report, and the 1978 EAB expired in 1980. Because it failed to appoint another EAB to consider additional research proposals, the Department effectively forestalled any attempts to support IVF research either internally or through its extramural grantees, and no experimentation involving human embryos was ever funded pursuant to EAB review.[14]

In early 1993, the new administration proposed, and Congress passed, sweeping legislation intended to repeal the existing executive moratorium on fetal tissue transplant research, along with related provisions.[15] The NIH Revitalization Act of 1993 also provided, inter alia, for the nullification of 45 CFR § 46.204(d), the regulatory provision mandating EAB review that had become a de facto ban on IVF and human embryo research.[16] Before allocating any funds under the new authority, NIH Director Harold Varmus convened the Human Embryo Research Panel (HERP) to distinguish fundable grants from those requests that, for ethical reasons, would be considered "unacceptable for Federal funding."[17] The panel identified several areas of potential research activity that it considered ethically appropriate for federal funds.[18] In a controversial decision that is still debated, the panel suggested that it might be ethically permissible to allow researchers to create human embryos for certain research purposes.[19]

The panel submitted its report to the Advisory Committee to the NIH Director (ACD) in September 1994.[20] Acting on this submission, the ACD formally approved the HERP recommendations (including provision for the deliberate creation of research embryos) and transmitted them to NIH Director Harold Varmus on December 1, 1994.[21] On December 2, preempting any NIH response, the President intervened to clarify an earlier endorsement of embryo research, stating that "I do not believe that Federal funds should be used to support the creation of human embryos for research purposes, and I have directed that NIH not allocate any resources for such requests."[22] Director Varmus proceeded immediately to implement those HERP recommendations not proscribed by the President's clarification, concluding that NIH could begin to fund research activities involving "surplus" embryos.[23] Before any funding decisions could be made however, Congress took the opportunity afforded by the DHHS appropriations process (then underway) to stipulate that *any* activity involving the creation, destruction, or exposure to risk of injury or death to human embryos for research purposes may not be supported with federal funds under any circumstances.[24] Three additional legislative riders were inserted into subsequent annual DHHS appropriating statutes, enacting identically worded provisions into law.[25]

II. Recent Developments

A number of recent developments have arisen as a direct result of the congressional ban on human embryo research. In January 1997, media controversy erupted when NIH-supported geneticist and former HERP panelist Mark Hughes from Georgetown University was found to have violated the federal restrictions governing the use of human embryos.[26] Hughes had included NIH-supported equipment and personnel in laboratory experiments on prenatal embryo diagnosis, violating strict segregation rules designed to implement the ban.[27] NIH Director Harold Varmus severed ties with the scientist and told a congressional committee investigating the incident that NIH had taken "several steps to further diminish the risk of subsequent violations."[28]

As controversial as the Hughes violation proved to be, considerably more public and media interest was generated by the recent announcements of Drs. James A. Thomson and John Gearhart regarding research advances in stem cell biology.[29] While Gearhart's work involved embryonic germ (EG) cells derived from fetal tissue, Thomson's team created embryonic stem (ES) cell lines from human blastocysts (the developing preimplantation embryo, four days after fertilization).[30] In the wake of Thomson's discovery, the issue of federal funding for human embryo research has been actively considered by NIH and other interested stakeholders. The National Bioethics Advisory Commission (NBAC) is presently examining the legal and bioethical implications of research on human embryos in the context of stem cell research[31] and was invited by Director Varmus at its January 1999 meeting to advise NIH on any relevant bioethical recommendations it chose to develop.[32] The invitation to NBAC was prompted by the NIH Director's own decision, supported by a favorable DHHS General Counsel's ruling, that federal funding for embryonic stem cell research (excluding derivation) was legally appropriate and could proceed.[33]

NIH Director Varmus has moved cautiously to implement his decision to fund ES and EG cell research, suspending any intra- or extramural grant funding in this area until an Ad Hoc Working Group of the ACD can develop guidelines for the ethical research use of human fetal and embryological materials.[34] The Ad Hoc Working Group met on April 8, 1999, to deliberate, but has not yet released a revised draft of its work for public comment (slated to last 60 days).[35] After comment and finalization, the Working Group has indicated that it will submit its proposed guidelines to the ACD to consider along with the underlying question of *whether* such research should receive federal funding at NIH.[36] The ACD's assessment will be transmitted to the Director for his review, and presumably some decision about the future funding of embryo research at NIH will be made at that time. Director Varmus has also asked his newly appointed Council of Public Representatives to consider these issues and to advise him accordingly.[37]

B. Federal Statutes, Presidential Directives, and CFR

A number of relevant federal statutes and Executive directives relating to human embryo research, along with DHHS departmental regulations included in the Code of Federal Regulations (CFR), are examined below.

I. Federal Human Embryo Research Ban

Public Law No. 105-277, Sec. 511(a) (1998), an appropriations rider restricting the DHHS and its subordinates, provides that "none of the funds made available in this Act may be used for (1) the creation of a human embryo or embryos for research purposes; or (2) research in which a human embryo or embryos are destroyed, discarded, or knowingly subjected to risk of injury or death greater than that allowed for research on fetuses *in utero* under 45 CFR § 46.208(a)(2) and section 498(b) of the Public Health Service Act (42 USC § 289g(b))."[38] The standard of risk referenced in 45 CFR § 46.208(a)(2) conditions *in utero* fetal research activity on the requirement that "risk to the fetus imposed by the research is minimal and the purpose of the activity is the development of important biomedical knowledge which cannot be obtained by other means."[39] The term "minimal risk" is defined at 45 CFR § 46.102(i) as "mean[ing] that the probability and magnitude of harm or discomfort anticipated in the research are not greater in and of themselves than those ordinarily encountered in daily life or during the performance of routine physical or psychological examinations or tests."[40] The standard of risk referenced in section 498(b) of the Public Health Service Act (42 USC § 289g(b)) provides that "in administering the regulations for the protection of human subjects research [at 45 CFR 46]...the Secretary [of Health and Human Services] shall require that the risk standard...be the same for fetuses which are intended to be aborted and fetuses which are intended to be carried to term."[41]

Congress has particularly broad authority to direct or withhold funds in its legislative discretion as the appropriating body.[42] It is extremely unlikely that the embryo research funding prohibition supra would be deemed defective or unconstitutional by a reviewing court.

II. Presidential Deliberate Research Embryo Directive

On December 2, 1994, the Office of the White House Press Secretary released the following Notice of Presidential Directive ("Statement by the President"):[43]

> The Director of the National Institutes of Health has received a recommendation regarding federal funding of research on human embryos. The subject raises profound ethical and moral questions as well as issues concerning the appropriate allocation of federal funds. I appreciate the work of the committees that have considered this complex issue and I understand that advances in *in vitro* fertilization research and other areas could derive from such work. However, I do not believe that federal funds should be used to support the creation of human embryos for research purposes, and I have directed that NIH not allocate any resources for such research. In order to ensure that advice on complex bioethical issues that affect our society can continue to be developed, we are planning to move forward with the establishment of a National Bioethics Advisory Commission over the next year.

The Presidential Directive has not been rescinded, but is technically superseded in its effect by Public Law No. 105-277, Sec. 511(a) (1998) supra, governing DHHS. The Presidential directive was not formally inscribed as an Executive Order.[44] Absent congressional action to nullify the directive, it is extremely unlikely that a reviewing court would interfere with Executive discretion in this area.[45]

III. Presidential Cloning Directives

On March 4, 1997, the Office of the White House Press Secretary released the following "Memorandum for the Heads of Executive Departments and Agencies" (quoted in relevant part):[46]

> Federal funds should not be used for cloning of human beings. The current restrictions on the use of Federal funds for research involving human embryos do not fully assure this result. In December 1994, I directed the National Institutes of Health not to fund the creation of human embryos for research purposes. The Congress extended this prohibition in FY 1996 and FY 1997 appropriations bills, barring the Department of Health and Human Services from supporting certain human embryo research. However, these restrictions do not explicitly cover human embryos created for implantation and do not cover all Federal agencies. I want to make it absolutely clear that no Federal funds will be used for human cloning. Therefore, I hereby direct that no Federal funds shall be allocated for cloning of human beings.

The Presidential directive was not formally inscribed as an Executive Order.[47] Absent congressional action to nullify the directive, it is extremely unlikely that a reviewing court would interfere with Executive discretion in this area.[48] On February 9, 1998, the Executive Office of the President, Office of Management and Budget, released a "Statement of Administration Policy" regarding S. 1601, the "Human Cloning Prohibition Act."[49] The statement provides, in relevant part:

> [T]he Administration supports amendments to S. 1601 that would...permit somatic cell nuclear transfer using human cells for the purpose of developing stem cells (unspecified cells capable of giving rise to specific cells and tissues) technology to prevent and treat serious and life-threatening diseases and other medical conditions, including the treatment of cancer, diabetes, genetic disorders, and spinal cord injuries and for basic research that could lead to such treatments.

The quoted statement, supra, does not have the force or effect of a Presidential Directive or Executive Order and does not modify the March 4, 1997, Presidential Directive prohibiting funding for human cloning by federal agencies (though it may reflect a departure by the administration from broad opposition to human cloning for any purpose—somatic cell nuclear transfer of this type is generally considered a form of human cloning that involves the research use of human embryos).[50]

IV. 45 CFR 46, Subpart B[51]

45 CFR 46, Subpart B of the DHHS version of the federal Common Rule is entitled "Additional Protections Pertaining to Research, Development, and Related Activities Involving Fetuses, Pregnant Women, and Human *In Vitro* Fertilization."[52] 45 CFR § 46.201 provides that "the regulations in this subpart are applicable to all Department of Health and Human Services grants and contracts supporting research, development, and related activities involving: (1) the fetus, (2) pregnant women, and (3) human *in vitro* fertilization."[53] At 45 CFR § 46.203(g), the term "*in vitro* fertilization" is defined as "any fertilization of the human ova which occurs outside the human body of a female, either through admixture of donor human sperm and ova or by any other means."[54] NIH has publicly affirmed that the term "*in vitro* fertilization" presently includes all categories of extracorporeal embryo research.[55] The NIH position would appear to mean that relevant IVF sections of 45 CFR 46, Subpart B also apply more broadly to regulate *any* DHHS-funded research involving human embryos not *in utero* (potentially including embryonic stem cells, if applicable).[56]

Under this interpretation, another EAB could be impaneled by the Secretary pursuant to 45 CFR § 46.204 to consider "policies and requirements...as to ethical issues" of stem cell derivation from human embryos, for example (the capacity to impanel a new EAB was not repealed with the *de facto* moratorium at section 46.204(d)).[57] Similarly, it would appear that 45 CFR § 46.205, specifying additional duties for Institutional Review Boards (IRBs) "in connection with *in vitro* fertilization, pregnant women, and fetuses," would also still apply to embryonic stem cell research.[58] The additional requirements are not onerous, but "no award may be issued until the applicant or offeror has certified to the [DHHS] Secretary that the Institutional Review Board has made the [necessary] determination[s]...and the Secretary has approved these determinations, as provided in [45 CFR § 46.120]."[59] Finally, 45 CFR § 46.206 specifies general limitations on "activit[ies] to which this subpart is applicable," including mandates for prestudies involving animals and nonpregnant women;[60] least possible risk;[61] restrictions on investigator involvement in the "timing, method, and procedures used to terminate the pregnancy;"[62] and a prohibition on the use of inducements "monetary or otherwise...offered to terminate pregnancy for purposes of the activity."[63] The newly revised 45 CFR § 46, Subpart B (not yet finalized) makes no substantive changes that would affect the regulations described above.[64]

V. 45 CFR 46, Subpart A (The Federal Common Rule)

Subpart A of the DHHS version of the federal Common Rule applies more broadly to federally sponsored human subjects research and requires protections such as IRB review, informed consent, and research subject privacy. The regulations define "human subject" as "*a living individual* about whom an investigator (whether professional or student) conducting research obtains (1) data through intervention or interaction with the individual, or (2) identifiable private information."[65] Like the research involving fetal tissue, this definition leaves considerable room for debate over whether embryo research, the derivation of embryonic stem cells, and research involving successor stem cells from embryonic sources are governed by the federal Common Rule.[66] A close analysis of the DHHS Common Rule reveals no persuasive basis for distinguishing whether or to what extent subpart A should apply in the present context of embryonic stem cell research.[67] NBAC's attempts to clarify its understanding of this subpart through correspondence with the NIH Office for Protection from Research Risks (OPRR) have not been successful.[68] It is strongly suggested that a clearer, more accessible Common Rule interpretation is needed for stem cell research involving human embryos.

C. State Statutes

I. State Legislation

State legislative activity on IVF and embryo research varies widely.[69] Issues specific to IVF are often ignored, and much legislative activity that affects human embryo research can be ascribed to overly broad statutory constructions that were intended to regulate fetal tissue research and disposal. In considering state legislation affecting the human embryo, there are at least six discrete areas of experimental activity that may potentially involve state statutory regulation (depending upon how broadly the underlying statutes are interpreted): embryo cryopreservation, preimplantation screening, gene therapy, cell line development, twinning, and basic research.[70] Further complicating this analysis, numerous states have enacted regulations affecting the commercialization of embryo research, and their impact will be examined in detail below.[71]

Two experimental activities, cell line development and basic research, are especially relevant to the embryonic stem cell research question now before NBAC. States that regulate cell line development from human embryos either prohibit the practice entirely, or restrict it substantially.[72] Commentator June Coleman observed that "all ten states that prohibit embryological research have vaguely worded statues which could encompass cell line development if the statutes were interpreted broadly...[although] some [activity] could be characterized as non-experimental, thus removing it from the scope of experimentation bans."[73] Issues inherent in cell line development will include the potential for restrictions on downstream commercialization and uncertainty over the extent to which gamete donors must be informed about the nature and potential commercial uses of the biological materials they donate.[74]

Basic research typically involves precommercial scientific activity designed to explore basic biological processes or to understand genetic and cellular control mechanisms. Currently, the practice of basic embryonic and fetal research is unrestricted in 24 states and the District of Columbia.[75] Of the remaining 26 states that regulate embryo or fetal research in one form or another, basic embryological research is prohibited or restricted in ten states.[76] New Hampshire's statute regulates the experimental use of human embryos to a degree not likely to impair research in that state.[77] However, the remaining nine states have legislated more broadly, effectively banning all research involving *in vitro* embryos.[78] Researchers should carefully consider the laws applicable to their laboratories; penalties for violation of state statutes regulating embryo research can include civil fines or imprisonment.[79] In the absence of a supervening federal regulatory scheme, state laws control.[80] Legal issues specific to basic research tend to be more theoretical and constitutional in nature.

The subject of commercialization is a potentially important one, affecting both researchers who must acquire embryos from for-profit IVF clinics or other sources, to downstream users who may develop derivative, commercial applications from basic embryological and stem cell research. Five states currently prohibit payment for IVF embryos for research purposes.[81] Eight additional states prohibit payment for human embryos for any purpose.[82] Five states apply ambiguous restrictions that may or may not prohibit sale of embryos, depending upon interpretation or, in some cases, action by state officials.[83] As Coleman noted, some statutes can be interpreted to prevent payment for cell lines derived from human embryos.[84] Commentator Lori Andrews suggests that "it is possible that because a cell line is new tissue produced from the genetic material of, but not originally a part of, the embryo, laws proscribing the sale of embryonic tissue may not apply."[85] By contrast, at least one state (Minnesota) "prohibits the sale of living [embryos] or nonrenewable organs but does allow 'the buying and selling of a cell culture line or lines taken from a non-living human [embryo].'"[86] Organ transplant statutes restricting sale of human organs or organ parts are common throughout the states, but none include language likely to impede research involving human embryos.[87]

II. Constitutional Implications of State Action

Previous discussion has focused on the regulatory structures enacted by the states affecting embryonic stem cell research. A broader question presents itself, however, in the extent to which any particular state regulatory action may involve constitutional issues or implications important to investigators. In effect, a thorough analysis of ES cell research regulations must consider whether state oversight of embryonic research is acceptable under the current articulation of constitutional law and under what circumstances. The question is not one that is easily answered. Litigation on the subject is comparatively minor (involving fetuses or fetal material), and at no point to date has a federal court considered the constitutionality of state statutes regulating research on embryos or their derivatives.[88] Commentary on the question is prospective and theoretical.

Three important cases deserve mention. In 1978, the state of Louisiana enacted La. Rev. Stat. Ann. § 40:1299.31 et seq., a statute that prohibited virtually all research or study involving the fetus or fetal tissue ("a live child or unborn child") not "therapeutic" to that child.[89] The statute protected the fetus *in utero*, but did not address *ex utero* fetal tissue except by implication. In 1981, the Louisiana legislature expanded the Act's scope to include aborted fetal tissue in its prohibition.[90] In its initial review, the court considered only the pre-1981 Act (without the fetal tissue amendment), holding that plaintiffs' failure to demonstrate the statute's negative impact on the right of privacy left the court to conclude that "no obstacle has been placed in the path of the woman seeking abortion."[91] Under the resulting rational basis analysis, the court found that the statute was rational because it tended to protect the public from the "dangers of abuse inherent in any rapidly developing field."[92] The statute was challenged again after the 1981 amendment and included a showing by plaintiffs that a prohibition on research did burden the right of privacy.[93] Reversing its earlier decision, the court found that the revised statute infringed on the fundamental right of privacy and applied strict scrutiny analysis. The court concluded that a research ban did not further the state's compelling interest in protecting the health of the woman and that its interest in the potential life of the unborn did not continue past the death of the fetus.[94] Finally, the district court addressed the statute's vagueness, noting that it was not possible, *ex utero*, to distinguish between fetal and maternal tissue or the products of spontaneous and induced abortions.[95] On appeal to the Fifth Circuit, the court ignored the district court's analysis entirely, finding instead that the term "experiment" as used in the statute's prohibition against fetal experimentation was unconstitutionally vague.[96] "The whole distinction between experimentation and testing, or between research and practice, is...almost meaningless," such that "experiment" is not adequately distinguishable from "test."[97] As a criminal prohibition without effective standards, the statute was deemed void.[98]

A less stringent Utah statute was examined by the Tenth Circuit in 1995 in *Jane L. v. Bangerter* and rejected on similar grounds.[99] Unlike the Louisiana law, Utah Code Ann. § 76-7-310 permitted discretionary experimentation aimed at acquiring genetic information about the embryo or fetus. A lower court upheld the statute by narrowly interpreting the term "experimentation" to mean "tests or medical techniques which are designed solely to increase a researcher's knowledge and are not intended to provide any therapeutic benefit to the mother or child."[100] The Tenth Circuit disagreed, arguing that the district court "blatantly rewrote the statute, choosing among a host of competing definitions for 'experimentation.'"[101] The court further concluded that the word "benefit" was itself ambiguous: "If the mother gains knowledge from a procedure that would facilitate future pregnancies but inevitably terminate the current pregnancy, would the procedure be deemed beneficial to the mother? Does the procedure have to be beneficial to the particular mother and fetus that are its subject?"[102] Without clear boundaries between permissible action and criminal conduct, the statute was deemed unconstitutionally vague and invalid.[103]

Finally, an Illinois district court in *Lifchez v. Hartigan*[104] considered a claim by a class of reproduction and infertility specialists seeking to invalidate a criminal misdemeanor statute that prohibited "experiment[ation] upon a fetus produced by the fertilization of a human ovum by a human sperm unless such experimentation is therapeutic to the fetus thereby produced."[105] Plaintiffs claimed that the words "experimentation" and "therapeutic" rendered the statute vague and unconstitutional, and the district court agreed. "[P]ersons of common intelligence will be forced to guess at whether or not their conduct is unlawful....[T]here is no single accepted definition of 'experimentation' in the scientific and medical communities."[106] The court observed that experimental procedures evolve quickly into routine diagnostic and therapeutic interventions, and tests to obtain information about a fetus' development often are not therapeutic to the fetus in the sense meant by that term.[107] The court was also troubled by the possibility that a term like "therapeutic" might prohibit assisted-reproduction technologies generally or impede the detection or novel treatment of disorders that are considered life-threatening to the mother.[108] Under this analysis, the court decided that a scienter (knowledge or intent) requirement included in the Illinois statute did not mitigate vagueness where the law "has no core meaning to begin with."[109] Rather, the court expanded its vagueness argument to conclude that potential restrictions on a woman's reproductive decision arising from the law's broad effect and definitional uncertainty were an encroachment on the essential right of privacy as outlined by the Supreme Court in *Griswold* and *Roe*.[110]

Addressing the three cases, commentator Lori Andrews suggests that "if an embryo stem cell researcher were to challenge the state statutory bans, it is unlikely that he or she would be as successful as the plaintiffs in Illinois, Louisiana, or Utah."[111] According to Andrews, "researchers presumably know when they are engaging in research," overcoming the inherent ambiguity identified by the courts between experimentation, prohibited under the statutes, and protected access to clinical care.[112] Since the possibility of therapeutic intervention would not be an issue in the research context, the investigator could be assumed to fully understand the scope of any state-imposed research regulation, even if, for all practical purposes, most of the state statutes only include embryonic material by implication (suggesting at least one possible alternative basis for challenging such statutes for ambiguity). Andrews concludes, moreover, that "the female patient's reproductive freedom was implicated in the [three] previous cases, providing another reason for overturning the statute. Stem cell researchers do not have those potential legal arguments" available to them since the reproductive freedom identified in *Roe* and later cases is attributed solely to the female patient.[113]

Despite Andrews' well-founded objections, there is at least some basis to consider whether the states may be limited under the federal constitution in their ability to regulate embryonic research. Given most statutes' origin in fetal regulation, significant ambiguity exists over whether or to what degree embryonic research may be implicated in their scope. Commentator John Robertson has observed that, "taken literally, [some statutes] would...appear to ban the destruction of extracorporeal embryos and might apply to other maneuvers with early embryos."[114] Robertson suggests, however, that "crucial questions of interpretation remain. For example, do they create an obligation to place embryos in the uterus or merely not to destroy them by active discard? What about removing embryos from [cryogenic] freezing [but] doing nothing further?"[115] Framed in these terms, the question of what affirmative duties may exist to preserve embryonic life under the regulating statutes is uncertain, even for investigators engaged in embryo research not involving clinical intervention.

Beyond the statutes' obvious and inherent ambiguity, there is also a clear sense that restrictive regulations may impair some component of the reproductive freedom of gamete or embryo donors, protected by implication under *Griswold* and *Roe*.[116] Andrews has suggested that female patients' separation from research activities undermines their claim for constitutional protection from overly restrictive state statutes (stem cell research involving the embryo or its derivatives would not provide the donor with immediate clinical or fertility benefit to which she may be constitutionally entitled). Commentator Christine Feiler suggests some obvious areas where protection may lie, noting that "research involving 'surplus' embryos may be more complex. Experiments using these embryos might be conducted to explore the reproductive health or status of the mother as part of an IVF protocol and, thus, could implicate reproductive autonomy."[117]

Even assuming the donor's separation from research (no direct beneficial effect, for example), some basis for constitutional protection may remain: "One might also argue persuasively that the couple has a constitutional right to make decisions concerning embryos formed from their gametes, including decisions concerning creation, storage, transfer, donation, manipulation, and research."[118] According to Robertson, "that right would be based on their right to determine whether they reproduce or avoid reproduction with their gametes and the embryos created therefrom."[119] Where germ or embryonic material is entrusted to a third party (like a research investigator), the capacity to reproduce biologically exists even though the specific protocol may not envision implantation or eventual live birth. Under such circumstances, it is possible to construct a theory to support the proposition that parental autonomy in the decision to reproduce or to refrain from reproduction would normally control (and by extension, parental autonomy in whom to entrust such material and for what purposes).[120]

Finally, commentator Christine Feiler identifies several plausible alternative arguments, including "the researchers' right to investigate and the general public's interest in obtaining beneficial information from such experiments...."[121] The Supreme Court in *Meyer v. Nebraska*, 262 U.S. 390 (1923), acknowledged that the Fourteenth Amendment protects the freedom "to acquire useful knowledge." In *Henley v. Wise*, 303 F. Supp. 62 (N.D. Ind. 1969), a district court indicated that the "right of scholars to do research and advance the state of man's knowledge" was a protected activity. Commentator June Coleman has observed that "basic research is specifically related to acquiring useful knowledge....Based on [the] statements by the courts, basic research clearly falls within...the definition of fundamental rights—those which society has traditionally protected—because the courts have protected this right since 1923."[122] Coleman also notes that "some commentators have argued that bans on research are a direct and indirect burden on the ability to make reproductive decisions" no matter what their scope.[123] Such prohibitions, Coleman suggests, "freez[e] information at an arbitrary point in time and halt the use and development of new medical procedures."[124] NIH Director Harold Varmus advanced a similar argument in support of embryo-derived stem cell research in his testimony before Congress on the human embryo research ban then under consideration by lawmakers.[125]

Notes

1 See e.g., 42 USC §§ 289g, 289g-1, 289g-2, 274e, and 45 CFR 46, Subpart B et al.

2 Members of Congress who have opposed stem cell funding maintain that "current law...also specifically covers cells and tissue obtained from embryos," citing as applicable 42 USC § 289g-1(b)(2)(ii) ("no alternation of the timing, method, or procedures used to terminate the pregnancy...made solely for the purposes of obtaining the [fetal] tissue"). See *Statement of Members of the House of Representatives to the Working Group of the Advisory Committee to the Director of the National Institutes of Health* 1 (8 April 1999) (hereinafter *Congressional Statement*). The apparent basis for this assertion is the definition of "human fetal tissue" at 42 USC § 289g-1(g) ("for purposes of this section, the term 'human fetal tissue' means tissue or cells obtained from a dead human embryo or fetus after a spontaneous or induced abortion"). Two elements render the congressional arguments unpersuasive: 1) neither 42 USC § 289g-1 nor 289g-2 is directed at embryo or IVF research; rather, both sections are exclusively centered in a conventional understanding of aborted fetal tissue and the issues arising from fetal tissue research; and 2) biological embryology, IVF, and ES cell research typically include only "live" embryos that are maintained in a living state for research purposes until they are either implanted, disaggregated for living unicellular components, or terminated upon the experiment's completion. A "dead human embryo" would, by definition, comprise a multicellular tissue mass in which all cellular functions associated with life activity had previously ceased (clinical cell death), and would be more in the nature of a stored pathology specimen. The draft guidelines of the NIH Ad Hoc Working Group of the Advisory Committee to the Director, *Draft NIH Guidelines for Research Involving Pluripotent Stem Cell Research* (8 April 1999) at 5 support this interpretation ("after implantation, embryos are legally considered fetuses and are governed by different laws and regulations—see Section III (B) [referencing 45 CFR 46 and 42 USC §§ 289g-1 and 289g-2 "for regulations regarding research involving human fetal tissue"]). Cf. 45 CFR § 46.203(c) ("fetus means the product of conception *from the time of implantation* (as evidenced by any of the presumptive signs of pregnancy, such as missed menses or a medically acceptable pregnancy test), until a determination is made, following expulsion or extraction of the fetus, that it is viable" [italics added]).

3 See Public Law No. 105-277, 112 Stat. 2461 (1998), as printed in H.R. Conf. Rep. No. 105-825 (1998) (federal human embryo research ban). No federal legislation or CFR currently regulates private sector experimentation in a manner specific to human embryo research. See also state law discussion, infra. NIH Director Harold Varmus has stated that "there are a number of advantages to using public funding for research. Perhaps the most important reason is the fact that Federal involvement creates a more open research environment—with better exchange of ideas and data among scientists—more public engagement and more oversight." *Statement of Harold Varmus, M.D., NIH Director, Before the Senate Appropriations Subcommittee on Labor, Health, and Human Services, Education, and Related Agencies* (26 January 1999) at 3. Note that some private sector biotechnology companies have independently self-regulated their stem cell research activities through the use of advisory boards and ethical protocols. See e.g., Geron, Inc., *A Statement on Human Embryonic Stem Cells by the Geron Ethics Advisory Board* (undated).

4 See note 1, supra, and 45 CFR 46.101 et seq. (human subjects protections).

5 See Gregory Gelfand and Toby R. Levin, "Fetal Tissue Research: Legal Regulation of Human Fetal Tissue Transplantation," *Washington and Lee Law Review* 50 (1993):668.

6 See New York State Task Force on Life and the Law, *Assisted Reproduction Technologies: Analysis and Recommendations for Public Policy* 382 (New York, N.Y.: 1998) (hereinafter *Task Force*).

7 See *National Research Act*, Pub. L. 93-348, Section 201(a), 88 Stat. 348 (1974).

8 National Commission for the Protection of Human Subjects of Biomedical and Behavioral Research, *Research on the Fetus: Report and Recommendations* (Washington, D.C.: 1975), reprinted in 40 Fed. Reg. 33,526 (1975); see 45 CFR § 46.201-.211 (1997) (Subpart B); 42 USC § 274e (1997); 42 USC § 289g(a) (1997).

9 Id. Note that 45 CFR Part 46 ("Protection of Human Subjects"), Subparts A–D comprise the full complement of human subjects regulation as it affects DHHS, but the term "Common Rule" is commonly used to refer to those regulations contained in Subpart A ("Federal Policy for the Protection of Human Subjects [Basic DHHS Policy for Protection of Human Research Subjects]").

10 Id.

11 See 45 CFR § 46.204(d), nullified by section 121(c) of the NIH Revitalization Act of 1993, Pub. L. 103-43, June 10, 1993 in 59 Fed. Reg. 28276, June 1, 1994. Former section 46.204(d) provided that "No application or proposal involving *in vitro* fertilization may be funded by the Department or any component thereof until the application or proposal has been reviewed by the Ethics Advisory Board and the Board has rendered advice as to its acceptability from an ethical standpoint."

12 See *Task Force* at 382.

13 Id. ("The EAB's report concluded that the federal government should support human embryo research, including research on embryos not intended to be transferred for implantation. According to the report, research not involving embryo transfer must comply with existing regulations governing research on human subjects, must be designed to establish the safety and efficacy of embryo transfer and to obtain scientific information not reasonably attainable by other means, and must stop fourteen days after fertilization. The report also urged researchers to obtain specific informed consent for the use of gametes in research and to inform the public of possible risks associated with the practice of IVF.").

14 Id.

15 Public Law 103-43 ("NIH Revitalization Act of 1993"), codified at *NIH Revitalization Act of 1993*, U.S. Code, vol. 42, secs. 289g-1 and 289g-2 (1997) as Section 498A of the Public Health Service Act.

16 See note 11, supra. DHHS affirms that human embryo research is included within the meaning of IVF and was subject to EAB review. 59 Fed.Reg. 45,293 (Sept. 1, 1994). "*In vitro* fertilization" is defined at 45 CFR § 46.203(g) as "any fertilization of human ova which occurs outside the body of a female, either through admixture of donor human sperm and ova or by any other means."

17 59 Fed. Reg. 28874, 28875 (June 3, 1994) (notice of meeting); National Institutes of Health, *Report of the Human Embryo Research Panel* ix (vol I 1994) (hereinafter *Panel I*).

18 The panel oriented its analysis around "three primary considerations": "[1] the [significant] promise of human [medical] benefit from research"; [2] the relative inappropriateness of according preimplantation human embryos "the same moral status as an infant or child"; and [3] the consensus that without "Federal funding and regulation in this area, preimplantation human embryo research that has been and is being conducted...would continue, without consistent ethical and scientific review." The HERP members identified a series of research categories that they deemed acceptable for federal funding, including "studies aimed at improving the likelihood of a successful outcome for a pregnancy"; "research on the process of fertilization"; "studies on egg activation and the relative role of paternally derived and maternally derived genetic material in embryo development (parthenogenesis without transfer)"; "studies in oocyte maturation or freezing followed by fertilization to determine developmental and chromosomal normality"; "research involving preimplantation genetic diagnosis with and without transfer"; "research involving the development of embryonic stem cells, but only with embryos resulting from IVF for infertility treatment or clinical research that have been donated with the consent of the progenitors"; "nuclear transfer into an enucleated, fertilized or unfertilized (but activated) egg without transfer for research that aims to circumvent or correct an inherited cytoplasmic defect"; "research involving the use of existing embryos where one of the progenitors was an anonymous gamete source who received monetary compensation"; and "a request to fertilize ova where this is necessary for the validity of a study that is potentially of outstanding scientific and therapeutic value" *Panel I* at x; xvii–xviii. See also *Panel I* at xviii–xx for research categories deemed to "warrant additional review" or "unacceptable for Federal funding."

19 See *Panel I* at xi, "Fertilization of Oocytes Expressly for Research Purposes" ("[it] would not be wise to prohibit altogether the fertilization and study of oocytes for research purposes...however, the embryo merits respect as a developing form of human life and should be used in research only for the most serious and compelling reasons....The Panel believes that the use of oocytes fertilized expressly for research should be allowed only under two conditions. The first condition is when the research by its very nature cannot otherwise be validly conducted....The second condition...is when a compelling case can be made that this is necessary for the validity of a study that is potentially of outstanding scientific and therapeutic value.").

20 See June Coleman, "Comment: Playing God or Playing Scientist: A Constitutional Analysis of State Laws Banning Embryological Procedures," *Pacific Law Review* 27 (1996):1339.

21 Id.

22 Id.; see also Christine L. Feiler, "Note: Human Embryo Experimentation: Regulation and Relative Rights," *Fordham Law Review* 66 (1998):2461–62 ("William Galston, deputy director of [President] Clinton's Domestic Policy Council, later confirmed that the Clinton administration had decided even before the ACD's meeting that deliberate creation of human embryos for experimentation exceeded the public's tolerance for 'exotic' research") citing R. Alta Charo, "The Hunting of the Snark: The Moral Status of Embryos, Right-to-Lifers, and Third World Women," *Stanford Law and Policy Review* 6 (1995):14. The administration's current position on human embryo research is not clear. While the President's Press Secretary Mike McCurry affirmed the earlier 1994 Executive Order

prohibiting deliberate creation of research embryos with federal funds in early 1997 (White House Press Briefing, 24 February 1997), the Executive Office of the President in February 1998 released a Statement of Administration Policy on S. 1601 ("Human Cloning Prohibition Act") that encourages Congress to refrain from prohibiting somatic cell nuclear transfer (SCNT), a potentially more radical form of deliberate embryo fertilization involving "the transfer of a cell nucleus from a somatic cell into an egg from which the nucleus has been removed....the resulting fused cell, and its immediate descendants, are believed to have the full potential to develop into an entire animal, and hence totipotent." Executive Office of the President, *Statement of Administration Policy [on] S.1601 - Human Cloning Prohibition Act* (9 February 1998): 1; definition of SCNT taken from NIH, *Fact Sheet on Stem Cell Research* (21 April 1999):1. This despite the President's March 4, 1997, ban on federal funding for cloning of human beings (SCNT is generally classified as a form of human cloning). Office of the White House Press Secretary, *Memorandum for the Heads of Executive Departments and Agencies* (4 March 1997): 1 ("Federal funds should not be used for cloning of human beings. The current restrictions on the use of Federal funds for research involving human embryos do not fully assure this result.").

23 See Feiler, "Human Embryo Experimentation," 2461; See NIH, *Background Information on the Impact of the Human Embryo Research Amendment* (30 January 1996) (hereinafter Background) at 2 ("Since 1993, nine applications that include research involving human *in vitro* fertilization have reached the stage in the NIH review process at which this research would have been funded. Of these, six applications would likely have received support if the NIH had been able to proceed according to the panel's recommendations and the President's directive.").

24 Id. Public Law No. 104-99, Title I, § 128, 110 Stat. 26, 34 (1996). (Preventing the use of any funds made available by Public Law 104-91 for the creation of human research embryos or embryo research in which the embryo is destroyed, discarded, or knowingly subjected to risk of injury or death greater than that allowed for research on fetuses *in utero* under 45 CFR § 46.208(a)(2) ["risk to the fetus imposed by the research is minimal and the purpose of the activity is the development of important biomedical knowledge which cannot be obtained by other means"] and section 498(b) of the Public Health Service Act (42 USC § 289g(b)) ["...the risk standard (published in [45 CFR §] 46.102(g))...[shall] be the same for fetuses which are intended to be aborted and fetuses which are intended to be carried to term"]; "human embryo" is defined as "any organism, not protected as a human subject under 45 CFR 46 as of the date of the enactment of this Act, that is derived by fertilization, parthenogenesis, cloning, or any other means from one or more human gametes or human diploid cells.") NIH has described the effect of the ban as prohibiting "*in vitro* fertilization of a human egg for research purposes where there is no direct therapeutic intent....as well as...research with embryos resulting from clinical treatment and research on parthenogenesis" (Background at 2).

25 Public Law No. 104-208, Div. A, § 101(e), Title V, § 512, 110 Stat. 3009, 3009-270 (1996); Public Law No. 105-78, Title V, § 513, 111 Stat. 1467, 1517 (1997); Public Law No. 105-277, 112 Stat. 2461 (1998), as printed in H.R. Conf. Rep. No. 105-825 (1998).

26 See Rick Weiss, "Georgetown Geneticist Admits Disobeying Test Ban on Embryos," *Washington Post*, 15 January 1997:3.

27 Id.

28 Id., *Congressional Statement* at 2; and *Letter from John J. Callahan, Assistant Secretary for Management and Budget, DHHS, to DHHS Institutional Officials* (February 1997) (reinforcing the legal requirements of the congressional human embryo research ban).

29 See e.g., James A. Thomson et al., Embryonic Stem Cell Lines Derived from Human Blastocysts, Science 282 (6 November 1998):1145; Michael J. Shamblott, John Gearhart, et al., "Derivation of Pluripotent Stem cells from Cultured Human Primordial Germ Cells," *Proceedings of the National Academy of Sciences* 95 (November 1998):13726.

30 Id.

31 See *Letter from President Clinton to Dr. Harold Shapiro, Chair, National Bioethics Advisory Commission* (14 November 1998); and *Letter from Dr. Harold Shapiro, Chair, National Bioethics Advisory Commission to President Clinton* (20 November 1998).

32 *Statement of Harold Varmus, M.D., NIH Director, Before the Senate Appropriations Subcommittee on Labor, Health, and Human Services, Education, and Related Agencies* (26 January 1999) at 4; *Testimony of Harold Varmus, M.D., NIH Director, Before the National Bioethics Advisory Commission* (19 January 1999) at xx.

33 Id. *Memorandum from Harriet S. Rabb, HHS General Counsel, to Dr. Harold Varmus, M.D., NIH Director* (15 January 1999); NIH, *Fact Sheet on Stem Cell Research* (21 April 1999) ("After a thorough analysis of the law, DHHS [General Counsel Rabb] concluded that the congressional prohibition on the use of DHHS funds for certain types of human embryo research does not apply to research utilizing human pluripotent stem cells because such cells are not embryos....Thus, research using pluripotent stem cells derived from human embryos can be funded by DHHS.").

34 Id. See also NIH, *NIH Position on Human Pluripotent Stem Cell Research* (undated) at 1 ("NIH funds [including equipment, facilities, and supplies purchased on currently funded grants] should not be used to conduct research using human pluripotent stem cells derived from human fetal tissue or human embryos until further notice....While the NIH proposes to support research utilizing these human pluripotent stem cells, it will not do so until public consultation has occurred, guidelines are issued, and an oversight committee has ensured that each project is in accord with these guidelines. Research on human stem cells derived from sources other than human embryos or fetal tissue will not be subject to these guidelines and oversight: this research will continue to be funded under existing policies and procedures."). The NIH Director's caution has not avoided public controversy, however. See e.g., *Letter from Seventy Members of the U.S. House of Representatives to Donna E. Shalala, Secretary of Health and Human Services* (11 February 1999); Seventy-Three Scientists, "Science Over Politics," *Science* 283 (19 March 1999):1849.

35 See generally NIH Ad Hoc Working Group of the Advisory Committee to the Director, *Draft NIH Guidelines for Research Involving Pluripotent Stem Cell Research* (8 April 1999).

36 Opening statement of Co-Chair Ezra C. Davidson, Jr., M.D., at 8 April 1999 meeting of the NIH Ad Hoc Working Group of the Advisory Committee to the Director.

37 See note 32, supra.

38 Public Law No. 105-277, 112 Stat. 2461 (1998), as printed in H.R. Conf. Rep. No. 105-825 (1998) (federal human embryo research ban).

39 45 CFR § 46.208(a)(2) (1997).

40 45 CFR § 46.102(i) (1997).

41 Pub. L. 99-158 ("Health Research Extension Act of 1985"), codified at *Health Research Extension Act of 1985*, USC, vol. 42, secs. 289g(b) (1997) as Section 498(b) of the Public Health Service Act.

42 U.S. Const. art. I, § 8; see e.g., *Fullilove v. Klutznick*, 448 U.S. 448, 100 S.Ct. 2758, 65 L.Ed.2d 902 (1980) ("The reach of the Spending Power within its sphere is at least as broad as the regulatory powers of Congress"—Burger, C.J. [citations omitted]).

43 Office of the White House Press Secretary, *Statement by the President* (2 December 1994):1.

44 No Executive Order exists on this topic as of 3 May 1999, see White House Virtual Library, Executive Orders (http://www.pub.whitehouse.gov/search/executive-orders.html).

45 U.S. Const. art. II, § 1.

46 Office of the White House Press Secretary, *Memorandum for the Heads of Executive Departments and Agencies* (4 March 1997):1.

47 No Executive Order exists on this topic as of 3 May 1999, see White House Virtual Library, Executive Orders (http://www.pub.whitehouse.gov/search/executive-orders.html).

48 U.S. Const. art. II, § 1.

49 Executive Office of the President, *Statement of Administration Policy [on] S.1601 - Human Cloning Prohibition Act* (9 February 1998):1.

50 National Bioethics Advisory Commission (NBAC), *Cloning Human Beings* (1997) at Appendix 3 ("Nuclear transplantation cloning: a type of cloning in which the nucleus from a diploid cell is fused with an egg from which the nucleus has been removed. The DNA of the transplanted nucleus thus directs the development of the resulting embryo.").

51 References to human embryo research are presently superseded by Public Law No. 105-277, 112 Stat. 2461 (1998), as printed in H.R. Conf. Rep. No. 105-825 (1998) (federal human embryo research ban), considered supra.

52 45 CFR § 46.201 et seq. (1997) ("Additional Protections Pertaining to Research, Development, and Related Activities Involving Fetuses, Pregnant Women, and Human *In Vitro* Fertilization").

53 45 CFR § 46.201 (1997).

54 45 CFR § 46.203(g) (1997).

55 See 59 *Fed.Reg.* 45,293 (Sept. 1, 1994), "National Institutes of Health (NIH); Notice of Meeting of Panel ("It is the enactment of [Public Law No. 103-43 repealing the *de facto* moratorium] that now enables the NIH to fund IVF proposals, as well as research involving human embryos that result from IVF or other sources"); but see John A. Robertson, "Article: Reproductive Technology and Reproductive Rights: In the Beginning: The Legal Status of Early Embryos," *Virginia Law Review* 76 (1990): n. 177 ("the federal regulations for fetal research [at 45 CFR 46, Subpart B] define the fetus as the entity that exists from the time of implantation, thus exempting preimplantation embryos from their reach.").

56 To the extent that DHHS/NIH maintains that ES cell research is "not embryo research," see note 33, supra, IVF/embryo research-relevant sections of Subpart B would also not be interpreted to apply. However, one or more sections of the general human subjects protections at 45 CFR 46, Subpart A, may be applicable to the extent that such research is or is interpreted to "involve human subjects." See 45 CFR §§ 46.101(a) and et seq. Further complicating this issue, Gary Ellis, Director of the NIH OPRR has addressed some issues relevant to 45 CFR 46 in his recent correspondence with NBAC, *Letter from Gary B. Ellis, Ph.D., Director, OPRR, to Eric M. Meslin, Ph.D., Executive Director, NBAC* (3 June 1999) (hereinafter *OPRR I*). According to Ellis, "although Subpart B does not apply to research involving an embryo *per se*, it does apply to research that involves the process of *in vitro* fertilization. An embryo formed by a means that does not involve *in vitro* fertilization would not be subject to Subpart B." Ellis' remarks do not address the status of embryos fertilized *in vitro* by private clinicians for assisted reproductive therapy and later declared "spare/available for research use" (the probable source for most embryonic stem cell derivation). Nor is it clear whether downstream research that includes ES cells would qualify for Subpart B regulation. Asked to describe OPRR's guidance to an Institutional Review Board (IRB) faced with embryo research, Ellis responded that "when the inclusion in proposed research of the process of *in vitro* fertilization invokes review of the proposed research by an IRB, OPRR routinely directs attention of IRBs to [45 CFR § 46.205, imposing additional duties for the IRB in research subject to Subpart B]," although Ellis admits that "Subpart B provides little, if any, specific direction to the IRB." Ellis predicts that OPRR will distribute any forthcoming guidelines promulgated by NIH to guide research involving pluripotent stem cell research, although "it would be inaccurate...to describe OPRR as 'enforcing' NIH guidelines on this or any other matter."

57 45 CFR § 46.204(a)-(c) (1997).

58 45 CFR § 46.205 (1997); but see n. 56, supra. Note that language in this section and 45 CFR § 46.201 ("Applicability") ("the requirements of this subpart [B] are in addition to those imposed under the other subparts [A, C, and D] of this part [46—Human Subjects Protections]") may imply that IRB and other provisions contained in Subpart A (the "federal Common Rule") are also applicable as against any research governed by Subpart B. Thus, while embryo derivation of stem cells and research on cells derived from embryo sources may not independently qualify for human subjects protections under Subpart A, they may impliedly receive some or all of those protections if Subpart B is held to apply to them. The Department may wish to clarify this question.

59 In addition to their other duties, IRB's reviewing research subject to Subpart B must "(1) Determine that all aspects of the activity meet the requirements of this subpart; (2) Determine that adequate consideration has been given to the manner in which potential subjects will be selected, and adequate provision has been made by the applicant or offeror for monitoring the actual informed consent process (e.g., through such mechanisms, when appropriate, as participation by the Institutional Review Board or subject advocates in: (i) Overseeing the actual process by which individual consents required by this subpart are secured either by approving induction of each individual into the activity or verifying, perhaps through sampling, that approved procedures for induction of individuals into the activity are being followed, and (ii) monitoring the progress of the activity and intervening as necessary through such steps as visits to the activity site and continuing evaluation to determine if any unanticipated risks have arisen); (3) Carry out such other responsibilities as may be assigned by the Secretary") 45 CFR § 46.205(a) (1997). See also 45 CFR § 46.205(c) (1997) ("Applicants or offerors seeking support for activities covered by this subpart must provide for the designation of an Institutional Review Board, subject to approval by the Secretary, where no such Board has been established under Subpart A of this part.").

60 45 CFR § 46.206(a)(1) (1997).

61 45 CFR § 46.206(a)(2) (1997).

62 45 CFR § 46.206(a)(3) (1997).

63 45 CFR § 46.206(b) (1997).

64 See 45 CFR §§ 46.201-210, Subpart B, "Additional DHHS Protections for Pregnant Women, Human Fetuses, and Newborns Involved as Subjects in Research, and Pertaining to Human *In vitro* Fertilization," *Fed. Reg.* 27794-27804 (20 May 1998).

65 45 CFR § 46.101(f) (1997).

66 Note 45 CFR §§ 46.101(c) (1997) ("Department or Agency heads retain final judgment as to whether a particular activity is covered by this [Common Rule]"); 45 CFR §§ 46.101(d) (1997) ("Department or Agency heads may require that specific research activities or classes of research activities conducted, supported, or otherwise subject to regulation by the Department or Agency but not otherwise covered by this [Common Rule], comply with some or all of the requirements of this policy."). DHHS has not acted to vary the terms of the Common Rule to apply its terms to fetal tissue transplant, stem cell derivation or research. See also *OPRR I*, at 1. Director Ellis, OPRR, observes that "a progenitor of an embryo may be a human subject in regulatory terms and covered by 45 CFR Part 46" when it satisfies the triggering language for Common Rule oversight at 45 CFR § 46.102(f) ("human subject" is defined as "a living individual about whom an investigator (whether professional or student) conducting research obtains (1) data through intervention or interaction with the individual, or (2) identifiable private information). Ellis also notes that such research "must be conducted in accordance with applicable State or local laws" [referencing 45 CFR § 46.210].

67 For a descriptive analysis of the various alternatives, see J. Kyle Kinner, "Bioethical Regulation of Human Fetal Tissue and Embryonic Germ Cellular Material: Legal Survey and Analysis," a paper commissioned for NBAC and available in this volume.

68 See note 56, supra.

69 See generally Lori B. Andrews, "State Regulation of Embryo Stem Cell Research," a paper commissioned for NBAC, available in this volume.

70 See Coleman, "Constitutional Analysis," 1352.

71 See e.g., Robertson, "Legal Status of Early Embryos," 512; Feiler, "Human Embryo Experimentation," n. 244.

72 See Coleman, "Constitutional Analysis," 1358; n. 177; Fla. Stat. Ann. § 390.0111(6); La. Rev. Stat. Ann. §§ 9:121-133 (West 1991); Me. Rev. Stat. Ann. tit. 22, § 1593 (West 1992); Mass. Ann. Laws ch. 112, § 12j(a)(I) (Law. Co-op. 1996); Mich. Comp. Laws Ann. §§ 333.2685, 333.2686, 333.2692 (West 1992); Minn. Stat. Ann § 145.422 subd. 1,2 (West 1989); N.H. Rev. Stat. Ann. § 168-B15(II) (1994); N.D. Cent. Code §§ 14-02.2-01, 14-02.2-02 (1991); 18 Pa. Cons. Stat. Ann. § 3216(a) (Supp. 1995); R.I. Gen. Laws § 11-54-1(a)-(c) (1994).

73 Id. at 1358.

74 Id. at 1358; see also Moore v. Regents of the University of California, 51 Cal.3d 120, 793 P.2d 479, 271 Cal. Rptr. 146 (Cal. 1990), cert. denied, 499 U.S. 936 (1991) (court recognized both a possessory and a commercial interest in patient's bodily tissues).

75 See Andrews, "State Regulation of Embryo Stem Cell Research," at 4; n. 5 ("In those states...embryo stem cell research is not banned"); but see D.C. Code § 6-2601 (1998) prohibiting sale of any part of human body (even cells), a restriction that may extend to human embryos.

76 See Feiler, "Human Embryo Experimentation," at n. 239. Note that a careful analysis of Andrews' work lists 32 states as regulating embryo research in some way (rather than the typically reported 26). This larger number reflects an aggregate of both research-restrictive statutes and those state statues limiting or prohibiting sale of embryo or human tissues that may indirectly include embryos.

77 N.H. Rev. Stat. Ann. § 168-B:15 (limiting the maintenance of *ex utero* pre-implantation embryos in a noncryopreserved state to under 15 days and prohibiting the transfer of research embryos to the uterine cavity).

78 See Andrews, "State Regulation of Embryo Stem Cell Research," at 4; n. 13–14. Louisiana broadly prohibits research involving IVF embryos. La. Rev. Stat. Ann. §§ 9:121-122 (West 1991). Eight other states restrict embryo research indirectly, banning all research on "live" embryos or fetuses. Fla. Stat. Ann. § 390.0111(6); Me. Rev. Stat. Ann. tit. 22, § 1593 (West 1992); Mass. Ann. Laws ch. 112, § 12j(a)(I) (Law. Co-op. 1996); Mich. Comp. Laws Ann. §§ 333.2685, 333.2686, 333.2692 (West 1992); Minn. Stat. Ann § 145.422 subd. 1,2 (West 1989); N.D. Cent. Code §§ 14-02.2-01, 14-02.2-02 (1991); 18 Pa. Const. Stat. Ann. § 3216(a) (Supp. 1995); R.I. Gen. Laws § 11-54-1(a)-(c) (1994).

79 Id. at 4; n. 16; see e.g., Me. Rev. Stat. tit. 22 § 1593 (maximum five-year term); Mass. Ann. Laws 112 § 12J(a)(V) (maximum five-year term); Mich. Comp. Laws § 333.2691 (maximum five-year term).

80 U.S. Const. art. VI, § 2; Amend. X.

81 See Andrews, "State Regulation of Embryo Stem Cell Research," at 8; n. 59; Me. Rev. Stat. Ann. tit. 22 § 1593; Mass Ann. Laws ch. 112 § 12(J)(a)(IV); Mich. Comp. Laws § 333.2609; N.D. Cent. Code § 14-02.2-02(4); and R.I. Gen. Laws § 11-54-1(f).

82 Id. at 8; n. 60; Fla. Stat. Ann. § 873.05; Georgia Code Ann. § 16-12-160 (a) (except for health services education); Ill. Stat. Ann. ch 110 _ para. 308.1; La. Rev. Stat. Ann. § 9:122; Minn. Stat. Ann. § 45.422(3) (live); 18 Pa. Cons. Stat. Ann. § 3216(b)(3) (forbids payment for the procurement of fetal tissue or organs); Texas Penal Code § 48.02; Utah Code Ann. § 76-7-311. But see Feiler, "Human Embryo Experimentation," at 2455 ("Although some state laws prohibit the sale of fertilized embryos, they do nothing to prevent the sale of gametes (sperm and eggs), which can easily be converted into research embryos through deliberate fertilization. Payment for sperm and eggs is widespread among American infertility clinics." [citations omitted]).

83 Id. at 9; n. 67; 75; 76; 80; Tenn. Code Ann. § 39-15-208 (1998) and Utah Code Ann. § 76-7-311 (1998) prohibit sale of an "unborn child"; D.C. Code § 6-2601 (1998) and Va. Code § 32.1-289.1 (1998) prohibit sale of all or a portion of the "human body" (D.C.) or a "natural body part" (Va.); two state statutes prohibit sale of specified organs (not including embryos), but permit state health officials to expand the list under prescribed conditions. N.Y. Pub. Health § 4307 (1998); W. Va. Code § 68.50.610(2) (1998).

84 See Coleman, "Constitutional Analysis," 1358.

85 See *Andrews*, at 8.

86 Id., Minn. Stat. Ann. § 145.422(3).

87 See e.g., *Uniform Anatomical Gift Act* secs. 1–17, 8A U.L.A. 29 (1994) (enacted in pre- or post-1987 revisions in all 50 states and DC; § 10 of the post-1987 revision prohibits organ sale). At least 20 states have enacted statutory prohibitions on the sale of organs or organ parts. See Andrews. Note also Minnesota statute § 525.921 restricting sale of decedent organ parts and defining "decedent" as "a deceased individual [that] includes a stillborn infant or an embryo or fetus that has died of natural causes *in utero*" (not applicable to IVF or embryo research).

88 The only relevant litigation is *Doe v. Shalala*, 862 F. Supp. 1421 (D.Md. 1994), overturned, sub nom., *International Found. for Genetic Research v. Shalala*, 57 F.3d 1066 (4th Cir. 1995) (unpublished table decision), available in 1995 WL 361174, at *1. Complainants brought a class-action suit on behalf of more than 20,000 embryos stored in fertility clinics nationwide, alleging material imbalance under the Federal Advisory Committee Act in the membership of the HERP. The District Court dismissed for lack of standing, citing *Roe* and *Casey* to support its conclusion that the unborn do not constitute "persons" with legal, assertable rights. In its review, the Fourth Circuit Court of Appeals vacated the lower court's decision (and its reasoning) but remanded for dismissal, determining instead that the publication of the panel's report rendered the claim moot.

89 La. Rev. Stat. Ann. § 40:1299.35.13 (1998).

90 La. Rev. Stat. Ann. § 40:1299.35.13 (1998). See Marilyn Clapp, "Note, State Prohibition of Fetal Experimentation and the Fundamental Right of Privacy," *Columbia Law Review* 88 (1988):1076–77 ("The Louisiana statute effectively prohibits any research, experimentation, or even observational study on any embryo, fetus, or aborted fetal tissue. The ban encompasses a range of activities, including studies of the safety of ultrasound and pathological study of fetal tissues removed from a woman for the purpose of monitoring her health. Research on *in vitro* fertilization is likewise barred. Since the aborted pre-viable fetus is not living or cannot survive for long, no procedure performed upon it could be considered 'therapeutic,' and therefore use of this tissue is likewise prohibited. If performed on tissues from a miscarriage, such experimentation would be acceptable under the statutory scheme." [footnotes omitted]).

91 *Margaret S. v. Edwards*, 488 F. Supp. 181, 220 n.124 (E.D. La. 1980) [hereinafter *Margaret S. I*]. This suit was a class action brought on behalf of pregnant women who sought abortions, three physicians who performed abortions, and five clinics that provided abortion facilities.

92 Id. at 221.

93 *Margaret S. v. Treen*, 597 F. Supp. 636 (E.D. La. 1984), aff'd sub nom. *Margaret S. v. Edwards*, 794 F.2d 994 (5th Cir. 1986) [hereinafter *Margaret S. II*]. See Clapp, "State Prohibition," 1078–79 (the court "specifically not[ed] that reproductive choice was 'not limited to abortion decisions...but extends to both childbirth and contraception.' Prohibiting experimentation on fetal tissues could deny a woman knowledge that would influence her own future pregnancies, as well as prohibit procedures of immediate medical benefit such as pathological examination of tissues. The court also found that the prohibition curtailed the development and use of prediagnostic techniques, including amniocentesis. This result constituted a 'denial of health care' and a 'significant burden' on choice made during the first trimester." [footnotes omitted]).

94 *Margaret S. II*, at 674–75. See Clapp, "State Prohibition," 1079, n. 48 ("The court further suggested the statute would fail even a rational relation test because it failed to serve its own stated purpose of treating the fetus like a human being, since it treated fetal tissue differently from other human tissue.") Id. at 674–75.

95 *Margaret S. II*, at 675–76.

96 *Margaret S. v. Edwards*, 794 F.2d 994, 999 (5th Cir. 1986) [hereinafter *Margaret S. III*]. Note the court's concurring opinion that "criticized the majority for avoiding the real constitutional issue raised—that any statutory ban on experimentation would inevitably limit the kinds of tests available to women and their physicians and thus could not help but infringe on fundamental rights." Id. at 999–1002 (Williams, J., concurring). Clapp, "State Prohibition," 1080, n. 50.

97 *Margaret S. III* at 999 ("every medical test that is now 'standard' began as an 'experiment'"). But see Clapp, "State Prohibition," 1080, n. 54 ("the court hypothesized that the statute was intended 'to remove some of the incentives for research-minded physicians...to promote abortion' and was therefore 'rationally related to an important state interest.' This language suggests that if the statute had not been vague, the court would have applied less than strict scrutiny to a ban on fetal research. The court also implied, in dicta, that the rationale was based on the 'peculiar nature of abortion and the state's legitimate interest in discouraging' it, relying on *H.L. v. Matheson*, 450 U.S. 398, 411–13 (1981).").

98 *Margaret S. III* at 999.

99 61 F.3d 1493 (10th Cir. 1995).

100 *Jane L. v. Bangerter*, 794 F. Supp. 1528 (D. Utah 1992).

101 Id. at 1501.

102 Id. at 1502.

103 Id.

104 735 F. Supp. 1361 (N.D. Ill. 1990).

105 Ill. Rev. Stat., Ch. 38 para 81-26, § 6(7) (1989).

106 *Lifchez* at 1364.

107 Id. at 1366–67.

108 Id. at 1367–70.

109 *Lifchez* at 1372.

110 Id. at 1376–77. See e.g., *Griswold v. Connecticut*, 381 U.S. 479 (1965); *Roe v. Wade*, 410 U.S. 113 (1973); *Planned Parenthood of Southeastern Pennsylvania v. Casey*, 505 U.S. 833 (1992).

111 Andrews, "State Regulation of Embryo Stem Cell Research," 6.

112 Id.

113 Id.

114 Robertson, "Legal Status of Early Embryos," n. 44.

115 Id.

116 See e.g., Feiler, "Human Embryo Experimentation," 2445–6 ("*Roe* and its successors make clear that a woman's liberty interests in reproductive autonomy and bodily integrity often outweigh the State's interest in protecting unborn life.").

117 Feiler, "Human Embryo Experimentation," n. 97. Robertson notes that "at least four of the areas curtailed by state laws—cryopreservation, preimplantation screening, gene therapy, and twinning—are associated with procreation, which is traditionally a protected area....[T]hese various procedures could be generally characterized as specific liberties or means by which one exercises the more general right of procreation" (Robertson, "Legal Status of Early Embryos," 1364).

118 Robertson, "Legal Status of Early Embryos," 460 and n. 62 ("The claim of a constitutional right to dispose of gametes and their products would be based on a right to procreate or to avoid procreation, rather than a more general right to control disposition of tissue or parts removed from the body. While a person would have a fundamental right to determine whether tissue, organs, or gametes would be removed from her body, it is less clear that she would have a fundamental right to dispose of those bodily parts....Although there may be no general right to control disposition of removed body parts, the argument for a special right in the case of gametes is much stronger because of their reproductive significance. If people have a fundamental right to decide to procreate or not, then control of their gametes, including their combined, embryonic form, is necessary to exercise that choice.... A state intent on limiting dispositional authority over gametes and embryos would thus have to show a compelling interest to sustain such restrictions."). Arguing along similar lines, Coleman suggests that "another constitutional area that might afford protection for various medical procedures that use human embryos is the Fourteenth Amendment's right to privacy" (Coleman, "Constitutional Analysis," 1362). Under Coleman's privacy argument, basic research involving embryos may be protected under a broader understanding of reproductive privacy that includes not only the immediacy of human reproduction, but also its necessary predicates. See e.g., *Lifchez*, 735 F. Supp. at 1377 ("it takes no great leap of logic to see that within the cluster of constitutionally protected choices that includes the right to have access to contraceptives, there must be included within that cluster the right to submit to a medical procedure that may bring about, rather than prevent pregnancy").

119 Id.

120 Robertson, "Legal Status of Early Embryos," 499–500 ("A cogent argument, based on Supreme Court contraceptive cases, exists for finding a fundamental right to avoid biologic offspring. Even if parenthood entails only psychological burdens...a person should still be free to decide whether an event of such paramount importance to personal identity will occur"); and id. at 506 ("Strong policy arguments exist for permitting and supporting with public funds many kinds of embryo research under the aegis of the IRB system of review, as most review bodies have concluded"); but see id. ("because there is no right to particular means of conducting scientific research, laws restricting the gamete providers' right to donate or use embryos in research are likely to be valid, unless they infringe procreative liberty. A desire to show respect or protect embryos would thus be a sufficient basis for a ban on embryo research unrelated to procreation, whatever the origin of the embryos"); but see id. ("a ban on embryo research that prevented studies designed to improve birth control or to enhance methods of achieving pregnancy might be more questionable. Such a ban would directly interfere with the person's ability to avoid or to achieve reproduction.").

121 Feiler, "Human Embryo Experimentation," 2446.

122 Coleman, "Constitutional Analysis," 1368.

123 Id. at 1368–69, citing Marilyn J. Clapp, Note, State Prohibition of Fetal Experimentation and the Fundamental Right of Privacy, *Columbia Law Review* 88 (1988):1073, 1086–90; Rebecca J. Cook, Human Rights and Reproductive Self-Determination, *American University Law Review* 44 (1995):975, 1002–04.

124 Id.

125 See e.g., *Departments of Labor, Health and Human Services, Education, and Related Agencies Appropriations for 1996: Testimony of Harold Varmus, M.D., NIH Director, Before the* Subcomm. *of the House* Comm. *on Appropriations*, 104th Cong., 1st Sess. 1480 (1995) ("The second issue that's of great interest concerns the nature of so-called embryonic stem cells. These cells have the capacity to differentiate into virtually any kind of tissue. All work of this sort proceeds with animal models with considerable vigor throughout the biomedical research enterprise. We know from studies of embryonic stem cells of the mouse that we now have the potential to consider using embryonic stem cells as potential universal donors, for example, for bone marrow transplantation and other kinds of therapeutic maneuvers. But we can't begin to pursue such studies with human materials until we're able to use early stage embryos.").

INTERNATIONAL PERSPECTIVES ON HUMAN EMBRYO AND FETAL TISSUE RESEARCH

Commissioned Paper
Lori P. Knowles
The Hastings Center

Introduction

The National Bioethics Advisory Commission (NBAC) is charged with the task of researching and analyzing the ethical implications of primordial stem cell research (human embryonic stem [hES]and germ [hEG] cells)[1] and recommending directions for future regulation of this research. In order to acquit itself of this task, the Commission must first answer one fundamental question, "How do we understand where primordial stem cell research fits in the existing regulatory scheme?"

hEG cell research, which uses tissue from aborted fetuses, is analogous to the use of fetal tissue in transplantation. This is not only because it involves the use of fetal tissue, but also because it is research directed at developing therapies for the benefit of people which are unrelated to the facilitation of human reproduction.[2] Consequently, hEG cell researchers and funders should be aware of safeguards and guidelines regarding the use of fetal tissue in research.[3] These guidelines will be discussed briefly later in this paper.[4]

Use of Human Embryos In Research

Existing federal law in the United States which prohibits federal funding of research using human embryos[5] can be interpreted to permit the use of stem cells derived from that research.[6] That legal interpretation, however, does not provide the answer to how we should understand and situate hES cell research within the existing policies on the use of reproductive tissue in research. It is clear that since deriving hES cells requires the use and destruction of human embryos, hES cell research involves embryo research. Ethical issues of hES cell research, therefore, encompass those issues that arise in embryo research.

If NBAC deems that hES cell research is ethically acceptable and should be publicly supported and privately permitted, then it would be appropriate to revisit the current ban on federal funding in embryo research. Recommending that hES cell research be permitted would require that the use of human embryos in research be sanctioned in certain circumstances.[7] A recommendation to lift the prohibition on federal funding of human embryo research requires that NBAC express its opinion on the principles and some of the restrictions and guidelines that should govern that research. Examination of international embryo research policies can provide guidance in this endeavor.

As most embryo research takes place in the context of assisted reproductive technology (ART), discussion of embryo research ethics and policy takes place primarily in that context. In addition, as embryo research involves the use of tissues of men and women, some issues of concern are also discussed in the context of human subjects research policy and regulation.

Numerous countries have grappled with setting policies in both the areas of ART and human subjects research; one of the areas of greatest conflict is the permissibility and regulation of the use of human embryos in such research. Controversy and diversity of opinion continue to dominate discussions about embryo research regulation to a much greater extent than the use of fetal tissue for therapy. For that reason, international perspectives on embryo research and their implications for hES cell research will be the main focus of analysis and discussion. This paper will examine policies and regulations on human embryo research in the following countries: Canada,[8] Australia,[9] the United Kingdom,[10] and France.[11] I will also study statements from the European Union.[12]

These countries have been chosen for a number of reasons. Canada (with the exception of Québec), Australia, and the United Kingdom share the same legal tradition in Common law as the United States. The United Kingdom produced the first policy statement of any European country,[13] which led to the *Human Fertilisation and Embryology Act 1990* (the HFE Act). The HFE Act has been the blueprint of successful, thorough ART legislation for other countries drafting policies in ART.

In contrast to the United Kingdom, which has a strong tradition of promoting scientific freedom resulting in very liberal regulation of embryo research, France currently has a very restrictive policy on embryo research. France also represents a different perspective; it is a predominantly Catholic country with a civil law tradition, and it has a long history of thoughtful and prescient leadership in the area of bioethics. Finally, I have chosen to analyze the policies of the European Union, as statements from the European Commission and the European Group on Ethics in Science and New Technologies (EGE) reflect the diversity of opinion in and among the member states of the European Community.

Each region under examination has struggled to develop clear policies with respect to embryo research. The task is made difficult in part due to confusion in terminology, but primarily due to the great diversity of opinions on the moral status of the embryo. From the determination of moral status flows the possible responses to the questions on the permissibility, restrictions, and prohibitions of embryo research. Despite the great cultural, social, and religious differences within and among the countries examined, it is possible to find commonalties in their responses.

The question of whether to permit embryo research is characterized everywhere by a tension between the desire for therapeutic benefits derived from that research and the need to prevent abuses. In addition, similarities in international human embryo research policy exist in guiding principles, recommendation strategies, limitations on permissible research, and uses subject to prohibition. The importance of examining and adopting standards on which there is international consensus must be respected. The reputation of human embryo research conducted in the United States depends on the standards under which it is performed. In addition, as research is increasingly conducted in multicenter trials and with international cooperation, agreement on appropriate standards will be necessary.

Context of Embryo Research Policy and Regulation

Regulation of the use of human embryos in research falls into ART or human subject research oversight and sometimes in both. Very occasionally, legislation is drafted that deals solely with embryo research. The majority of international policies on embryo research are developed in the context of ART regulation, as is illustrated in the table below.

Table 1. International Policies in the Context of ART Regulation

Assisted Reproductive Technology	Human Subjects Research	Specific Embryo Research Legislation
Canada[14]	Canada[22]	Germany[24]
Australia[15]	Australia[23]	(Belgium)*
United Kingdom[16]	(Finland)*	
Austria[17]		
Denmark[18]		
Spain[19]		
Sweden[20]		
France[21]		
(Italy)*		
(Netherlands)*		
(Portugal)*		

*legislation in progress

Regulatory Background

Examination of the background of embryo research regulation in many countries shows that scientific developments, resulting public interest and concern, and considerable diversity of opinion have prompted the appointments of national commissions to explore and discuss the issues surrounding the use of ART. These commissions have traditionally been multidisciplinary and have been appointed to study the social, legal, and ethical issues of concern in ART, including human embryo research.

The mandates of the national commissions have included the pursuit of knowledge, identifying current areas of public concern and ethical problems, making recommendations for oversight, outlining guiding principles, and updating norms about ART and human subjects research. Other common mandates include:

- Identification of future developments and issues of concern.

- Outlining of guiding principles and basic standards of practice.

- Encouragement of continued reflection and thoughtful consensus around more contentious ethical issues.

- Advancement of knowledge and understanding.

Generally, the commissions required a period of two to four years, marked by numerous public meetings, calls for submissions, issuance of drafts for comment, and consultation, before submitting a final report.[25] Most commissions stated that they would not offer definitive answers to contentious ethical issues, but would outline the issues and elucidate guiding principles and the application of those principles in specific contexts.

For the most part, the national commissions have produced reports that are wide-ranging in scope and exhaustive, examining all issues surrounding ART, including embryo research.[26] In discussing embryo research, the reports include an examination of the uses of embryos in research, the sources of embryos (including creation of embryos for that research), and prohibitions and limitations on embryo research. In many countries, once a report was issued by the national commission, legislation was drafted and implemented.[27]

NBAC does not have the necessary time to examine thoroughly the issues and make comprehensive recommendations that balance public and scientific opinion concerning the regulation of embryo and primordial stem cell research. Consequently, a partial response to the President's request for clarification of the ethical issues may be more appropriate, to be followed by a more thorough examination of the issues surrounding embryo research. Such a partial response might set out principles and strategies to guide policy and regulation, areas of special concern, and areas requiring further inquiry.

The Need for Clear Definitions

Rapidly changing technology, resulting public anxiety, and diversity of firmly held beliefs make thoughtful, intelligent analysis of ART and embryo research regulation extremely difficult and politically sensitive. One further difficulty in developing domestic policy and in understanding international policy stems from the lack of precise or consistent use of terminology. In many countries the term "embryo" is not defined in the legislation that regulates embryo research. In those countries that do define the term, the definitions vary greatly. For example, in Australia (Victoria), the embryo is understood as differing from the zygote and as coming into being at syngamy (the alignment of the chromosomes on the mitotic spindle) approximately 22–24 hours after fertilization.[28] Consequently, as only creation of embryos for research is prohibited, ova may be fertilized and research conducted until syngamy.

By contrast, the United Kingdom HFE Act defines embryo as "a live human embryo where fertilisation is complete, references to an embryo include an egg in the process of fertilisation, and fertilisation is not complete until the appearance of a two-cell zygote."[29] Canada's Royal Commission on New Reproductive Technologies underlines this terminology problem:

In the language of biologists, before implantation the fertilized egg is termed a 'zygote' rather than an 'embryo.' The term 'embryo' refers to the developing entity after implantation in the uterus until about eight weeks after fertilization. At the beginning of the ninth week after fertilization, it is referred to as a 'fetus,' the term used until time of birth. The terms embryo donation, embryo transfer, and embryo research are therefore inaccurate, since these all occur with zygotes, not embryos. Nevertheless, because the terms are still commonly used in the public debate, we continue to refer to embryo research, embryo donation, and embryo transfer. For accuracy, however, we also refer to the developing entity during the first 14 days as a zygote, so that it is clear that we mean the stage of development before implantation and not later.[30]

Clearly, how a commission decides to define "embryo" greatly affects the resulting interpretation and impact of any recommendations. There is a danger if the terminology is manipulated to achieve certain ends indirectly that could not be achieved directly. For example, in the United States, attempts to define "embryo" as coming into existence at syngamy or 14 days after fertilization would surely be criticized as sophistic. The appearance of "skirting the issue" by an appeal to mechanistic approaches or legalistic interpretation should be avoided; transparency of findings and reasoning is required. Terminology should respond to public understandings and concerns. Whether embryos are viable or nonviable, hybrid or human, exist at fertilization or sometime thereafter, the fertilized human egg and developing human form is the locus of ethical concern regardless of its name.

Guiding Principles, Vision, and Strategy

Guidance on framing the issues involved in human embryo research can be found by examining the commonalities in guiding principles and recommendation strategies outlined by other countries. Many national commissions articulated guiding principles and values that informed their policy decisions and provided a framework for their recommendations on embryo research. Common principles in the reviewed reports, policies, and laws include:

- Respect for human life and dignity.
- The quality, including safety, of medical treatment.
- Respect for free and informed consent.
- Minimizing harm and maximizing benefit.
- The relief of human suffering.
- Freedom of research.
- Noncommercialization of reproduction.

In making decisions about using embryos in research or ART, most commissions adopted a long-term vision. This means that recommendations should be drafted in general terms and allow for flexibility and adaptability in the face of future developments. For example, the *Warnock Report* adopted the following recommendation strategy:

- Frame recommendations in general terms, leaving matters of detail to be worked out by government.
- Indicate what should be matters of good practice.
- Indicate what recommendations, if accepted, would require legislation.
- Apply any proposed changes equally throughout the United Kingdom.[31]

Treatment of Moral Status Arguments

The central finding from public consultation about embryo research is that there is a great diversity of opinion on the acceptability of that research. The diversity of opinion reflects a lack of consensus on the moral status of the embryo. This lack of consensus is acknowledged openly in most reports. For example, the Canadian Royal Commission states:

> Canadians have differing views on the moral status of the zygote and embryo. Although there is strong agreement on a commitment to the principle of respect for human life, Canadians differ about what form that respect should take and what level of protection is owed to human life at its different stages of development. There is also a wide range of answers to these questions in the history of moral philosophy.[32]

The EGE states that, despite the diversity of views on the moral status of the human embryo among its member states, one can find two conflicting tendencies emerging with respect to the moral status of the embryo and the legal protection that should be afforded the embryo with respect to scientific research. These two positions are:

1. Human embryos have the same moral status as human persons and consequently are worthy of equal protection.
2. Human embryos do not have the same moral status as human persons and consequently have a relative worth of protection.[33]

As the European Commission Working Group on human embryos and research stated in its 1992–1993 report, "these views are fundamentally different and it is difficult to see how, at these extremes, the differences can be reconciled."[34]

There is another common formulation of the two positions which presents the conflicting positions as greater extremes. In this formulation, the moral argument against the use of human embryos in research is essentially the following: The human embryo has the same status as a child or an adult by virtue of its potential for human life; as it is unacceptable to make use of children and adults in research that could harm or kill them, it is also wrong to use human embryos in such research. The argument on the other side of the debate is essentially the following: It is only *human persons* that must be respected, and human embryos are not persons, or even potential persons, but simply a collection of cells; therefore, there is no reason to accord embryos any protected status.

No matter how the conflicting opinions are expressed, in the face of this fundamental disagreement the most common response has been to state that no definitive answer can yet be given to the question of when life begins or whether the human embryo is a "person" using scientific information.[35] The *Warnock Report* states,

> Although the questions of when life or personhood begin appear to be questions of fact susceptible of straightforward answers, we hold that the answers to such questions in fact are complex amalgams of factual and moral judgments.[36]

After highlighting the unsolvable nature of the problem, most reports adopted an intermediate position or compromise between the two positions. This pragmatic approach seeks to balance the scientific and medical cost of not pursuing embryo research with the moral cost of permitting such research. However, it is important to note that if the conflicting positions are expressed as they are by the EGE, the decision to permit embryo research is less a compromise and more a choice. Just as those countries which prohibit embryo research are choosing between the two positions in favor of the former position (embryos are full human persons, entitled to the same protections against harmful research), those countries that sanction any embryo research[37] are also making a choice between the two positions—rejecting the position that human embryos have the same status and rights of full human persons.

Where the opinions of a population can accurately be described as greater extremes (human persons versus collection of cells), the intermediate position is a true compromise. Such a dichotomy permits a commission to choose a compromise based on rejecting both extreme positions and stating that the embryo must have special status—less then full personhood and more than simply a mass of cells. The *Warnock Report* provides a good example of this reasoning.

Interestingly, the *Warnock Report* and others appeal to the legal status of the human embryo, which is less than that of a legal person in nearly every country in the Western world. This fact is used to bolster the argument that embryos are not moral persons, but without much explanation why the legal interpretation informs the question of moral status. It would seem that the widespread legal agreement that embryos ought not to be accorded the same rights as children and adults could be used as evidence of an international norm; however, this is not explicitly stated.

Once the question of definitively determining the moral status of the embryo has been abandoned, most national reports turn to identifying ways in which to respect the human embryo within the context of scientific research. For example, the *Warnock Report* states that the definitive determination of the status of the embryo is not open to resolution, lays out the conflicting positions of those opposed and in favor or embryo research, and states:

> Instead of trying to answer these questions [about the moral status of the embryo] directly we have therefore gone straight to the questions of *how it is right to treat the human embryo*. We have considered what status ought to be accorded to the human embryo, and the answer we give must necessarily be in terms of ethical or moral principles.[38]

Where embryo research is permitted, a common solution is to provide some protection to human embryos in the form of certain prohibitions and limits on research.

Limits to Embryo Research

Although there is no consensus about the moral status of the embryo, there is agreement that if embryo research is permissible, limitations are necessary and appropriate.[39] As such, limitations on research reflect a compromise between the acceptability and unacceptability of embryo research and are a means of allaying public anxiety. Many of the fears of abuse in embryo research are widely shared and have resulted in considerable consensus about what uses should be prohibited. There is less consensus, although some commonalties, about what limitations on embryo research are required to allay public concerns, promote beneficial research, and respect the connection between human embryos and the rest of the human community. These limits represent both acknowledgments that public fears are respected and are a "sign of respect for the special status of the embryo without the cost of an outright ban."[40]

The following restrictions exist in most countries that permit embryo research:

- ***Informed Consent of Gamete Donors***

  This condition reflects the principle of respect for individual autonomy, as well as a desire to protect vulnerable people, including patients undergoing infertility treatment who may be subject to emotional stresses. Of course, a formal requirement of consent may not be enough to protect female patients from feeling subtle coercion to consent to the removal of ova for research at the same time that ova are being removed for IVF. This is especially true where it is the physician treating the woman or couple who makes the request to remove additional ova for research. It is easy to imagine that patients may consent to the removal of extra ova for research in order to appear compliant and establish or maintain a positive doctor/patient relationship. For this reason, the Canadian Royal Commission makes the recommendation that:

> [a] woman's or couple's consent to donate zygotes generated but not used during infertility treatment for research never be a condition, explicit or implicit, of fertility treatment. Potential donors must be informed that refusal to consent does not jeopardize or affect their continuing treatment in any way.[41]

An additional recommendation worth consideration is the recommendation by the Royal Commission that no additional surgical procedure be permitted to retrieve eggs for the creation of embryos strictly for research.[42]

■ *Time Limits Within Which Research Must Be Conducted*

In keeping with the changing physical status of the embryo, many countries have stipulated that as embryonic development progresses, greater protections are required. A common line drawn is 14 days after fertilization, the point believed to represent the last opportunity for twinning to occur; the point in time beyond which the primitive streak (precursor to the central nervous system) begins to develop; and the point before sentience is attained. While recognizing that any line is, to some extent, arbitrary, this is a line which is adopted by most countries permitting embryo research.[43] Broad international adoption of the 14-day limit is another reason for adopting such a standard. The Canadian Royal Commission concedes that "it is appropriate to agree to a standard that enjoys broad international support, if only to ensure that research done [nationally] will be as respected as that done in the rest of the world."[44]

■ *Embryos Must Be Necessary for the Research*

Limits on the necessity and often the significance[45] of the research involving human embryos simply underline the need to ensure that protocols that use human embryos have scientific validity. The Australian *NHMRC Guidelines* suggest that where embryos are destroyed in research, the number used in such a protocol must be restricted.[46] In addition, the Canadian Royal Commission has indicated that "necessary" means that no other animal research or model is available or appropriate to conduct the experiment.[47] The Warnock Commission explicitly links this limitation to the special status and consequent protection of the human embryo: "the embryo of the human species ought to have a special status...no one should undertake research on human embryos the purposes of which could be achieved by the use of other animals or in some other way.[48]

■ *Protocol Review*

Australia, Canada, the United Kingdom, France, and the European Union all require review of the protocol by a local or national body, or both.[49]

■ *Regulatory Oversight*

In addition to protocol review, several countries have recommended the establishment of a regulatory board or national commission to license and regulate infertility treatments and embryo research. Although national oversight is desirable, the use of law to regulate (rather than set limits) in this area would be inappropriate, given the rapid development in uses of technologies. A national commission or authority coupled with subcommittees responsible for various areas of ART would provide flexibility and adaptability and relieve the need to campaign for removal of legislative bans and prohibitions as technologies and attitudes change.[50] In addition, national regulation ensures more consistent application of safeguards and can ensure greater public accountability and transparency.[51]

The need for national as well as local oversight of primordial stem cell research has recently been echoed by the Australian Academy of Science. No such system currently exists in the United States, with reliance placed on a system of review by local institutional review boards (IRBs). The ability of IRBs to adequately assess the merits and ethics of primordial stem cell protocols, given their time and resources, is surely limited. A national review mechanism that reviewed not only primordial stem cell research but also research protocols using human embryos would ensure strict adherence to guidelines and standards across the country.

■ *Use of Only Spare Embryos from IVF for Research*

The issue of whether to create embryos for research purposes or to limit the supply of embryos to surplus embryos from *in vitro* fertilization treatment (IVF) is one on which there is no consensus. Very few of the national commissions discuss the ethics of creating embryos for research in a detailed manner. For example, the opinion of the EGE states simply, "There is a debate on the distinction between research on donated spare embryos and on eggs donated for research and subsequently fertilised *in vitro*."[52]

The primary objection to creating embryos specifically for the purposes of research is based on the notion that there is a qualitative difference between creating an embryo that may have a chance of implantation and creating embryos without that chance. Objections about creating embryos for research often appeal to arguments about respecting human dignity by avoiding instrumental use of human embryos. Creation of embryos without the intention of implanting them is argued by some to be disrespectful. The other side of the argument does not accept the difference between creating embryos for the purposes of reproduction, in which case there is still the chance that they will not be implanted, and creating them for research.

The *Warnock Report* acknowledges that the committee was divided on this issue. Some committee members argued for and some against this qualitative difference. Those against argued that "if research on human embryos is to be permitted at all, it makes no difference whether these embryos happen to be available or were brought into existence for the sake of research....In both cases, the research would be subject to the limitations outlined above and the moral status of the embryo would be the same."[53] Despite the dissent of four of the committee members, the use of embryos created *in vitro* for research purposes was endorsed, and under the HFE Act, creation of research is permitted within the regulatory scheme of the Act.[54]

In Canada, where the Royal Commission endorsed the creation of embryos for research, the commission expressed the belief that the limitations on embryo research, such as the necessity of human embryos to conduct the research, the 14-day limit, and the importance of the research could ensure appropriate respect for the embryos. With regard to this point, the following statement is worth reproducing:

> On the one hand, we believe that [the creation of embryos for research purposes] would create the danger of promoting instrumentalization of zygotes, thereby potentially undermining commitment to respect for human life and dignity. On the other hand, it is not clear whether we can distinguish effectively between zygotes that become available because they are 'surplus' to the needs of couples undergoing IVF treatment and zygotes created specifically for research. Some commentators argue that the distinction is unworkable, since doctors can stimulate the maturation of more eggs than are needed for purposes of IVF by using fertility drugs. According to one submission to the Australian Senate Select Committee, 'any intelligent administrator of any IVF program can, by minor changes in his ordinary clinical way of going about things, change the number of embryos that are fertilized. So in practice there would be no purpose at all in enshrining in legislation a difference between surplus and specially created embryos.'[55]

The Canadian Royal Commission endorsed creation of embryos for research purposes if conducted within the strict regulatory scheme proposed and subject to the commission's proposed restrictions, such as those on use and time limits.[56]

When deciding how to deal with this issue, there are a number of points to keep in mind. First, most of the reports acknowledge that the creation of embryos provides the only way to conduct certain research, such as research on the process of human fertilization. Second, as techniques for IVF improve, it is possible that the need to create surplus embryos will be eliminated; one of the frequently approved uses of embryo research is the improvement of IVF techniques.[57] At the same time that legislation permits embryo research, it is advocating that research improve IVF techniques and that fertility experts attempt to reduce the surplus of embryos created for infertility treatment.[58] As this happens, and if embryo research is dependent on the existence of spare embryos donated with informed consent, it is possible that the supply of embryos for research will dwindle.

If the research supply is limited to surplus embryos from IVF, two other results are possible. First, in light of the tremendous interest in hES cell research, there is little doubt that demand for embryos and ova for research will increase.[59] Increased demand will only augment incentives for fertility clinics and physicians to ensure a supply of ova is available. This is particularly true where physicians and clinics are also engaged in embryo research themselves, and also if ova and embryos can be bought and sold between clinics and research institutions. Demand for embryos could translate into pressure on women undergoing IVF to donate ova or embryos specifically for research when having ova removed for IVF.[60] Alternatively, clinicians might simply create more embryos than are necessary for treatment purposes to ensure a surplus.

Second, if the IVF supply ever were to dwindle, and creation of embryos for research purposes had been explicitly prohibited, it would be necessary to revisit the issue of creating embryos for research, which could prove difficult both logistically and politically. Finally, since much hES cell research is aimed at the creation of tissues and organs to replace damaged or diseased tissues, it is likely that autologous tissue transplants will be the desired procedure to reduce risk of rejection in transplantation. This treatment would require the creation of embryos using the patients' genetic material. If this procedure is perfected, a prohibition on the creation of embryos could eliminate the possibility of autologous organ or tissue transplantation.

In light of the foregoing, NBAC ought to consider endorsing the use of spare embryos where possible and to permit the creation of embryos for research when research depends on that creation to achieve its objectives, or in situations in which access to spare embryos is not possible. This does not fully address the concerns over possible coercion of infertility patients; these concerns require greater analysis and specific guidelines or recommendations.

Appropriate Uses of Human Embryos in Research

The search for appropriate limits in embryo research regulation can also be seen in the regulation of the scientific ends to which the research must be directed in order to be acceptable for funding or licensing. Upon examining international polices, it becomes clear that how a country determines the uses for which embryo research may be approved is a crucial issue when determining the implications for embryonic stem cell research.

Therapeutic and Nontherapeutic Research

Confusion similar to that over the definition of "embryo" exists with respect to the definition of "research." Again, many countries simply do not define the term. A few countries draw a distinction between "nontherapeutic" and "therapeutic" research on embryos.[61] For example, the Australian *NHMRC Guidelines* define "therapeutic research" as "research aimed at benefiting the well-being of the embryo" and "nontherapeutic research" as research "not intended to benefit the embryo and which may or may not be destructive."[62]

This distinction results in part from the fact that in the field of ART there is considerable overlap between clinical practice and research. For example, research on new techniques for cryopreservation and fertilization has been used in clinical practice for years. It is difficult to draw a clear line between innovative clinical practice and research since much of this area is based on technologies which are new or developing. Both the Canadian Royal Commission and the Australian NHMRC note this overlap and recommend that in ART, innovative or experimental therapies fall under the rubric of research in order to be regulated.[63]

The distinction between human embryo research which is therapeutic and that which is nontherapeutic is particularly unhelpful and should be avoided in the context of ART. The EGE and the Canadian Royal Commission have suggested that this distinction is not only unhelpful but may even be unethical. As the EGE suggests, some countries where the distinction is used

> only allow an IVF embryo to be used for research if the research is intended for the benefit of that particular embryo, and if the embryo is subsequently replaced in the uterus. If there existed the possibility that procedures might damage the embryo, this would amount to experimentation on the fetus or the baby and mother, and would be clearly unethical.[64]

The Canadian Royal Commission echoes this view:

> [T]he only way to develop therapeutic embryo research is to allow for some nontherapeutic embryo research. Allowing therapeutic research while at the same time prohibiting nontherapeutic research would not be workable, nor would it be ethical, because of the risks it would create for women and children.[65]

Drawing distinctions between appropriate and inappropriate uses of human embryos, such as the distinction between therapeutic and nontherapeutic research, is emblematic of the ambivalence about the permissibility of human embryo research.

Therapy Unrelated to Human Reproduction: Primordial Stem Cells

As most policies or laws regulating embryo research are directed at regulating ART, the closer the relationship to human infertility and reproduction, the more acceptable the research is likely to be regarded. Many countries sanction embryo research aimed at:

- Improvement of infertility treatments.
- Development of contraceptive technologies.
- Improvement of detection of genetic/chromosomal anomalies in embryos before implantation.
- Advancement of knowledge about congenital diseases, causes of infertility, and human development.

Conversely, the more attenuated the relationship to human infertility, the more controversial the research. So, for example, where research is aimed at therapeutic approaches to disease or tissue damage, many laws or policies make no provision for these uses, particularly as most policies or acts are specifically directed at reproductive technologies. This lacuna is also a function of recent scientific developments; possibilities like those presented by primordial stem cell research were not envisaged when most of the acts were drafted.

The British HFE Act, arguably the most liberal act, makes no explicit provision for research of this sort. However the British provided a mechanism to add research not currently available for licensing through amendment of the regulations to the Act. Consequently, the Human Genetics Advisory Commission/Human Fertilisation and Embryology Authority (HGAC/HFEA) statement of December 1998 states:

> [W]hen the 1990 HFE Act was passed, the beneficial therapeutic consequences that could potentially result from human embryo research were not envisaged. We therefore recommend that the Secretary of State should consider specifying in regulations two further purposes to be added to the list [of approved purposes], being:
>
> - developing methods of therapy for mitochondrial diseases
> - developing methods of therapy for diseased or damaged tissues or organs.[66]

In addition, the HGAC/HFEA specifically recommends permitting research of this sort, advising that it would be unwise to rule out absolutely research using cell nucleus replacement involving embryos "that might prove of therapeutic value."[67]

The Australian Academy of Science (the Academy) issued a position statement, *On Human Cloning*, on February 4, 1999. This statement is aimed at distinguishing between "*reproductive cloning* to produce a human fetus" and "*therapeutic cloning* to produce human stem cells, tissues and organs."[68] The Academy, which speaks for the Australian scientific community, states that reproductive cloning of humans is unethical and should be prohibited, but that it must be distinguished from therapeutic cloning which holds the potential of "great benefit to mankind."

> For Australia to participate fully and capture benefits from recent progress in cloning research, it is necessary to review the [*NHMRC Guidelines*] and repeal restrictive legislation in some States. This could be done in the context of establishing a national regulatory arrangement, taking into account recent advances in biomedical research and advocated best practice elsewhere. Human cells, whether derived from cloning techniques, from ES cell lines or from primordial germ cells should not be precluded from use in approved research activities in cellular and developmental biology.[69]

The Academy goes on to recommend that, "if Australia is to capitalize on its strength in medical research, it is important that research on human therapeutic cloning is not inhibited by withholding federal funds or prevented by unduly restrictive legislation in some States."[70] The recommendations also suggest that primordial stem cell research should be subject to a two-tiered regulatory approval process, passing first before a local ethics committee and then requiring national approval.

The World Health Organization (WHO) *Draft Bioethics Guidelines*, 1999, also assert that the use of cloning techniques for nonreproductive means should not be foreclosed:

> As recognized by the World Health Organization, nonreproductive, *in vitro* cloning research, with the clinical objective of repairing damaged tissues and organs has important potential benefits. Relevant animal research would be acceptable provided it was carried out in accordance with the CIOMS ethical guidelines on the use of animals in biomedical research. Guidelines addressing the possible involvement of human gametes or embryos must be developed.[71]

Like the Australian guidelines, the German *Embryo Protection Act*, 1991, also protects human embryos from harmful research. However, unlike the Australians, in discussing the use of hES cells, the German government came to the conclusion that there was no need to relax the strict embryo protection laws to permit hES research, since hEG cell research is permitted under laws relating to the use of fetal tissue.[72] This is, of course, an option open to NBAC; however, whether significant research differences exist between hES and hEG cells is not currently known.

Perhaps the most interesting statements directly on the use of primordial stem cells are those that issue from the French statement on bioethics. The French have banned nontherapeutic human embryo research, which effectively bans all research. Since destruction of human embryos is not possible, creation of embryonic stem cell lines is also not possible. The French National Commission says the following:

> We are approaching a paradoxical situation as a result of legislation....experimentation or therapeutic research on [stem cells] from embryos *in vitro* are banned, but it is possible to import cells from collections established without any observance of specific ethical law applicable in France to embryonic cells.

The French commission has suggested that, taking into account the prospects for therapeutic research, the ban may be modified this year when the existing law is up for review to permit hES cell research.

A similar paradox exists in the United States. In this country, there is a ban on federal funding for research that would destroy an embryo, which therefore bans funding for the creation of hES cell lines, but permits the use of hES cell lines created in the private sector or overseas without reference to national protections and oversight. NBAC should take steps toward eliminating this paradoxical situation and outline a consistent set of protections with national application. There is room for leadership on this issue both here and around the world—other counties will be watching the response of NBAC on this issue.[73]

Prohibitions on Human Embryo Research

One of the most important facts that can be gleaned from an examination of international embryo research polices is that near unanimity exists with respect to practices that should be prohibited as unethical. The following practices are widely regarded as unacceptable, and many are deemed to be offensive to human dignity:

Cloning for Reproduction
All the countries under examination have prohibited the use of cloning techniques for the purposes of human reproduction either in law or in policy recommendation.[74]

Creation of Hybrids/Chimeras (also described as cross-species fertilization)
There is ambiguity over whether the prohibition on creating hybrids and chimeras refers to the creation of individuals or embryos. For a number of years, hamster ova have been fertilized with human sperm as a test for human sperm motility. It can be argued, therefore, that it is the creation of hybrid or chimeric *individuals* which is the prohibited practice.[75] However, some countries explicitly exempt the fertilization of hamster ova from the prohibition, which would indicate that creation of all other hybrid embryos is prohibited.[76] This issue remains unclear, but given the conservative legislation in many countries, it is arguable that any creation of hybrid embryos would be considered unethical and therefore would be prohibited in many countries.

Cross-Species Implantation; Germline Interventions; and Sex-Selection for Other Than Prevention of Hereditary Disease
All three of these prohibitions are widely adopted in the countries under examination and in international organization statements.[77]

Transfer of Embryos Used in Research into a Woman
This practice is clearly unacceptable, as it would amount to conducting research on women and any resulting children.

Commercialization of Embryos/Gametes
Most countries also abhor the commercialization of embryos and fetal tissue: This has lead to prohibitions on sales of ova, sperm, embryos (both nationally and internationally)[78] and fetal tissue[79] and to recommendations

that research on embryos not be conducted for commercial gain. In fact, the WHO *Draft Guidelines* suggest that countries that have not already done so should take steps to regulate the patenting of genetic materials and life forms, in keeping with the stated guiding principles:

> Patents are designed to protect intellectual property and stimulate innovation and they are part of the product development process. The private sector, however, also has public responsibilities. A balance must be sought between the need for patent protection and the obligation to ensure society's access to the health benefits of new knowledge and technology.

With respect to the use of primordial stem cells which are already subject to patent protection, the issue is particularly complex. The French National Ethics Committee Opinion on embryonic stem cells notes that prohibitions exist on the sales and patentability of embryonic and fetal cell collections, but not explicitly on embryonic stem cell collections. Given the strong endorsement of the principle of noncommercialization of genetic and reproductive materials, it is likely that many countries will extend these prohibitions to embryonic stem cells. However, in the United States where private research is often funded by pharmaceutical companies lured by lucrative patent rights, this provides a complex and difficult area of regulation. The British HGAC/HFEA expresses this tension:

> A significant number of respondents expressed fears and reservations about the possible commercialisation of therapeutic uses of [cloning] techniques....There is an understandable desire on the part of the public that curative process should not simply be exploited as sources of financial gain for their developers, but that there should be respect for the public good and corresponding access to these techniques for those who would benefit from them. A balance has to be struck between affording a reasonable recompense to those who have exercised initiative...and ensuring that the needs of the sick are properly met.[80]

While the issue of recommending specific changes to the United States patenting system is beyond the scope of NBAC's current mandate with respect to primordial stem cells, the issue of patenting of human body materials needs to be revisited and modified in the United States.

Use of Fetal Eggs and Eggs from Female Cadavers

This prohibition is discussed below in the context of prohibitions on the use of fetal tissue for research.

Use of Fetal Tissue in Research

The use of fetal tissue to isolate hEG cells is less problematic than the similar use of human embryos for three reasons. First, the removal of the fetal germ cells does not occasion the destruction of a live fetus. Second, there is no question of creating fetal tissue specifically for research. Third, the use of fetal tissue to develop therapies for people unrelated to reproduction has been raised before in the context of fetal tissue transplantation, and therefore a number of laws and policies exist regarding this use.[81]

Regulatory Background

The fact that fetal tissue can be derived only from aborted fetuses means that the ethical dilemma that marks the debate on the permissibility of using fetal tissue in therapy is the issue of complicity with the abortion. Due to the ferocity of the abortion debate in the United States, legal restrictions were enacted that blocked the use of fetal tissue in federally funded research on transplantation therapy. Accordingly, the only permissible source of tissue for such research was tissue from spontaneously aborted fetuses or ectopic pregnancies. As little of that tissue proved suitable for such research, the ban on using other aborted fetal tissue effectively brought that research to a halt.

In 1993, President Clinton lifted the ban on the use of fetal tissue from elective abortions for federally funded fetal tissue transplantation research. Consequently, there are no federal legal prohibitions that would inhibit the use of that tissue for hEG cell research.[82] In addition, there is considerable agreement in the international community that the use of fetal tissue in therapy for people with diseases, such as Parkinson's disease, is acceptable.

Policies that address the use of fetal tissue for therapy indicate that consensus exists with respect to the following:

Guiding principles

- Respect for human life.
- Respect for the woman's dignity and integrity.

Limitations

- Decision to terminate pregnancy made before donation of fetal tissue discussed.

  In order to avoid a decision to terminate pregnancy being influenced by potentially beneficial consequences that could result from donating fetal tissue for transplantation, the decision to terminate must be made without reference to possible donation of fetal tissue. Clearly, with respect to hEG cell research, the same restriction should be in place. A woman's decision to terminate her pregnancy should be neither induced nor coerced by the possibility that the resulting fetal tissue can be used for research or therapy that will benefit others.

- Informed consent.
- Establishment of a regulatory and licensing scheme.[83]

  The establishment of a regulatory scheme that licenses the use of both fetal and embryonic tissue in research has been suggested with respect to primordial stem cell research.

Prohibitions

- Donation of the tissue to a designated recipient.

  The prohibition on directed donation reflects a fear that women will get pregnant and seek abortions with the aim of donating the fetal tissue to a loved one or relative in need of fetal tissue for therapy. Although this largely misunderstands the motivations of women making choices to terminate pregnancies, it is not outside the realm of possibility that some situations of this type could arise. Consequently, to avoid this scenario, many countries have foreclosed the possibility by removing the woman's ability to designate that a particular person receive the tissue donated for therapy.

  The World Medical Association's *Fetal Tissue Transplantation Statement* nicely sums up the ethical justifications for the above limits and prohibitions:

  > Prominent among the currently identified ethical concerns is the potential for fetal transplants to influence a woman's decision to have an abortion. These concerns are based, at least in part, on the possibility that some women may wish to become pregnant for the sole purpose of aborting the fetus and either donating the tissue to a relative or selling the tissue for financial gain. Others suggest that a woman who is ambivalent about a decision to have an abortion might be swayed by arguments about the good that could be achieved if she opts to terminate

the pregnancy. These concerns demand the prohibition of: (a) the donation of fetal tissue to designate a recipient; (b) the sale of such tissue; and (c) the request for consent to use the tissue for transplantation before a final decision regarding abortion has been made.[84]

- Commercialization of fetal tissue.

Most countries explicitly prohibit the commercialization of human fetal tissue. The Canadian Royal Commission states that the noncommercialization of reproduction is one of its guiding principles and recommends that no for-profit trade be permitted in fetal tissue and that the "prohibition on commercial exchange of fetuses and fetal tissue extend to tissues imported from other countries."[85] This prohibition is in place to prevent the exploitation of poor women, especially in developing countries, who might be persuaded to begin and end pregnancies for money.[86] With respect to patenting, the Royal Commission states:

> Commissioners believe strongly that fetuses should never be an appropriate subject for patents. However, if they are intended to benefit human health and if the safeguards we have recommended for obtaining and using fetal tissue are in place, innovative products and processes using fetal tissue as a source may warrant some limited form of patent protection.[87]

This limited patent protection should be considered with respect to primordial stem cells and other living tissue patents, especially patents involving reproductive tissue.

- Use of fetal eggs.

Both Australia and Canada have prohibited the use of fetal eggs for the creation of embryos.[88] The Canadian Royal Commission states:

> We would object strongly to fertilisation of eggs obtained from female fetuses, even if it becomes technically feasible to retrieve and mature them. We find this suggestion deeply offensive to all notions of human dignity and have recommended that it be among the activities prohibited outright in the *Criminal Code* of Canada.[89]

By contrast the United Kingdom HFE Authority has stated that it is acceptable to use fetal eggs for the creation of embryos or in embryo research.[90] The use of fetal eggs provides an unlimited supply of immature eggs, and, of course, embryos created from fetal eggs could also provide an unlimited source of hES cells.

It is likely that the use of fetal eggs would be unacceptable to the majority of Americans including not only those who oppose the use of fetal eggs, but also those opposed to the creation of embryos for research, and, of course, those who oppose embryo research at all. If adequate safeguards are in place for the creation of embryos for research, there are good policy reasons to argue that the use of fetal eggs should be prohibited.

Conclusion

Primordial stem cell research offers the potential for life-saving technology. Restricting research, therefore, on these cells has both scientific and moral costs; such restrictions must not be imposed in ignorance of the costs involved. Deriving hES cells means sanctioning the destruction of human embryos for that purpose. This action also has moral costs. With respect to the derivation of hEG cells, the use of human fetal tissue is less controversial. Although human fetal tissue is obtained from aborted fetuses, the process of deriving the hEG cells is not itself implicated in the death of the fetus and is legally sanctioned.

The possibilities presented by primordial stem cell research serve to illustrate the great need for comprehensive and thoughtful regulation of ART in the United States. Although NBAC has not been asked to undertake that task, it is clear that the recommendations of NBAC with respect to primordial stem cell research must be designed with reference to the regulation of human embryo and fetal tissue research. Consequently, the recommendations made by NBAC will also lay the groundwork for regulation of embryo research and fetal tissue research in the United States. This is a great responsibility.

Commonalities in international policies on human embryo and fetal tissue research clearly exist. Countries with different religions and with diverse social and cultural backgrounds share views on the principles and strategies which should guide regulation of this research. Much can be gained from following the lead of those countries that have examined these issues with in-depth public and scientific consultation.

Those responsible for developing policy in this area need to address the rapidly changing techniques in genetics and ART. The *WHO Draft Guidelines* state:

> Hurried and premature legislation in the rapidly-evolving field of genetics can be counterproductive. Legislation and guidelines should be based on full and sound scientific and ethical assessment of the techniques concerned. They should be general enough to accommodate new developments, and they should be reviewed periodically.[91]

It is imperative to provide mechanisms for accommodating change within the regulatory structure and to anticipate the wider application of human embryo research by looking at the state of relevant animal research. The need to anticipate changes within the near future is crucial; the goal is to build a framework that anticipates rather than reacts.

Notes

1 See *Nature* 391, 1998, at 325, James A. Thomson et al., "Embryonic Stem Cell Lines Derived from Human Blastocysts" *Science*, 282, 6 November 1998, pp. 1145–1147, and Michael J. Shamblott et al., "Derivation of Pluripotent Stem Cells from Cultured Human Primordial Germ Cells," *95 Proc. Nat'l. Acad. Sci. USA* 13726, November 1998.

2 It is anticipated that the benefits of primordial stem cell research will be aimed at the replacement of damaged or diseased tissues, such as is currently the goal of skin grafting and organ transplantation.

3 In the United States see *DHHS Regulations for the Protections of Human Subjects*, 45 CFR 46 § 46.210 and *NIH Reauthorization Act* (1993) Ss. 111, 112 amending *Public Health Service Act*, 42 USC 289 et seq. For international regulations see, for example: United Kingdom, Committee to Review the Guidance on the Research Uses of Fetuses and Fetal Material. *Report (The Polkinghorne Report)* London: HMSO, 1989 (hereinafter *Polkinghorne Report*); World Medical Association, *Statement on Fetal Tissue Transplantation*, Adopted by the 415 World Medical Assembly, Hong Kong, September, 1989 (hereinafter *WMA Fetal Tissue Transplantation Statement*); Medical Research Council of Canada, Natural Sciences and Engineering Research Council of Canada, Social Sciences and Humanities Research Council of Canada, *Tri-Council Policy Statement: Ethical Conduct for Research Involving Humans*, August 1998 (hereinafter the *Tri-Council Policy Statement*) at 9.4; *Proceed with Care: Final Report of the Royal Commission on New Reproductive Technologies*, Vols. 1 and 2, Minister of Government Services, Canada, 1993 (hereinafter *Proceed with Care*) Vol. 2 at 967–1015; Australia, National Health and Medical Research Council *Statement on Human Experimentation and Supplementary Notes, 1992, Supplementary Note 5: The Human Fetus and the Use of Human Fetal Material*; and France, *Opinion No. 53* (see note 11).

4 See infra, at 32–36.

5 *Departments of Labor, Health and Human Services, and Education, and Related Agencies in the Omnibus Consolidated and Emergency Supplemental Appropriations Act*, Fiscal Year 1999, Public Law 105-277, S.511.

6 See "Embryonic Stem-Cell Research Exempt from Ban, NIH is Told," *Nature* 397, 21 January 1999, at 185.

7 I do not entertain the option of permitting public funding of embryo research on existing hES cell collections but maintaining that embryo research is unsuitable for public funding. That position is not only disingenuous but ethically untenable. See the French statement on the paradox of forbidding research on embryos yet permitting human embryo research. (See note 11, *Opinion No. 53*.)

8 *Proceed with Care; Tri-Council Policy Statement*. Legislation was drafted based on the Royal Commission's Report; however, it died on order paper in April 1997. Although there has been no comprehensive legislation regulating ART passed in Canada to date, at the time of press, a working group had been formed to begin the drafting process once more.

9 The National Health and Medical Research Council, *Ethical Guidelines on Assisted Reproductive Technology*; 1996 (hereinafter *NHMRC Guidelines*) Supplement to *NHMRC Statement on Human Experimentation*, 1992; The Australian Academy of Science, *On Cloning Humans: A Position Statement*; 4 February 1999.

10 Report of the Committee of Inquiry into Human Fertilisation and Embryology, HMSO London, 1984 (hereinafter the *Warnock Report)*; *Human Fertilisation and Embryology Act 1990* (hereinafter HFE Act).

11 Lois 94-653 and 94-654 29 July 1994 (hereinafter Loi 94-654) French National Consultative Ethics Committee for Health and Life Sciences, *Opinion No. 53*, "Opinion on the Establishment of Collections of Human Embryo Cells and Their Use for Therapeutic or Scientific Purposes," 11 March 1997 (hereinafter, *Opinion No. 53*).

12 European Commission, European Group on Ethics in Science and New Technologies, *Opinion No. 12, Ethical Aspects of Research Involving the Use of Human Embryo in the Context of the 5th Framework Programme*, 23 November, 1998 (hereinafter *EGE Opinion*).

13 The *Warnock Report*.

14 *Proceed with Care*.

15 *NHMRC Guidelines*. Although these guidelines are supplementary to the Australian *Guidelines on Research Involving Human Subjects,* they stand alone as guidelines with respect to ART.

16 HFE Act.

17 *Reproductive Medicine Law*, Federal Law of 1992 (Serial No. 275).

18 Law No. 503, 24 June 1992 amending Law No. 353 3 June 1987, Order No. 650, 22 July 1992.

19 Spanish Law 35, *Health: Assisted Reproduction Techniques* 24 November 1988 (NUM.282).

20 Swedish *In Vitro* Fertilisation Law, 1988.

21 Loi 94–654.

22 *Tri-Council Policy Statement*.

23 *NHMRC Guidelines*, see note 15 above.

24 German *Embryo Protection Act*, 1991.

25 For example, the *Warnock Report* required approximately two years to complete, and the Canadian Royal Commission on New Reproductive Technologies took four years from Order in Council until submission of its final report, *Proceed with Care,* from 1989 to 1993.

26 *Cf.* The German *Embryo Protection Law*, 1991.

27 For example, the *Warnock Report* led to the HFE Act in the United Kingdom. Similarly, the Committee to Consider the Social, Ethical and Legal Issues Arising from *In Vitro* Fertilisation (Report 3: Report on the Disposition of Embryos Produced by *In Vitro* Fertilisation) (hereinafter the *Waller Report*) in Victoria, Australia, was followed by the *Infertility (Medical Procedures) Act,* 1984 (replaced by the 1995 Act).

28 Victorian *Infertility Treatment Act*, 1995, at S.3 (1).

29 HFE Act, S. 1(1)(a)(b).

30 *Proceed with Care*, Vol. 1, at 581.

31 The *Warnock Report*, at 6–7.

32 *Proceed with Care*, at 631.

33 *EGE Opinion*, at para 1.13.

34 *EGE Opinion*, at para 1.25.

35 The Canadian Royal Commission states, "Commissioners recognize that no amount of deliberation on our part will definitively answer the question of the moral stature of the embryo. Philosophers and theologians have grappled with the issue for centuries" *Proceed with Care*, at 632. The *EGE Opinion* states, "The legislations of the EU Member States differ considerably from one another regarding the question of when life begins and about the definitions of 'personhood.' As a result, no consensual definition, neither scientifically nor legally, of when life begins exists" Para. 1.13.

36 The *Warnock Report*, at 60.

37 Here I am not referring to those countries in which only "therapeutic research" is permitted, as this effectively prohibits true embryo research. See the discussion below.

38 The *Warnock Report*, at 60.

39 The *EGE Opinion* states that the prerequisite for European funding of embryo research must be the respect of strict legal and ethical principles. In addition it states that when discussing and signing the Council of Europe *Convention on Human Rights and Biomedicine,* many European countries "failed to reach a consensus concerning the definition of 'person' or the admissibility of embryo research. It accepted however the principle of research on embryos *in vitro*, and moreover, provided that if national laws permit research on human embryos, they shall ensure that such laws provide 'adequate protection of the embryo' (Art. 18, 1)" *EGE Opinion*, at para 1.14.

40 Lori B. Andrews and Nanette Elster, *Cross Cultural Analysis of Policies Regarding Human Embryo Research* (Background paper for National Institutes of Health Final Report of the Human Embryo Research Panel, 27 September 1994) (unpublished), at 11 quoting John Robertson.

41 *Proceed with Care*, at 640.

42 Recommendation 188: Zygotes be created for research purposes only if gametes for this purpose are available without conducting any additional invasive procedures. *Proceed with Care*, at 640.

43 See *Warnock Report*, at 65 explaining that although no one single identifiable line exists, a time limit on keeping embryos alive must be imposed to allay public anxiety. Perhaps the best evaluation of the 14-day limit is contained in the Canadian Royal Commission *Proceed with Care*, at 632–637.

44 *Proceed with Care*, at 636 citing Law Reform Commission of Canada. *Biomedical Experimentation Involving Human Subjects*. Working Paper 61. Ottawa: Minister of Supply and Services, Canada, 1989.

45 *NHMRC Guidelines* mandate that approval of research involving destruction of embryos requires "a likelihood of significant advance in knowledge or improvement in technologies for treatment as a result of the proposed research" Art. 6.4.

46 Ibid.

47 *Proceed with Care*, at 630.

48 *Warnock Report*, at 63.

49 Spain, Finland, Sweden, and the United Kingdom all require approval by a national authority. Canada, the United Kingdom, Australia, Denmark, France, and the European Union require national or local ethics committee approval. See *EGE Opinion*, at para 1.20.

50 This is the recommendation of the Canadian Royal Commission and the system used by the United Kingdom Human Fertilisation and Embryology Authority.

51 *EGE Opinion* at para. 2.11. See also the recommendation of the Australian Health Ethics Committee advocating complementary ART legislation which reflects the national guidelines be enacted in the Australian States and Territories without ART legislation. *NHMRC Guidelines*, at v.

52 *EGE Opinion*, at para 1.29.

53 The *Warnock Report*, at 66–69.

54 The *Warnock Report*, at 67–68. HFE Act Schedule 2 S.3.

55 *Proceed with Care*, at 636, citing Australia, Senate Select Committee on the Human Embryo Experimentation Bill 1985. Human Embryo Experimentation In Australia, Canberra: Australian Government Publishing Service, 1986 para. 3.31.

56 *Proceed with Care*, at 638.

57 The *Warnock Report* echoes this thought: "A further argument for the generation of embryos for research is that as the techniques of freezing become more successful there would be fewer spare embryos available for research," at 68.

58 *NHMRC Guidelines*, Art. 6.5.

59 The British HGAC advises that given the primordial stem cell research conducted in the United States, it is likely that there will be an increase in the number of research projects involving cell nucleus replacement in human oocytes and that such research would require a source of oocytes donated for research which at present are not widely available. At para. 5.5.

60 The Canadian Royal Commission recognizes this by saying "Doing research on the zygotes could put women enrolled in IVF programs under pressure to consent to donate unused eggs or zygotes. This pressure could be particularly acute if the creation of zygotes for research purposes were prohibited" *Proceed with Care,* at 639.

61 See the laws of France, Victorian *Infertility Treatment Act*, 1995 at S.3 (1) and Australian *NHMRC Guidelines*.

62 *NHMRC Guidelines*, at vi.

63 *NHMRC Guidelines*, at iv; *Proceed with Care*, at 614.

64 *EGE Opinion*, at para 1.3.

65 *Proceed with Care*, at 614.

66 *HGAC/HFEA Statement*, at para. 9.3.

67 *HGAC/HFEA Statement*, at para. 5.4.

68 Australian Academy of Science, *On Human Cloning: A Position Statement*, 4 February, 1999 (hereinafter *On Human Cloning*) at 4.

69 *On Human Cloning*, at 5.

70 Ibid.

71 World Health Organization *Draft Bioethics Guidelines 1999* Art. 28 (hereinafter *WHO Draft Guidelines*).

72 DFG *Statement Concerning the Question of Human Embryonic Stem Cells*, March 1999.

73 *EGE Opinion*, at para 2.12.

74 For a comprehensive account of current legal and policy statements prohibiting the cloning of human beings for reproduction, see Elisa Eiseman, *Cloning Human Beings: Recent Scientific and Policy Developments*, RAND, Science and Technology Products, 1999 (unpublished).

75 This is explicit in Danish law. See *Reproductive Medicine and Embryological Research: A European Handbook of Bioethical Legislation*, C. MacKellar ed. (Edinburgh: European Bioethical Research) 1997, at 6.

76 The *Warnock Report* expressly exempts the use of cross-species fertilization for the purposes of alleviating infertility or assessment of subfertility from the prohibition on creation of hybrids, at 71. A similar exception is written into Spanish law.

77 See Council of Europe, *Convention on Human Rights and Biomedicine* (1997): "An intervention seeking to modify the human genome may only be undertaken for preventative, diagnostic or therapeutic purposes and only if its aim is not to introduce any modification in the genome of any descendants" Article 13; "The use of techniques of medically assisted procreation shall not be allowed for the purpose of choosing a futures child's sex, except where serious hereditary sex-related disease is to be avoided" Art. 14 (hereinafter the *Convention on Human rights and Biomedicine*); *WHO Draft Guidelines*, "Sex is Not a Disease. Except for Severe Sex-Linked Genetic Disorders, the Use of Genetic Services for the Purpose of Sex-Selection is Not Acceptable."

78 *NHMRC Guidelines*, Art. 11.9 "Commercial trading in gametes or embryos (is prohibited)" and S. 11.10; *Convention on Human Rights and Biomedicine*, "Paying donors of gametes or embryos beyond reasonable expenses (is prohibited)." Art. 21. France, Loi 96-327, 16 April 1996, extending the principles of noncommercialization and nonpatentability to human embryo cell collections. In Canada, the noncommercialization of reproduction is a guiding principle to the Royal Commission's report, "As we have emphasized throughout this report, it is inappropriate for decisions involving human reproduction to be motivated by the prospect of financial gain" *Proceed with Care*, at 643. In Italy, current regulation prohibits embryo research for commercial purposes.

79 *WMA Fetal Tissue Transplantation Statement*.

80 *HGAC/HFEA Statement*, at para. 5.11.

81 See note 3.

82 There may be state restrictions that inhibit this research, although most states permit other research involving fetal tissue. Use of fetal tissue in research is also permitted in Canada, the United Kingdom, Australia, and in most countries in the European Union. Germany, for example, permits no embryo research but does permit the use of fetal tissue for the derivation of germ cells. The DFG statement concerning human embryonic stem cells upholds the ban on destructive embryo research effectively banning the derivation of hES cells because the option of deriving hEG cells exists in that country. See DFG *Statement Concerning Question of Human Embryonic Stem Cells*, March 1999, at 8–10.

83 See for example *Tri-Council Policy Statement* at 9.4 and *NHMRC Guidelines*.

84 *WMA Statement on Fetal Tissue Transplantation*.

85 *Proceed with Care*, at 1003.

86 *Proceed with Care*, at 1001.

87 *Proceed with Care*, at 1003.

88 *NHMRC Guidelines*, Art. 11.4.

89 *Proceed with Care*, at 594.

90 See Human Fertilisation and Embryology Authority, *Donated Ovarian Tissue in Embryo Research and Assisted Conception*, Public Consultation Document and Final Report, (London: 1994).

91 *WHO Draft Guidelines*, at Art. 6.

What Has the President Asked of NBAC? On the Ethics and Politics of Embryonic Stem Cell Research

Commissioned Paper
Erik Parens
The Hastings Center

Introduction

In November of 1998, President Clinton wrote to Dr. Harold Shapiro, Chair of the National Bioethics Advisory Commission (NBAC), requesting that NBAC turn its attention to emerging issues in embryonic stem (ES) cell research. In that letter, the President raises the general question concerning ES cell research in the current context of a congressional ban on embryo research. He had, of course, made his interest in embryo research generally known several years earlier, in his response to the 1994 Human Embryo Research Panel's (HERP's) report. In that response the President said that while he could endorse research on embryos originally created by means of *in vitro* fertilization (IVF) for the purpose of reproduction, he could *not* endorse using IVF to create embryos for research. In early 1998, however, he indicated that he *could* endorse using somatic cell nuclear transfer (SCNT) to create embryos for research. Indeed, in February of 1998, the administration announced that it could support a bill to prohibit using SCNT to produce children only if it met four conditions, the second of which was that the bill should "permit [SCNT] using human cells for the purpose of developing stem-cell…technology to prevent and treat serious and life-threatening diseases."[1]

Not only has the President made clear his interest in the questions concerning ES cell research in general and the creation of embryos for research in general, but the very first sentence of his November 1998 letter to Dr. Shapiro makes clear that he is also interested in the particular question concerning Advanced Cell Technology's (ACT's) work. He wrote, "This week's report of the creation of an embryonic stem cell that is part human and part cow raises the most serious of ethical, medical, and legal concerns." Despite the confusion about the fact that ACT claimed to have produced a hybrid *source* of embryonic stem cells (rather than hybrid stem cells), it is clear the President thinks that SCNT involving a human somatic cell and an enucleated bovine ovum raises important ethical questions that he wants addressed. In other words, the President's letter to Dr. Shapiro raises a complex and interrelated set of issues concerning ES cell research and the creation of embryos as sources of ES cells.

In this paper, I aim to delineate the issues raised by the President's request for a report from NBAC. By showing the interrelationships among those issues, I intend to show that an intellectually honest and adequate response to the President's request will acknowledge, if not fully address, each of them. First, how should policy makers view and talk about the relationship between ES cell research and embryo research? For example, is it reasonable to attempt to cordon off the public conversation about ES cell research from the public conversation about embryo research? Second, what is the current state of the policy conversation concerning embryo research? Specifically, what was HERP's argument for limited embryo research and how could it have been made more persuasively? Third, if in general it were acceptable to do limited research on embryos, then would the original *intention* of the maker of the embryo make a moral difference? For example, was it reasonable for the President to endorse research on "discarded" embryos but to oppose research on "created" embryos? Fourth, if there were agreement that under carefully circumscribed conditions it is acceptable to create embryos for the purpose of research, then would it make a moral difference which *means* are used to create them? For example, was it reasonable for the President to endorse using SCNT to create embryos for research but not to endorse using IVF for the same purpose? Finally, if it were acceptable to use SCNT with human cells to produce embryos for research, then would it be acceptable to use SCNT with human and nonhuman cells for the same purpose? That is, how should policy makers view the ACT experiment that moved the President to write his letter to Dr. Shapiro?

1. The Relationship Between ES Cell Research and Embryo Research

The director of the National Institutes of Health, Dr. Harold Varmus, requested a legal opinion "on whether federal funds may be used for research conducted with human pluripotent cells derived from embryos [i.e., ES cells]…." In January of 1999, Dr. Varmus's counsel, Harriet Rabb, rendered an opinion. In it Ms. Rabb acknowledges that federal funding may not be used for "research in which a human embryo or embryos are destroyed." Thus she acknowledges that insofar as isolating ES cells requires destroying embryos, using federal funds to isolate ES cells is prohibited. According to Ms. Rabb's reading of the law, however, insofar as ES cells themselves are not embryos, research on them is not prohibited.[2]

Though according to the law ES cell research is not embryo research (and thus is eligible for federal funds), people speaking about these matters generally recognize that ES cell research and embryo research are inextricably entwined. For example, when HERP discussed embryo research that was acceptable for federal funding, one area it identified was "research involving the development of [ES] cells."[3] In his letter to Dr. Shapiro, the President made it clear that he also understands how inextricably entwined are ES cell research and embryo research. He did so by suggesting that since the potential medical benefits of stem cell research are now less hypothetical than when the HERP wrote its report, he and the Congress might have to rethink the current ban on embryo research.

Indeed, not only are ES cells isolated by dismantling embryos, but in principle it seems that human ES cells could be transformed into embryos by fusing ES cells with "disabled" blastocysts. In mice, ES cells have been fused with tetraploid host blastocysts to form embryos (and mature animals) that are "solely derived from the ES cells."[4] In the era of SCNT, however, when presumably all somatic cells can be transformed into embryos, it may seem that the capacity of ES cells to be so transformed does not make their relationship to embryos especially significant.

But that view overlooks an important characteristic of ES cells. Whereas the public policy conversation about ES cells thus far has focused on their *pluripotentiality*,[5] it has largely ignored their so-called *immortality*— or, more accurately, their capacity for "prolonged undifferentiated proliferation."[6] Because ES cells "grow tirelessly in culture....they give researchers ample time to add or delete DNA precisely."[7] Because it is easier to make precise gene insertions in ES cells than it is to make such insertions in other kinds of cells (including zygotes[8] and somatic cells), ES cells are potentially a powerful tool with which to produce germline interventions.

Thus ES cells and embryos are importantly related: With *relative* ease, ES cells can be genetically altered, those genetically altered ES cells can be fused with a disabled blastocyst to give rise to an embryo derived solely from the ES cells, and that embryo can give rise to a genetically altered organism. While there are large practical (not to mention ethical) obstacles in the way of using ES cells to produce germline alterations in humans,[9] Geron's Ethics Advisory Board (EAB) has quietly acknowledged that these obstacles may not be in place forever. After asserting that at this time Geron has no intention of using ES cells for reproductive purposes, the EAB states: "Should Geron consider initiating [activities involving genetic manipulation for reproductive purposes], the EAB will undertake the necessary ethical analysis."[10] That is, it is at least theoretically possible that in the future, the practical obstacles now in the way of using ES cells to produce genetically altered human embryos will no longer stand in the way. A comprehensive analysis of ES cell research should acknowledge this theoretical possibility. More to the point here, a careful analysis will avoid too quickly asserting that there is nothing special about the capacity of ES cells to be transformed into embryos.

Ms. Rabb may be accurate to say that insofar as ES cells are not embryos, the letter of the law against embryo research does not apply. However, insofar as ES cells are harvested by destroying embryos and ES cells can in principle be used to produce not just embryos but altered embryos (i.e., embryos with added or deleted genes), and insofar as the spirit of the law aims to prevent such destruction and production, the spirit of the law does "apply."

Though I believe that the current congressional ban against all embryo research is not in the public interest, I also believe that public policymakers are obliged to respect that ban or make the arguments to lift it. A legalistic end run around the spirit of the law is contrary to what we might call a basic rule of making public policy with regard to publicly funded scientific research: Makers of such policy are obliged to speak openly and clearly about what that research entails. Insofar as research on human ES cells entails the destruction of embryos (and ultimately could entail their production), and insofar as many members of the public and their representatives feel passionately about such activities, the public should be involved in the public policy conversation about that research. Because topics like ES cell research are discussed in a language that is foreign to many members of the public, those who conduct those conversations are obliged to attempt to translate. An end run around the public will is not acceptable, no matter how great the potential medical benefits to be garnered. Medical progress is a very great good. But in a democracy, transparent and respectful public conversation may be an even greater good.

2. The State of the Policy Argument About Embryo Research

The major argument for conducting embryo research is that it promises to reduce human suffering and promote well-being, from helping overcome infertility to curing disease. The major argument against using embryos for research is that they have the moral status of persons and thus should not be destroyed, no matter how great the human benefit.

HERP rejected the argument that embryos have the same moral status as persons. "That is because of the absence of developmental individuation in the preimplantation embryo, the lack of even the possibility of sentience and most other qualities considered relevant to the moral status of persons, and the very high rate of natural mortality at this stage."[11] Though the panel denied that embryos have the moral status of persons, it did state that "the human embryo warrants serious moral consideration as a developing form of human life."[12]

In suggesting that the preimplantation embryo "warrants serious moral consideration," but not the same moral consideration as persons, HERP suggested a way between two radical alternatives. Again, the panel could not accede to the view that embryos are persons. Indeed, that view is persuasive only if one proceeds from a particular set of beliefs that citizens in a democratic society are not obliged to accept. Nor, however, could the panel accede to the view that embryos are mere property. That view is persuasive only if one chooses to ignore that *if* someone wanted to transfer these entities to a uterus, they might become human beings. In light of the determination that embryos have an intermediate moral status (neither persons nor property), HERP suggested that appropriate respect could be showed for embryos by limiting the timeframe in which research is done on them and by limiting the purposes to which they can be put.[13]

In my estimation, HERP's line of argument is as reasonable as any policy group is likely to make. Nonetheless, both friends and foes of embryo research have raised objections to it. Next, I would like to suggest how the portion of NBAC's document dealing with embryo research might respond to some of those objections, thereby making for a clearer and perhaps more persuasive argument. As already stated, my view is that federal money should be used to fund such research if and only if a majority of the public's representatives understands what is at stake and has been persuaded by the arguments.

Objections to and concerns about HERP's line of argument. Alta Charo has suggested that the panel's report is significantly flawed insofar as it claims to have made a determination of the moral status of the embryo. According to Charo, "it is impossible for a governmental body to determine the moral status of the embryo."[14] In one sense, that is surely true. No body, governmental or otherwise, can "determine" the moral status of the embryo in the way we can, say, determine the time it will take an object dropped from a given height to reach the ground. There is no "correct" answer to the question, What is the moral status of the embryo? Human beings cannot determine—in the sense of *discovering* through simple empirical investigation—what the moral status of embryos is.

In another sense, however, government bodies cannot *not* determine the moral status of embryos. They have to make a determination, in the sense of *implicitly or explicitly making an interpretation of* their moral status.

What we think is appropriate to do with things is to a large extent a function of what we think they are. When, for example, an advisory body makes a policy concerning the disposition of embryos, it has to rely on an interpretation of—it has to make a "determination" about—their moral status. If an advisory body wants to make recommendations about how to treat embryos, it can choose among many interpretations of what they are. However, no matter how keenly such a body might be aware that the interpretation it relies on is tentative and potentially divisive, it cannot choose not to choose an interpretation. Thus, pace Charo, I would suggest that the HERP should *not* have "abandoned any effort to determine the moral status of the embryo."[15] The panel should not have attempted to avoid making a determination, because no such attempt could succeed. Indeed, while the panel technically may have been correct to assert that it "was not called upon to decide which of [of the many views on the moral status of the embryo] is *correct*," it should have more clearly acknowledged that it nonetheless had to base its recommendations on its *interpretation* of the embryo's moral status.

The panel's technically accurate but unfortunately worded assertion about not being called upon to decide which view of the moral status was correct was reinforced by the also technically accurate but unfortunately worded assertion that it "conducted its deliberations in terms that were independent of a particular religious or philosophical perspective." Someone in a hurry might have missed the work that "particular" does in that sentence. The sentence could be read to mean that commissioners thought they were somehow free of *all* philosophical and religious commitments, just offering commitment-free analysis from on high. That is clearly not what the commissioners meant. The report was free of particular religious commitments in the sense that it did not appeal to a particular biblical tradition or religious authority to support its interpretation of the moral status of embryos. Similarly, it was free of particular philosophical commitments in the sense that it did not appeal to any particular philosophical "school" such as deontology or utilitarianism. On the other hand, however, the very idea of democracy has deep roots in commitments that are arguably religious and surely philosophical. Although, for example, the idea that all human beings are "created equal" can be given a philosophical account, its religious roots are obvious. Moreover, the idea that the government should, to the extent possible, allow a plurality of life projects to flourish is rooted in fundamental philosophical commitments.

NBAC should try to be clearer than HERP was about the difference between *particular* philosophical and religious commitments that are not essential to the idea of democracy and more general philosophical (and arguably "religious") commitments that undergird the idea of democracy itself. The so-called pluralistic approach does not come from nowhere; it is not value-free; on the contrary, it grows out of a commitment to and tradition of giving reasons that are accessible to all. Rather than emphasize that no particular philosophy was appealed to, NBAC should specify some of the essential democratic and philosophical commitments to which it did appeal—such as the commitment to giving reasons that do not decisively depend upon particular schools of religion or philosophy.

This nation's founders, of course, understood that sometimes disagreements about policy matters would be rooted in deep religious and philosophical commitments. The founders thought that such disagreements would have to be resolved through the political process. Even if the reality of political and economic power is often otherwise, in principle, those arguments prevail that persuade the majority. In accordance with that process, the government will sometimes have to implement determinations that conflict with the fundamental values of some citizens. It is utopian to imagine that at all times all deep commitments will be able to flourish. As John Rawls puts it: "There is no social world without loss: that is, no social world that does not exclude some ways of life that realize in special ways certain fundamental values."[16] Inevitably, some citizens will sometimes feel the pain of such exclusion. We are all obliged to notice and try to respond to the pain of that loss.[17] Yet as Rawls points out, there is no social world without it.

The founders believed, however, that those whose values were not embraced in a given case could take solace in understanding that the procedure that produced that result was rooted in a shared fundamental value: the value of relegating such disagreements to a public arena in which those with the most persuasive arguments prevail. The founders were aware that history is strewn with examples of bad arguments persuading the majority. But in a democracy, the remedy for bad arguments is not religious fiat; it is better arguments.

Again, because one cannot avoid making an interpretation of the moral status of embryos, in a democracy one has to instead give reasons to support one's interpretation and show how one's interpretation recognizes competing interpretations to the greatest extent possible. Proponents of the "intermediate status" view cannot claim to be without an interpretation of the moral status of embryos; they should be clear and open about that, and they should feel no need to apologize for it. They should acknowledge the pain and/or frustration that those holding minority views will experience. Advocates of the "intermediate moral status" interpretation can and should point to how their interpretation acknowledges minority claims: Acknowledging advocates of the embryos-are-persons view, limits are placed on the timeframe in which embryos can be used for research as well as on the purposes to which they can be put; acknowledging those who hold the embryos-are-property view, much, but not all, research is allowed.

It was perhaps because HERP was not sufficiently clear about how it understood the intermediate status of embryos that another objection was leveled against its report. One observer suggested that it is incoherent to say that we can both "respect" embryos and accept their dismemberment in the research process.[18] That would be true if, for example, one assumed that embryos are persons and thus deserving of the same respect as persons. But if, as did HERP, one conceives of the moral status of embryos differently, then respecting them differently could be altogether coherent. Again, what we think we should do with things—and how we think we should respect things—is a function of what we think they are. For example, we think we can consistently accord cadavers the respect they are due and allow medical students under carefully circumscribed conditions to dismember them. If one accepts the middle way interpretation of the moral status of embryos, then limited (but appropriate) respect for them is consistent with limited research on them.

3. The "Discarded-Created" Distinction: On the Intentions of Embryo Makers

Let us assume NBAC has successfully argued that, in general, federal funding should be available for research on embryos under certain circumstances and for certain purposes. The next question is, should the intention of the maker of an embryo at the time of its creation make a difference for how we evaluate the ethical acceptability of doing research on it? Thoughtful people have suggested that there is an important moral difference between doing research on embryos originally created with the intention of using them for reproduction and doing research on embryos originally created with the intention of using them for research. The former class of embryos becomes available for research only when it is discovered that members of it are no longer needed for reproduction; only then are they "discarded" and only then do they become available for research. The latter class of embryos would be "created" specifically for the purpose of research. According to this view, doing research on embryos originally created for reproduction ("discarded") is far easier to justify than is doing research on embryos originally created for research ("created").

There is of course a large practical problem with investing much intellectual capital in the created-discarded distinction: It is altogether unclear how oversight bodies will be able to discern the intentions of embryo makers. Though regulations could perhaps impose limits on the number of embryos allowed to be created in the IVF context, there will always be room for creative overestimation of the need for embryos for reproduction. As HERP reported, the Australian Senate Select Committee that took up these issues wrote that "any intelligent administrator of any IVF program can, by minor changes in his [sic] ordinary clinical ways of going about things, change the number of embryos that are fertilized"[19]

While members of a bioethics commission need to take into account such practical concerns, it is ethical concerns that should drive their analysis. One ethical intuition that seems to motivate the discarded-created distinction is that whereas the act of creating an embryo for reproduction is respectful in a way that is commensurate with the moral status of embryos, the act of creating an embryo for research is not. Because the first class of embryo was brought into being under moral circumstances—because the intentions of its makers were moral—research on them is deemed acceptable.[20] Because the second class of embryo was *not* brought into being under equally moral circumstances—because the intentions of its makers were not equally respectable—research on them is deemed unacceptable. According to this view, the moral status of the embryo (and thus the moral status of research on it) is a function of the intention of its maker. The problem with this intuition is that it is difficult to see what the intention of the maker of something has to do with the moral status of that thing once it has come into being. We do not think, for example, that the moral status of children is a function of their parents' intention at the time of conception. If what something *is* obliges us to treat it some ways and not in others, then *how it came into being* is usually thought to be morally irrelevant.

It may be that another and closely related motivation for taking the discarded-created distinction seriously is the intuition that whereas in creating embryos for reproduction scientists are helping nature along toward a natural purpose, in creating embryos for research they are not. According to this intuition, whereas helping nature along is praiseworthy, doing something different from what happens "naturally" is not. In other words, whereas intending to create embryos for the purpose of reproduction is *natural*, intending to create them for the purpose of research is *artificial*. The problem with this intuition is that both projects (reproduction and research) entail the intentional creation of embryos in the highly "artificial" context of an IVF clinic. Thus it is difficult to see why policymakers should give credence to the natural/artificial distinction in attempts to delineate the moral difference between doing research on embryos originally intended for reproduction and those originally intended for research.

But perhaps what motivates the distinction is not a view about the intention or purpose of the maker of the embryo at the time of creation, but, more pragmatically, a view about what to do with embryos once they are already here. Perhaps the motivation for the distinction is simply the view that it would be wasteful *not* to use embryos that are already here (regardless of their origin). Whereas this view about wastage may support the claim that using embryos that are already here is ethically acceptable, it sheds no light on whether creating embryos for research is acceptable. The holder of this view assumes that creating embryos for research is wrong. But that assumption is rejected by those who hold the "intermediate moral status" view of embryos. That is, by itself, the intuition about wastage cannot alone justify the created-discarded distinction.

It may be that another thing at work in taking the distinction seriously is the intuition that the good of helping an infertile couple become pregnant is *greater than* the good of doing embryo research. But insofar as most of that research aims at helping *many* couples overcome infertility and become pregnant, it is difficult to see why that good is of lesser moral weight than the good of helping an individual couple. If the good of helping an individual couple become pregnant is great enough to justify the creation of embryos, then it would seem that the good of helping many couples to become pregnant is an equally strong justification.

Another thing that clearly motivates taking the distinction seriously is a concern about instrumentalization.[21] The concern is that, different from creating embryos for the purpose of reproduction, creating them for the express purpose of research could make us increasingly think of them as mere means to our ends rather than as ends in themselves. In one sense, this concern seems off the mark to those who hold an "intermediate status" view of embryos. While it is clear to holders of that view that embryos deserve respect commensurate with their intermediate moral status, it does not seem to them that embryos are "ends in themselves" the way persons are. Nonetheless, holders of the intermediate status view, too, would be concerned if creating embryos in the research context leads to a more general degradation of the respect due to entities that, if transferred, might become human beings. Thus, the worry about instrumentalization strikes me as worthy of further reflection.

A last thing that may motivate the created-discarded distinction is a concern that allowing the creation of embryos for research will increase pressure on women to donate ova for that purpose. It is interesting to note, however, that the Canadian Royal Commission on New Reproductive Technologies (1993) suggested that *not* allowing the creation of embryos for research would increase pressure on women; the Canadians suggested that allowing researchers to create embryos for research would *decrease* pressure on women in IVF programs to donate unused eggs or zygotes.[22] Though the Canadian Commission's strategy might decrease pressure on women who already have undergone IVF procedures, there remains the question concerning when and where else researchers will get the eggs they need to create embryos. It is entirely plausible that that perceived need will create subtle or not-so-subtle pressure on women to donate eggs. Thus, like the concern about instrumentalization, the concern about pressure on women is not unreasonable. Unlike the concern about instrumentalization, however, the concern about pressure on women might be mitigated by the use of nonhuman ova (see Part 5 below).

In Part 2 above I attempted to suggest reasons to suppose that the way we show respect for embryos commensurate with the middle-way interpretation of their "moral status" is by placing limits on the timeframe in which researchers work on them and on what researchers can do with them. In this part, I have suggested reasons to believe that the attempt to show respect for embryos by means of distinguishing between the intentions of the makers of embryos is fraught with practical and conceptual difficulties. Indeed, neither the British Human Fertilisation and Embryology Authority (1993), the Canadian Commission, nor the U.S. Ethics Advisory Board (1979) put much stock in the distinction; all three approved the fertilization of ova for research purposes.[23] I have also, however, tried to signal the concerns about instrumentalization and pressure on women that deserve further consideration.

4. IVF Versus SCNT: On the Different Means Used to Create Embryos

If the *intentions* of the maker of embryos do not make a moral difference (or at least do not make the sort of clear moral difference suggested by some proponents of the created-discarded distinction), then do the *means* used to make the embryo make a moral difference? This question arises from the observation that whereas IVF as a means to achieve the purpose of reproduction is widely accepted, SCNT as a means to achieve the same purpose has been widely rejected.

Aside from concerns about risk, the rejection of "reproductive cloning" is based upon a widespread worry about the psychological consequences of producing children with means that replicate an extant genotype rather than creating a new one.[24] However, since here we are talking about using SCNT for research (*not* reproduction), worries about reproducing an extant genotype (worries about psychological consequences for children) are not relevant. If, in general, we accept the limited creation of embryos for research, and if by definition the harms-to-children concerns do not apply to using SCNT to produce embryos for research, then is there another reason to object to or worry about using SCNT for that purpose?

One reason to object to using SCNT to produce embryos for research might be that SCNT will significantly increase the supply of embryos—and thereby decrease respect or awe before them. This worry overlooks two facts. First, both "traditional" IVF and SCNT are limited by the number of available human ova; I am not aware of a reason to think that that number is going to grow fast. Second, at this point, it is more difficult to produce embryos with SCNT than with IVF; it is not reasonable to assume that researchers will rush to use SCNT. Thus, it does not seem reasonable to worry that SCNT will significantly increase the number of, and thereby decrease the respect accorded, embryos in general.

There may, however, be another more substantial worry in this context. This is the worry that since embryos created by means of SCNT are not genetically unique,[25] and since genetic uniqueness is one of the valued properties of embryos created by IVF, embryos created by means of SCNT may be respected less than those created by IVF. That is, one might worry that producing embryos by means of SCNT will contribute to an instrumental or cheapened view of them.

This worry strikes me as important and worthy of further reflection. To put it in "humanistic" terms, it is not implausible that the more we imagine ourselves to be the masters of nature, the more we will forget our fundamental indebtedness to nature. While I believe that that worry deserves further reflection, I acknowledge that it is not obvious that such an increasingly instrumental attitude must emerge. We have been able to maintain our awe before IVF embryos created for the purpose of reproduction. There is no prima facie reason why we cannot also maintain awe before SCNT embryos created for the purpose of doing research to promote human well-being more generally.

5. Using SCNT to Create Nonviable, Hybrid "Entities"

As I observed in the beginning of this paper, the first line of President Clinton's letter requesting an NBAC report mentions ACT's attempt to use SCNT to create hybrid "entities." As we noticed (in Part 1) above, in the current context of a ban on embryo research, much is at stake in whether we call ACT's entities "hybrid embryos" or something else, such as "embryonic cells."[26]

People at ACT have suggested that we should not call their entities "embryos." But if, as ACT's CEO, Michael West, has suggested, we do not know what ACT's "entities" are, then how can we say they are not embryos? And if we do know that ACT's entities are not embryos (and thus not capable of implanting in a uterus), then why do researchers say that they would not transfer such entities in an experiment to see if they would implant?

I would suggest that to most speakers of English who followed the Dolly story, if you take a sheep somatic cell and fuse it with an enucleated sheep egg, you get a sheep embryo. If for some reason that embryo is not viable, we would call it a nonviable sheep embryo. To most speakers of English, if you take a human somatic cell and fuse it to an enucleated cow egg, you get a hybrid embryo. If for some reason that hybrid embryo is not viable, most would call it a nonviable hybrid embryo.

Once again, in deference to what above I called a basic rule of public conversation, I would recommend that we call ACT's entities *hybrid embryos*. (If it becomes clear that these hybrid embryos are not viable, we should call them *nonviable hybrid embryos*.) Though that requires facing hard questions about the production of embryos by means of SCNT for the purpose of research, facing those questions is preferable to violating the obligation to engage in public conversation in terms that attend to rather than obfuscate the concerns of many citizens.

If we can agree that we should call ACT's entities hybrid embryos, then what ethical questions arise? The question concerning the risk that results from mixing mitochondrial DNA from one species and nuclear DNA from another is large. On the one hand, it has been shown that the mitochondrial DNA from common chimpanzees, pigmy chimpanzees, and gorillas is compatible enough with human nuclear DNA for one of the cell's basic functions, oxidative phosphorylation, to occur. On the other hand, mitochondrial DNA from orangutans, New World monkeys, and lemurs has been shown to be *not* compatible enough with human nuclear DNA for oxidative phosphorylation to occur.[27]

If the risk question were resolved, the more complicated question concerning the ethics of creating hybrid organisms would remain. In the past, anxiety about at least some forms of hybridity has rested on deep but utterly indefensible intuitions: Perhaps the best example is the "intuition" that "miscegenation" is criminal. Yet, it would seem both practically and theoretically unwise for a bioethics commission to dismiss a general worry with such a long and powerful history. For example, it was so obvious to HERP that producing chimeras is "unacceptable" that they did not think it necessary to give reasons for that decision. In principle at least, it would seem that while some "anxiety" may harbor nothing more than ignorance, some may harbor "insight." This question concerning hybridity deserves further exploration.

In the meantime, fortunately for NBAC, ACT's work does not entail the production of "hybrids" in the way that seems to motivate most of the worry about hybridity. If concerns about hybridity are really about the production of "chimeras," and if ACT only wants to use an enucleated cow egg as a way station for human nuclear DNA destined to become ES cells, then concerns about hybridity would in general appear not to apply. Insofar as we are willing to place genes from one species into another to produce things like insulin or transplantable organs, it is not easy to see on what grounds we might object to temporarily housing a nuclear genome from one species in the cytoplasm of another.

If the risk and hybridity questions were resolved, then there would, of course, be two very great benefits to using ACT's strategy. First, ACT's technique provides a way around the problem of histoincompatibility; the person needing the transplanted tissue provides the somatic cell from which the tissue is produced. Perhaps more important, insofar as ACT's strategy does not involve human ova, it eliminates one of the largest concerns about creating embryos for research: pressure on women to donate ova.

Concluding Thoughts

In this essay I have delineated the issues that I believe an intellectually honest and adequate response to President Clinton's letter must address—or at least acknowledge. Indeed, because of time constraints, NBAC will not be able to do much more than acknowledge some issues. But if NBAC wants to both answer the President and serve the public interest, then it must squarely place the question concerning ES cell research in the context of embryo research. It must at least help the President and the public see the intellectual work that will need to be done after NBAC has submitted its report.

Before taxpayer money is spent on research involving embryos (whether human or hybrid), arguments should be made that persuade the majority of taxpayers and Congress that such research is ethically acceptable. Experience of the last few years makes it clear that this will not be easy. Nonetheless, it will not be impossible. Americans are mightily and appropriately impressed by the potential medical benefits associated with embryo research. That HERP's argument for limited embryo research did not prevail should not keep others from trying to make that same argument more persuasively.

Medical progress is a very great good, but it does not trump all others. In particular, it does not trump the good that is transparent and respectful public debate. It is ultimately (if not immediately) in everyone's best interest to be as clear as possible about the facts. One of those is that ES cell research cannot be done without dismantling embryos, whether they are hybrid or from a single species, viable or nonviable, created by IVF or SCNT.

If NBAC so narrows the scope of its report as to exclude, for example, the general question concerning SCNT to create embryos for research and the particular question concerning SCNT to create hybrid embryos, then it will have failed to adequately respond to the President's request. Beyond the fact that the President asked for a broader analysis, it is what the public needs and deserves. If NBAC will not at least begin to address the issues raised by the President's request, who will?

Acknowledgments

I thank Marguerite Strobel for valuable research assistance and conversation about this paper. For thoughtful comments on earlier drafts, I am grateful to Lori Andrews, Erika Blacksher, Andrea Bonnicksen, Cynthia Cohen, Dena Davis, Mark Frankel, Theodore Friedmann, Greg Kaebnick, Lori Knowles, Eric Meslin, Tom Murray, Andy Siegel, Paul Silverman, Bonnie Steinbock, and Carol Tauer. If only I had responded to all of their insightful observations!

Notes

1 Statement of Administration Policy (Concerning S. 1601 - Human Cloning Prohibition Act), February 9, 1998.

2 January 15, 1999, Memo from Ms. Rabb to Dr. Varmus. Ms. Rabb wrote, "federally funded research that utilizes human pluripotent stem cells would not be prohibited by the HHS appropriations law prohibiting human embryo research, because such stem cells are not human embryos."

3 National Institutes of Health, *Report of the Human Embryo Research Panel* (HERP), September 27, 1994, at 10.

4 David Solter and John Gearhart, "Putting Stem Cells to Work," *Science* 283 (March 5, 1999):1468–70, at 1469.

5 Insofar as ES cells (different from embryos) do *not* develop into the cells of the placenta, they are only *pluripotent*. Insofar as ES cells do develop into *all* the cells of the fetus, they are *totipotent*. Whether one describes ES cells as totipotent or pluripotent may be a function of whether one wants to emphasize the respect in which they are different from or similar to embryos. An adequate analysis will recognize both the similarity and difference.

6 James A. Thomson et al., "Embryonic Stem Cell Lines Derived from Human Blastocysts," *Science* 282 (November 6, 1998):1145–47, at 1145.

7 Antonio Regalado, "The Troubled Hunt for the Ultimate Cell," *Technology Review* 101(4) (July/August 1998):4–41, at 40.

8 Jon W. Gordon, "Genetic Enhancement in Humans," *Science* 283 (March 26, 1999):2023–24, at 2024.

9 On the practical and ethical obstacles, see Solter and Gearhart, at 1469; Thomson et al., at 1145; and Gordon, at 2024.

10 Geron Ethics Advisory Board, "Research with Human Embryonic Stem Cells: Ethical Considerations," *Hastings Center Report* 29 (2) (1999):31–36, at 34.

11 HERP, at 2.

12 HERP, at 2.

13 Bonnie Steinbock, "Ethical Issues in Human Embryo Research," background paper for HERP.

14 Alta Charo, "The Hunting of the Snark: The Moral Status of Embryos, Right-to-Lifers, and Third World Women," 6 *Stanford Law and Policy Review* 11 (1995).

15 Charo, at 18.

16 John Rawls, *Political Liberalism* (New York: Columbia University Press, 1993):197.

17 Though earlier I disagreed with Charo's "Hunting the Snark," here I am indebted to that same essay.

18 Daniel Callahan, "The Puzzle of Profound Respect," *Hastings Center Report* 25(1) (1995):39–41.

19 HERP, at 56.

20 "For most people it is the intention to create a child that makes the creation of an embryo a moral act." George Annas, Arthur Caplan, and Sherman Elias, "The Politics of Human-Embryo Research—Avoiding Ethical Gridlock," *New England Journal of Medicine* 554(20) (May 16, 1996):1329–32, at 1331.

21 HERP, at 53.

22 HERP, at 56.

23 HERP, at 53; additionally, the state of Victoria, Australia, permits fertilization for research purposes through the first 24 hours following fertilization.

24 National Bioethics Advisory Commission, *Cloning Human Beings* (Rockville, MD: U.S. Government Printing Office, 1997).

25 To be more precise, the *nuclear* DNA of embryos produced by SCNT is not unique; because of the mitochondrial DNA contributed by the enucleated ovum, an embryo produced by SCNT *is* genetically unique. This minor technical point does not change the fact that many may worry about the moral significance of replicating (nuclear) genotypes, even in the research context.

26 Nicholas Wade, "Researchers Claim Embryonic Cell Mix of Human and Cow," *New York Times* (November 12, 1998).

27 L. Kenyon and C.T. Moraes, "Expanding the Functional Human Mitochondrial DNA Database by the Establishment of Primate Xenomitochondrial Cybrids," *Proceedings of the National Academy of Sciences* 94(17) (August 19, 1997):9131–35.

Locating Convergence: Ethics, Public Policy, and Human Stem Cell Research

Commissioned Paper
Andrew W. Siegel
The Johns Hopkins University

I. Introduction: Benefits and Constraints

The principal moral justification for promoting research with human pluripotent stem cells is that such research has the potential to lead to direct health benefits to individuals suffering from disease. Research that identifies the mechanisms controlling cell differentiation would provide the foundation for directed differentiation of pluripotent stem cells to specific cell types. The ability to direct the differentiation of stem cells would, in turn, advance the development of therapies for repairing injuries and pathological processes. The great promise of human stem cell research inspired 33 Nobel laureates to voice their support for the research and to lay down the gauntlet against those who oppose it: "Those who seek to prevent medical advances using stem cells must be held accountable to those who suffer from horrible disease and their families, why such hope should be withheld" (Letter to Congress 1999).

While the invocation of the potential benefits of stem cell research furnishes strong moral grounds for supporting the research, considerations of social utility are not always sufficient to morally justify actions. There are moral constraints on the promotion of the social good. For example, considerations of justice and respect for persons often trump considerations of social utility. Those who oppose research involving the use of stem cells derived from embryos and fetal tissue argue that the research is morally impermissible because it is implicated in the unjust killing of human beings who have the moral status of persons. Opponents of the research maintain that the constraints against killing persons to advance the common good apply to fetuses and embryos. On this view, one cannot approve of a policy supporting such research without holding that elective abortion and the destruction of embryos are morally permissible.

In this paper, I consider whether it is possible to offer a justification for federal support of research with stem cells derived from fetal tissue and embryos that is not premised on a view about the morality of abortion and the destruction of embryos. The aim of seeking such a justification is to provide a foundation for consensus on public policy in the area of stem cell research. I will argue that people with divergent views about the moral status of fetuses and embryos should be able to endorse some research uses of stem cells derived from these sources.

II. Research with Stem Cells Derived from Fetal Tissue

The ethical acceptability of deriving stem cells from the tissue of aborted fetuses is, for some, closely connected to the morality of elective abortion. For those who believe that abortion is permissible because the fetus has no moral standing, there are no significant moral barriers to research using stem cells derived from fetal tissue. Restrictions that separate decisions to donate fetal tissue from decisions to abort might be thought necessary, but the purpose of the restrictions would be to protect the mother against coercion and exploitation rather than to protect the fetus.

What is less clear is whether one can both morally oppose abortion and support this method of deriving stem cells. A common view in the literature on the ethics of human fetal tissue transplantation research is that we can support the research without assuming that abortion is morally permissible. As long as guidelines are in place that ensure that abortion decisions and procedures are separated from considerations of fetal tissue procurement and use in research, using aborted fetuses for research is no more problematic than using other cadavers donated for scientific and medical purposes.

Opponents of the research use of fetal materials obtained from induced abortions dispute the claim that we can dismiss the relevance of the morality of abortion. They appeal to two grounds in asserting the relevance of the moral question: 1) those who procure and use fetal material from induced abortions are complicit with the abortions which provide the material, and 2) it is impossible to obtain valid informed consent for the use of fetal materials. Each of these claims merits consideration.

Complicity

There are two general ways in which one can maintain that those involved in the research use of stem cells derived from aborted fetuses are complicit with abortion: 1) They bear some causal responsibility for abortions, or 2) they symbolically align themselves with abortion.

1. Causal Responsibility

A researcher or tissue procurer may bear direct or indirect causal responsibility for abortions. The former kind of responsibility exists where one directly motivates a woman to have an abortion (e.g., by offering financial incentives) or is directly involved in the performance of an abortion from which fetal tissue is procured. There are several measures that can help prevent direct responsibility for abortions (Fetal Tissue Transplantation Research Panel 1988):

- A requirement that the consent of women for abortions be obtained prior to requesting or obtaining consent for the donation of fetal tissue.

- A requirement that those who seek a woman's consent to donate not discuss fetal tissue donation prior to her decision to donate.

- A prohibition against the sale of fetal tissue.

- A prohibition against directed donation of fetal tissue.

- A separation between tissue procurement personnel and abortion clinic personnel.

- A prohibition against any alteration of the timing of or procedures used in an abortion solely for the purpose of obtaining tissue.

Those involved in research uses of stem cells derived from fetal tissue would be indirectly responsible for abortions if the perceived benefits (or promise of benefits) of the research contributed to an increase in the number of abortions. Opponents of fetal tissue research argue that it is not realistic to suppose that we can always keep a woman's decision to abort separate from considerations of donating fetal tissue, as many women facing the abortion decision are likely to have already gained knowledge about fetal tissue research through widespread media attention to the issue. The knowledge that having an abortion might promote the common goodwill, opponents argue, tips the balance in favor of going through with an abortion for some of the women who are ambivalent about it. More generally, some also argue that the benefits achieved through the routine use of fetal tissue will further legitimize abortion and result in more socially permissive attitudes and policies concerning abortion.

Although there has been one empirical study examining whether the potential for fetal tissue transplantation is likely to influence abortion decisions, the issue remains largely speculative. The study involved a survey which asked, "If you became pregnant and knew that tissue from the fetus could be used to help someone suffering from Parkinson's disease, would you be more likely to have an abortion?" (Martin et al. 1995, 548). Twelve percent of the women responded in the affirmative. The authors conclude that the option to donate tissue may influence some women's abortion decisions. The main deficiency of the study, however, is its reliance on a hypothetical that is stripped of the complexities of the actual circumstances a pregnant woman considering abortion might be operating under. As Dorothy Vawter and Karen Gervais have noted: "[G]iven how situation-specific women's abortion decisions are, it is unclear what useful information can be obtained from asking women global hypothetical questions about whether they believe the option to donate would affect their decision to terminate a 'generic' pregnancy sometime in the future" (Vawter and Gervais 1993, 5).

It is difficult to deny that there is a risk that knowledge of the promise of research on stem cells derived from fetal tissue will play a role in some abortion decisions, even if only very rarely. However, it is not clear that much moral weight ultimately attaches to this fact. One might be justified in some instances in asserting that "but for" research using fetal tissue a particular woman would not have chosen abortion. But one might assign this kind of causal responsibility to a number of factors which figure into abortion decisions without making ascriptions of complicity. For example, a woman might choose to have an abortion principally because she does not want to slow the advancement of her education and career. She might not have had an abortion in the absence of policies that encourage career development. Yet, we would not think it appropriate to charge those who promote such policies as complicit in her abortion. In both this case and that of research, the risk of abortion is an unintended consequence of a legitimate social policy. The burden on those seeking to end such policies is to show that the risks of harm resulting from the policies outweigh the benefits (Childress 1991). This minimally requires evidence of a high probability of a large number of abortions that would not have occurred in the absence of those policies. There is, however, no such evidence at present; nor is there any reason to think that it is forthcoming.

2. Symbolic Association

Agents can be complicit with wrongful acts for which they are not causally or morally responsible. One such form of complicity arises from an association with wrongdoing that symbolizes an acquiescence in the wrongdoing. As James Burtchaell characterizes it, "It is the sort of association which implies and engenders approbation that creates moral complicity. This situation is detectable when the associate's ability to condemn the activity atrophies" (Burtchaell 1991, 9). Burtchaell maintains that those involved in research on fetal tissue enter a symbolic alliance with the practice of abortion in benefiting from it.

A common response to this position is that there are numerous circumstances in which persons benefit from immoral acts without tacitly approving of those acts. For example, transplant surgeons and patients may benefit from deaths resulting from murder and drunken driving but nevertheless condemn the wrongful acts (Robertson 1988; Vawter et al. 1990). A researcher who benefits from an aborted fetus need not sanction the act of abortion any more than the transplant surgeon who uses the organs of a murder victim sanctions the homicidal act.

This response has not, however, been satisfactory to opponents of fetal tissue research. They maintain that fetal tissue research implicates those involved in a different and far greater evil than is the transplant surgeon in the example above. Unlike drunken driving and murder, abortion is an institutionalized practice in which a certain class of humans (which pro-lifers regard as the moral equivalent of persons) are allowed to be killed. In this respect, some foes of abortion suggest that fetal tissue research is more analogous to research which benefits from victims of the Holocaust (Bopp 1994).

But whatever one thinks of comparisons between the victims of Nazi crimes and aborted fetuses—and many are understandably outraged by them—one could concede the comparisons without concluding that fetal tissue research is morally problematic. There are, of course, some who believe that those who use data derived from Nazi experiments are morally complicit with those crimes. For example, Seidelman writes: "By giving value to (Nazi) research we are, by implication, supporting Himmler's philosophy that the subjects' lives were 'useless.' This is to argue that, by accepting data derived from their misery we are, *post mortem*, deriving utility from otherwise 'useless' life. Science could thus stand accused of giving greater value to knowledge than to human life itself" (Seidelman 1989). But one need not adopt this stance. Instead, one can reasonably hold that the symbolic meaning of scientists' actions must be divined solely from their intentions. As Benjamin Freedman argues:

> A moral universe such as our own must, I think, rely on the authors of their own actions to be primarily responsible for attaching symbolic significance to those actions...[I]n using the Nazi data, physicians and scientists are acting pursuant to their own moral commitment to aid

patients and to advance science in the interest of humankind. The use of data is predicated upon that duty, and it is in seeking to fulfill that duty that the symbolic significance of the action must be found (Freedman 1992, 151).

One could likewise maintain that the symbolic significance of support for research using stem cells derived from fetal tissue lies in the desire to promote public health and save lives. This research is allied with a noble cause, and any taint that might attach from the source of the stem cells arguably diminishes in proportion to the potential good that the research may yield.

One who opposes the use of fetal tissue might reply, however, that those who use Nazi data can symbolically disassociate themselves from the immoral acts that produced the data only because the Nazi regime no longer exists. Those who use fetal tissue cannot—so the argument runs—divorce themselves from the immoral act that produces it because elective abortion is legally protected and largely socially accepted in our society. Opponents of fetal tissue research might seek to generate different intuitions about the moral permissibility of research using fetal tissue by offering a different analogy. Imagine the following: You live in a society where members of a particular racial or ethnic group are regularly killed when they are an unwanted burden. The practice is legally protected and generally accepted. Suppose that biological materials obtained from these individuals subsequent to their deaths are made available for research uses. Would it be morally problematic for researchers to use these materials?

It is likely that many who consider it permissible to use Nazi data would think it problematic to use materials derived from the deceased individuals in the hypothetical scenario. Arguably, what underlies the judgment that the research in the hypothetical is morally problematic is the belief that there is a heightened need to act in ways that protest moral wrongs where those wrongs are currently socially accepted and institutionalized. Attempts to benefit from the moral wrong in these circumstances may be viewed as incompatible with mounting a proper protest.

On the pro-life view, the hypothetical case is analogous to the case of research uses of fetal material. Hence, if we concede (at least for the sake of argument) that the fetus has the moral status of a person, it is not clear that an investigator can engage in fetal tissue research without implicitly sanctioning abortion. However, it is not obvious that this conclusion cuts against a public policy which supports fetal tissue research. Federal funding of research may be problematic where taxpayers who oppose the research would regard themselves as complicit with abortion as a result of the funding. There would, for example, be a basis for concern if fetal tissue research played an important causal role in an increase in the number of elective abortions. But funding research in which investigators are complicit with abortion through their symbolic association with it does not render opponents complicit with abortion. That a person's tax dollars are being used to support fetal tissue research does not imply that he or she acquiesces in abortion, nor does it diminish his or her ability to condemn abortion.

Consent

Many people hold that women should not be allowed to terminate a pregnancy solely for the purpose of donating fetal material. There are two consent requirements that can help insulate the decision to donate from the decision to abort: a) informed consent for an abortion must be provided prior to an informed consent to donate the fetal tissue, and b) in the consent process for abortion, there must be no (unsolicited) mention of the possibility of using fetal materials in research and transplantation.

The most serious charge against these restrictions on the consent process is that it is disrespectful of the autonomy of women considering abortion to withhold information from them regarding the donation of fetal tissue. Because this information might be important to a woman's abortion decision, the failure to disclose the information would render the consent for the abortion ethically invalid (Martin 1993).

There are, however, a number of difficulties with this argument. First, it is not clear that information about the possibility of donation is materially relevant to the abortion decision, since, as discussed above, there is not adequate evidence that the option of donation (where financial incentives and directed donations are prohibited) would ever function as a reason for a woman to abort a fetus. Second, assuming the possibility of donation is materially relevant to some women's abortion decisions, there is an obligation not to disclose the option if it is unethical for women to abort for this purpose. Finally, if clinic personnel are permitted to discuss donation prior to obtaining a woman's consent for abortion, women may be (or feel) pressured to have an abortion, in which case the voluntariness of the consent will be in doubt (Vawter and Gervais 1993).

Another problem about consent concerns the matter of who has the moral authority to consent to donate fetal tissue. Some object that, from an ethical standpoint, a woman who chooses abortion forfeits her right to determine the disposition of the dead fetus. Burtchaell, for instance, argues that "the decision to abort, made by the mother, is an act of such violent abandonment of the maternal trusteeship that no further exercise of such responsibility is admissible" (Burtchaell 1989).

John Robertson argues that this position mistakenly assumes that the disposer of cadaveric remains acts as the guardian or proxy of the deceased. Instead, "a more accurate account of their role is to guard their own feelings and interests in assuring that the remains of kin are treated respectfully" (Robertson 1988, 6). But even if we suppose that a woman does forfeit her moral authority to determine the disposition of her aborted fetus, it is not clear that informed consent is always ethically required for the use of cadaveric remains. The requirement of informed consent in medical practice is largely meant to protect the autonomy of persons. In the absence of any person whose autonomy must be respected, it does not seem that the failure to obtain consent violates anyone's rights (Jones 1991).

III. Research with Stem Cells Derived from Embryos

Research with stem cells obtained from human embryos poses moral difficulties that do not arise in the case of fetal tissue. Whereas researchers using fetal tissue are not responsible for the death of the fetus, researchers using stem cells derived from embryos will often be implicated in the destruction of the embryo. Researchers using stem cells derived from embryos are clearly implicated in the destruction of the embryo where they a) derive the cells themselves, or b) enlist others to derive the cells. However, there may be circumstances in which opponents of embryo research could not properly deem researchers who use embryonic stem cells complicit with the destruction of embryos. Suppose, for example, that X creates a cell line for his own study and later makes an unsolicited offer to share the cell line with Y so that Y may pursue her own research. Is Y implicated in X's act of destroying the embryo from which the cell line was derived? It does not seem that Y is implicated in X's act. As Robertson argues, it does not appear one can assign causal or moral responsibility for the destruction of an embryo to an investigator where his or her "research plans or actions had no effect on whether the original immoral derivation occurred" (Robertson 1999). Nonetheless, it does seem evident that much research with embryonic stem cells will be morally linked to the derivation of the cells (and the resulting destruction of the embryo), especially in the early stages of the research. Thus, an analysis of the ethics of research with embryonic stem cells, as well as the ethics of the funding of this research, must address the issue of the moral permissibility of destroying embryos.

The Moral Status of Embryos

The moral permissibility of destroying embryos turns principally on the moral status of the embryo. The debate about the moral status of embryos has traditionally revolved around the question of whether the embryo has the same moral status as children and adult humans, with a right to life that cannot be sacrificed for the benefit of society. At one end of the spectrum of positions is the view that the embryo is a mere cluster of cells which has no more moral standing than any other human cells. From this perspective, there are few, if any, limitations on research uses of embryos. At the other end of the spectrum is the view that embryos have the moral status of persons. On this view, research involving the destruction of embryos is absolutely prohibited. An intermediate position is that the embryo merits respect as human life, but not the level of respect accorded persons. Whether research using embryos is acceptable on this account depends upon just how much respect the embryo is thought to deserve.

While the moral permissibility of research using embryonic stem cells turns upon the status of the embryo, the prospects of mediating the stand-off between opposing views on the matter are dim. A brief consideration of the competing positions will reveal some of the difficulties of resolving the issue.

The standard move made by those who deny the personhood of embryos is to identify one or more psychological or cognitive capacities that are thought essential to personhood (and a concomitant right to life) but which embryos lack. The capacities most commonly cited include consciousness, self-consciousness, and reasoning (Warren 1973; Tooley 1983; Feinberg 1986). The problem faced by such accounts is that they seem either under- or over-inclusive, depending on which capacities are invoked. If one requires self-consciousness or reasoning, most early infants will not satisfy the conditions for personhood. If sentience is regarded as the touchstone of the right to life, then nonhuman animals will also possess this right. Since most of those who reject the personhood of the embryo believe that newborn infants do possess a right to life and animals do not, these capacities cannot generally be accepted as morally distinguishing embryos from other human beings.

Of course, those who reject that embryos have the standing of persons can maintain that the embryo is simply too nascent a form of human life to merit the kind of respect that we accord more developed humans. However, in the absence of an account which decisively identifies the first stage of human development at which destroying human life is morally wrong, one can hold that it is not permissible to destroy embryos.

The fundamental argument of those who oppose the destruction of human embryos is that these embryos are human beings, and as such, have a right to life. The humanity of the embryo is thus thought to confer the status of a person upon it. The problem is that for some, the premise that all human beings have a right to life (i.e., are persons in the moral sense) is not self-evidently true. Indeed, some believe that the premise conflates two categories of "human beings"—namely, beings which belong to the species *homo sapiens*, and beings which belong to the moral community (Warren 1973). According to this view, the fact that a being belongs to the species *homo sapiens* is not sufficient to confer on it membership in the moral community. While it is not clear that those who advance this position can establish the point at which human beings first acquire the moral status of persons, those who oppose the destruction of embryos likewise fail to establish that we should ascribe the status of persons to human embryos.

Those constructing public policy on the use of embryos in research would do well to avoid attempting to settle the debate over the moral status of embryos. Ideally, public policy recommendations should be formulated in terms which individuals with opposing views on the status of the embryo can accept. As Thomas Nagel argues, "In a democracy, the aim of procedures of decision should be to secure results that can be acknowledged as legitimate by as wide a portion of the citizenry as possible" (Nagel 1995, 212). Amy Gutmann and Dennis Thompson similarly argue that the construction of public policy on morally controversial matters should involve a "search for significant points of convergence between one's own understanding and those of citizens whose positions, taken in their more comprehensive forms, one must reject" (Gutmann and Thompson 1996, 85).

Locating Convergence

R. Alta Charo suggests an approach for informing policy in this area that seeks to accommodate the interests of individuals who hold conflicting views on the status of the embryo. Charo argues that the issue of moral status can be avoided altogether by addressing the proper limits of embryo research in terms of political philosophy rather than moral philosophy:

> The political analysis entails a change in focus, away from the embryo and the research and toward an ethical balance between the interests of those who oppose destroying embryos in research and those who stand to benefit from the research findings. Thus, the deeper the degree of offense to opponents and the weaker the opportunity for resorting to the political system to impose their vision, the more compelling the benefits must be to justify the funding (Charo 1995).

In Charo's view, once we recognize that the substantive conflict among fundamental values surrounding embryo research cannot be resolved in a manner that is satisfactory to all sides, the most promising move is to perform some type of cost-benefit analysis in considering whether to proceed with the research. Thus, one could acknowledge that embryo research will deeply offend many people, but argue that the potential health benefits for this and future generations outweigh the pain experienced by opponents of the research.

It is, however, questionable whether Charo's analysis successfully brackets the moral status issue. One might object that placing the lives of embryos in this kind of utilitarian calculus will only seem appropriate to those who already presuppose that embryos do not have the status of persons. After all, we would expect most of those who believe—or who genuinely allow for the possibility—that embryos have the status of persons, to regard such consequentialist grounds for sacrificing embryos as problematic.

An acceptable political approach must seek to develop public policy around points of convergence in the moral positions of those who disagree about the status of the embryo. Of course, as long as the disagreement is cast strictly as one between those who think the embryo is a person with a right to life and those who think it has little or no moral standing, the quest for convergence will be an elusive one. But there are grounds for supposing that this is a misleading depiction of the conflict. Once this is recognized, it will become clear that there may be sufficient consensus on the status of embryos to justify some research uses of stem cells derived from them.

In his discussion of the abortion debate, Ronald Dworkin maintains that, despite their rhetoric, a large faction of the opposition to abortion does not actually believe that the fetus is a person with a right to life. This is revealed, he claims, through a consideration of the exceptions that they permit to their proposed prohibitions on abortion:

> It is a very common view, for example, that abortion should be permitted when necessary to save the mother's life. Yet this exception is also inconsistent with any belief that a fetus is a person with a right to live. Some people say that in this case a mother is justified in aborting a fetus as a matter of self-defense; but any safe abortion is carried out by someone else—a doctor—and very few people believe that it is morally justifiable for a third party, even a doctor, to kill one innocent person to save another (Dworkin 1994, 32).

It can further be argued that one who regards the fetus as a person would seem to have no reason to privilege the mother's life over the fetus's life. Indeed, inasmuch as the mother typically bears some responsibility for putting the fetus in the position in which it needs her aid, one who regards the fetus as a person might well hold that the fetus's life should be privileged over the mother's life.

Some abortion conservatives further hold that abortion is morally permissible when a pregnancy is the result of rape or incest. Yet, as Dworkin comments, "it would be contradictory to insist that a fetus has a right to live that is strong enough to justify prohibiting abortion even when childbirth would ruin a mother's or a family's life, but that ceases to exist when the pregnancy is the result of a sexual crime of which the fetus is, of course, wholly innocent" (Dworkin 1994, 32).

The importance of these exceptions in the context of research uses of embryos is that they suggest we can identify some common ground between liberal and conservative views on the permissibility of destroying embryos. Conservatives who allow these exceptions implicitly hold with liberals that very early forms of human life can sometimes be sacrificed to promote the interests of other humans. While liberals and conservatives disagree about the range of ends for which embryonic or fetal life can ethically be sacrificed, they should be able to reach some consensus. Conservatives who accept that killing a fetus is permissible where it is necessary to save a pregnant woman or spare a rape victim additional trauma could agree with liberals that it is also permissible to destroy embryos where it is necessary to save people or prevent extreme suffering. Of course, the cases are different inasmuch as the existence of the fetus directly conflicts with the pregnant woman's interests, while a particular *ex utero* embryo does not threaten anyone's interests. But this distinction does not bear substantial moral weight. It is the implicit attribution of greater value to the interests of adult humans over any interests the fetus may have that informs the judgment that it is permissible to kill it in the cases at issue. Thus, the following would seem a reasonable formulation of a position on embryo research that liberals and conservatives could agree upon: *Research that involves the destruction of embryos is permissible where there is good reason to believe that it is necessary to cure life-threatening or severely debilitating diseases.*

Given the great promise of stem cell research for saving lives and alleviating suffering, this position would appear to permit research uses of stem cells derived from embryos. Some might object, however, that the benefits of the research are too uncertain to justify a comparison with the abortion exceptions. But the lower probability of benefits from research uses of embryos is balanced by a much higher ratio of potential lives saved for lives lost. Another objection is that it is unnecessary to use embryos for stem cell research because there are alternative means of obtaining stem cells. This reflects an important concern. The derivation of stem cells from embryos is justifiable only if there are no less morally problematic alternatives for advancing the research. At present, there appear to be strong scientific reasons for using embryos. But this is a matter that must continually be revisited as science advances.

It is important to further note that the abortion exceptions that serve as the basis for the type of convergence identified above are exceptions to the law banning federal funding for abortions (Title V, Labor, HHS, and Education Appropriations, 112 Stat. 3681-385, Sec. 509(a) (1)&(2)). Thus, federal funding of research uses of embryos within such limitations as those specified above can reasonably be viewed as consistent with current federal funding practices in the abortion context.

There is, however, one further qualification that must be made here. The justification developed above for federal funding of research that involves the destruction of embryos does not appear to extend to research in which embryos are created expressly for research purposes. While there is wide agreement that it is acceptable to produce embryos for *in vitro* fertilization (IVF), there is much controversy about the moral permissibility of creating embryos for research. Indeed, many who approve of research using embryos remaining from IVF treatments oppose the creation of embryos solely for research purposes (Annas, Caplan, and Elias 1995; Edwards 1990). The main basis for the distinction is that, whereas embryos created for procreative purposes are originally viewed as potential children, embryos created for research are meant to be treated as mere objects of study from the outset. Some dispute that there is a morally relevant distinction between the use of spare IVF embryos and embryos created for research (Harris 1998). But, regardless of which view is more sound, the foundation for agreement on embryo research identified above does not appear to support the use of embryos that are created expressly for research purposes. This limitation on federal funding of stem cell research could ultimately prove to be a significant impediment to the advancement of the research. However, it should come as little surprise that parties on both sides of the issue may incur substantial costs when constructing public policy around points of convergence between competing visions of the meaning and value of life.

References

Annas, G., Caplan, A.L., and Elias, S. 1996. "The Politics of Human-Embryo Research: Avoiding Ethical Gridlock." *New England Journal of Medicine* 334(20):1329–32.

Bopp, J. Jr. 1994. "Fetal Tissue Transplantation and Moral Complicity with Induced Abortion." In *The Fetal Tissue Issue: Medical and Ethical Aspects*, P. Cataldo and A. Moraczewski, eds., 61–79. Braintree, MA: Pope John Center.

Burtchaell, J.T. 1988. "University Policy on Experimental Use of Aborted Fetal Tissue." *IRB: A Review of Human Subjects Research* (10)4:7–11.

Burtchaell, J.T. l989. "The Use of Aborted Fetal Tissue in Research: A Rebuttal." *IRB: A Review of Human Subjects Research* 11(2):9–12.

Charo, R.A. 1995. "The Hunting of the Snark: The Moral Status of Embryos, Right-to-Lifers, and Third World Women." *Stanford Law and Policy Review* 6:11–27.

Childress, J.F. 1991. "Ethics, Public Policy, and Human Fetal Tissue Transplantation Research." *Kennedy Institute of Ethics Journal* 1(2):93–121.

Dworkin, R. 1994. *Life's Dominion: An Argument About Abortion, Euthanasia, and Individual Freedom*. New York: Vintage.

Edwards, R. 1990. "Ethics and Embryology: The Case for Experimentation." In *Experiments on Embryos*, A. Tyson and J. Harris, eds., 42–54. London: Routledge.

Feinberg, J. 1986. "Abortion." In *Matters of Life and Death*, T. Regan, ed., 256–293. New York: Random House.

Freedman, B. 1992. "Moral Analysis and the Use of Nazi Experimental Results." In *When Medicine Went Mad*, A.L. Caplan, ed., 141–154. Totowa, NJ: Humana Press.

Gutmann, A., and Thompson, D. 1996. *Democracy and Disagreement*. Cambridge, MA: Belknap Press.

Harris, J. 1998. *Clones, Gene and Immortality: Ethics and the Genetic Revolution*. New York: Oxford University Press.

Jones, D.G. 1991. "Fetal Neural Transplantation: Placing the Ethical Debate Within the Context of Society's Use of Human Material." *Bioethics* 3(1):22–43.

Letter to Congress and President Clinton, March 4, 1999.

Martin, D.K. 1993. "Abortion and Fetal Tissue Transplantation." *IRB: A Review of Human Subjects Research* 15(3):1–3.

Martin, D.K., et al. 1995. "Fetal Tissue Transplantation and Abortion Decisions: A Survey of Urban Women." *Canadian Medical Association Journal* 153(5):548.

Nagel, T. 1995. "Moral Epistemology." In *Society's Choices*, R.E. Bulger, E.M. Bobby, and H.V. Fineberg, eds., 201–214. Washington DC: National Academy Press.

National Institutes of Health (NIH). 1988. *Report of the Fetal Tissue Transplantation Research Panel*. Bethesda, MD: NIH.

Robertson, J. 1988. "Fetal Tissue Transplant Research Is Ethical." *IRB: A Review of Human Subjects Research* (10)6:5–8.

Robertson, J. 1999. "Ethics and Policy in Embryonic Stem Cell Research." *Kennedy Institute of Ethics Journal* (9)2:109–136.

Seidelman, W.E. 1988. "Mengele Medicus: Medicine's Nazi Heritage." *The Milbank Quarterly*. 66:221–239.

Tooley, M. 1983. *Abortion and Infanticide*. New York: Oxford University Press.

Vawter, D.E., and Gervais, K.G. 1993. "Commentary on Abortion and Fetal Tissue Transplantation." *IRB: A Review of Human Subjects Research* 15(3).

Vawter, D., Kearney, W., Gervais, K., Caplan, A.L., Garry, D., and Tauer, C. 1991. "The Use of Human Fetal Tissue: Scientific, Ethical and Policy Concerns." *International Journal of Bioethics* 2(3):189–196.

Warren, M.A. 1973. "On the Moral and Legal Status of Abortion." *The Monist* 57:43–61.

ETHICAL ISSUES IN HUMAN STEM CELL RESEARCH

VOLUME III
Religious Perspectives

Rockville, Maryland
June 2000

National Bioethics Advisory Commission

Harold T. Shapiro, Ph.D., Chair
President
Princeton University
Princeton, New Jersey

Patricia Backlar
Research Associate Professor of Bioethics
Department of Philosophy
Portland State University
Assistant Director
Center for Ethics in Health Care
Oregon Health Sciences University
Portland, Oregon

Arturo Brito, M.D.
Assistant Professor of Clinical Pediatrics
University of Miami School of Medicine
Miami, Florida

Alexander Morgan Capron, LL.B.
Henry W. Bruce Professor of Law
University Professor of Law and Medicine
Co-Director, Pacific Center for Health Policy and Ethics
University of Southern California
Los Angeles, California

Eric J. Cassell, M.D., M.A.C.P.
Clinical Professor of Public Health
Cornell University Medical College
New York, New York

R. Alta Charo, J.D.
Professor of Law and Medical Ethics
Schools of Law and Medicine
The University of Wisconsin
Madison, Wisconsin

James F. Childress, Ph.D.
Kyle Professor of Religious Studies
Professor of Medical Education
Co-Director, Virginia Health Policy Center
Department of Religious Studies
The University of Virginia
Charlottesville, Virginia

David R. Cox, M.D., Ph.D.
Professor of Genetics and Pediatrics
Stanford University School of Medicine
Stanford, California

Rhetaugh Graves Dumas, Ph.D., R.N.
Vice Provost Emerita, Dean Emerita, and
    Lucille Cole Professor of Nursing
The University of Michigan
Ann Arbor, Michigan

Laurie M. Flynn
Executive Director
National Alliance for the Mentally Ill
Arlington, Virginia

Carol W. Greider, Ph.D.
Professor of Molecular Biology and Genetics
Department of Molecular Biology and Genetics
The Johns Hopkins University School of Medicine
Baltimore, Maryland

Steven H. Holtzman
Chief Business Officer
Millennium Pharmaceuticals Inc.
Cambridge, Massachusetts

Bette O. Kramer
Founding President
Richmond Bioethics Consortium
Richmond, Virginia

Bernard Lo, M.D.
Director
Program in Medical Ethics
The University of California, San Francisco
San Francisco, California

Lawrence H. Miike, M.D., J.D.
Kaneohe, Hawaii

Thomas H. Murray, Ph.D.
President
The Hastings Center
Garrison, New York

Diane Scott-Jones, Ph.D.
Professor
Department of Psychology
Temple University
Philadelphia, Pennsylvania

CONTENTS

Testimony of
Ronald Cole-Turner, M.Div., Ph.D.
Pittsburgh Theological Seminary A-1

Testimony of
Father Demetrios Demopulos, Ph.D.
Holy Trinity Greek Orthodox Church B-1

Testimony of
Rabbi Elliot N. Dorff, Ph.D.
University of Judaism C-1

Testimony of
Margaret A. Farley, Ph.D.
Yale University D-1

Testimony of
Gilbert C. Meilaender, Jr., Ph.D.
Valparaiso University E-1

Testimony of
Edmund D. Pellegrino, M.D.
Georgetown University F-1

Testimony of
Abdulaziz Sachedina, Ph.D.
University of Virginia G-1

Testimony of
Rabbi Moshe Dovid Tendler, Ph.D.
Yeshiva University H-1

Testimony of
Kevin Wm. Wildes, S.J., Ph.D.
Georgetown University I-1

Testimony of
Laurie Zoloth, Ph.D.
San Francisco State University J-1

Nancy J. Duff, Ph.D., Princetown University Theological Seminary, also provided testimony to the National Bioethics Advisory Commission on May 7, 1999. Her written testimony was not available for this publication.

Testimony of

Ronald Cole-Turner, M.Div., Ph.D.
Pittsburgh Theological Seminary

I want to thank the Commission for the attention that you are giving today to religious perspectives regarding human stem cell research and for the opportunity to speak before you. I come as a member of a mainline Protestant denomination, the United Church of Christ, and although no one individual speaks for our church, I will try to represent the positions we have taken and the concerns that we hold.

Let me begin by saying that we do not have an official position regarding the status of embryos. That is not to say we have no opinion or do not care about their rightful status before God. But officially, we have never declared that we regard embryos as persons. Some of our members would agree with that declaration; many—perhaps most—would not agree, believing instead that embryos have an important but lesser status. But we have, deliberately, I think, avoided any such declarations. On the contrary, we have made statements in which we express our openness to embryo research, given certain conditions, which I will come to in a moment.

I quote at length from a report that served as the background to a 1997 General Synod resolution on the question of human cloning:

> Beginning with the 8th General Synod in 1971, various General Synods of the United Church of Christ have regarded the human pre-embryo as due great respect, consistent with its potential to develop into full human personhood. General Synods have not, however, regarded the pre-embryo as the equivalent of a person. Therefore, we on the United Church of Christ Committee on Genetics do not object categorically to human pre-embryo research, including research that produces and studies cloned human pre-embryos through the 14th day of fetal development, provided the research is well justified in terms of its objectives, that the research protocols show proper respect for the pre-embryos, and that they are not implanted. We urge public discussion of current research and future possibilities, ranging from pre-implantation genetic screening of human pre-embryos to nuclear transfer cloning to human germline experimentation. We do not categorically oppose any of these areas of research, but we believe they must be pursued, if at all, within the framework of broad public discussion. In 1989, the 17th General Synod of the United Church of Christ stated that it was 'cautious at present about procedures that would make genetic changes which humans would transmit to their offspring (germline therapy)....We urge extensive public discussion and, as appropriate, the development of federal guidelines during the period when germline therapy becomes feasible'....We on the United Church of Christ Committee on Genetics are opposed to the idea that human pre-embryo research, such as germline experimentation or research involving cloned pre-embryos, should be permitted but left largely unregulated if funded privately, or that there is no federal responsibility for the ethics of such research if federal funds are not used. We believe that this approach merely seeks to avoid the difficult public deliberation that should occur prior to such research. We believe that all such research should be subject to broad public comment and that it should only proceed within a context of public understanding and general public support.[1]

And so when it comes to the specific questions before you regarding the ethics of pluripotent human stem cell research and federal policy in this area, my view is that it is broadly consistent with the views of the United Church of Christ that human stem cell research go forward with federal funds. In fact, we go further and encourage reconsideration of the ban on federal funding for embryo research. We are open to the possibility that somatic cell nuclear transfer be used to create embryos for research, but not implantation, under highly defined research protocols, and that this research, too, be done with public funding.

One of the conditions that we attach to the possibility of this research is that a clear and attainable benefit for science and for medicine be indicated in advance. And it is reasonable to think that now, with pluripotent stem cell technology, such benefit is becoming clearer.

Another condition that we attach is that this research should follow a period of intense and open public discussion. In fact—and let me be as clear as I can about this—all that I have said about our support for research in these areas depends upon meeting the condition of advanced public discussion. I believe this is especially important for this Commission, because it represents one of very few places in our national life where such a conversation can begin.

We stipulate this condition for two reasons. First, we believe that although enormous advances for medicine lie ahead in these areas of research and that we are obliged to work to achieve these advances, our efforts could be undermined, and it could be very bad for science if research proceeds in the short term without broad public understanding and support. Public misunderstanding and public exclusion from discussion could result in public rejection of this and related forms of research.

The second reason why we set forth the condition of advanced public discussion and support is that we value living in a society that makes basic public moral decisions based on the deliberations of informed citizens. As a historic church, our congregational forebears extended congregational decisionmaking to the public square. As a church today, we believe that our views are not the only views worth hearing, but that public policy on morally problematic issues should be the result of honest and sustained discourse during which all views are brought forward in public. This view of a public society is an article of faith with us.

As a commission, you are, of course, under time constraints and must offer your report on specific policy questions. As a church, we offer at least some support for the view that federally funded research in embryonic stem cells, and possible even in embryos, should move forward as quickly as possible. But on the basis of the condition our Church has set on this support, I ask you to do whatever you can in your report to satisfy our condition by helping to bring about a new, open, and sustained national discussion of these difficult questions and issues. Such a sustained discussion may be well beyond your mandate and may require some new institutional platform, but you are one of the key voices in our national life that can urge that this challenge be taken on for the good of research, for the good of public support of research, and for the good of the kind of society in which we want to live.

I will conclude by noting two concerns, both of which involve contextual factors that a church such as ours will bring to the discussion table that I am urging you to help create. The first is social justice. Precisely because this research promises such great benefit, we worry that the benefit will be distributed unevenly and therefore that it will further the position of the rich and the powerful at the expense of the poor and the weak. We believe that the moral test of any system, including our system of medical research and treatment, is how well it treats the least privileged members of society, first of all within our own nation, but also globally. And so we would challenge those who fund and develop these therapies by asking the following question: How will the benefits be shared universally? We are aware that the difficult problems of delivery and cost recovery must be considered, but in offering our support for this research based on the promise of medical benefit, we do not mean that the benefit should be distributed only by market means.

The second concern involves the broader scientific and medical context of research. It is impossible for any of us to offer a moral assessment of human stem cell technology in isolation from other current or pending areas of research, among them somatic cell nuclear transfer and human germline modification. Through these technologies and the combination of these technologies, we are about to acquire a wholly unprecedented level of control over our health, our longevity, and our offspring. And so I urge you to do whatever is in your power not only to create a broad public discussion, but also to define its agenda broadly as involving this wide but inter-related set of emerging technologies.

I conclude with this simple observation: If the question before us is narrowly defined as involving embryos and stem cells, the various religious traditions will take different positions. But if the question is framed in terms of concern for social justice or of our ability to chart our common future in view of the overwhelming changes that lie ahead, the various religious traditions will find there is much upon which to agree. If this is correct, then we might find greater understanding on the narrow issues as we move along the pathway of greater engagement on the contextual issues.

Note

1 "'Statement on Cloning' by the United Church of Christ Committee on Genetics," in Ronald Cole-Turner ed., *Human Cloning: Religious Responses* (Louisville, KY: Westminster John Knox Press, 1997), 149–151.

Testimony of

Father Demetrios Demopulos, Ph.D.
Holy Trinity Greek Orthodox Church

An Eastern Orthodox View of Embryonic Stem Cell Research

I would like to thank the Commission for providing me with an opportunity to present an Orthodox view of the ethical problems and challenges associated with human embryonic stem cell research. I would like to emphasize that I do not speak for the Greek Orthodox Church but instead offer comments that I believe are consistent with the teachings and tradition of the Orthodox Church.

The Orthodox Church has a long tradition of encouraging the "medical art" that alleviates unnecessary pain and suffering and restores health. The Church, however, also has reminded us that this art is given to us by God to be used according to His will, not our own, since "the medical art has been vouchsafed us by God, who directs our whole life, as a model for the cure of the soul" and "we ought not commit outrage against a gift of God by putting it to bad use."[1] What constitutes bad use is what has brought us together here today. An important consideration for the Orthodox is based on our understanding of what it is to be a human person.

Humans are created in the image and likeness of God and are unique in creation because they are psychosomatic, beings of both body and soul—physical and spiritual. We do not understand this mystery, which is analogous to that of the Theanthropic Christ, who at the same time is both God and a human being. We do know, however, that God intends for us to love Him and grow in relationship to Him and to others until we reach our goal of theosis, or deification, participation in the Divine Life through His grace. We grow in the image of God until we reach the likeness of God. Because we understand the human person as one who is in the image and likeness of God, and because of sin we must strive to attain that likeness, we can say that an authentic human person is one who is deified. Those of us who are still struggling toward theosis are human beings, but potential human persons.[2]

We believe that this process toward authentic human personhood begins with the zygote. Whether created *in situ* or *in vitro*, a zygote is committed to a developmental course that will, with God's grace, ultimately lead to a human person. The embryo and the adult are both potential human persons, although in different stages of development. As a result, Orthodox Christians affirm the sanctity of human life at all stages of development. Unborn human life is entitled to the same protection and the same opportunity to grow in the image and likeness of God as are those already born.

Given this Orthodox understanding of human personhood and life, I cannot condone any procedure that threatens the viability, dignity, and sanctity of that life. In my view, the establishment of embryonic stem cell lines[3] was done at the cost of human lives. Even though not yet a human person, an embryo should not be used for or sacrificed in experimentation, no matter how noble the goal may seem. For me, then, the derivation of embryonic stem cell lines is immoral because it sacrificed human embryos, which were committed to becoming human persons. That the embryos donated for this work were not going to be implanted and had no chance of completing their development cannot mitigate the fact that they should not have been created. *In vitro* fertilization techniques that routinely result in "surplus" embryos that are eventually discarded is immoral for the same reasons I have already mentioned. I believe, then, that the prohibition of research using human embryos should be continued and, if possible, extended to the private sector as well.

Wishing that something had not been done will not undo it. Established embryonic stem cell lines exist, and their use has great potential benefits for humanity, which need not be reviewed here. The Orthodox Church, as I mentioned before, has a long tradition of encouraging the medical arts. We have a long list of healer-saints—physicians who became authentic persons through the practice of medicine. Invariably, they obeyed the commandment of Christ to his apostles, "Heal the sick, raise the dead, cleanse lepers, cast out demons. You received without paying, give without pay."[4] Without going into an extensive exegesis of the verse, the intention is clear: Attend not to profit, but to the medical needs of others.

Using our healer-saints as a paradigm, I am concerned about how the existing stem cell lines will be used. Will they be used to heal, or will they be used to maximize profits? Market forces are very strong, and, in my opinion, are often contrary to the general good. Allowing the cell lines to be owned by private companies that are responsible first to their stockholders and investors rather than to the general welfare may compromise the use of the lines. It is imperative that steps be taken to ensure that the lines be used only for therapeutic procedures that will benefit those in need and not be limited to the few who will be able to afford them. I want to emphasize that the lines must be used only therapeutically, to restore health and to prevent premature death. They must not be used cosmetically or to further any eugenic agenda. None of us is physically perfect, but all are called to be perfected in Christ. Part of our challenge to participate in the Divine Life is to overcome our deficiencies. We must not attempt to re-create ourselves in our own image.

Because stem cell lines have such great potential for healing, efforts should be made to encourage discovery of more morally acceptable sources. A recent report suggests that adult stem cells may be less restricted than previously thought.[5] It may be possible to develop techniques to culture such cells without the need to sacrifice the donor. Alternatively, because organ donation is viewed favorably by many (but not all) Orthodox Christians, I would accept cell lines derived from fetal primordial germ cells, but only in cases of spontaneous miscarriage. A fetus cannot be killed for an organ, just as an adult cannot. Also, great care must be taken to assure that the mother's consent is truly informed.

In summary, the Orthodox Church promotes and encourages therapeutic advances in medicine and the research necessary to realize them, but not at the expense of human life. The Church considers human life to begin with the zygote and to extend beyond our physical death, as we were promised eternal life by our God and Savior. Recognizing that we are all in a sinful and imperfect state, the Church admonishes us to strive for perfection through God's grace as we strive to become authentic human persons in communion with God. Because we tend to follow our own will rather than God's, we are reminded to be discerning so that we do not commit outrages by putting a gift of God to bad use.

Notes

1 St. Basil the Great, *The Long Rules* 55, M.M. Wagner, tr., *St. Basil: Ascetical Works, The Fathers of the Church*, vol. 9 (Washington, D.C.: The Catholic University of America Press, 1962), 330–37.

2 See also, Nellas, P., *Deification in Christ: The Nature of the Human Person*, Chapters 1 and 2 (Crestwood, New York: St. Vladimir's Seminary Press, 1987) and Breck, J., *The Sacred Gift of Life: Orthodox Christianity and Bioethics*, Chapters 1 and 3 (Crestwood, New York: St. Vladimir's Seminary Press, 1998).

3 Thomson, J.A., Itskovitx-Eldor, J., Shapiro, S.S., Waknitx, M.A., Swiergiel, J.J., Marshall, V.S., and Jones, J.M., "Embryonic Stem Cell Lines Derived from Human Blastocysts," *Science* 282 (1998):1145–1147.

4 Matthew 10:8.

5 Bjornson, C.R., Rietze, R.L., Reynolds, B.A., Magli, M.C., and Vescovi, A.L., "Turning Brain into Blood: A Hematopoietic Fate Adopted by Adult Neural Stem Cells In Vivo," *Science* 283 (1999):534–537.

Testimony of
Rabbi Elliot N. Dorff, Ph.D.
University of Judaism

Stem Cell Research

A. Fundamental Theological Convictions[1]

1. The Jewish tradition uses both theology and law to discern what God wants of us. No legal theory that ignores the theological convictions of Judaism is adequate to the task, for any such theory would lead to blind legalism without a sense of the law's context or purpose. Conversely, no theology that ignores Jewish law can speak authoritatively for the Jewish tradition, for Judaism places great trust in law as a means to discriminate moral differences in similar cases, thus giving us moral guidance. My understanding of Judaism's perspective on stem cell research will, and must, draw on both theological and legal sources.

2. Our bodies belong to God; we have them on loan during our lease on life. God, as owner of our bodies, can and does impose conditions on our use of our bodies. Among those conditions is the requirement that we seek to preserve our lives and our health.

3. The Jewish tradition accepts both natural and artificial means for overcoming illness. Physicians are the agents and partners of God in the ongoing act of healing. Thus, the mere fact that human beings created a specific therapy rather than finding it in nature does not impugn its legitimacy. On the contrary, we have a duty to God to develop and use any therapies that can aid us in taking care of our bodies, which ultimately belong to God.

4. At the same time, all human beings, regardless of level of ability and disability, are created in the image of God and are to be valued as such.

5. Moreover, we are not God. We are not omniscient, as God is, and so we must take whatever precautions we can to ensure that our actions do not harm ourselves or our world in our very effort to improve them. A certain epistemological humility, in other words, must pervade whatever we do, especially when we are pushing the scientific envelope, as we are in stem cell research. We are, as Genesis says, supposed to work the world *and* preserve it; it is the achievement of that *balance* that is our divine duty.[2]

B. Jewish Views of Genetic Materials

1. Because doing research on human embryonic stem cells involves procuring them from aborted fetuses, the status of abortion within Judaism is a subject that immediately arises. Within Judaism, by and large, abortion is forbidden. The fetus, during most of its gestational development, is seen as "the thigh of its mother," and neither men nor women may amputate their thigh at will, because that would be injuring their bodies, which belong to God. On the other hand, if the thigh turns gangrenous, both men and women have the positive duty to have their thigh amputated in order to save their lives. Similarly, if a pregnancy endangers a woman's life or health, an abortion *must* be performed to save her life or protect her physical or mental health, for she is without question a full-fledged human being with all the protections of Jewish law, while the fetus is still only part of the woman's body. When there is an elevated risk to the woman beyond that of normal pregnancy, but insufficient risk to constitute a clear threat to her life or health, abortion is permitted, but it is not required. That is an assessment that the woman should make in consultation with her physician. Some recent authorities also would permit abortions in cases where genetic testing indicates that the fetus will suffer from a terminal disease such as Tay-Sachs or from serious malformations.[3]

The Jewish stance on abortion, then, is that *if* a fetus was aborted for legitimate reasons under Jewish law, it may be used to advance our efforts to preserve the life and health of others. In general, when a person dies, we must show honor to God's body by burying it as soon as possible after death. To benefit the lives of others, however, autopsies may be performed when the cause of death is not fully understood, and organ transplants are allowed to enable other people to live.[4] The fetus, as I have said, does not have the status of a full-fledged human being. Therefore, if we can use the body of a human being to enable others to live, how much the more so may we use a part of a body—in this case, the fetus—for that purpose. This all presumes that the fetus was aborted for good and sufficient reason within the parameters of Jewish law.

2. Stem cells for research purposes also can be procured from donated sperm and eggs mixed together and cultured in a petri dish. Genetic materials outside the uterus have no legal status in Jewish law, for they are not even a part of a human being until implanted in a woman's womb, and even then, during the first 40 days of gestation, their status is "as if they were simply water."[5] Abortion is still prohibited during that time, except for therapeutic purposes, for in the uterus such gametes have the potential of growing into a human being. Outside the womb, however, at least at this time, they have no such potential. As a result, frozen embryos may be discarded or used for reasonable purposes and so may the stem cells that are procured from them.

C. Other Factors in This Decision

1. Given that the materials for stem cell research can be procured in permissible ways, the technology itself is morally neutral. It gains its moral valence on the basis of what we do with it.

2. The question, then, is reduced to a risk-benefit analysis of stem cell research. The articles in the most recent *Hastings Center Report*[6] raise some questions to be considered in such an analysis, but I will not rehearse them here. I want to note only two things about them from a Jewish perspective:

   a. The Jewish tradition views the provision of health care as a communal responsibility, and so the justice arguments in the *Hastings Center Report* have a special resonance for me as a Jew. Especially because much of the basic science in this area was funded by the government, the government has the right to require private companies to provide their applications of that science at reduced rates, or if necessary, at no cost, to those who cannot afford them. At the same time, the Jewish tradition does not demand socialism, and for many good reasons we in the United States have adopted a modified, capitalistic system of economics. The trick, then, will be to balance access to applications of the new technology with the legitimate right of a private company to make a profit on its efforts to develop and market those applications.

   b. The potential of stem cell research for creating organs for transplant and cures for diseases is, at least in theory, both awesome and hopeful. Indeed, in light of our divine mandate to seek to maintain life and health, one might even argue that from a Jewish perspective we have a *duty* to proceed with that research. As difficult as it may be, we must draw a clear line between uses of this or any other technology for cure, which are to be applauded, as opposed to uses of this technology for enhancement, which must be approached with extreme caution. Jews have been the brunt of campaigns of positive eugenics both here, in the United States, and in Nazi Germany,[7] and so we are especially sensitive to creating a model human being that is to be replicated through the kind of genetic engineering that stem cell applications will involve. Moreover, when Jews see a disabled human being, we are not to recoil from the disability or count our blessings for not being disabled in that way; rather, we are commanded to recite a blessing thanking God for making people different.[8] Thus, in light of the Jewish view that all human beings are created in the image of God, regardless of their levels of ability or disability, it is imperative from a Jewish perspective that the applications of stem cell research be used for cure and not for enhancement.

D. Recommendation

My recommendation is that we take the steps necessary to advance stem cell research and its applications in an effort to take advantage of its great potential for good. We should do so, however, in such a way that we provide access to its applications to all Americans who need them and at the same time prohibit the development of applications intended to make all human beings fit any particular model of human excellence. Through this technology, we should seek to cure diseases and to appreciate the variety of God's creatures.

Notes

1 For more on these and other fundamental assumption of Jewish medical ethics, and for the Jewish sources that express these convictions, see Dorff, E.N., *Matters of Life and Death: A Jewish Approach to Modern Medical Ethics*, Chapter 2 (Philadelphia: Jewish Publication Society, 1998).

2 Genesis 2:15.

3 For more on the Jewish stance on abortion, together with the biblical and rabbinic sources that state that stance, see Dorff, *Matters of Life and Death*, 128–133, and Feldman, D.M., *Birth Control in Jewish Law: Marital Relations, Contraception, and Abortion as Set Forth in the Classic Texts of Jewish Law* (New York: New York University Press, 1968), reprinted under the title *Marital Relations, Abortion, and Birth Control in Jewish Law*, Chapters 14 and 15 (New York: Schocken, 1973).

4 For classical sources on this, see Dorff, *Matters of Life and Death*, Chapter 9.

5 Babylonian Talmud, *Yevamot* 69b. Rabbi Immanuel Jakobovits notes that "40 days" in talmudic terms may mean just under two months in our modern way of calculating gestation, since the rabbis counted from the time of the first missed menstrual flow while we count from the time of conception, approximately two weeks earlier. See Jakobovits, I., *Jewish Medical Ethics: A Comparative and Historical Study of the Jewish Religious Attitude to Medicine and Practice* (New York: Bloch Publishing Company, 1959, 1975), 275.

6 *Hastings Center Report*, March-April (1999):30–48.

7 See Gould, S.J., *The Mismeasure of Man* (New York: W.W. Norton and Company, 1996) and Annas, G.J., and Grodin, M.A., *The Nazi Doctors and the Nuremberg Code: Human Rights in Human Experimentation* (New York: Oxford, 1992).

8 For a thorough discussion of this blessing and concept in Jewish tradition, see Astor, C., "...*Who Makes People Different:" Jewish Perspectives on the Disabled* (New York: United Synagogue of America, 1985).

Testimony of

Margaret A. Farley, Ph.D.
Yale University

Roman Catholic Views on Research Involving Human Embryonic Stem Cells

The Roman Catholic moral tradition offers potentially significant perspectives on questions surrounding research on human embryonic stem cells. I use the plural, "perspectives," because there is no simple, single voice from the Catholic community on such questions. There is, however, a shared "community of discourse," so that one can easily identify common convictions expressed in a common language as well as specifically divergent views on this and other particular moral issues.

First, then, the common convictions: The Catholic tradition is undivided in its affirmation both of the goodness of creation and the importance of human agency in the ongoing processes within creation. God is actively present in the world, and human persons are called to discern the sacredness of creation and their own responsibilities as, in a sense, co-creators with God. With one mind, Catholics also affirm the importance of both the individual and the community, seeing these not finally as competitors but as essentially in need of each other for the fulfillment of both. It is never possible from this tradition to justify, in an ultimate sense, the sacrifice of an individual to the community or to forget the common good when thinking about the individual. It is also clear to everyone in the Catholic tradition that human persons are responsible for their offspring in ways particular to humans and that future generations matter both in this world and in a hoped-for unlimited future.[1] The Catholic tradition is unified in its belief in God's active and intimate care for the world and each person in it and in our own correlative obligations to care for those who are in need—preventing unjustified harm, alleviating pain, and protecting and nourishing the well-being of individuals and the wider society. There are deep roots in the Catholic tradition that anchor a commitment to the poorest, the most marginalized, and the most ill, and that in doing so sustain a commitment to human equality in its most basic sense.

At the same time, there are clear disagreements among Catholics (whether moral theologians, Church leaders, or ordinary members of the Catholic community) on, for example, particular issues of fetal and embryo research, assisted reproductive technologies, and the prospects for morally justifiable human stem cell research. These disagreements include conflicting assessments of the moral status of the human embryo and the use of aborted fetuses as sources of stem cells.

That there is so much agreement on fundamental approaches to human morality yet disagreement on specific moral rules is not surprising. For one thing, affirmations of the goodness of creation, human agency, and principles of justice and care do not always yield directly deducible recommendations on specific questions such as stem cell research. Or again, genuine concerns for the moral fabric of society do not by themselves settle empirical questions regarding possible good or bad consequences of the development of particular technologies. There is, for example, often no easy and direct way to determine whether a particular set of choices regarding scientific research will violate the rights of some persons to basic medical care or undermine respect for the dignity of each individual.

At the heart of the Catholic tradition, however, there is a conviction that creation is itself revelatory and knowledge of the requirements of respect for created beings is accessible at least in part to human reason. This is what is at stake in the Catholic tradition's understanding of natural law. For most of its history, a Catholic natural law theory has not assumed that morality can simply be "read off" of nature, not even with the important help of Scripture. Nonetheless, what natural law theory does is tell us where to look—that is, to the concrete reality of the world around us, to the basic needs and possibilities of human persons in relation to one another, and to the world as a whole. "Looking" (to concrete reality) means a complex process[2] of discernment and deliberation, and a structuring of insights and determination of meaning, from the fullest vantage point available, given a particular history—one that includes the illumination of Scripture and the accumulated wisdom of the tradition. The limits, yet necessity, of this process account for many of the disagreements about specific matters, even within the faith community.

This brings us, then, to disagreements regarding human embryonic stem cell research. Those who stand within the Catholic tradition tend to "look" to the reality of stem cells and, what is more relevant in this instance, to the realities of the sources of stem cells for current research—that is, human embryos and fetuses. Within the Catholic tradition, a case can be made both against and for such research—each dependent upon different interpretations of the moral status of the human embryo and the aborted human fetus. There are, first, a significant number of Catholics, including present spokespersons for the American bishops,[3] who make the case *against*. They argue that human embryos must be protected on a par with human persons—at least to the extent that they should not be either created or destroyed merely for research purposes. Moreover, the use of aborted fetuses as a source for stem cells, while not in one sense different from the harvesting of tissue from any human cadavers, nonetheless should be prohibited because it is complicit with and offers a possible incentive for elective abortion. (If the fetuses in question have been spontaneously aborted, however, some opening is allowed for their use in this research.[4]) Part of the case against embryo stem cell research also rests on the identification of alternatives (the use of adult cells, dedifferentiated and redifferentiated into specific lineages[5]). One can also presume that the case against embryo stem cell research includes a case against cloning, if and insofar as this research incorporates steps involved in procedures for cloning.[6]

But on the other hand, a case *for* human embryo stem cell research can also be made on the basis of positions developed within the Catholic tradition. A growing number of Catholic moral theologians, for example, do not consider the human embryo in its earliest stages (prior to the development of the primitive streak or to implantation) to constitute an individualized human entity with the settled inherent potential to become a human person. The moral status of the embryo is, therefore (in this view), not that of a person, and its use for certain kinds of research can be justified. (Because it is, however, a form of human life, it is due some respect—for example, it should not be bought or sold.) Those who would make this case argue for a return to the centuries-old Catholic position that a certain amount of development is necessary in order for a conceptus to warrant personal status.[7] Embryological studies now show that fertilization ("conception") is itself a process (not a "moment"), and such studies provide support for the opinion that in its earliest stages (including the blastocyst stage, when stem cells would be extracted for purposes of research) the embryo is not sufficiently individualized to bear the moral weight of personhood.[8] Moreover, some of the concerns regarding the use of aborted fetuses as a source for stem cells can be alleviated if safeguards (such as ruling out "direct" donation for this purpose[9]) are put in place—not unlike the restrictions articulated for the general use of fetal tissue for therapeutic transplantation. And finally, concerns about cloning may be at least partially addressed by insisting on an absolute barrier between cloning for research and therapeutic purposes on the one hand and cloning for reproductive purposes on the other (the latter, of course, raising many more serious ethical questions than the former).

We have, then, two opposing cases articulated within the Roman Catholic tradition. It would be a mistake to conclude that what this tradition has to offer, however, is only a kind of "draw." It offers, rather, an ongoing process of discernment that remains faithful to a larger set of theological and ethical convictions, that takes account of the best that science can tell us about some aspects of reality, and that aims to make one or the other case persuasive on the basis of reasons whose intelligibility is open to the scrutiny of all. I myself stand with the case *for* embryonic stem cell research, and I believe this case can be made persuasively both within the Catholic tradition and in the public forum. The newest information we have from embryological studies supports this case, and I would argue that it can be made without sacrificing the tradition's commitments to respect human life, promote human well-being, and honor the sacred in created realities. Further, to move forward with human embryonic stem cell research need not soften the tradition's concerns to oppose the commercialization of human life and to promote distributive justice in the provision of medical care.[10]

Our tradition's ongoing conversation on such matters yields more light than I have time to show here. It is also a reminder to all of us of the importance of epistemic humility, especially if and as we decide to open more and more room for the human control of creation.

Notes

1 This implies that for those in the Roman Catholic tradition, a goal of longer and longer life spans is not an unqualified or in itself absolute good. This has some relevance for arguments for stem cell research that suggest a major goal of a greatly expanded human life span.

2 Hence, the intelligibility of "realities" is not such that their meaning is immediately obvious. What is given to our understanding through experience is not only always partial, but it must always be *interpreted*.

3 See Doerflinger, R., "Destructive Stem-Cell Research on Human Embryos," Origins 28 (1999):769–773; see also "Donum Vitae (Respect for Human Life)," Origins 16 (1987):697–711; and Grisez, G., "When Do People Begin?" *Proceedings of the American Philosophical Association* 63 (1990).

4 The difficulty often noted regarding this option, however, is that spontaneously aborted fetuses are frequently not a source for healthy cells or tissue (there is a reason why they spontaneously aborted).

5 See, for example, Pittenger, M.F., et al., "Multilineage Potential of Adult Human Mesenchymal Stem Cells," *Science* 284 (1999):143–147; and Wade, N., "Discovery Bolsters a Hope for Regeneration," *New York Times* (2 April 1999). This alternative could prove to be extremely important precisely because it does not involve the harvesting of stem cells either from embryos or from aborted fetuses. Many scientists, however, consider this alternative as too distant (in terms of the research still needed to develop it) to be a realistic competing possibility.

6 There is insufficient time to expand on the relevance of this point. But some stem cell research, at least, does involve the first stages of cloning—although the goal is not to bring a clone to birth.

7 See, for example, Donceel, J., "Immediate and Delayed Hominization," *Theological Studies* 31 (1970):76–105. The early views on this matter were, of course, based on inadequate knowledge of reproductive biology, and twentieth-century views that hold the presence of potential for personhood from the "moment" of conception are based on more adequate knowledge. The contemporary position on delayed "hominization," however, is argued on the basis of more recent embryological studies. For the Catholic tradition, science is extremely important for theology, though it is not determinative in every case.

8 See, for example, Shannon, T.A., and Walter, A.B., "Reflections on the Moral Status of the Pre-Embryo," *Theological Studies* 51 (1990):603–626; McCormick, R.A., "Who or What Is the Preembryo?" *Corrective Vision: Explorations in Moral Theology* (Kansas City, MO: Sheed and Ward, 1994), 176–188; and Cahill, L.S., "The Embryo and the Fetus: New Moral Contexts," *Theological Studies* 54 (1993):124–142.

9 That is, ruling out the possibility of a woman who elects abortion directly donating fetal stem cells for therapeutic treatment of someone she knows. Other safeguards insist that the investigator also not be the attending physician for an abortion.

10 These and other concerns are urgent in regard to the overall question of human stem cell research. However, there is insufficient time to pursue them, or even articulate them, here.

Testimony of
Gilbert C. Meilaender, Jr., Ph.D.
Valparaiso University

Thank you for the invitation to address your Commission. I had such an opportunity on one previous occasion, and, at the risk of simply repeating myself, I would like at the outset to make clear just a few of the qualifications that must be applied to everything I say here today.

As I understand it, I have been invited to speak specifically in my capacity as a Protestant theologian, and I will try to do so. At the same time, I cannot claim to speak for Protestants generally—alas, no one can. I will, however, try to draw on several theologians who speak from within different strands of Protestantism. I think you can and should assume that a significant number of my co-religionists more or less agree with the points I will make. You can, of course, also assume that other Protestants would disagree, even though I like to think that, were they to ponder these matters long enough, they would not.

Moreover, I have tried not to think of what I am doing as an attempt by some Protestant "interest group" to put its oar into your deliberations. Although I will begin as best I can from somewhere rather than nowhere, that is, from within a particular tradition, its theological language seeks to uncover what is universal and human. It begins epistemologically from a particular place, but it opens up ontologically a vision of the human. You might, therefore, be interested in it not only because it articulates the view of a sizable number of our fellow citizens but also because it seeks to uncover a vision of the life that we share.

Finally, I confess at the outset that the topic before you—human embryonic stem cell research—raises for me complexities that I do not fully understand. As I have tried to follow recent developments, they have often seemed bewildering. You, no doubt, understand them better than I, but perhaps I can also bring an angle of vision that will enrich your deliberations.

To that end I will make three points. For each of the three points, I will take as my starting point a sentence from a well-known Protestant thinker—not in order to claim that theologian's authority for or agreement with what I have to say, but simply to provide some "texts" with which to begin my reflections.

First, a passage from Karl Barth, perhaps the greatest of twentieth-century theologians, who writes from the Reformed (Calvinist) tradition: "No community, whether family, village or state, is really strong if it will not carry its weak and even its very weakest members."[1]

This sentence invites us to ponder the status of the human embryo—the source of many, though not all, of the stem cells that would be used in research. One of the complexities that I do not fully understand involves the question of whether stem cells are not themselves and cannot develop into embryos. I will assume that they are not and cannot, although perhaps I need to be instructed further on this matter. Even in making this assumption, however, we face the fact that procuring embryonic stem cells for research requires the destruction of the embryo. Hence, we cannot avoid thinking about its moral status.

No doubt in our society it is impossible to contemplate this question without feeling sucked back into the abortion debate, and we may sometimes have the feeling that we cannot consider any other related question without always ending up arguing about abortion. Perhaps there is something to that, and I will not entirely avoid it myself before I am done, but the question of using (and destroying) embryos in research is a separate question. The issue of abortion, as it has been framed in our society's debate and in Supreme Court decisions, has turned chiefly on a conflict between the claims of the fetus and the claims of the pregnant woman. It is precisely that conflict, and our seeming inability to serve the woman's claim without turning directly against the life of the fetus, that has been thought to justify abortion. But there is no such direct conflict of lives involved in the instance of embryo research.

Here, as in so many other areas of life, we must struggle to think inclusively rather than exclusively about the human species, about who is one of us, and about whose good should count in the common good we seek to fashion. The embryo is, I believe, the weakest and least advantaged of our fellow human beings, and no community is really strong if it will not carry its weakest members.

This is not an understanding shaped chiefly in the fires of recent political debate; rather, it has very deep roots in Christian tradition, and, invited as I have been to address you from within that tradition, I need to explore briefly those roots. We have become accustomed in recent years to distinguishing between persons and human beings, to thinking about personhood as something added to the existence of a living human being—and then to debating where to locate the time when such personhood is added. There is, however, a much older concept of the person—for which no threshold of capacities is required—that was deeply influential in Western history and that had its roots in some of the most central Christian affirmations. The moral importance of this understanding of the person has been noted recently by the Anglican theologian, Oliver O'Donovan.[2]

Christians believed that in Jesus of Nazareth, divine and human natures were joined in one person, and, of course, they understood that it was not easy to make sense of such a claim. For if Jesus had both divine and human natures, he would seem to be two persons, two individuals, identified in terms of two sets of personal capacities or characteristics—a sort of chimera, we might say, in terms appropriate to this gathering.

So Christian thinkers turned in a different direction that was very influential in our culture's understanding of what it means to be an individual. In their view, a person is not someone who has a certain set of capacities; a person is simply, as O'Donovan puts it, a "someone who"—a someone who has a history. That story, for each of us, begins before we are conscious of it, and, for many of us, may continue after we have lost consciousness of it. It is nonetheless our personal history even when we lack awareness of it, even when we lack or have lost certain capacities characteristic of the species.

This is, as I noted, an insight that grew originally out of intricate Christological debates carried on by thinkers every bit as profound as any we today are likely to encounter. But starting from that very definite point, they opened up for us a vision of the person that carries deep human wisdom, that refuses to think of personhood as requiring certain capacities, and that therefore honors the time and place of each someone who has a history. In honoring the dignity of even the weakest of living human beings—the embryo—we come to appreciate the mystery of the human person and the mystery of our own individuality.

Second, a sentence from the late John Howard Yoder, a well-known Mennonite theologian: "I am less likely to look for a saving solution if I have told myself beforehand that there can be none, or have made advance provision for an easy brutal one."[3]

Stem cell research is offered to us as a kind of saving solution, and it is not surprising therefore that we should grasp at it. Although I suspect that promises and possibilities could easily be oversold, none of us should pretend to be indifferent to attempts to relieve or cure heart disease, Parkinson's and Alzheimer's diseases, or diabetes. Suffering, and even death, are not the greatest evils of human life, but they are surely bad enough—and all honor goes to those who set their face against such ills and seek to relieve them.

The sentence from Yoder reminds us, however, that we may sometimes need to deny ourselves the handiest means to an undeniably good end. In this case the desired means will surely involve the creation of embryos for research—and then their destruction. The human will, seeing a desired end, takes control, subjecting to its desire even the living human organism. We need to ask ourselves whether this is a road we really want to travel to the very end. Learning to think of human beings as will and freedom alone has been the long and steady project of modernity. At least since Kant, ethics has often turned to the human will as the only source of value. But C. S. Lewis, an Anglican and surely one of the most widely read of twentieth-century Christian thinkers, depicted what happens when we ourselves become the object of this mastering will:

> We reduce things to mere Nature in order that we may 'conquer' them. We are always conquering Nature, because 'nature' is the name for what we have to some extent conquered. The price of conquest is to treat a thing as mere Nature....The stars do not become Nature till we can weigh and measure them: the soul does not become Nature till we can psycho-analyse her. The wresting of powers from Nature is also the surrendering of things to Nature. As long as this process stops short of the final stage we may well hold that the gain outweighs the loss.

> But as soon as we take the final step of reducing our own species to the level of mere Nature, the whole process is stultified, for this time the being who stood to gain and the being who has been sacrificed are one and the same. This is one of the many instances where to carry a principle to what seems its logical conclusion produces absurdity. It is like the famous Irishman who found that a certain kind of stove reduced his fuel bill by half and thence concluded that two stoves of the same kind would enable him to warm his house with no fuel at all....[I]f man chooses to treat himself as raw material, raw material he will be.[4]

What Yoder reminds us is that only by stopping, only by declining to exercise our will in this way, do we force ourselves to look for other possible ways to achieve admittedly desirable ends. Only by declining to use embryos for this research do we awaken our imaginations and force ourselves to seek other sources for stem cells—as may be possible, for example, if recent reports are to be believed, by deriving stem cells from bone marrow or from the placenta or umbilical cord in live births. The discipline of saying no to certain proposed means stimulates us to think creatively about other, and better, possibilities.

One such possibility will, however, be almost as controversial as deriving stem cells from embryos, and it must, therefore, be noted here. I refer to the possibility of deriving stem cells from the germ cells of aborted fetuses. I have opposed the use of embryos for stem cell research, and I also want, in the last analysis, to oppose this method of acquiring the cells, but the reasons are not immediately apparent. On the face of it, after all, this is simply another form of tissue or organ donation from a cadaver. It does not use—or create and then use—a living human being solely for research purposes. Obviously, though, it threatens to suck us back into the situation I described earlier: where every problem becomes, ultimately, the abortion problem. And here, I fear, we cannot so easily separate the issues, although there are, of course, various procedural safeguards that can be put in place in order to try to assure ourselves that the promised benefits of research do not in any way encourage abortion.

We can clarify our own judgments on the matter by two simple thought experiments that aim to distinguish the several moral issues interwoven here. Would we object to research using tissue acquired only from spontaneously aborted (miscarried) fetuses? I cannot see why we should—though, of course, it is not really very helpful to propose such a source. Would we object to research using tissue acquired only from those abortions which, though induced and intended, were abortions we thought permissible (however large or small that class might be)? This, at least in my view, is a harder call. But to use for the benefit of others those whom we have already (even if legitimately) condemned to die is so clearly an example of the strong using the weak that I think we should draw back and say no. The life of a human being has been sacrificed in abortion, legitimately, by hypothesis, for the good of someone else. And, as Kathleen Nolan once put it, "a moral intuition insists that being used once is enough."[5] We need to challenge ourselves to look for other, better solutions.

Third, a passage from Stanley Hauerwas, a Methodist theologian: "The church's primary mission is to be a community that keeps alive the language and narrative necessary to form lives in a truthful manner."[6]

Hauerwas does not mean that Christians are necessarily more truthful than other people. He means that, when they are doing what they ought to be doing, they worry lest we deceive ourselves, lest we fail to speak the truth about who we are individually and communally, and about what we are doing. This is certainly important for our larger society, and I am quite sincere when I say that—whatever this Commission decides to recommend—you can do us all an enormous service if you will speak truly and straightforwardly and if you will help us avoid euphemism and equivocation, so that we may together think clearly about who we are and wish to be.

What, more precisely, do I have in mind? I have in mind matters such as the following: that we avoid sophistic distinctions between funding research on embryonic stem cells and funding the procurement of those cells from embryos; that we not deceive ourselves by supposing that we will use only "excess" embryos from infertility treatments, having in those treatments created far more embryos than are actually needed;[7] that we speak simply of embryos, not of the "pre-embryo" or the "pre-implantation embryo" (which is really the unimplanted embryo); and that, if we forge ahead with embryonic stem cell research, we simply scrap the language of "respect" or "profound respect" for those embryos that we create and discard according to our purposes. Such language does not train us to think seriously about the choices we are making, and it is, in any case, not likely to be believed. You can help us to think and speak truthfully, and that would be a very great service indeed.

I have pressed these three points with some reluctance, because I have the sense—as you may well imagine—that I will be taken to be standing athwart history and yelling, "Stop!" But it is a risk worth taking. We may easily deceive ourselves about what we do—especially when we do it in a good cause, with a good conscience. We need help if we are to learn to speak truthfully and to face with truthfulness the choices we make, and, whatever this Commission's precise determinations, I hope you will give us such help.

Notes

1 Barth, K., *Church Dogmatics*, III/4 (T. & T. Clark, 1961), 424.

2 O'Donovan, O., *Begotten or Made?* (Clarendon Press, 1984), 49–66.

3 Yoder, J.H., "What Would You Do If...? An Exercise in Situation Ethics," *Journal of Religious Ethics* 2 (1974):91.

4 Lewis, C.S., *The Abolition of Man* (Macmillan, 1947), 82–84.

5 Nolan, K., "*Genug ist Genug*: A Fetus Is Not a Kidney," *Hastings Center Report* 18 (1988):14.

6 Hauerwas, S., *Truthfulness and Tragedy* (University of Notre Dame Press, 1977), 11.

7 That this is not simply my private suspicion can be seen from the following passage from Andrews, L.B., "Legal, Ethical, and Social Concerns in the Debate Over Stem-Cell Research," *Chronicle of Higher Education* (29 Jan 1999):B5. "Moreover, as embryos become valuable to biotech companies as sources of cell lines, doctors may increase the dose of fertility drugs to insure that multiple embryos are created—in effect, to manufacture more 'excess' embryos."

Testimony of
Edmund D. Pellegrino, M.D.
Georgetown University

I am grateful for the opportunity to appear before this Commission to present a Roman Catholic perspective on the moral issues involved in stem cell research. I speak as an individual physician, ethicist, and former clinical and laboratory investigator. Because of limited time, I shall confine myself to a summary of the moral issues. I have read and agree with the testimony regarding the legal and moral issues presented by Mr. Richard Doerflinger of the National Conference of Catholic Bishops on April 16 of this year.

I will argue against the moral acceptability of research involving embryonic stem cells obtained from *in vitro*-fertilized blastocysts and embryonic primordial germline cells obtained from aborted fetuses. My objections are grounded in the following: 1) my understanding of the teachings of the Roman Catholic Church about the moral status of the fetus and embryo, 2) the insufficiency of the utilitarian arguments that would justify destruction or discarding of embryos, and 3) the practical difficulties of effectively regulating the practice even if it were morally defensible.

I recognize, as do Roman Catholics generally, the great potential for human therapeutics in stem cell research. I do not oppose stem cell research per se if the cells are obtained from sources such as adult humans, miscarriages, or placental blood. What is morally unsustainable is the harvesting of stem cells by either of two currently proposed methods: 1) the creation and destruction of human embryos at the blastocyst stage by removal of the inner cell mass or 2) the harvesting of primordial germ cells from aborted fetuses. Both cases involve complicity in the direct interruption of a human life, which Roman Catholics believe has a moral claim to protection from the first moments of conception. In both cases, a living member of the human species is intentionally terminated.

In the Roman Catholic view, human life is a continuum from the one-cell stage to death. At every stage, human life has dignity and merits protection. Upon conception, the biological and ontological individuality of a human being is established. Human development unfolds in an orderly way, and each stage of that development must be treated as an end in itself, not as a mere means to other ends, however useful they might be to others.

The Roman Catholic perspective, therefore, rejects the idea that full moral status is conferred by degrees or is achieved at some arbitrary point in development. Such arbitrariness is liable to definition more in accord with experimental need than ontological or biological reality. Terms such as "pre-embryo" or "pre-implantation embryo" seem to be contrivances rather than biological or ontological realities.

Also rejected are socially constructed models that leave moral status to definition by social convention. In this view, moral status may be conferred at different times, or taken away, depending on social norms. This is a particularly perilous model for the most vulnerable among us: fetuses, embryos, the mentally retarded, or those in permanent vegetative states. The horrors of genocide in current events force us to recognize how distorted social convention can become, even in presumably civilized societies.

There is, admittedly, a difference in moral gravity in harvesting cells from aborted fetuses if the act of terminating life is clearly separated from the use of the harvested cells. The moral problem of complicity remains, however, because Roman Catholics believe abortion to be intrinsically wrong. To use tissue from an aborted fetus is morally akin to using the data from unethical human experimentation under dictatorial regimes. For most Roman Catholics, both the fetus and the embryo have the same moral claim to protection of their lives, even though the moral gravity of use of their respective tissues may be different.

In addition to objections to the current sources for stem cells, the moral arguments for permitting embryonic stem cell research are faulty. Only a few of these arguments can be mentioned here. One argument is that the so-called spares (fertilized ova) that result from *in vitro* fertilization will be discarded anyway, so why not use them? But the facts are otherwise: Many spare embryos have been frozen; all have not been destroyed, even though permission may have been given. The fate of spare embryos is, therefore, not as certain as we may suppose.

Even if parents were to consent to use of their spare embryos, this would not change the inherent moral status of the embryo itself. Embryos created specifically for research do not have a different moral status than embryos created for reproductive purposes. In both cases, the embryo would be treated as a means to an end. Its inherent moral status is violated because it must be killed in order to obtain stem cells. There is no moral or legal basis for subjecting any member of the human species to harm or death in nontherapeutic research based on the prediction that it will die anyway, no matter how certain that prediction may be.

The Department of Health and Human Services has argued that funds can be used for research on cells obtained by the destruction of embryos so long as the act of destruction, itself, is not federally funded. I will not address the question of whether this reasoning distorts the intent of Congress in prohibiting use of federal funds for embryo destruction. But it is reasonable to question the logic of moral cleansing of the act of destruction by this artificial separation of killing the embryo and using its cells.

An issue of complicity as well as justice lies in the use of tissues from aborted fetuses or therapies developed from the destruction of embryos. Many Catholics, and probably others, would object, as some already do, to vaccines and transplants derived from the use of those sources in ways they take to be immoral. Catholic hospitals could not on principle use such therapies. Supporting such research from federal funds would impose an injustice on Catholics, who would be forced into complicity through taxation even though they perceive grave moral harm in the practice, however legal it may be.

Even in the general public, there is as yet no overwhelming moral consensus for approval of the destruction of embryonic human life for experimental purposes. Even if there were such a consensus, the moral dilemmas would still exist for many members of our society. Opinion polls and plebiscites do not per se establish moral norms.

Those who favor embryonic stem cell research, like the Human Embryo Research Panel, grant, as have legal opinions, that the embryo should be treated with "respect." When we inquire into what this means, it seems to be merely an assurance that these embryos will be destroyed only in "...research that incorporates substantive values such as reduction of human suffering."[1] This is a fragile form of respect, since it makes the embryo's dignity and value conditional on something other than its intrinsic value.

Even if these and many other ethical issues were surmountable—as I think they are not—much of the argument for embryonic stem cell research rests on the promise to control abuses by appropriate legal regulation. How is it possible to separate "spare embryos" from embryos intentionally produced as stem cell sources? The temptation to make "spares" is obvious. In any case, we cannot, and should not, post monitors in every laboratory. Morality has always depended on character, not on legal regulation.

The temptation to stretch the moral envelope is already apparent. Clearly, a major biological problem is how to direct pluripotential stem cells to take a desired direction—let us say, towards myocytes rather than osteocytes. The question of whether cells a little further along in differentiation might not be more successful already has been raised. The pressure to use somewhat more mature cells will mount, if only to test the hypothesis. Furthermore, it is not at all certain that frozen spare cells will actually function the same way as "fresh" cells. The temptation to create, or "find," spare cells during *in vitro* fertilization will be strong. Finally, it is still uncertain that pluripotential cells are not totipotential and capable of developing into a complete human embryo.

There are also the obvious complications of profits and patents and the close association of the current research with the biotechnology industry. It is not unfair to question the "protection" provided by ethics review boards appointed by and serving corporate entities. This is not to impugn their motives, but only to recognize the conflicts of interest that occur when profit and prestige are at stake.

I believe the Commission would serve the public welfare and the cause of morality best if it were to reject any attempt to legitimize embryonic stem cell research from *in vitro* fertilized-blastocysts or from aborted fetuses, the moral, legal, and practical impediments of which are of such great magnitude and complexity. The Commission should instead strongly encourage the funding and development of alternate sources of stem cells—those that do not depend on the destruction of living human embryos or make use of cells from induced abortions. In light of the rapid developments in this field, the possibility and probability of the development of morally acceptable sources of stem cells is a reality. Therefore, both scientific and ethical prudence would dictate a delay in the implementation of any policy covering such research.

Like all scientific research, stem cell research has tremendous potential for human benefit. But without ethical constraints, it can easily overshadow the very humanity it purports to benefit. As presently conceived, human stem cell research goes beyond the boundaries of moral acceptability.

Note

1 Geron Ethics Advisory Board, "Research with Human Embryonic Stem Cells: Ethical Considerations," *Hastings Center Report* March/April (1999):32–33.

Testimony of
Abdulaziz Sachedina, Ph.D.
University of Virginia

Islamic Perspectives on Research with Human Embryonic Stem Cells

Thank you very much for inviting me to give an Islamic perspective on human stem cell research. I do not represent a particular school of thought ("church") in Islam; rather, I speak for the Islamic tradition in general, which is a textual tradition. I have been able to examine a number of primary and secondary sources that have been produced by scholars representing different schools of thought. Two major sects or schools of thought, the Sunni, who form the majority in the Muslim community, and the Shi`ite, who form the minority, do not represent an Orthodox/Reform divide; instead, they are both "orthodox" in the sense that both base their arguments on the same set of texts that are recognized as authoritative by all of their scholars. And yet, it is important to keep in mind the plurality of interpretations displayed by the "traditionalists" and "conservatives" on the one hand, and the "liberals" on the other.

The ethical-religious assessment of research uses of pluripotent stem cells derived from human embryos in Islam can be inferentially deduced from the rulings of the Shari`a, Islamic law, that deal with fetal viability and the sanctity of the embryo in the classical and modern juristic decisions. The Shari`a treats a second source of cells, those derived from fetal tissue following abortion, as analogically similar to cadaver donation for organ transplantation in order to save other lives, and hence, the use of cells from that source is permissible. For this presentation, I have researched three types of sources in Islamic tradition to assess the legal-moral status of the human embryo: commentaries on the Koranic verses that deal with embryology; works on Muslim traditions that speak about fetal viability; and juridical literature that treats the question of the legal-moral status of the human fetus (*al-janin*).

Historically, the debate about the embryo in Muslim juridical sources has been dominated by issues related to ascertaining the moral-legal status of the fetus. In addition, in order to provide a comprehensive picture representing the four major Sunni schools and one Shi`ite legal school, I have investigated diverse legal decisions made by their major scholars on the status of the human embryo and the related issue of abortion in order to infer religious guidelines for any research that involves the human embryo.

Let me repeat here, as I did when I testified to the Commission about Islamic ethical considerations in human cloning, that since the major breakthrough in scientific research on embryonic stem cells that occurred in November 1998, I have not come across any recent rulings in Islamic bioethics regarding the moral status of the blastocyst from which the stem cells are isolated.

The moral consideration and concern in Islam have been connected, however, with the fetus and its development to a particular point when it attains human personhood with full moral and legal status. Based on theological and ethical considerations derived from the Koranic passages that describe the embryonic journey to personhood developmentally and the rulings that treat ensoulment and personhood as occurring over time almost synonymously, it is correct to suggest that a majority of the Sunni and Shi`ite jurists will have little problem in endorsing ethically regulated research on the stem cells that promises potential therapeutic value, provided that the expected therapeutic benefits are not simply speculative.

The inception of embryonic life is an important moral and social question in the Muslim community. Anyone who has followed Muslim debate over this question notices that its answer has differed at different times and in proportion to the scientific information available to the jurists. Accordingly, each period of Islamic jurisprudence has come up with its ruling (*fatwa*), consistent with the findings of science and technology available at that time. The search for a satisfactory answer regarding when an embryo attains legal rights has continued to this day.

The life of a fetus inside the womb, according to the Koran, goes through several stages, which are described in a detailed and precise manner. In the chapter entitled "The Believers" (24), we read the following verses:

We created (*khalaqna*) man of an extraction of clay, then We set him, a drop in a safe lodging, then We created of the drop a clot, then We created of the clot a tissue, then We created of the tissue bones, then we covered the bones in flesh; thereafter We produced it as another creature. So blessed be God, the Best of creators (*khaliqin*) (K. 24:12–14)!

In another place, the Koran specifically speaks about "breathing His own spirit" after God forms human beings:

Human progeny He creates from a drop of sperm; He fashions his limbs and organs in perfect proportion and breathes into him from His own Spirit (*ruh*). And He gives you ears, eyes, and a heart. These bounties warrant your sincere gratitude, but little do you give thanks (K. 41:9–10).

And your Lord said to the angels: 'I am going to create human from clay. And when I have given him form and breathed into him of My life force (*ruh*), you must all show respect by bowing down before him' (K. 38:72–73).

The commentators of the Koran, who were in most cases legal scholars, drew some important conclusions from this and other passages that describe the development of an embryo to a full human person. First, human creation is part of the divine will that determines the embryonic journey developmentally to a human creature. Second, it suggests that moral personhood is a process and achievement at the later stage in biological development of the embryo when God says: "*thereafter* We produced him as another creature." The adverb "thereafter" clarifies the stage at which a fetus attains personhood. Third, it raises questions in Islamic law of inheritance as well as punitive justice, where the rights and indemnity of the fetus are recognized as a person, whether the fetus should be accorded the status of a legal-moral person once it lodges in the uterus in the earlier stage. Fourth, as the subsequent juridical extrapolations bear out, the Koranic embryonic development allows for a possible distinction between a biological and moral person because of its silence over a particular point when the ensoulment occurs.

Earlier rulings on indemnity for homicide in the Shari`a were deduced on the premise that the life of a fetus began with the appreciation of its palpable movements inside the mother's womb, which occurs around the fourth month of pregnancy. In addition to the Koran, the following tradition on creation of human progeny provided the evidence for the concrete divide in pre- and post-ensoulment periods of pregnancy:

Each one of you possesses his own formation within his mother's womb, first as a drop of matter for forty days, then as a blood clot for forty days, then as a blob for forty days, and then the angel is sent to breath life into him (*Sahih al-Bukhari* [d. 870] and *Sahih al-Muslim* [d. 875], The Book of Destiny [*qadar*]).

Ibn Hajar al-`Asqalani (d. 1449) commenting on the above tradition says:

The first organ that develops in a fetus is the stomach because it needs to feed itself by means of it. Alimentation has precedence over all other functions for in the order of nature growth depends on nutrition. It does not need sensory perception or voluntary movement at this stage because it is like a plant. However, it is given sensation and volition when the soul (*nafs*) attaches itself to it (*Fath al-bari fi sharh al-Sahih al-bukhari, kitab al-qadar*, 11:482).

A majority of the Sunni and some Shi`ite scholars make a distinction between two stages in pregnancy divided by the end of the fourth month (120 days) when the ensoulment takes place. On the other hand, a majority of the Shi`ite and some Sunni jurists have exercised caution in making such a distinction because they regard the embryo in the pre-ensoulment stages as alive and its eradication as a sin. That is why Sunni jurists in general allow justifiable abortion within that period, while all schools agree that the sanctity of fetal life must be acknowledged after the fourth month.

The classical formulations based on the Koran and the Tradition provide no universally accepted definition of the term "embryo." Nor do these two foundational sources define the exact moment when a fetus becomes a moral-legal being. With the progress in the study of anatomy and in embryology, it is confirmed beyond any doubt that life begins inside the womb at the very moment of conception, right after fertilization and the production of a zygote. Consequently, from the earliest stage of its conception, an embryo is said to be a living creature with sanctity whose life must be protected against aggression. This opinion is held by Dr. Hassan Hathout, a physician by training, who was unable to be here today. This scientific information has turned into a legal-ethical dispute among Muslim jurists over the permissibility of abortion during the first trimester and the destruction of unused embryos, which would, according to this information, be regarded as living beings in the *in vitro* fertilization clinics. Some scholars have called for ignoring the sanctity of fetal life and permitting its termination at that early stage.

A tenable conclusion held by a number of prominent Sunni and Shi`ite scholars suggests that aggression against the human fetus is unlawful. Once it is established that the fetus is alive, the crime against it is regarded as a crime against a fully formed human being. According to these scholars, science and experience have unfolded new horizons that have left no room for doubt in determining signs of life from the moment of conception. Yet, as participants in the act of creating and curing with God, human beings can actively engage in furthering the overall good of humanity by intervening in the works of nature, including the early stages of embryonic development, to improve human health.

The question that still remains to be answered by Muslim jurists in the context of embryonic stem cell research is, When does the union of a sperm and an ovum entail sanctity and rights in the Shari`a? Most of modern Muslim opinions speak of a moment beyond the blastocyst stage when a fetus turns into a human being. Not every living organism in a uterus is entitled to the same degree of sanctity and honor as is a fetus at the turn of the first trimester.

The anatomical description of the fetus as it follows its course from conception to a full human person has been closely compared to the tradition about three periods of 40-day gestation to conclude that the growth of a well-defined form and evidence of voluntary movement mark the ensoulment. This opinion is based on a classical ruling given by a prominent Sunni jurist, Ibn al-Qayyim (d. 1350):

> Does an embryo move voluntarily or have sensation before the ensoulment? It is said that it grows and feeds like a plant. It does not have voluntary movement or alimentation. When ensoulment takes place voluntary movement and alimentation is added to it (*Ibn al-Qayyim, al-Tibyan fi aqsam al-qur'an*, 255).

Since there is no unified juridical-religious body representing the entire Muslim community globally, different countries have followed different classical interpretations of fetal viability. Thus, for instance, Saudi Arabia, might choose to follow Ibn Qayyim; while Egypt might follow Ibn Hajar al-`Asqalani. We need to keep in mind that the same plurality of the tradition is operative in North America when it comes to making ethical decisions on any of the controversial matters in medical ethics. Nevertheless, on the basis of all the evidence examined for this testimony, it is possible to propose the following as acceptable to all schools of thought in Islam:

1. The Koran and the Tradition regard perceivable human life as possible at the *later* stages of the biological development of the embryo.

2. The fetus is accorded the status of a legal person only at the later stages of its development, when perceptible form and voluntary movement are demonstrated. Hence, in earlier stages, such as when it lodges itself in the uterus and begins its journey to personhood, the embryo cannot be considered as possessing moral status.

3. The silence of the Koran over a criterion for moral status (i.e., when the ensoulment occurs) of the fetus allows the jurists to make a distinction between a biological and a moral person, placing the latter stage after, at least, the first trimester of pregnancy.

Finally, the Koran takes into account the problem of human arrogance, which takes the form of rejection of God's frequent reminders to humanity that God's immutable laws are dominant in nature and that human beings cannot willfully interfere to cause damage to others. "The will of God" in the Koran has often been interpreted as the processes of nature uninterfered with by human action. Hence, in Islam, research on stem cells made possible by biotechnical intervention in the early stages of life is regarded as an act of faith in the ultimate will of God as the Giver of all life, as long as such an intervention is undertaken with the purpose of improving human health.

Testimony of

Rabbi Moshe Dovid Tendler, Ph.D.
Yeshiva University

Stem Cell Research and Therapy: A Judeo-Biblical Perspective

I. Pre-Introduction

There is something unreal about this National Bioethics Advisory Commission meeting at which the focus is on religious views on the humanhood of stem cells or of pre-embryos. The death-dealing silence of the leaders of organized religion as to the humanhood of men, women, and children subjected to murder, rape, exile, torture, and hunger, raises the question of "who cares?" Do the views of religious leaders mean anything to anybody anymore? Ambiguity and indecision in the face of crimes against humanity is an admission that organized religion has failed in its mission to humanize animal-man. The amorality of this silence may indeed be more destructive of the moral fabric of society than the immorality of the perpetrators of the crimes.

The prophet Elija cried out in exasperation (Kings I 18:21), "How long will you vacillate between G-d and idolatry?" If you would but take a stand I could then hope to influence you with the truth and beauty of our faith. Ambiguity and indecision are fatal to any social order based on morals and ethics.

II. Introduction

The chronology of American laws that affect the funding of fetal cell research represents a record of ambiguity regarding the "humanhood" of the embryo.

1) 1973 - *Roe v. Wade* conferred a constitutional right on women who decide to have an abortion during the first trimester of gestation.

2) 1988 - In March, a National Institutes of Health (NIH) panel concluded that existing regulations permitted federal support for fetal transplantation research, such as using fetal brain cells to treat Parkinsonism. The Reagan administration imposed a moratorium on the grounds that fetal cell research would encourage abortion by ameliorating the heinous nature of the crime if some good would come from the abortus.

3) 1992 - President Clinton enacted into law the NIH panel recommendations with the provisions that the decision to have an abortion is determined independently from the research; that no fees are paid to women to donate the abortuses; and that no selection of the recipient by the donor is permitted.

4) 1998 - On October 21, Congress passed an Appropriations Act ordering that no funds can be used for any research in which embryos are destroyed or exposed to significant risk.

5) 1999 - On January 15, The Department of Health and Human Services Counsel advised that federal law permits NIH to support research conducted with stem cells obtained from embryos despite the ban on destroying or injuring embryos.

III. The Moral Status of the Embryo

There are two sources of human embryonic stem cells:

1) The inner cell mass of the blastocyst, consisting of approximately 140 cells 14 days post-conception. The Judeo-biblical tradition does not grant moral status to an embryo before 40 days of gestation. Such an embryo has the same moral status as male and female gametes, and its destruction prior to implantation is of the same moral import as the "wasting of human seed." After 40 days—the time of "quickening" recognized in common law—the implanted embryo is considered to have humanhood, and its destruction is considered an act of homicide. Thus, there are two prerequisites for the moral status of the embryo as a

human being: implantation and 40 days of gestational development. The proposition that humanhood begins at zygote formation, even *in vitro*, is without basis in biblical moral theology.

2) Human embryonic germ cells are derived from gamete ridge tissue removed from an approximately eight-week-old aborted fetus. These cells, like human embryonic stem cells, are assumed to be pluripotent and immortal. The theological and secular moral concerns expressed in the long-running abortion drama that has for years disregarded the principle of separation of church and state is relevant to the use of human embryonic germ cells.

Biblical law as practiced for 3,500 years views the destruction of an eight-week-old fetus as tantamount to homicide. It is permitted only to save the endangered life of the mother. However, an abortus from a spontaneous or medically necessitated abortion may be used to further research for the benefit of mankind.

In the United States, the right of a woman to obtain a first trimester elective abortion is protected by our constitution. Surely, if all concerned give "fully informed consent," the use of cells from such abortuses to provide life-saving therapy must be declared legal. In the Jewish tradition, which declares abortion after 40 days of gestation to be homicide, the use of the abortus in life-saving therapy is not precluded. An illicit act does not necessarily result in a prohibition to use the product of that act. For example, biblical law prohibits cross-breeding any two species of animal, such as a horse and a donkey. The product of such an illicit mating, the mule, may be used for the benefit of the owner, even though a biblical prohibition was transgressed (Maimonides Laws of Cross-Breeding [Klayim] 9:3).

In stem cell research and therapy, the moral obligation to save human life, the paramount ethical principle in biblical law, supersedes any concern for lowering the barrier to abortion by making the sin less heinous. Likewise, the expressed concern that this research facilitates human cloning is without merit. First, no reputable research facility is interested in cloning a human, which is not even a distant goal, despite the pluripotency of stem cells. Second, those on the leading edge of stem cell research know that the greater contribution to human welfare will come from replacement of damaged cells and organs by fresh stem cell products, not from cloning. Financial reward and acclaim from the scientific community will come from such therapeutic successes, not from cloning.

IV. Fences Around the Law

Jewish law consists of biblical and rabbinic legislation. A good deal of rabbinic law consists of erecting "fences" to protect biblical law. Surely our tradition respects the effort of the Vatican and fundamentalist Christian faiths to erect fences that will protect the biblical prohibition against abortion. But a fence that prevents the cure of fatal diseases must not be erected, for then the loss is greater than the benefit. In the Judeo-biblical legislative tradition, a fence that causes pain and suffering is dismantled. Even biblical law is superseded by the duty to save lives, except for the three cardinal sins of adultery, idolatry, and murder.

The commendable effort of the Catholic citizens of our country to influence legislation that will assist in preventing the further fraying of the moral fabric of our society must not impinge on the religious rights and obligations of others. Separation of church and state is the safeguard of minority rights in our magnificent democracy. Life-saving abortion is a categorical imperative in Jewish biblical law. Mastery of nature for the benefit of those suffering from vital organ failure is an obligation. Human embryonic stem cell research holds that promise. The recently announced joint effort between Geron Corporation and Roslyn Labs of Scotland (of Dolly fame) has its focus on the use of human embryonic stem cells to bolster a failing heart or liver, without need for immunosuppressive drugs or dependency on organ donors.

V. Ethical Concerns

If abortion and cloning are excluded from the analysis of the moral and ethical status of stem cell research and therapy—as they should be—what ethical concerns must be addressed? A succinct listing should suffice for those conversant with medical ethics literature.

1) Risk/Benefit Evaluation

If immunosuppressive therapy is needed after stem cell transplantation, the dangers inherent in "transplantation sickness" must be weighed against the benefit to be achieved. In general, only cases of fatal disease, such as vital organ failures, would justify such therapy.

The hope that human embryonic stem cells can provide cells or organs that are genetically identical with the patient via the Roslyn-Dolly technique would greatly expand the list of those eligible for such therapy.

2) Allocation of Scarce Resources

The decision to fund stem cell research must be weighed against other demands for governmental financial support. What funding must be redirected to this project? How many will benefit from such research, and how many will be deprived of funding for competing services such as preventive medicine, drug rehabilitation, and assisted reproductive technology?

Failure to fund human embryonic stem cell research will not stop this research. The private sector has already declared its readiness to support this research in return for the great profits that may accrue to investors. Decisions that prevent government support guarantee the financial success of the private sector and will surely result in hundreds of patents that will restrict other scientists from free access to the developing technology. Without government involvement in this area of research, "Wild-West" ethics will prevail. It is government guidelines that establish ethical principles for research that affect the private sector.

3) Fully Informed Consent

Fully informed consent can be a "can of worms," given the different educational and social backgrounds of the parents who decide to abort, especially if permission for research use is requested in addition to any clinical application.

A broad-based educational effort directed at the clergy of the religious denominations that are represented in our great country, as well as at the schools, can help even this field.

4) Financial Reward

Should the donors share in the financial gains that may result from the research done with the donated stem cell lines that will be established? Surely altruism should not be demanded only of the cell line donor!

However, the critical contribution to any clinical application of human embryonic stem cell research is that of dedicated biomedical researchers. Cell lines can be obtained from any number of individuals; however, brilliant and dedicated biomedical research is the domain of the gifted few.

Testimony of

Kevin Wm. Wildes, S.J., Ph.D.
Georgetown University

Let me begin by thanking the Commission for undertaking its important work in the area of human embryonic stem cell research. As a Georgetown University faculty member, it is an honor to welcome you to Georgetown University, which was founded, in part, to foster dialogue between religious faiths and civil society on important matters.

In my brief testimony I would like to identify two important, though different, areas of profound moral concern of the Roman Catholic community regarding stem cell research using human embryos: the source of the stem cells used in the research and issues of social justice.

The Roman Catholic Bishops of the United States have made known their opposition to stem cell research, opposition that is based on the need to destroy human embryos in order to conduct this type of research.[1] Because the Bishops work from an assumption that the human embryo should be treated as a human person, destruction of the embryo to conduct research is morally problematic. If one begins with this assumption, then many of our commonly held views on research ethics come into play. Research ethics are grounded in an understanding of respect for persons that views the consent of the research subject as essential to the moral appropriateness of the research itself. Furthermore, any research that is undertaken should minimize the risks and harms to research subjects. In research involving human stem cells, consent cannot be obtained, and it is certain that harm will come to the embryos because they must be destroyed so that the research might take place.

The use of embryos in stem cell research, whether they be "spare" embryos or embryos created for research, presents a moral roadblock to that research, because the use of the embryos involves the destruction of human life for the sake of the research itself. Although the status of the embryo is clear in hierarchical statements about the embryo, this is a far-from-settled matter in our society, which is deeply divided over the moral standing of early human life. As Glenn McGee and Arthur Caplan have noted, "Embryonic and germ cell status is not a scientific matter. There is neither consensus nor fact from which to deduce the social meaning of different embryonic or fetal tissue."[2]

Another possibility for obtaining stem cells for research is to develop them from fetal tissue. However, if the tissue comes from an aborted fetus, this, of course, leads to an immediate problem in the Roman Catholic tradition, because such a situation puts the research and the researcher in a compromised position. Here we have traditionally used the language of cooperation or complicity with evil to describe such situations. Since abortion is viewed as the destruction of human life, one cannot "profit" from evil or immoral actions. Indeed, this has been the position held on the use of fetal tissue in other types of experimentation. As an alternative, fetal tissue from spontaneous abortions could be used as a source for stem cell research. However, I am led to think that such tissues have not proven to be good sources for this type of research.

This latter point leads me to make clear something that may be too easily lost. That is, I do not think one can argue that there is, in Roman Catholic thought, opposition to stem cell research itself. The crucial moral issues and stumbling blocks are the problems of the derivation of the stem cells used in the research itself. That is, the destruction of embryos or the use of fetal tissue from abortion are the key moral problems. If you think that embryos should be treated as human persons, then it makes sense to argue that they should not be destroyed for purposes of research. However, if there were a way to conduct stem cell research without destroying human life, either embryonic or fetal, I do not think the Roman Catholic tradition would have a principled opposition to such research. Indeed, Richard Doerflinger closed his testimony before this Commission by saying: "This commission should urge the National Institutes of Health to devote its funds to stem-cell techniques and other promising avenues of research that in no way depend upon such killing."[3]

It is important to point out, however, that there is no single Roman Catholic "position" on this topic or many moral topics. Like many issues in Catholic moral thought, there has been a long line of reflection on the moral standing of early human life.[4] It is hard to see how one can speak of human personhood in the totipotent stage. Within the Roman Catholic tradition, how one views the status of the early embryo is often tied to one's views about authority within the Church. The assumptions made about authority shape the arguments, positions, and premises one holds.

The second area of moral concern that comes from the Roman Catholic tradition is the concern that questions involving morality cannot be asked in isolation. Rather, such questions must be situated in the larger context of society and its just organization. That is, if we were to proceed with stem cell research, what type of review and oversight would be in place (in the way that we now review the use of human subjects in research and experimentation)?[5] In addition, if one thinks with a Roman Catholic imagination, one must also ask about the questions of justice in devoting resources, especially national resources, to such research when there are so many other basic medical and health needs that are unmet. Issues of social and distributive justice are not easy to discuss in American society. Nonetheless, I would argue that the Roman Catholic tradition would say that such questions must be included in any discussion about how we organize our medical research and delivery.

Notes

1 Doerflinger, R., "Destructive Stem-Cell Research on Human Embryos," *Origins* 18 (1999):770–773.

2 McGee, G., and Caplan, A., "What's in the Dish?" *Hastings Center Report* 29 (1999):36–38.

3 Doerflinger, 773.

4 See, for example, Donceel, J., "Immediate and Delayed Hominization," *Theological Studies* 31 (1970):76–105; Shannon, T.A., and Walter, A.B., "Reflections on the Moral Status of the Pre-Embryo," *Theological Studies* 51 (1990):603–626; and Cahill, L., "The Embryo and the Fetus: New Moral Contexts," *Theological Studies* 54 (1993):124–142.

5 Tauer, C.A., "Private Ethics Boards and Public Debate," *Hastings Center Report* 29 (1999):43–45.

Testimony of

Laurie Zoloth, Ph.D.
San Francisco State University

The Ethics of the Eighth Day: Jewish Bioethics and Genetic Medicine A Jewish Contribution to the Discourse

Introduction

When the first serious work in genetics became possible, the initial public reaction to the exploration of the genetically coded structure of the human being and the research on that code was a curious mixture of fascination and fear. The fascination has driven an intense public interest in each new genetic advance, project, or claim, as well as support for research projects that appear to have the potential for therapeutic use. The fear has prompted both initial caution and legislation to limit that very research.

Limits on the clinical use of genetic interventions and limits on research and testing were created for three reasons. First, technical barriers themselves made the successful manipulations of genetic material for reliable medical use highly risky endeavors. Hence, regulators and theorists focused on the issues of safety and avoidance of the clearly foreseeable chances for harm, understanding that even with the best of intentions, unforeseen error and unintended consequence were unavoidable. Second, the mere activity of intervention into human DNA was viewed with alarm, since such research seems to tamper with what are understood as the basic building blocks of human life itself. Third, since much of the proposed research on human molecular biology called for the use of gametes or early embryonic tissue and since such use involved questions at the heart of the most volatile issue in American political life, that of abortion, the scientific use of the human embryonic tissue to explore and manipulate human DNA was seen as a violation of essential moral limits.

Hence, political and legal "bright lines" were erected to prohibit certain experimental trajectories, and political and religious pressure was exhorted to prohibit research and curtail intellectual discourse in various aspects of the field. Additionally, the problems of informed consent and refusal that are raised by all sorts of medical research seemed uniquely daunting in this research, surrounded by both great and desperate hopes and sobering uncertainties. Given this, public funding sources available to researchers were limited, as Congress and the administration reflected on the issues, curtailing active searches for new techniques in this area. In particular, research on human embryos has been limited by federal bans on funding, and research on the manipulation and alteration of germline DNA for therapeutic medical purposes has been constrained by federal law.

But the enormous potential of such genetic intervention is a powerful incentive for this research, particularly as our understanding of disease has unfolded and the claim strengthened for a genetic etiology of many complex human disorders. Furthermore, as the practical technical skills of genetic scientists have improved, the ethical issues at the margins of genetic research have been raised for reconsideration. Private companies have continued to fund university researchers, and work on the human genome and embryonic cellular manipulation has continued. In fact, the research in human embryonic stem (ES) cells (the search for the "Holy Grail" of genetic research, itself a religiously freighted term[1]) and the possibility of successful germline intervention have proceeded swiftly, and recent breakthroughs in this technology have again raised questions about the ethical implications for such interventions in the clinical context.

I have been asked to comment on the *halachic* and moral and historical context for such genetic research and to provide an analysis of how the field of Jewish bioethics might respond to the technologies being proposed in order to conduct this research. The National Bioethics Advisory Commission hearings are particularly important in reflecting on the use of human ES (hES) cells, and this testimony will focus on both the specific ethical issues regarding the use of hES cells and human embryonic germ (EG) cells, as well as the emerging and congruent issues that such cells allow us to consider. These are notes toward such a consideration that will serve to delineate the types of questions (rather than specific answers) that further work in bioethics will need to address.

Note on Methodology

In the Jewish ethical-legal tradition (*halachah*), which functions methodologically as a discursive community in which the justification is created by the force of moral suasion, no single authoritative voice or one particular council of authority speaks for the entire tradition or the community. Judaism itself is divided into four distinctive movements, each with a varying degree of allegiance to rabbinic and textual authority.[2] Hence, in confronting emerging ethical issues, what will serve best in beginning to frame a coherent Jewish understanding of these issues is the widest possible call for inquiry and the widest possible response.

This paper is a preliminary contribution in that direction, in which I raise what I argue are the framing questions for further debate. This is a broader step than a strictly halachic review, because I want to raise a wider set of questions in addition to the halachic ones, although my allegiance is to a halachic sensibility in my research. At stake in the halachic method of reasoning is the finding of cases that, while not having all of the same features as the case before us, have distinguishing moral appeals that might be similar to our case. Thinking broadly about the multiple dimensions of genetic research will allow us to capture and creatively debate which features will be the relevant ones that will then allow us to create normative outlines for social policy. In this, halachic reasoning is a form of linguistic, definitional analysis in which the parties to the debate seek epistemological commonality as a first step. However, as an important caveat, it is critical that we remember that the new terrain upon which we now find ourselves bears scant analogy to the terrain of the rabbinic world. The biology of the Talmud was still couched in terms later retracted, gamete reproduction was still not fully understood, and microbiological techniques were not even imagined.

There is another critical methodological point at which Jewish thought can said to be distinctive. For Jewish ethics, the framing questions will be those of obligations, duties, and just relationships to the other, rather than the protection of rights, privacy, or ownership of the autonomous self. Because much of our thinking in contemporary American bioethics is rights based and relies on a model of intricate semi-legal contracts carefully made between autonomous and anonymous strangers, the idea of centering our obligations rather than worrying about our rights can seem simple-minded or naive. But the other-regarding binding gesture, this commanded act of justice, responsibility itself, is the first premise of Jewish ethics.[3]

In general, there are three categories we need to consider in thinking the issues through. The first is the general issue of whether the act that we are considering—that of allowing for the research, manipulation, and use of hES cells—is itself a good act. The research on stem cells—on the possibility of manipulating them, pushing them toward differentiation, or from pluripotency to totipotency, away from differentiation, and growing and collecting vast amounts of them—raises issues of use and meaning. Are human persons collections of potentially deconstruct-able and dismantle-able other parts, or even other selves?[4] Here we need to address issues of goal, meaning, moral status, and the ontological nature of the person; the meaning and scope of medical intervention; the question of what constitutes disease and what normalcy; the relationship between God and human partners; the tension between faith and science; and the issue of safety. In general, these are problems of *telos*.

The next genre of question, important in a religious legal system such as Judaism, is whether the technical aspects of the complex manipulation required are themselves permitted. Here we need to address questions of origin, informed consent, the use of assisted reproductive technology (ART), such as *in vitro* fertilization (IVF), cell harvesting, the use of third parties, extra-coital reproduction, the perimeters of the family, contracts, the effect on the character of the researchers, and the issue of limits on the applications and participants. In general, these are problems of *process*.

The last category of questions, and one that is, I argue, critically important in Jewish thought, includes the issues of justice, access, and distribution, and the implications of the work on the human community in which we will share an altered medical and social universe. In general, these are problems of *context*.

Part I: Issues in the Use of Primordial Stem Cells—hES Cells and Human EG (hEG) Cells

A Statement of the Issue: Technological Summary

ES cells are the cells that are present in the early stages of all animal development. After fertilization, the fertilized egg splits into two and then into several identical cells. In the human, this occurs as the egg tumbles toward the uterus for implantation. In the first few days, it changes into a hollow ball of approximately 140 cells, called a blastocyst. The outside of this ball will form the placenta and chorionic villi of the animal, with cells on the inside and at the top of the ball eventually developing into the organism itself. Each of these ES cells has certain characteristics critical to our discussion. Each cell is pluripotent, meaning that it can develop into any of the body parts that will be required for development (for example, some will develop into cardiac cells, some into neural cells, and some into gametes). Each cell is immortal, meaning that it can theoretically continue its development indefinitely. A critical feature of hES cells is that they can repopulate themselves while remaining in their undifferentiated state, thus creating colonies that once begun are self-sustaining. Each cell is malleable, meaning that its DNA can be manipulated and the cell function will continue. Each cell is re-insertable. It can be moved into another blastocyst, where it will continue its development within the organism. Finally, each cell is telomerase expressive. The cells produce high levels of telomerase, which is the enzyme that stimulates cells to grow continually. When telomerase production ceases, the cell will stop growing (the aging process) and will die, and when telomerase production is overactivated, the cell will proliferate (cancer). All of these qualities make the study of these cells and their potential use extremely important for medical science. At an early stage, the entire blastocyst can be divided into multiple parts (resulting in spontaneous twinning/multiple births), and each part would develop into a complete individual being. At this point, what causes individual ES cells to differentiate, form into organs, and specialize is not understood. One of the reasons, in fact, that research is called for so forcefully in the scientific community is this very lack of knowledge of this basic science.

Given these unique characteristics, ES cells have attracted significant attention for therapeutic uses. If ES cells could be taken from the blastocyst and allowed to grow and differentiate into specific tissue in a culture medium, sheets of specific tissue could be grown. Such tissue could then be implanted into humans to correct conditions in which tissue is lacking or damaged. For example, individual cardiac cells might be placed into a heart in which cardiac tissue is damaged, islet cells could be introduced into a pancreas that is unable to correctly produce insulin, or stem cells could be introduced into bone marrow that lacks the immune system response (as in Severe Combined Immunodeficiency). Tissue cultures instead of human subjects could also be used to test drugs or could be used to study little-understood areas of cell death, as in Parkinson's disease and Alzheimer's disease, for example.

Before 1998, researchers were able to harvest ES cells from rats, mice, rhesus monkeys, chickens, cows, and pigs. Using this and other technologies, researchers were able to alter the DNA in a few cells, replace the cells into a developing animal embryo, and create a chimera animal, with DNA from two species. When these fused-DNA animals are mated, after several generations they become purebred carriers of the new genetic material, able to pass the characteristics reliably on to their offspring. Thus, permanent genetic changes are introduced into the animal. Applications include altering milk in cows to produce useful proteins needed for certain drugs and altering organs of animals in such a way that when they are transplanted into a human, the proteins that identify them as "pig," for example, would no longer be present, reducing the problem of graft-versus-host disease.

But collecting, harvesting, isolating, and growing hES cells are difficult processes. In part, this has been because obtaining human blastocysts has been ethically and technically problematic. Two recent developments have altered this landscape, however. One is the growth of IVF clinics. When human eggs are fertilized *in vitro*, many blastocysts are created, only some of which will be implanted into the womb of the recipient. Others may be frozen for future use, and still others are deemed unusable or damaged and are discarded. These blastocysts are "graded" by physicians into ranked groups according to their presumed viability. Grades I and II are used for therapeutic purposes, but grades III and IV are discarded.[5] It is to these blastocysts designated to be discarded that researchers initially turned their attention in developing this technology.

A second change is the insight that the cells that *would* have developed into germ cells in human fetuses, hEG cells, are very much like ES cells They are pluripotent, possibly totipotent, and immortal. These cells can be collected from the primordial genital ridge that is destined to develop into the testes or ovaries of aborted fetuses from five to nine weeks of age.

The harvesting of this fetal tissue raises other ethical issues. The first step in the harvesting process is obtaining the informed consent from a woman who is having an early elective abortion to use the tissue of her aborted fetus.[6] Then, the abortion is performed, the age of the fetus is established, and its physical integrity is checked. The fetus must be between the ages of 5 to 8 weeks postconception (which means 7 to 10 weeks since the last menstrual period or 3 to 5 weeks since the woman become aware that she missed a cycle). That is, the fetus must be between 35 and 56 days old (a precise number that will have implications for our later halachic discussion). The fetus must be large enough and well developed enough that its primordial genital ridge can be located, and yet not so large that the abortion itself could damage the very small fetal parts. Human EG cells have been collected and grown in tissue culture in the same manner as hES cells, and they have undergone laboratory testing that indicates there may be promise in using them as stem cell origin culture.

Our understanding of these cells has grown rapidly since they have been successfully grown in culture. One feature of this research is that the basic science is itself rapidly evolving, and hence the ethical reflections that follow the science are changing as well. For example, the lower grade embryos can now be more successfully used in reproduction, making the distinction between embryos that would be discarded and ones that are viable for reproductive use a false one. Since the embryos that were used for research could technically be used for implantation, our halachic reasoning based on their status as "doomed" is no longer satisfactory. This technology then raises the ethical issue of whether we should be permitted to create embryos solely for research. Most recently, there have been questions regarding the distinction between thinking of these cells as totipotent or pluripotent. In the work of Nagy, et al., in mouse stem cells,[7] for example, mice were successfully gestated from a cluster of stem cells placed in a trophocytic matrix (to simulate the placenta-forming cells) and then into a mouse uterus. Hence, in theory, if given the correct matrix, stem cells can make all of the parts of a living organism, at least in mice. If all stem cells could potentially become embryos if given the right sort of cellular environment, then what exactly does one have when one has a stem cell cluster? Is it a canonical cell line? Or is it actually also a potential human fetus? Finally, if stem cell technology develops in tandem with nuclear transfer technology (cloning), then cells could potentially be programmed not only to differentiate into specific tissue types, but could be histocompatible, tailored for each transplant recipient. It is a stunningly important technology, potentially eliminating graft-versus-host/host-versus-graft disease and saving millions of lives. However, such technology raises the problem of reproductive uses, moral status, and instrumentality. And, if such nuclear transfer could be done, surely the DNA of the transferred nucleus could be manipulated as well, raising the complex problem of inadvertent or deliberate germline intervention, yet another ethical issue that will require reflection.

In part, this difficulty in understanding the moral meaning of these implications is a problem with the language we use, which is one of discourse about human reproduction that emerges from the classic understanding of gametes, families, sexual reproduction, debates about abortion, birth control, privacy, rights, and sanctity. However, as we contemplate a world of cellular replication and reproductive potential without gametes, we will need new language to describe what we intend, its moral meaning, and what we find fitting.

Part II: Ethical Questions in the Pursuit of hES Research—The Secular Discourse

Secular concerns are of concern to Jewish discourse for several reasons. The first is that secular anxieties create a specific social context to which religious communities respond. Cultural zeitgeist, cultural practices, and aesthetic sensibilities create the landscape upon which the locus of Jewish discourse, the *Beit Midrash*, meets. Medical theory creates the horizon of possibility for the ethical consideration of the issue at hand. Ethical theory emerges as a result of casuistic debate. Jews confront the dilemmas of all aspects of modernity, and when questions or cases emerge in the ordinary activity of the world, Jewish laypersons are directed to turn to their rabbinic leadership for advice based on traditional texts. If the question is a new one, or one concerning an emerging or contended topic, the question is framed as a formal one, a *poskin*. Such *poskinim* are then responded to by leading national authorities with expertise in a particular area. These may be answered with formalized written *responsa*. Such debates are then further discussed and debated. The questions themselves, however, as in this case, are generated not only by ritual transgressions, but by all ethical concerns.

It is in this spirit that we turn to the specific problems raised by the research that might be raised in any secular or civic context.

A. The Problem of Telos

1. What Is the Meaning and the Goal of This Process?

One stated goal of this research is the replacement of diseased, damaged, or absent tissue. This end appears to be benevolent and straightforward. Yet, the use of genetically altered tissue could also be desired for conditions that are only marginally defined as "illness." In the current bioethics discourse, many authors have focused on precisely this difficulty. If we understand that the telos of this work allows consideration of enhancement of human capacities, then how are the ethical considerations altered? The secular issues focus on two premises that have marked discontent about nearly every medical advance. (In these debates, one is reminded of the struggles surrounding the use of anesthesia, in which the essential link between childbirth and pain or surgery and pain was challenged, thus raising ontic questions for religious thinkers.) A second use for hES/hEG cells is the need for basic research in human embryonic development. It is for this reason that adult stem cells cannot be substituted for hES/hEG cells derived from embryonic tissue. With a clearer understanding of human development, we will be able to more clearly understand how to prevent many disorders.

2. Could Evil Uses Be Made of hES/hEG Cells?

Ends cannot be controlled without close regulation and enforcement of research. Is this feasible? It is difficult to imagine how it could be made so. In this technology, one is not intending to create new persons, only new personal parts. Yet, all genetic alteration is surrounded by public fear of such alteration, marginalization, and use of unwarranted power in the hands of a malevolent state. Is this the first step in an unacceptable alteration of human species by genetic means? Because the hES cells can become anything, they could also be altered, and with technological issues resolved, they could be used in the construction of new embryos. Furthermore, it is conceivable that new blastocysts could be manufactured using hES cells and nuclear transfer. If this is the case, then this technology is close to the process of cloning.

In the sober consideration of this technology, it is critical that we distance ourselves from such concerns and linkages. Here, the parallel use of transplants is a useful analogy. As ethicists, it will be a key part of our shared discourse not only to worry about possible abuses of power, but also to raise concerns about unwarranted fears that might unduly block research efforts.

3. The Problems of Origins and Moral Status

What is the moral status of the human stem cell? By moral status, we mean how we describe the standing of an entity relative to other moral agents and the obligations and relationships that other moral agents have toward this entity. In the work of philosopher Mary Ann Warren, if a being has moral status, then "we cannot just treat it any way we please."[8] If hES cells are understood as tissue or as organic nonhuman life forms, then it might be permissible to use them even instrumentally for very compelling reasons and just ends. But if hES cells are to be considered human entities, our obligation toward them shifts sharply.

One suggestion about the determination arises from some arguments that derive from some commentators within the Catholic moral theological tradition.[9] In this formulation, one can differentiate between the embryo at the time before and after the appearance of the primitive streak, a line of division in the embryo that is the first step in the formation of a spinal cord of one individual. It is at this point in embryology that the blastocyst can no longer split into two identical portions and become monozygotic twins. It is argued by some scholars (Shannon) that it is at this moment that one individual with a particular subject-life narrative begins. Further, it is this moment in development that represents a limit, the line beyond which we cannot take cells from the new organism (the child-to-be) unless we are subject to the same kind of constraints that we would place on pediatric research. The definition of moral status is important in medical research in large measure not only because of the nature of our obligation to another entity, but because of the issue of consent of the subject itself, if in fact we are dealing with a "subject." And this line of reasoning creates substantive problems in regulation. Research on human subjects is carefully regulated by international codes, national legislation, and considerable case law. In the case of research on embryos, to whom should we turn for guidance if we have a "subject?" Then, in what sense is that subject an entity to which one can claim patent rights? What is the meaning of a subject blastocyst generated by nuclear transfer, with no parents in the normative sense?

Another genre of questions of origin arises in considering whether there is a difference when the tissue that is generated is obtained as ES or EG cells. Each site of tissue origin raises significant problems. ES cell collection will destroy what might potentially become a human person; EG cell collection uses aborted fetuses, with the entire attendant debate about abortion at play.

4. Can We Breach Essential Creaturely Limits?

The next problem of the moral meaning of this work is how it challenges the idea that there are firm limits to what humans should do. It raises the question of whether there are limits in the creation of life or of other essential biological boundaries that now define the meaning, scope, and purpose of human existence. Such limits are often marked by our uneasiness when certain biological "bright lines" are crossed, for example, the change in the meaning of aging after a "normal" human life span, the replacement of human parts, the ability to reproduce after a certain age, the number of infants in a pregnancy, or the extension of life with sophisticated machinery. Many of these bright lines have been broached in the last decade as a result of the steady development of medical technology. We accept revolutionary challenges to creaturely limits,[10] and, after initial concerns about safety and efficacy and the standard Food and Drug Administration challenges, each medical change in our creaturely limits is ultimately first feared, then celebrated, and then seen as an entitlement, becoming the new norm for community medical practice.

5. The Problem of Interpretation: Language and Meaning of the Self

The issue of exactly what we are doing when we "touch" DNA or alter cellular development is of concern at the level of meaning and language. We lack a coherent theory that allows broad philosophic agreement on the issues of definition of disease and normalcy. What is a coherent self? In framing the issue in this way, we encounter other issues, such as that of the permissibility of altering fundamental features of human reproduction. But is such work a change in kind or in efficacy? In other words, is our ability to create *de nova* tissue for use in disease intervention or repair so significantly different from our use of transplants of cadaver organs? Such a problem is compounded by the interlocking issues of the marketplace as raised above, such as owning, patenting, and selling human DNA patterns or processes.

6. The Question of Interpretation

Over all of this technology lies the complex social surround of the marketplace, in particular the pharmaceutical industry. At the present time, the medical model supports the ideological construction of the self as an entity that is basically intact, but that is besieged by alien germs that must be confronted with drugs that will kill them (antibiotics), or as a self that merely lacks a chemical that could be also nicely be supplied by a drug company (insulin). Hence, the marketplace endeavors to supply these commodities. But if the self-qua-self can be altered to change the underlying proteins that control the immune system, for example, or the production of enzymes, then externally offered drugs will not be needed daily. Such a shift represents both a significant marketplace change and a significant change in the meaning of the self.

We understand the "self" in philosophic and in religious terms as a creature with specific boundaries and specific obligations based on this creaturely fragility and wiliness. But all genetic speculation raises deep anxieties and corresponding hopefulness about the way that this notion of the self could be altered. For example, Jewish ethics calls for a self that is extraordinarily "other-regarding" (relative to Western secular traditions). This ethical stance derives its power from the constancy of the need of the stranger. But what of this obligation in a world free—or freer—from the constancy of illness or the reality of disabling difference?

7. Immortality and Meaning

The telomerase-expressive ability of these cells remains one of the most compelling and fascinating aspects of their nature and the one that could most fundamentally alter how we think about aging, illness, and death. More reflection will be needed on the various issues this raises in our essential framing of what it is to be a self at play in a limited mortally and morbidly bound universe. But Jewish thought is framed in terms of both the reality of aging and the limits of mortality. Changing the embodied experience of aging has important and interesting implications for intergenerational obligations.

B. The Problems of the Process

1. Is the Informed Consent Process Adequate for the Donors of the Fetal Tissue/Blastocyst?

Classic issues of informed consent for adult donors are traditionally addressed by using the formal consent process, including discussion and a carefully constructed consent form. But is this adequate? The women from whose bodies the eggs come are already participating in a heavily freighted dilemma and are in many cases at the end of a year-long process of increasingly intricate and invasive technology to achieve a desperately yearned for pregnancy. The women from whom the aborted fetuses come are similarly at risk, only they are enmeshed in the medical care system because of (at least) stressful and (at most) tragic options. It is at this junction that they are confronted with signing this consent form and speaking about research protocols. Can any decision that is made under these circumstances really be a considered, reflective decision?

2. How Can We Address the Inevitability of Errors?

All medical interventions are fraught with error. Errors in diagnosis, prognosis, and treatment occur far more often than is commonly supposed by the lay public, high rates of error and failure exist in genetic research, and there may be special problems associated with this particular technology. In agriculture, ES cell research has been used to created advantageous mutations, beginning with the stem cells and allowing them to grow and divide in the compromised and altered environment of the laboratory. A higher-than-normal level of mutagenic effects would be expected, some of which will develop into useful organisms. Thus, in this application, the mutation is useful and desired. But in medical therapy, we want stable and predictable cellular division, although some mutations might not reveal themselves until the organism is fully mature. How are we to evaluate this problem? Some mutations reveal themselves only later in the process. It is unknown exactly how cellular replication works or what causes cells to differentiate. Telomerase is implicated in cancer and in tetratomas in laboratory animals. How will this affect the use of these tissues in transplantation? How can we quantify such risks?

We can anticipate that despite our best efforts, we are sure to err and that human loss will be the result of such error. What we cannot know is exactly what will occur or when. Now, we know that this is the case in many situations in ordinary life, but we can calculate the relative benefits and proceed.

3. The Problem of Unintended Consequences
The process is one of experimentation, in which a product is developed whose uses are unclear, but profound. We cannot know what will happen as we develop this technology, but we do know that it will have both intended and unintended consequences. While we can debate the intended consequences, are we willing to accept the uncertainty of the unknown? Even if we do not err badly, even if we carefully follow all of our protocols and do not "fail," we could still create a chain of events that leads to complex interactions and social changes that we do not intend but that we cannot ignore. By definition, human experimentation will provoke both intended and unintended events.

C. The Problem of Context

1. The Problem of Application and Distribution: Justice and the Marketplace's Use of the Process
Medical intervention in the United States takes place within an elaborate system that is a construction of class, race, and geography, creating deep problems of justice in the very construction of the context of our questions. Here we can see clearly the intersection between the civil commons of medicine and the religious concerns for justice for the vulnerable. The issues of access to a new technology and of simultaneous protection from it are considerably freighted. Such a technology, if it becomes practical after enormously expensive development, will create paradoxical shifts offering, for example, a one-time solution to chronic illness or disabling life-long conditions that now require large quantities of social, human, and financial resources.

The essential issue is that this technology is highly technical and, at least initially, very expensive. How would the developed world's access to this material benefit the emerging economies of the underdeveloped world? Linked to this are the issues of the broad general marketplace, such as the patenting authority of the process, the ownership of the tissue, and the relationship of the veracities of profit, corporation, shareholders, and other third-party sponsors of the enterprise itself. These issues become more acute precisely because the federal government is not the primary sponsor of the research, as noted above. To what extent does the private sector, whose fiduciary responsibility is to the shareholders, have to be held to humanitarian concerns? Why is this different from pharmaceutical companies? (Note, Viagra.)

2. The Specter of History: What of the Problem of Evil Uses of This Technology?
Of all the considerations in medicine that evoke the specter of the Holocaust (*Shoah*), including physician-assisted suicide, abortion policy, treatment of the disabled, and research policy, none raise the issue more definitively than the idea of genetic engineering to create an altered human self. The historical link to "race" enhancement, the nomenclature of eugenics, and the marking of some as genetically "inferior" is unavoidable and lead us to sober consideration of the role of state power in medical ideology. In many ways, the gross indignities of the state's use of genetic technology seem less a hazard than the temptations of medicine itself. The link between somatic improvement, class standing, and subsequent power has been made in other work.[11]

But critical issues, such as the meaning of difference, the meaning of ethnicity, and the responsibility of a whole society to bear the vulnerability that illness and disability carries, will be raised by the possibilities inherent in this technology. How will the dynamics of power drive, for example, research funding for interventions? How will private uses of emerging technology be controlled?

Further questions of context and norms arise: How will aesthetics, physical progress, and advantage create a climate of approval for genetic changes that allow or disallow regulations? (Consider here the link between funding and mandated testing for genetically borne diseases of the neonate, such as PKU, and the lack of fully funded support for a child bearing the disease itself in many states.)

D. Commodification and Commercialization

Critical issues of how the marketplace treats the "products" that are generated by stem cell research need to be more fully addressed. In the sphere of organ donation, we do not allow the buying and selling (in this country) of body parts. But gametes have been treated differently, with a marketplace approach guiding their sale. Can we avoid the pressures to sell, patent, and barter this tissue to the highest bidder? Or should we allow for a spontaneous organization of the market in this field? How will each decision change our research interests?

Part III: Jewish Halachic Responses Raise Other Questions

Jewish consideration of issues in bioethics is, of course, textually based and is based in the casuistry of halachah in which specific considerations are addressed by textual recourse. Halachic reflection on all innovative scientific research is constrained by the fact that none of the specific issues raised by new technology is directly addressed by Talmudic conversations compiled in the first centuries of the common era, nor in the elaborate medieval commentary that carries the most considerable weight in the classic tradition. Moreover, in researching the halachic conversations that touch upon this arena, we can note that what the rabbinic culture understood as central is not necessarily what moderns consider most salient. For example, the rabbis were concerned that we act *more* like God might in many ethical and social/political arenas, as in helping the poor, creating justice, and healing the sick, rather than having a re-occurring horror of "acting like God." Sexuality and procreativity were cheerfully and enthusiastically promoted by social and chemical means, and the use of all available means to aid health was promoted.

What follows is an account of the essential halachic concerns and questions that might be primary in assessing hES cell research. Some of these issues will seem odd in the context of modern discussions, but I have raised these topics to provide a comprehensive account of the kinds of issues that might be raised. Some of the questions that we as moderns consider important will be absent from the medieval debates found in the rabbinic and responsa literature, such as the issue of the autonomy as a part of moral status, or the consideration of saliency, or the considerations of individual rights, all of which have a tangential or weak moral appeal in the tradition. This is to be expected. Other issues, such as the primary value placed on relationships, family, or community; the considerations of justice; and the obligations to the poor and the stranger, I will address as we continue our work together.

A further note on method: Jewish reasoning does not entail simply setting out of a list of principles and then deciding whether they are applicable or not in a facile binary sense. Rather, it is a series of open-ended arguments intended to include the broad and creative use of history, text, and culture, with many interrupting voices representing competing narratives. What I have done here is to provide a series of such framing questions to elicit such responses from a range of perspectives.

Of note here and critically important to our thinking is the essential premise of the halachic system itself—that of full, frank, and, I might add, contentious discourse. To that end, I have come to understand that the widest possible discourse is a vital part of that process.

Modern concerns shape the context for the contemporary use of classic sources in the current deliberation. However, the rabbinic discourse on medicine raises substantively different concerns, and hence, particular responses, than do secular ones. What follows are the results of my first discussions and a review of the published literature that bears on this problem. Since this research has not been the focus of medical questions that have arisen for patients, and since Jewish law is case driven (no cases, no responsa), the literature is as yet thin. The intent of this work is to direct specific attention to this emerging issue and to stimulate serious inquiry in this direction.

Of importance to note is that Jewish law, unlike American secular law, in which something is permitted or prohibited, describes four categories for possible action that are based on the relationship between morality, halachah norms, and the laws of the secular nation-state. An action may be permitted, or at least unpunishable under the halachic code, but morally undesirable; an action may be permitted and desirable; an action may be prohibited (even if desirable); and an action may be permitted by Jewish law, but prohibited by the secular state (and thus not be permitted in Jewish law, since "the law is the law of the land," "*dinah d'malchuta dinah*").

A. The Problems of Telos

1. The Prominence of Life-Saving or Life-Extending Medical Intervention

The first responses to hES/hEG cell research seem to indicate a general sanguinity with the procedure when it is framed as breakthrough medical therapy for life-threatening conditions. This general response is based on the clear mandate to save life whenever possible, even if the saving of life requires the violation or suspension of other commanded acts. This entire category of response stems largely from the defining moment in the Talmud in which the rabbinic authorities debate whether one can violate the mandate to rest and to sanctify the Sabbath in order to rescue a man trapped in the rubble of a collapsed building. From this vivid (and, I might add, graphically obvious) source text springs a whole set of cases that are then defined as like being trapped—by illness, catastrophe, hunger, war, or threat. This has provided the warrant text for virtually all experimental therapy, including genetic research. (Limiting factors include the calculus of risk—if the therapy itself is more likely to threaten a life than to save it, as in the first organ transplants, then the case is altered, and the intervention not permitted.) Hence, even if otherwise proscribed actions are involved (taking the organs of the dead, for example), the use is permitted if the life can be reliably saved. Jewish medical ethics is nearly entirely constructed around the principle of *pikuach nefesh*, to save a life. To save even one life, the halachah states, it is permissible, and in fact mandated, that all other *mitzvot* can be abrogated (except for the case of the prohibitions against murder, adultery, and idolatry). Using this consideration alone, the technology could be considered ethical, since, as we have demonstrated above, it does not involve the mere taking of one life to save another, but the use of the cells of one, albeit special, type of tissue.

This is a consideration upon reflection that can be advanced about nearly all the technologies that are suggested by this research. If the full use were possible for this tissue, millions of persons would be afforded years of productive life. While no technological fix should be regarded as, in the words of Christian ethicist Stanley Hauerwas, enabling us to "get out of life alive,"[12] the work of repair, patching, transfusing, and replacing damaged tissue would alleviate human suffering without altering the essential self of the recipient. Moreover, the use of this tissue as a front-line test for newly developed drugs would be a remarkable advance. In speaking to several commentators about this issue, this consideration was the trumping issue in all subsequent discussions.

Some have suggested (Karen Lebacqz[13]) that allowing longer life expectancy or allowing some to live who might otherwise die of, say, fatal cardiac dysfunction, has disturbing implications. It should be noted that classic halachic considerations would not address such concerns.

2. Moral Status and the Issue of Temporality: What Age Is the Embryo?

While moral status of the embryonic tissue is the threshold question for many religious traditions, I will argue that it is of secondary importance to the question of the life-saving consequences of this technology, given the textual tradition and the Jewish position on the developmental status of the embryo and fetus. Like nearly all discourse in this field, Jewish understanding of moral status derives from the abortion debate. At stake is whether the fetus is an independent entity or a part of the body of the mother (*ubar yerickh imo*[14]). The biblical text that grounds the literature is as follows:

> If two men fight, and wound a women who is pregnant (and is standing nearby) so that her
> fruit be expelled, but no harm shall befall (her) then he shall be fined as her husband assesses,
> and the matter placed before the judges. But if harm befall her, then shalt give life for life
> (Exodus 21:22).

By this, the text is understood to mean that if the women herself is not harmed, then the only harm, that of the miscarriage and loss of the pregnancy, is a loss of lessor importance. It can be made whole by monetary compensation, unlike the taking of a human life.[15]

The moral status of the embryo in Jewish considerations of abortion, the main textual location for discussion of embryos in the Talmud, is based on age and proximity to independent viability. Central to all understanding of embryology in the Talmud and subsequent halachic responsa is that before the 40th day after conception, the embryo and fetus are to be considered "like water."[16] The rabbis were close observers of fetal development because it fell within their purview as decisors to examine genital emissions, including spontaneous abortion, to answer questions of *niddah* (the period during the monthly cycle when a husband and wife are not permitted sexual relations) and the use of the *mikvah* (the ritual immersion following the niddah). At stake here is the understanding that the relationship of a woman to her community was closely tied to the moral status of the delivered fetus: Was this a stillbirth or a late menstrual period? Would the women be in niddah for 14 days or 6 weeks?

In that capacity, there are discussions about the nature and character of the contents of the womb at various stages of embryonic and fetal development. There are other considerations, such as quickening (the development of a spinal cord) and the external visual changes in a woman's body that also warrant differing social responses and a different consideration of the pregnancy. This developmental understanding of moral status is not limited to how the halachah considers moral status of fetuses. There is ample precedence for the rabbinic understanding of changing obligations, even life-saving obligations, based on the temporal standing of the human person. Liminal times exist not only at the beginning, but also at the end of life, and there are well-established norms that permit the instrumental consideration of an entity, clearly a human person, and clearly alive, based solely on this understanding of developmental moral statutes.

Let us turn to classic examples. When a person is in a state called *terefah*, one who is inevitably dying, our obligations to save his life and his life relative to others is altered. For example, there is a discussion in the texts about categories of persons to whom one might be differently obligated to protect in a crisis, such as a siege or a hostage taking. A category exists in rabbinic thought for the person who is already condemned by a court to death. Elliot Dorff, who noted the importance of this category in the work of Daniel B. Sinclair,[17] found it of importance in the problem of both discontinuation of treatment and allocation of the scarce resource, and by analogy, I would argue its use here.

> The term *gavra ketila* occurs four times in the Babylonian Talmud. According to Sanhedrin
> 71a, once a person has been sentenced to death, he is immediately a *gavra ketila*, a killed
> person. Because of that, Sanhedrin 81a deals with the possibility that one might think a person
> sentenced to one of the more lenient forms of execution, since immediately presumed dead,

could not be subsequently sentenced to a harsher form of execution for another crime. It rejects that conclusion, but in the meantime reaffirms the description of a doomed person as a dead one. Sanhedrin 85a adds the consideration that one sentenced to death, since considered an already killed person, is no longer 'abiding among your people' in the terms of Exodus 22:27. And, perhaps the most relevant for our purposes, Pesahim 110b says that a person who drinks more than 16 cups of wine is a *gavra ketila*. There it is medical, rather than judicial, factors that make the person thought of as dead.[18]

Dorff linked to this another category of person—the terefah, a person with an incurable, fatal organic disease. This category emerges both in Talmudic texts to describe persons and animals that have a fatal organic defect or a diagnosis thought to lead to death in 12 months. Dorff noted that since the death of a terefah is by definition inevitable, the killing of this person does not "count" as murder in quite the same way, nor is the civil obligation toward him exactly the same as if he were not doomed to die. The rabbis allow for the deserted wife of a terefah to remarry, for example, when the deserted wife of a man may not remarry unless there is clear proof of his death, or for the killer of a terefah to be exempt from the death penalty. The intentional killer of such a person is still subject to divine punishment and moral sanctions, but the legal status of the terefah is analogous to that of a dead person.

The view of this critical liminal state involves a definition of the period at the end of life and a highly nuanced view of personhood. The terms are different linguistic attempts at characterizing the process of dying. Why this is extremely difficult to understand is due in part to the modern attachment to the "moment of death." It is important in clinical medicine to have such a moment, because it is at that moment that everything about the person as patient changes. The language of rabbinic ethics reflects sensitivity to death as process, with the person as "person" retaining certain aspects of personhood even beyond the cessation of cardio-respiratory function and losing other aspects as personhood prior to clinical death. The rabbis struggled to define these states, just as moderns do, and used different vocabularies in an attempt to describe with accuracy a difficult and essentially mysterious boundary of human life.

A parallel in rabbinic categorization exists in the case of the beginning of personhood and the debate that surrounds abortion. Here too, Jewish law suggests a liminal status for the fetus and, exempt from the death penalty, its destruction. If a women is injured by standing near two men fighting, for example, the *Mishnah* records the penalties for her injury (life for a life, eye for an eye) differently than the death of the embryo or fetus who is regarded as property and whose loss is compensated as is all lost property, by money. The fetus is further considered property of the husband in this text in the *Mishnah*, but specifically not the property of the husband in a later text, *Arakin*, but a part of the woman's body—an interesting technicality in our debates.

Subsequent rabbinic commentary regards the fetus as "a part of the woman's body" until the moment at which the head or the greater part of the breech is delivered out of the birth canal. Until that moment, a pregnancy can be terminated and the fetus allowed to die to save the life or health (mental or physical) of the women. In fact, a classic early text allows the killing of a fetus to prevent it from being born in the grisly and queer case described if a near-term pregnant women is scheduled to be hung, to prevent the woman the "shame" of delivery and blood in her death.

After infants are born, their moral status is still in a process of development, albeit of a less dramatic nature. Children are not named or admitted to community (public) membership until the eighth day of life. And if a child dies before the thirtieth day of life, the necessary rituals of death are not performed, *Shiva* is not observed, and the *Kaddish* is not said for the requisite year of mourning. All of these considerations frame our ability to consider the moral status of the pre-implantation embryo as a nonensouled entity that deserves special consideration and respect and is not a human person within the mutually binding halachic system.

This is defensible on two counts. First, since the blastocyst only exists in the pre-40 day period, its status is as "water." Further, if we can determine a distinction in the moral status of the pre- and post-40-day embryo, then surely we can determine a distinction between the pre- and postimplantation embryo. Thus, the considerations that evolve around the primal streak are consistent with this genre of moral reasoning. Since the entity that we are discussing could at this point in its normal development still be a variety of individuals, it is hardly the case that we are altering the life narrative of a specific individual. It is of great importance that the intervention occurs after the first restriction in this developmental view and before the time when the cells can no longer split into several individuals. The cells have already divided into blastomere and placental cells; hence the cells are not totipotent (each cell cannot be a whole person on its own), but pluripotent. Other considerations in the developing moral status of the fetus include the development of a spinal cord or the time of quickening.

Second, the blastocysts that are being used to initiate this process, when first considered by researchers, were ones that had been deemed by physicians to be "nonviable" or not useful for reproductive purposes. There were visible differences between embryos, and many thought that these differences would render some embryos unsuccessful candidates for implantation and development into human persons in any case. It was the initial understanding of this author that the researchers who are involved in this technology use only blastocysts that are in any case not viable and hence that are analogous to a terefah or a gavra ketila. Paradoxically, this allowed for an argument that "we are allowing a particular genetic expression and an immortality" to the embryo that we would destroy. However, as this paper is written, the technology that might enable even grade III and IV blastocysts to be used from reproduction is advancing. In some cases, grade I and II blastocysts might be used for successful implantation. However, at the same time, other research proceeded that made implantation with fewer eggs, or even one egg, far more successful. This means that many more eggs and embryos might be generated for the purpose of treatment of infertility than could be used by any one couple. This raises further questions: These are "extra" blastocysts in that they are considered as such by the couples no longer needing them to achieve pregnancy. Should they be discarded? Donated to other couples? Should they be donated for research? In what sense are these socially nonuseful, "excessive" blastocysts considered "not viable" and thus likened to a person doomed to die? This problem needs more study. What if these entities could be donated for implantation in other women? Would this change their moral status?

It is important to clear up the impression that the technology involves the use of a few cells found in the blastocyst that if taken during a normally developing pregnancy would not endanger the life of the organism (as they would be, for example, in genetic testing of the chorionic villi at nine weeks of development). This technology is based on destroying the blastocyst by removing the outer membrane (preplacental cells) and taking the blastomere (inner cell mass) and placing it in the cell culture to grow.

In reflecting on the possible abuses of a blastocyst that might be created by the use of nuclear transplant (cloning) and then the subsequent use of stem cell techniques, theologian and legal scholar Ze'ev Falk raised the issue of whether one could even consider the entity that would be such a blastocyst created by nuclear transplant or the stimulation of the EG cells to be a human, since it was not created by sexual intercourse at all. In his reflections on this topic, Falk noted that the origin of this tissue would make it distinctive from a naturally occurring pregnancy in a womb, interrupted by science and taken out if its course of development.[19]

Further issues emerge about the legal status of the tissue cell lines themselves: Are they to be regarded as part of a woman's body for as long as they exist? How does ownership accrue to them? Since the rabbis did not have a halachic category for cells that can live ceaselessly and are, perhaps, capable of asexual reproduction, it will require further research to claim anything about the halachic status of this tissue.

3. Is the Pursuit of Genetic Research a Mandated Healing?

The task of healing in Judaism is not only permitted, it is mandated. This is supported and directed not only in early biblical passages ("you shall not stand idly by the blood of your neighbor," and "you shall surely return what is lost to [your neighbor]"), but in numerous rabbinic texts as well.[20] The general thrust of Jewish response to medical advances has been positive, even optimistic, linked to the notion that advanced scientific inquiry is a part of *tikkun olam*, the mandate to be an active partner in the world's repair and perfection. Judaism is not, after all, a nature-based religion; the very assertion of circumcision rests on the notion that the body is not sacred or immutable. There is no part of the body that is sacred, or untouchable. But disfigurement of the body (for example, piercing and tattoos) is not permitted, and the belief that the personal body is a property that belongs to the "self" alone is a late and nontraditional response to medical decisionmaking. But nearly all commandments can be abrogated to permit acts of lifesaving intervention or healing. Characteristically, "Judaism does not interfere with physicians' medical prerogative, providing his considerations are purely medical in character."[21]

The permission and the obligation to heal come directly from the Torah text of Exodus and Deuteronomy, as interpreted by the Talmud:[22]

> The school of R. Ishmael taught and heal he shall heal (Exodus 21:19). This is [the source] whence it can be derived that the authorization was granted [by God] to the physician to heal.

And further:

> How do we know [that one must save his neighbor from] the loss of himself? From the verse: And thou shall restore him to himself (Deut. 22:2).

There is a positive attitude toward medicine, which stresses that the recourse to prayer and faith alone is incomplete without the complete resourcefulness of which humans are capable. This capability is a God-given gift, part of the work of stewardship to which persons are entasked in Genesis.

As many commentators have noted, there is another text that directs the general attitude of Jewish theologians toward the medical endeavor. The physician's work is legitimate, and in fact, obligatory, as can be seen in the following story. Rabbi Akiva and Rabbi Ishmael are walking in Jerusalem and encounter an ill person who asks for their expertise in finding a cure. They tell him, but the man is puzzled: After all, are not the rabbis transgressing the will of God who made him sick in the first place by curing him? They answer by asking him about his work. He is a farmer, who works in the vineyard created by God: Does he not alter the world that God created by his work? The text continues as follows:

> [He answers to them] 'If I did not plow, sow, fertilize, and weed, nothing would sprout.' Rabbi Akiva and Rabbi Ishmael said to him, 'Foolish man.…Just as if one does not weed, fertilize, and plow, the trees will not produce fruit, and if fruit is produced but is not watered or fertilized, it will not live but die, so with regard to the body. Drugs and medicaments are the fertilizer, and the physician is the tiller of soil.'[23]

Hence, there is a mandate for humans to be partners with God in creation, and Dorff, for example, generally acclaims genetic engineering as "one of the wonders of modern medicine." While he notes the potential for eugenic uses, "the potential benefits to our life and our health are enormous," and hence research should continue.

There are no specific texts that address the issues of the use of research science specifically, although the Talmud is replete with stories about the general ability of the rabbis to examine closely the abortus itself or to observe closely specific medical conditions. On the other hand, there are no halachic texts that forbid basic research either. David Bliech notes that these phenomena are characteristic of several modern problems in medicine, ones where there are no clear textual referents.[24] In recent work, he has used texts that refer to the necessity to build fortification around cities. The community must build walls in the face of danger, but the obligation that the community has to protect itself against "imminent danger" does not extend to danger that exists in the not-yet-existing future. Hence, by extrapolation, genetics work that promises the very real chance of saving a life is an obligation to pursue even in the face of other theoretical dangers. In Bliech's view, the premise is clear. The science as promised offers enormous potential to cure horrific and fatal diseases. Further, scientists who do medical research should be assumed to be working for the welfare of humanity. Bliech notes that in terms of Jewish responsa literature, the possibility of "hard science" was a relatively recent one. No significant commentary emerges until the mideighteenth-century, and here only vague reference to basic research is found.

Given such positive halachic responses, the nearly universal communal response to all genetic advances that can promote health and increase fertility has been enthusiastically positive in the Jewish world. The absolute mandate to heal and the firm rejection of the claim that to intervene would counter God's will are clear features of rabbinic Jewish thought. Further, it is mandated to use the best and most advanced methods available as soon as they are proven to be efficacious and not dangerous to the patient. Using this argument, prohibiting the exploration of this field might create legal concerns as well.

4. Does It Assist in the Mitzvah of Procreativity?

Much of the impulse for genetic research, and in large part its justification, has rested on the premise that working on the edge of permitted research is allowed to assist in the *mitzvah* of procreativity and fecundity. "Where there is a rabbinic will, there is a rabbinic way" is a folkloric summary of the methodology of halachic discourse itself; nowhere are the rabbinical decisions more creative in their use of text than to support new reproductive technologies. At stake here is how this use of the IVF process will be affected by this technology. Here, the compelling reasons for the use of the technology are even stronger, since they may allow us to save a specific life. But how far can we allow this reasoning to carry us? Is there any risk of a purely instrumental creation and then destruction of blastocysts for this use? How is this better or worse as an instrumentality than the idea of the use of blastocysts created for fertility and then not used? This would take careful new research in Jewish law, history, and context.

5. What Do We Mean by Normalcy and Disease?

For Jews, the ideas of the normal have been historically used to mark Jews (Jewish blood, Jewish noses, Jewish "gaze" and gait) as different, deviant, and dangerous. Hence, mapping, marking, and altering the physicality of difference were linked to altering the social and psychological situation and finally the mental health of the Jew.[25] Is the alteration of the diseased "type" of the Ashkenazi Jew, now used as a marker population in a number of genetic diseases, a similar case? What are the implications if that is the case? How does the specific history of the Jew and the fate of the Jewish community at the hands of a state-supported German scientific community inform our discourse on this point—a position that has been explored in a number of European countries?

Classically, permission to alter the physical body has been linked to the way that the "disfigured" body has affected the mental health of the person—linking the intervention to a medicalized end. Will this be the warrant for genetic intervention? If this is the case, then a far larger window of possibility might open for the use of this technology.

B. Problems of Process

A different sort of ethical concern is raised by the process of the research.

1. Can We Use Drugs to Stimulate Ovulation?

This issue has been debated in early questions about the development of ART and has been resolved. Medications to stimulate fertility (mandrakes) are spoken of in biblical literature approvingly. The problem of biblical infertility is resolved on the spiritual level, but there is no prohibition against the use of all medical interventions that could help a couple achieve the commandment to raise at least two children.

2. Can We Harvest Eggs from a Woman for IVF?

This question, too, has been raised. Eggs are part of a body, but they do not have the status of fully moral entities even when fertilized *in vitro*, since before 40 days, the embryo is "like water." (See the lengthier discussion of this point in the section on hES cell research.)

3. Can We Use Donor Sperm to Perform IVF?

Here we find the first problems in the use of ART as a part of the process. Two issues are of concern: The first is whether it is adultery if the sperm of another man is used inside a woman's body (as is done in artificial insemination), and the second is whether a prohibited marriage could occur through the offspring of two families marrying by chance. (Prohibited marriage would include marriage to one's half-sibling, a remote but interesting theoretical possibility.) For this reason, some orthodox rabbinic sources prohibit the use of donor sperm for artificial insemination. Even for the use of the husband's sperm, or in some cases, the use of a mixture of sperm sources to meet the halachic requirements, there are special considerations. Sperm is not to be wasted (the sin of Onan), so elaborate collection devices have been created to allow for coital stimulation and collection of sperm.[26] But on this point there is sharp disagreement, even among Orthodox rabbinic commentators. ("You cannot commit adultery with a hypodermic syringe. Even if a woman uses donor sperm against the will of her husband it has no consequences for the child."[27])

4. Can We Use DNA Splicing Technique?

As we consider the future implications that this research suggests, we turn to other considerations. Responses to the question of using DNA splicing technique have focused on animals, where the prohibitions concern interspecies genetic transfer. Alteration within species, or enhancement of certain characteristics within species, has been accepted. (See above.) Hence, the concerns have been outcome driven. But in looking at the process, contemporary rabbinic response has been largely theological and not legal. Discussion refers to the mandates of Genesis 1-3, in which issues of creation, stewardship, and limits are described.

Immanuel Jakobovitz suggests a general response in his reflections on the problems of human cloning. Jakobovitz recalls that human holiness for Jews rests on cessation and not merely creation. Here he argues that Shabbat, not only for humans, but for God, represents this limit on production, creativity, and alteration of the world. Except for action needed to save lives in an immediate sense, even good human work is suspended in recognition of God's sovereignty.[28] Cognate cases include the theological limits on the boundaries of the *mishkan*, or tabernacle, the restrictions of *kashrut*, or kosher norms, and the general idea in Jewish thought that appetite, desire, business, and acquisition are to be limited and constrained by the social realities of a particular situation. The tension between unlimited freedom and social imperatives is discussed repeatedly in rabbinic debate.

5. Is It Disrespectful of the Dead?

If hES cells originate from the germ cells of aborted fetuses, then the halachic consideration is a separate one. Here, we have two questions. The first is when the abortion is actually performed. If it is performed within the 40-day halachic limit, it is considered differently than if it is done after. (See above.) Timing is essential for both the researcher and the halachic authority, since the cell must be collected prior to its differentiation. A second question arises about the use of the body parts of a fetus who is the subject of an abortion and who is past 40 days after (what the rabbis would consider) conception if the use is for medical purposes. To address this problem, I turned to the protracted debate about autopsy in the halachic literature. It seems clear here that the cutting, dissecting, and use of fetal tissue borders on the prohibitions about desecration of the dead. But several factors mitigate this problem. First, is the fetus given the same consideration as the stillborn child? If not, why not? Next, in the case of the permitted autopsy, the procedure is permitted in the case described above, *puchach nefesh*. While some decisors need to be assured that a specific life will be saved by the medical information derived from the procedure, many allow autopsies for the understanding of a disease process that affects a category of ill persons.

Unlike the hES cells, the issue of moral status for hEG cells is less troubling, since the cells are taken from the gamete ridge of an already dead fetus. Hence, the use of this tissue is closer to the use of other sorts of human cadaver tissue, such as the use of cadaver skin for grafting in burn victims or cadaver kidneys for transplantation. This autopsy model yields important results in our moral theory as well. While we may have qualms about the origins of the aborted fetus and while we may not like or may even abhor the circumstances of the death of the fetus, we understand that we may use the tissue for important and good ends. In thinking about this, we may make an extreme comparison by considering the use of tissue from the aborted fetus as exactly the same as we might allow the use of the kidneys or skin of a victim of a drive-by shooting. The use of the tissue is in no way seen in the second case as an endorsement of drive-by shootings, and the use of the tissue in the first case is not an endorsement of abortion.

6. Is It an Improper Mixing of Two Kinds?
The biblical prohibitions against mixing two types of things, or *shatnes,* might be another concern. For example, it is not permitted to use two animals of different species to plow a field or two kinds of animal hair to make a garment. This issue is seen in legal arguments about grafting, in particular, certain types of grafting that create new species. This was at first prohibited, then permitted (nectarines, for example), and now the deliberate breeding of interspecies is permitted. Some justify this as an extension of biblical tricks of breeding, as in the eugenic use of goats by Jacob (to create his new herd of spotted, robust goats.) It is the kind of question that does not readily occur to a modern audience (we tend to use the technical term for unease: the "yuck factor"), but it is an important halachic one: Is the stimulation of hEG cells to initiate a process using nuclear transfer such a mixing? Is the use of this and related technology that used human DNA inserted into animal cells to create chimerical species prohibited under this consideration? Would the prohibition against animal-human sexual liaisons stand in the case of the use of interspecies nuclear transplant?

In this specific technology, the transplantation is limited to human-to-human "mixing." Here, we might raise questions of further problems in the use of animals in further considerations and advances. If saving a life can trump all other factors, we are on slippery ground. This was a consideration in the early reflections on transplantation and has been resolved using the principle of puchach nefesh.

7. Does Its Collection Shame the Woman?
The dignity, reputation, and integrity of her body and the risk of immodest exposure to the woman who carries a fetus were all significant considerations for the rabbinic authorities, who, as in the text mentioned above, were deeply concerned about the protection of a woman's body from any event that would force her into shame. In this discussion, the consideration is close to the feminist stance that understands the gametes as a part of the self of a woman and not as her property to be sold. In fact, the texts make this clear even in the most extreme cases.

In this way, we need to reflect carefully on the informed consent process. The later texts are clear that the embryo and fetus are not the property of the husband. As such, since the fetus is considered part of the woman's body, the woman's mental status must be carefully considered, as well as the circumstances surrounding the collection of the egg.

In thinking about the steps of such therapy, we might also raise concerns here: Is the process of having one's pregnancy manipulated externally in the extreme way envisioned by this technique a violation of the bounds of dignity ("modesty") for a woman?

8. Does Informed Consent Involve Making a "Nonbinding Contract"?
As noted above, special consideration must be given to the mental state of the woman who donates the tissue. But also important for our careful reflection is the problem of a coerced and therefore unenforceable contract—a specific entity in Jewish law. It has been noted for centuries of legal debate that some contracts are not valid because they require an unnatural (in the rabbinic sense) act of imagination or will and cannot be enforced justly. Contracts that are clearly not in the best interest of the person who makes it are not valid. Is the agreement to donate blastocysts or fetuses such a contract? Does the consent for the use of fetal tissue or of embryos involve such contracts?

Linked to this are other possibilities for source texts: Perhaps in the laws regarding the use of slaves and limits on the use of female slaves or of captives one might find useful ideas for how we think of instrumental relationships between persons of differing levels of power. Here again, careful work will need to be done to research such relational issues.

9. Can an Analogy Be Made Between This Case and Any Other Case in History?
Rabbinic reasoning works by analogy. In thinking about any new case—for example, the invention of electricity, the exploration of America, the use of anesthesia in surgery—the rabbinic authorities had to seek parallel cases that offered precedent. In this case, the framing of the analogous case will be of central importance. For example, is the development of hES cell technology more like cloning or more like transplantation? Is the relationship of hES cell growth to other forms of tissue cultivation and transfer like the relationship between fire and the light switch? In other words, are they enough alike so that the same halachic prohibitions apply (as in the Orthodox view), or different enough that their use is permitted in situations where fire is not (as in the Conservative view)?

Should we look at moments in history in which new medical advances were developed or at periods of the discovery of new geographic terrain? Such a search will allow us to explore the range of limits and challenges in the halachic method. For example, Judaism presents a long history of possibilities and many possible textual venues for this work. However, Jewish responsa literature works like American legal systems, by considering all questions asked. If no one thought of an advance as problematic, then there are, perhaps, no textual precedents. But all such avenues need to be explored, a task beyond the scope of this paper. Often the rabbinic law works by an analysis of the separation of each part—if each step in the process is permitted, then the whole is allowed.

Part IV. Further Questions from Contemporary Jewish Bioethics
Not all questions that have arisen in the field of Jewish bioethics are ones that are raised by the normative halachic problems. Other considerations have emerged in the academic field of bioethics that have stimulated debate and then a distinctive response from the Jewish location(s).

I have chosen to address the critical issues of halachic regard of moral status and other immediate concerns. In the halachic method, it will be the framing of the question that will determine the critical reflection that will emerge. Hence, it will be critical to lay claim to these considerations as well as the standard discourse on the use of gametes and DNA that is found in the literature. In this section, I turn methodologically toward what we might call narrative ethics, or in Jewish ethics, what I call "*aggadic ethics*." In this, we are justifying our approach with derived ethical norms suggested by extra-legal sources (narrative, literature, and history).

A. Is This a Just Use of Technology?
Much of Jewish law and codes is concerned with the problem of justice in an unjust world. In this, the problem of how to create a world of just order is a clear preoccupation of the biblical and rabbinic argument about

the meaning and goals of a society that lives in a covenantal relationship with God. For justice to have real meaning, the civilization that is constructed will need to account for the primacy of this relationship.

B. What Will Be the Effect on the Poor?

The poor are to be protected not only out of a vague sense of compassion, but as a part of how the natural and agricultural world is structured. Our texts remind us that the structure of harvest is understood to include the provision of parts of the field and parts of the yield for the poor. In fact, essential economic decisions (such as how to plant, what to harvest, and when to refrain from planting) are mediated by this consideration. Limits are placed on the entire society to ensure that the widow, orphan, and stranger are provided for with full dignity. Hence, the concern for the sabbatical year, in which all production is suspended to allow for the use of the field by the disadvantaged, for the harvest to be organized to allow for gleaning, and for the corners of the field to be proscribed for one's own use and to be reserved for the use of the poor. Technological advances, even clever and expedient ones, cannot be permitted if persons or even animals might be unjustly used—hence the concern for the yoking of unlike animals for plowing.

C. Is This a Good Instance of Tikkun Olam or an Over-Reaching of Human Power? Is the Intuitive Uneasiness a Measure of a Greater Sense of Inequity? (Is This Use Like the Creation of the Golem?)

There are two important texts that recall a broad general concern for all of technology. The first is the creation of the *Golem*, a humanoid creature, by the manipulation of text and spells. This theme recurs frequently in the tradition. According to Moshe Idel, its most influential mention is in the Talmud:

> Rava said: If the righteous wished, they could create a world, for it is written: 'Your inequities have been a barrier between you and your God.' For Rava created a man and sent him to R. Zeira. The Rabbi spoke to him but he did not answer. Then he said: 'You are from the pietists: Return to dust.'[29]

What is occurring here? Rava demonstrates that creation of some type of human life is possible: The man moves and walks, but he does not talk. The work is flawed because of some inequity that must exist in Rava, the creator. The creation is undone, sent back to dust. In commentary on the text, Rashi notes that this sort of magical enterprise (in a way, the basic science of the time) was achievable by the manipulation of the letters of the name of God, the building blocks, as it were, of the Creator as known by humans.[30] The commerce is language: the word, the letters. In fact, in later Golem tales (the legend persists), the Golem has the Hebrew letters of the word "truth," *emet*, carved on his forehead. By removing the aleph, the first letter in one of God's names, the Hebrew word "death," *met*, is formed instead, and the Golem vanishes. Further legends link the Golem not only with the chimera of "truth" but play with the Golem: all body, no spirit. The Golem in later tales is a revenging and powerful force: Unlike the caricatured and vilified Jewish body—small, stooped, and awkward—the Golem of Prague legend is tall, muscular, and powerful, wreaking havoc on the gentile enemy. The Golem of Prague emerges to protect the Jews from the wrath of the Gentiles on the eve of Passover 1580, when the blood libel charges historically increased and lead to pogroms.

Yet, as appealing as this image is to a persecuted people, we are warned of the essential error in the pursuit of this particular type of creationist research: The manipulation of the whole by pieces of the whole does not lead to "truth," but to the excesses of spiritless power, unguided by faith, and ultimately dangerous. The texts are cautionary, but apparently not absolutely prohibitive: Otherwise the persistence of the story would not be evident.[31]

The second text is the *midrash* (metaphorical narrative) on the construction of the Tower of Babel. Here the rabbis struggle with the problem of why the construction of a joint human project is seen as problematic, even when the ostensible reason for the construction is to "reach up to God." Finding nothing in the direct text, they describe a theoretical scene:

> When a worker was killed, no one wept, but when a brick fell, all wept.[32]

What is occurring here? The rabbinic caution was that the use of humans instrumentally in a technologically impressive human project leads to a dismantling of the distinction between persons and things. It took a long time to make a brick, which then became more precious than the human self. It was perhaps this de-centering of the human and the reification of the thing that was the catastrophe that felled the enterprise, suggests this text, as much as the hubris of trying to "pierce heaven." It is not just that they have breached a limit between what is appropriate to create and what is not. It is that the process of the creation must be carefully mediated, with a deep respect for persons, over the temptations of the enterprise. Such a text elaborates on the tension between repairing the world in acts of tikkun olam and acts that claim that the world is ours to control utterly.

D. Do We Have an Obligation to Pursue This Research?

Given all this caution, should we halt or ban such research? What if the halachic considerations lead us toward supporting a ban on genetic research on human embryos? What would this mean for public policy? What would be lost and gained by such an approach? By the same token, what if we understood the Jewish position as mandating this research in an uncertain political climate? Would our stand imply an activist role for our leadership? Does a general obligation to "heal" include all possible avenues, and are we obligated even if the consideration of justice would mandate that other research be pursued? In other words, it is not enough for us to consider the question theoretically. If the work is a mandated healing, then the correct role would be to consistently argue and advocate for such a position, for to do any less might be the neglect of a commanded act. In so doing, we must recall that this action, that of mandated healing, is surely not the only place for commanded acts toward the health of our fellow Americans.

We must, for example remember that we speak out of our context of limited access to research funding and lack of health care for all Americans, much less to the needs of a wanting world in which infant diarrhea is still a leading cause of (preventable) death.

E. What of the Problem of Safety to the Persons on Whose Behalf We Intervene?

When Steptoe and Edwards first advanced the idea of IVF for purposes of reproduction, initial Jewish British reaction warned against the possibility of creating monster children or children who would suffer later effects of interventions that were thought to be life-saving at the time. For the Jewish American community, which was heavily affected by the "life-saving" reproductive intervention of 1950–1970, DES, the lessons are particularly acute. Further work will have to be explored regarding the issue of possible harm to specific others.

The issue of safety is critical. The hES/hEG cells are by their nature unstable, and we are only now learning about their unique properties. But clearly, some of what makes them interesting could make them dangerous in ways that may not be expressed for generations. For example, the highly telomerase expressive quality of these cells means that they can proliferate and are "immortal." But this is a quality shared with cancer cells. Will these cells retain this characteristic in higher percentages when used *in vivo*? Another question arises relative to the mutability of the stem cell. Will implanted hES/hEG cells have a far higher rate of mutability? How will we be able to test for such effects?

F. Will This Negatively Affect the Researcher Who Performs the Action?

In the consideration of genetic interventions in all cases, we must reflect on what in Hellenistic theory would be called "virtue theory." (Jewish sources include Maimonides and others.) How would the performance of the act of "harvesting" aborted fetal cells and all that this entails affect the scientists involved? What must be considered to protect the researcher from becoming indifferent to the human tissue involved in the use of the blastocyst? How can research scientists, by design removed from patients to protect the informed consent process, still act as though they are healers, motivated in the ways that must inform and direct the research? How will the significant monetary incentive affect this commitment? What is the effect on society if we create a "bank" of canonical cell lines, considering the potential of each cell and its special status? Here we raise the problem of the commodification of the tissue, or the denigration of its moral status.

G. Will This Have Implications for Evil Uses of Genetics (the Question of the Shoah)

The Shoah (Holocaust) changed the entire landscape of genetic research. While not only Jews have reason to raise deep concern about the evil specter of genetics, Jews certainly must do so as a primary consideration. Our firmness in remembering history and our disciplined stand to avoid any chance of repetition cannot overcome all efforts at new genetic research. However, the associations with genes, Jews, difference, and danger are extraordinarily strong. Here, we see the broadness of method that will be required. Halachic norms do not directly address this question, yet history matters for our account of whether such action should be permitted or prohibited.

H. What of the Marketplace Pressures on This Technology?

The field of ART is marked by its unrestrained use of the marketplace. Without oversight, fees, the nature of the contract, the standards for clinics, and the lengths that are permitted for individuals to pursue are unlimited. With new technology that will powerfully extend human life and potentially alter moral meaning necessarily, can we offer ethical guidelines to inform policy in this arena? How can the use of contract law, or the rabbinic prohibitions on marketplace exchanges, or rabbinic limits on the instrumental use of the body of another be used to regulate this arena? In other words, what rabbinic norms that are found in sources removed from medical consideration, but related to civil law and justice, might be mobilized to assist our thinking about the just use of technology?

I. Is Genetic Manipulation and Use of hES Cells an Instrumental Use of the Potential Human?

Here we have a classic example of the limits of halachic method. Like many scientific advances in medicine, we will simply not find a clear law to tell us, in a linear fashion, what to do when science discovers new capacities.

There are three possible responses: One is simply to describe classic commitments and familiar texts in Jewish medical ethics (saving life above all else, the duty to heal, the partnership with God to act as stewards in the world, or the general mandate to produce children). Using this mandate might allow nearly all technology that can be defended as "life-saving" to be permitted. In fact, this has operated in medical technology in a general, genial way.

Another response is to allow the general sensibilities of liberal theory to be bootlegged into Jewish thought, justifying such a move as necessary since no specific halachic norm exists for this new case. Another would begin an open and creative process of reflection on the several issues that I have raised. This would first entail a commitment to asking if these four basic texts are enough for us as we consider a response. It is my contention that, as medicine faces a critical junction, they do not go nearly far enough.

Part V: Next Steps

Remarkable advances in biotechnology and genetic medical interventions are fundamentally changing our most basic understanding of what it means to be human, of what the proper limits should be on research, and on the moral status of the essential components of human biology itself. For example, when cloning the human person became even remotely possible, the American press and public immediately understood the event to be one of religious importance and turned to its religious leadership for their responses. Such fundamental shifts call for significant reframing of what cases and narratives will stand for us as we reflect on the theological meaning and ethical choices within different faith traditions.

For scholars within the Jewish tradition, new science presents us with new challenges. We will be increasingly asked in the next few years about the permissibility, the telos, the moral meaning, and the appropriate limits of genetic intervention. For scholars within the field of bioethics, such calls are already emerging from scientists at the brink of new discoveries.

A Call for Research and Discourse Within the Jewish Tradition

Judaism is distinctive in foregrounding the text as rationale of its normative judgments. Hence, the casuistry that supports ethical response is based on exegetical reasoning that is debated over a prolonged temporal period. New developments in the field, however, call for innovations within this tradition. First, the presence of four major denominations has historically offered significant challenges to how Jewish academics and theologians can create scholarly discourse that is critical to developing and considering how we should reflect on the truly historic changes in science that will reconstruct not only medicine but also the basic view of the self. Second, scientific developments we are currently facing call for an imperative and a deeply informed discussion that is responsive to the rapidity with which new advances in genetic research emerge.

The textual method of Jewish thought takes place within a particular context—that of a larger discussant or discursive community. For centuries, Jewish law was based on a Beit Midrash—a discursive international community in which competing narrative justifications were offered to community leaders. Our careful and collaborative reflection on the topic will allow us to take a broader view of the issues before us.

Scholars of religion, theologians, and bioethicists have been asked to carefully reflect on the breathtaking and sweeping changes in medicine and research science. But our role, if prudently undertaken, cannot occur without a thoughtful and contextual account of the field of genetics as a whole. Learning about and approving each technology is akin to studying the elephant in small, and blinded groups—feeling trunk, legs, and tusk—each part understandable, but the larger whole incomprehensible. We will need to ask tough questions about how the use of any specific technology will relate to other pieces of research, such as reproduction technology, nuclear genetic transfer, and inheritable human genetic manipulation ("germline" intervention). As bioethicists, exploring the new technological beasts will take both courage and moral imagination. By this I mean the courage to resist a rush toward a swiftly moving future and the courage to believe that ethical and justice considerations must be taken into account at all stages of research, as well as the moral imagination to see beyond the perimeters of what we are given to what we might do and who we should become.

The Jewish textual tradition insists on the notion that the whole of the intellectual proposition of ethics is linked both to practicality and to prophesy, which means that one's epistemology must be sound, but one's vision intact. An Exodus tradition insists on the idea that what is given and what is now a fixity can be changed, healed, and imagined beyond. It is the act of moral imagination that this research calls us to make. But the leap from the present to the possible future will require, in that same tradition of Exodus, certain conditions.

First among these is the passion for just citizenship, for the idea that broad social liberation must take place in a responding and listening community. Next is the consideration for the vulnerable stranger. Finally, Jewish thought reminds us that the world we stand in now is ours only as stewards—and we will need to reflect carefully beyond the rhetorical flourish of that phrase to core issues of regulation and tough standards of enforcement. How will we set limits on research? How will a large public and plural discourse be assured? How will public justice, the passion for science, and the competing needs of the marketplace contend for our attention?

Our first, careful thinking about this new technology and our sober reflections and our tendency toward caution—which I argue are good and prudent responses—should not blind us to the extraordinary event that this discovery has been. This is a stunningly important moment in the history of medicine, one with the potential to save and sustain human life. The work that I have seen—the cardiac cells beating steadily in the laboratory, the nerve cells spinning out their tendrils—is impressive and bold work that challenges us to imagine beyond what is into what is possible. It challenges our moral sensibilities and our moral imaginations. It is work that reminds us that there is a special blessing that is said when one sees a wise secular scholar pass by, in praise of a Creator who makes human wisdom tangible:

> Blessed are You, Ruler of the Universe, who has given of Your knowledge to human beings.

In our cautionary deliberations of telos, process and meaning, and justice, we will need to foreground the essential ethicist's question of whether this is a "right act" and what makes it so, the essential Jewish question of how this act can repair a broken world, or the question of justice, of whether this research might ultimately not find a place in a world so broken. But at this moment in the process of scientific potential, we cannot forget our conjoint responsibilities: first, to be careful and thoughtfully reflective about the ends, goals, and norms of research, and second, to honor and support the extraordinary gesture of creative science that such a discovery represents. Both deliberation and a certain degree of radical awe frame the moral response of Jewish ethics in the assessment of human stem cell research.

Notes

1 Regalado, A., "The Troubled Hunt for the Ultimate Cell," *Massachusetts Institute of Technology Magazine, Technology Review* July/August (1998):35–41.

2 For the purposes of disclosure, this author is a member of a Modern Orthodox community.

3 For more on this topic, see E. Levinas, in numerous works, such as *Difficult Freedom: Essays on Judaism* (Baltimore, MD: Johns Hopkins University Press, 1990).

4 It raises the very interesting concept of persons as "text" with multiple "embedded narratives." Note here how we then conceptualize the human person in a postmodern way: a text with the potential for alternative narratives.

5 Biotechnology is a moving target. As this paper is being written, there are verbal reports of changes in techniques that might allow the use of these "lower" grade blastocysts for implantation.

6 The process of obtaining an abortion in Maryland, where this research is located, always involves a physician who will state that the abortion is undertaken for therapeutic reasons, either the physical or mental health of the woman being at stake. The consent for use of the fetal tissues is carefully worded. The researchers in the hEG cell project have no contact with the physicians treating the women. For a precise wording of the form, see "Research with Human Embryonic Stem Cells: Ethical Considerations," *Hastings Center Report*, March-April (1999):31–35, by the Geron Ethics Advisory Board.

7 Nagy, et al., "Derivation of Completely Cell Culture-Derived Mice from Early-Passage Embryonic Stem Cells," *Proceedings of the National Academy of Science USA* 90 (1993):8424–8428.

8 Warren, M.A., *Moral Status: Obligations to Persons and Other Living Things* (New York: Oxford University Press, 1998).

9 See the work of T. Shannon in *Genetics: Issues of Social Justice*, T. Peters, ed. (Cleveland, OH: Pilgrim Press, 1998). This alternative perspective is not the official position of the Vatican on this matter.

10 Here sits the ethicist, wearing glasses, having had children in her 40s, after having visited her 80-year-old father after his angioplasty.

11 Gilman, S.L., *The Jew's Body* (Bloomington: Routledge Press, 1995).

12 Hauerwas, S., *Dispatches from the Front: Theological Engagements with the Secular* (Durham, NC: Duke University Press, 1997).

13 See background paper on Jewish ethics and issues of moral status: unpublished, for internal use of the Geron Ethics Advisory Board, 1998. Each board member was asked to reflect on this issue from her/his particular faith perspective.

14 Literally, "part of the thigh of the mother." See Hulin 58a, Gittin, 23b. Talmud Balvi. (London: Soncino Press, 1958).

15 The Greek translation, from which the Christian tradition emerges, is different and assumes the exact opposite. The word in question is *ason*, which we have rendered as "harm." But the Greek renders the word as "form," yielding something like "if there yet be no form, he shall be fined, but if there be form, shalt thou give life for life." The "life for life" clause was thus applied to the fetus instead of the mother. This is of critical importance here. See Aptowitzer, V., as noted in Feldman, D.M., *Birth Control in Jewish Law: Marital Relations, Contraception, and Abortion as Set Forth in the Classic Texts of Jewish Law* (New York: New York University Press, 1968), 257.

16 See Feldman, 270–273, for an extensive reconstruction of this argument.

17 Sinclair, D., a doctoral dissertation, as quoted in Dorff, "A Jewish Approach to End-Stage Medical Care," *The Journal of Conservative Judaism* (1991):43.

18 Talmud Balvi. Pesahim 110b., as noted in Dorff.

19 In a discussion with the author, July 11, 1998. Philadelphia.

20 For an extensive listing, see Dorff, 1999, *Jewish Medical Ethics: Life and Death Decision Making*, Chapters 1–3. Jewish Publication Society.

21 Ibid. Also Jacobovitz, I., in *Jewish Medical Ethics: A Comparative and Historical Study of the Jewish Religious Attitude to Medicine and Practice* (New York: Bloch Publishing Company, 1997).

22 Dorff, "A Jewish Approach to End-Stage Medical Care."

23 Ibid.

24 In phone conversation with the author, November 1994. This conversation was in reference to the Human Genome Project.

25 Zoloth-Dorfman, "Mapping the Normal Human Self" in *Genetics*, T. Peters, ed. (Cleveland, OH: Pilgrim Press, 1998). See the extensive and masterful work of Sander L. Gilman here, for example, *The Jew's Body* (Bloomington: Routledge Press, 1995), or *Making the Body Beautiful: A Cultural History of Aesthetic Surgery*, by the same author (Princeton, NJ: Princeton University Press, 1999).

26 Such practices lead Orthodox feminist Blu Greenberg to observe, "where there is a rabbinic will, there is a rabbinic way." Greenberg, B., *How to Run a Traditional Jewish Household* (New York: Simon and Shuster, 1985).

27 The Jerusalem Report, April 16, 1998.

28 Ibid.

29 Talmud Sanhedrin, 65b, as quoted in Idel, M., *Golem* (New York: State University of New York Press, 1990).

30 Idel, op cit. It is a subject of a novel by Marge Piercy, *He, She and It* (New York: Ballentine Publishing Group, 1991). The tales reoccur in the eighteenth century, and in the texts of the responsa literature (Zevi Askenazi, She'elot u-Teshuvot, no. 93). In one such text, the question is raised about whether the Golem can be included in a prayer quorum, minyon. At stake is the issue of murder. If the Golem is a man, then is it not killing to "return him to dust?" (One thinks here of the legal cases involving the destruction of embryos.) The text resolves this in an odd way, not by claiming the humanity or countable status of the Golem, but of decrying the waste of a creature with "a purpose." Sherwin also comments on this text and notes that Askenazi's son, Jacob Emden, argues with this distinction. Emden is an important commentator on other issues in medical ethics.

31 This entire section is from Laurie Zoloth-Dorfman, "Mapping the Normal Human Self."

32 *Midrash Rabbah*, Bereshit, Soncino Edition, London. This is a collection of traditional narratives redacted from rabbinic sources, including the Babylonian Talmud.